ABIOTIC AND BIOTIC STRESS

ABIOTIC AND BIOTIC STRESS

Editor

Melitta Salzwedel

scitus academics

Abiotic and Biotic Stress
Edited by **Melitta Salzwedel**

This Book is Printed on Acid-Free Paper. ♾

ISBN: 9781681179049
Library of Congress Control Number: 2017948932

SCITUS Academics LLC,
616, Corporate Way, Suite 2, 4766,
Valley Cottage, NY 10989

www.scitusacademics.com

Notice

Printed in United States of America.

Preface

Environmental stresses play crucial roles in the productivity, survival and reproductive biology of plants as well as crops. Plants are subjected to many forms of environmental stress, which can be included into two broad areas: abiotic and biotic. The non-living variable must influence the environment beyond its normal range of variation to adversely affect the population performance or individual physiology of the organism in a significant way. Whereas a biotic stress would include such living disturbances as fungi or harmful insects, abiotic stress factors, or stressors, are naturally occurring, often intangible, factors such as intense sunlight or wind that may cause harm to the plants and animals in the area affected. Abiotic stress is essentially unavoidable. Abiotic stress affects animals, but plants are especially dependent on environmental factors, so it is particularly constraining. Abiotic stress is the most harmful factor concerning the growth and productivity of crops worldwide. Research has also shown that abiotic stressors are at their most harmful when they occur together, in combinations of abiotic stress factors. Epigenetic regulation is essential for growth and development in eukaryotic organisms and is also responsible for the establishment, maintenance, and reversal of non-genetic cellular memory that records developmental and environmental cues, including those arising from biotic and abiotic stress. Understanding of plant responses in relation with abiotic and biotic stresses is an important issue of concern not only because of yield loss but also because quality and longevity of plants are equally important for some crops not grown for food. In agriculture system loss of yield is taken as the final parameter of plant stress that shows the economic impact of stress. Even in natural systems it is more important to understand the physiology of plants in response to various types of abiotic and biotic stresses.

This volume represents a general overview of the major mechanisms developed by plants to tolerate environmental stresses, both abiotic (drought, high temperature, chilling and freezing, UV-B radiation, salinity and heavy metals) and biotic stress which includes physiology of plant-pathogen interaction, various ways through which pathogen invade the host tissues and many others. Clear understanding of physiological interaction of an organism with its environment is necessary to answer how competition, herbivory, and disease affect individual plants, plant processes, plant population and finally crop productivity. Consequently, basic knowledge of plant physiology leads to the foundation of developing plant species that are capable of withstanding stressful environment.

Table of Contents

CHAPTER 1

Plant Responses to Simultaneous Biotic and Abiotic Stress: Molecular Mechanisms

Ines Ben Rejeb, Victoria Pastor and Brigitte Mauch-Mani

Sciences, Institute of Biology, University of Neuchâtel, Rue Emile Argand 11, 2000 Neuchâtel, Switzerland

ABSTRACT

Plants are constantly confronted to both abiotic and biotic stresses that seriously reduce their productivity. Plant responses to these stresses are complex and involve numerous physiological, molecular, and cellular adaptations. Recent evidence shows that a combination of abiotic and biotic stress can have a positive effect on plant performance by reducing the susceptibility to biotic stress. Such an interaction between both types of stress points to a crosstalk between their respective signaling pathways. This crosstalk may be synergistic and/or antagonistic and include among others the involvement of phytohormones, transcription factors, kinase cascades, and reactive oxygen species (ROS). In certain cases, such crosstalk can lead to a cross-tolerance and enhancement of a plant's resistance against pathogens. This review aims at giving an insight into cross-tolerance between abiotic and biotic stress, focusing on the molecular level and regulatory pathways.

Keywords: cross-tolerance; biotic stress; abiotic stress; plant hormones

1. INTRODUCTION

Plants have to deal with various and complex types of interactions involving numerous environmental factors. In the course of evolution, they have evolved specific mechanisms allowing them to adapt and survive stressful events. Exposure of plants to biotic and abiotic stress induces a disruption in plant metabolism implying physiological costs [1,2,3,4], and thus leading to a reduction in fitness and ultimately in productivity [5]. Abiotic stress is one of the most important features of and has a huge impact on growth and, consequently, it is responsible for severe losses in the field. The resulting growth reductions can reach >50% in most plant species [6]. Moreover, biotic stress is an additional challenge inducing a strong pressure on plants and adding to the damage through pathogen or herbivore attack [7,8,9,10,11].

A crucial step in plant defense is the timely perception of the stress in order to respond in a rapid and efficient manner. After recognition, the plants' constitutive basal defense mechanisms [12] lead to an activation of complex signaling cascades of defense varying from one stress to another [13,14]. Following exposure to abiotic and/or biotic stress, specific ion channels and kinase cascades [15] are activated, reactive oxygen species (ROS) [16], phytohormones like abscisic acid (ABA), salicylic acid (SA), jasmonic acid (JA), and ethylene (ET) [17] accumulate, and a reprogramming of the genetic machinery results in adequate defense reactions and an increase in plant tolerance in order to minimize the biological damage caused by the stress [18].

In recent years, research has mainly concentrated on understanding plant responses to individual abiotic or biotic stresses [19,20,21,22], although the response to simultaneous stresses is bound to lead to a much more complex scenario [18]. From the perception of the stimulus (stress) to the final response in cells, plants use various signaling pathways depending on the challenge(s). It seems that plants respond in a specific manner when they have to face more than one stress simultaneously, and the response cannot be predicted based on the plant's response to the individual single stresses [23]. Research on multiple stresses has been trying to simulate natural conditions, but in the field, conditions are not controlled, and one stress can strongly influence the primary stress defense response of the plants [18]. Moreover, plants can show different degrees of sensitivity depending on the field condition and the developmental stage of the plant [24]. Additional factors that can influence an interaction are the intensity of the stress and the plant species. Various interactions can take place between the defenses induced after perception of the stresses. They depend on the specific combination of stresses and even on the degree of simultaneity [15,25,26]. It is not clear whether simultaneous stresses are rather antagonistic, synergistic or additive, inducing more or less susceptibility to a specific kind of stress [27,28]. Combination of two stressors can have a negative and additive effect on plants, the second stress being the one that leads to a greater damage [29]. On the other hand, the combination of stresses can also lead to antagonistic responses in the plants [30,31]. Common beans exposed to drought stress display more symptoms when infected by *Macrophomina phaseolina* [29] and treatment of detached tomato leaves with exogenously applied ABA increases the susceptibility of wild type plants to *Botrytis cinerea* [32].

Interestingly, one possible outcome of multiple stress exposure is that plants that are able to defend themselves facing one stress can become more resistant to other stresses [33]. This phenomenon is called cross-tolerance, showing that plants possess a powerful regulatory system that allows them to adapt quickly to a changing environment [33,34,35]. Wounding, for instance, increases salt tolerance in tomato plants [34]. Furthermore, in tomato plants again, localized infection by *Pseudomonas syringae* pv. tomato (*Pst*) induces systemic resistance to the herbivore insect *Helicoverpa zea* [36]. The association between abiotic and biotic stress is also possible [13], as demonstrated by the reduced infection of tomato by *Botrytis cinerea* and *Oidium neolycopersici* following the application of drought stress [37]. Ozone exposure can induce resistance to virulent *Pseudomonas syringae* strains in *Arabidopsis* [38]. Conversely, biotic stress can also interfere to increase the resistance to abiotic stress. This effect is visible when plants are under pathogen attack. Infection may cause stomatal closure to hinder pathogen entry and as a consequence water loss is reduced and leads to an enhanced plant resistance under abiotic stress [39]. Xu and colleagues [40] show that viral infection protects plants against drought stress. Verticillium infection in Arabidopsis plants induced the expression of the Vascular-Related No Apical meristem ATAF and Cup-Shaped Cotyledon (NAC) domain (VND) transcription factor *VND7*. VND7 induced *de novo* xylem formation ensuring the water storage capacity and as a consequence, increased plant drought tolerance [41]. Stress combination induces different signaling pathways, which share some components and common outputs [14,15,16,17,18,19,20,21,22,23,24,25]. This could help plants to minimize energy costs and create a flexible signaling network [42].

Resistance to both biotic and abiotic stress has been well documented in a variety of crops through priming of defenses. This component of induced resistance can be achieved through specific chemical stimuli like the resistance inducers BABA (beta-aminobutyric acid) or BTH (benzothiadiazole) [43,44], genetic manipulation of genes and proteins

[45] or by previous contact with a pathogen [46]. Due to the complexity of interactions in defense, in the present review, we aim to focus on the cross-tolerance between abiotic and biotic stress as a part of induced resistance for defense.

2. CROSS-TOLERANCE BETWEEN ABIOTIC AND BIOTIC STRESS

Plants are able to manage simultaneous exposure to abiotic and biotic stress, and there is evidence for a link between the responses to these two stressful situations [23,47,48,49]. Usually, environmental pressure by abiotic and biotic stress can induce plant resistance. However, some plants confronted with each stress individually have also been reported to be more susceptible compared to a simultaneous exposure to two different stresses [50,51]. In addition, certain environmental stresses have the possibility to predispose the plant in order to allow it to respond faster and in a resistant manner to additional challenges. Therefore, cross-tolerance between environmental and biotic stress may induce a positive effect and enhanced resistance in plants and have significant agricultural implications. Interestingly, abiotic stress regulates the defense mechanisms at the site of pathogen infection as well as in systemic parts, thus ensuring an enhancement of the plant's innate immunity system [31]. Likewise, osmotic and proton stress are inducers of resistance in barley against powdery mildew. This induced resistance depends on the formation of callose-containing papillae capable of blocking fungal growth [48]. This kind of resistance is similar to the chemically induced resistance by BTH and INA (isonicotinic acid) [52]. Achuo *et al.* [37] demonstrated that drought stress increased the ABA content of tomato leaves, concomitantly with increasing the resistance against the necrotophic fungus *Botrytis cinerea* and that salt stress reduced susceptibility towards the biotrophic fungus *Odium neolycopersici* but not against *Botrytis cinerea*. This difference between drought and salt stress is in accordance with the observation that they both induce different gene expression patterns [53]. Additionally, the acclimation of *Nicotiana benthamiana* to moderate drought stress (60% of field capacity) reduced the growth of *P. syringae* pv. *tabaci* [26]. Recently, Atkinson and Urwin [23] reviewed the interaction of abiotic and biotic stress where they showed the common threads in pathways leading to a regulation of plant responses. Therefore, in order to prepare the plant for the battle, the activation of various detoxifying enzymes, control hormones, signaling pathways, and gene expression are indispensable [4,42,54].

The defense response of plants exposed to different stressors is expected to be complex including the interconnection of various signaling pathways regulating numerous metabolic networks [55].

3. SIGNALING PATHWAYS INDUCED BY MULTIPLE STRESS RESPONSES

The interaction between abiotic and biotic stress induces complex responses to the different stressors. Under stress, the accumulation of certain metabolites positively affects a plant's response to both stresses and therefore protects it from multiple aggressors [25,47]. Callose accumulation, changes in ions fluxes, ROS, and phytohormones are the first responses induced to combat the stress and the resulting signal transduction triggers metabolic reprogramming towards defense [31,56].

3.1. Reactive Oxygen Species

A rapid generation of ROS is observed after stress sensing [57,58]. One of the major roles of ROS is to serve as signaling molecules in the cells [58,59,60,61,62]. The production of ROS is fine-modulated by the plant to avoid tissue damage [58,63,64,65,66,67,68,69,70,71]. ROS have long been known to be destructive and harmful compounds in stressed organisms. However, it has been shown that while high levels of ROS lead to cell death, lower levels are mostly responsible to regulate the plant's stress responses [67,68,69]. In biotic stress, ROS are mainly involved in signaling. This again might attenuate the oxidative stress caused by abiotic stress [70]. Furthermore, ROS could interfere in cross-tolerance [33]. ROS are involved in stress-induced tolerance in *Arabidopsis thaliana* after infection with the vascular pathogen *Verticillium* spp. by increasing drought tolerance due to *de novo* xylem formation and the resulting enhanced water flow [68]. Additionally, the production of ROS can help in cell-to-cell communication by amplifying the signal through the *Respiratory Burst Oxidase Homologue D* (*RBOHD*; [72]) and can act as a secondary messenger by modifying protein structures and activating defense genes [61,73]. ROS respond to abiotic and biotic stress, but differently from one stress to another [47]. Davletova *et al.* [74] showed that the transcription factor *Zat12* was involved in both abiotic and biotic stress and that *Zat12* could be a regulator in ROS scavenging. ROS may possibly be the central process mediating cross-tolerance between abiotic and biotic stress responsive networks [23]. In Arabidopsis, ROS production can be sensed by ROS-sensitive transcription factors [75,76] leading to the induction of genes participating in the stress responses.

Gechev *et al.* [77] proposed that ROS were inducers of tolerance by activating stress response-related factors like mitogen-activated protein kinases (MAPKs), transcription factors, antioxidant enzymes, dehydrins, and low-temperature-induced-, heat shock-, and pathogenesis-related proteins.

Priming for stress tolerance induced after application of specific chemicals is responsible for certain modifications in ROS signaling [70,71,72,73,74,75,76,77,78]. Treatment of cucumber plants with brassinosteroids lead to a rise in H_2O_2 levels and primed the plants for both biotic and abiotic stress tolerance [68]. H_2O_2 priming for salt tolerance in citrus moderately increased the abundance of oxidized and *S*-nitrosylated proteins, and the level remained the same after stress application, however, non-treated plants were more sensitive to the stress [78].

3.2. Mitogen-Activated Protein Kinase (MAPK) Cascades

Following perception and recognition of stress stimuli, Mitogen-Activated Protein Kinase (MAPK) cascades are activated. They control the stress response pathways [79,80]. MAPKs are highly conserved in all eukaryotes and are responsible for the signal transduction of diverse cellular processes under various abiotic and biotic stress responses, and certain kinases are involved in both kind of stress [18,81,82]. Since MAPKs are involved in different stress responses, they could have a role in the combination of abiotic and biotic stress [83,84]. For instance, in cotton the kinase *GhMPK6a* negatively regulates both biotic and abiotic stress [85]. MAPK pathways activated by pathogen attack are mediated by SA, and the resulting expression of *PR* genes induces defense reactions [86]. The Arabidopsis protein VIP1 is translocated into the nucleus after phosphorylation by MPK3 and acts as an indirect inducer of *PR1* [87]. Chinchilla *et al.* [88] showed that pathogen associated molecular patterns (PAMPs) like flagellin trigger MAPK cascades in order to establish pathogen response signaling. In addition, MAPK such as MPK3, MPK4, and MPK6 also responded to various abiotic stresses [89,90]. MAPK cascades are important in controlling cross-tolerance between stress responses [12]. MPK3 and MPK6 are essential to show full primed defense responses [91], therefore, these two kinases could be important for mediating tolerance to further stresses. Over-expression of the *OsMPK5* gene and also kinase activity of OsMPK5 induced by ABA contributes to increased abiotic and biotic stress tolerance. *OsMPK5* seems to play a double role in the rice stress response, one as a positive regulator of resistance to the necrotrophic brown spot pathogen *Cochliobolus miyabeanus* and the second as a mediator of abiotic stress tolerance [81,92]. Tomato plants activate MPK1 and MPK2 against UV-B, wounding, and pathogens in order to enhance their defense reactions [93]. MAPK signaling also interacts with ROS and ABA signaling pathways leading to enhanced plant defense and induction of cross-acclimation to both abiotic and biotic stress [94,95,96].

3.3. Relevance of Hormone Signaling under Stress Interaction

The control of every kind of stress by specific hormones allows defense responses against defined environmental conditions. ABA is considered the primary hormone involved in the perception of many abiotic stresses [97]. Increases in ABA concentration modulate the abiotic stress-regulation network [98] while biotic stress responses are preferentially mediated by antagonism between other stress hormones such as SA and acid JA/ET [99]. In certain cases, ABA has been shown to accumulate after infection [18,27,100,101]. For instance, higher levels of ABA were observed after *Pst* DC 3000 infection [102], and this provoked a suppression of other defense responses [103]. However, recent findings show a positive effect of ABA on biotic stress resistance [30,104,105]. This dual effect makes ABA a controversial molecule that can switch from "good to bad" depending on the environmental conditions (type and timing of the stress; [105]). Moreover, under combination of abiotic and biotic stress, ABA mostly acts antagonistically with SA/JA/ethylene inducing a susceptibility of the plant against disease and herbivore attack [28,31,32,106,107]. However, since an increase of ABA under the effect of abiotic stress induces stomatal closure, as a "secondary effect", the entry of biotic assailants through these passive ports of the plant is prevented. Hence, under such circumstances, the plant is protected from abiotic as well as from biotic stress [108]. There are three different phases showing the influence of ABA on pathogen infection [23,30]. The first effect of ABA on the combination of both, abiotic and biotic stress is related only to abiotic stress because an infection takes more time to establish itself and the plants react therefore later to it [30,31,32,33, 34,35,36,37,38,39,40,41,42,43,44,45,46,47,48,49,50,51,52,53,54,55,56,57 ,58,59,60,61,62,63,64,65,66,67,68,69,70,71,72,73,74,75,76,77,78,79,80,8 1,82,83,84,85,86,87,88,89,90,91,92,93,94,95,96,97,98,99,100,101,102,10 3,104,105,106,107,108,109]. At this moment, ABA induces stomatal closure [110], which allows a reduction in water loss and, as a consequence, the maintenance of a beneficial water potential. In this first phase, SA, JA and ethylene might not yet be activated and ABA can antagonize their induction. In this situation, future responses against potential pathogens are modified. The second phase concerns the post-infection reactions. Callose is an important inducible defense that can prevent pathogen invasion [111]. After infection, an intact ABA signaling pathway is required to increase callose accumulation in attacked plants [44,112], and the presence of ABA can induce or repress additional callose accumulation [98] depending on the environmental conditions. Therefore, ABA variation by a previous stress can influence the final output following biotic stress, such as strengthening the resistance phenotype through accumulation of callose or by inducing other defense pathways [96,108]. The third phase finally starts when PAMPs stimulate the accumulation of specific hormones that are SA, JA, and ethylene in order to regulate the defense reaction [27,96,113]. In summary, the exact role of ABA as a regulator of the dialogue between abiotic and biotic stress strongly depends on the timing of the stress perception: does the infection hit a plant that had already been exposed previously to abiotic stress or does an infected plant become additionally exposed to abiotic stress [30,97,114]?

The beneficial role of SA in the relationship between plants and pathogens has been extensively studied. What is known is that ABA and SA have an antagonistic role in plant defense against stressors [31]. However, Miura and Tada [88] have shown that in addition to ABA, SA seems to also be important in plant responses to drought stress. Furthermore, SA increased barley resistance against water deficit [115].

3.4. Transcription Factors and Molecular Responses in Cross-Tolerance

Changes in gene expression occur after detection of a given stress, and the reprogramming of the molecular machinery is regulated by the action of transcription factors. The altered expression of certain genes is a key event in helping plants to set up an effective defensive state, and there is convincing evidence that many genes are multifunctional and able induce tolerance in plants towards more than one stress [49,50,51,52,53,54,55,56,57,58,59,60,61,62,63,64,65,66,67,68,69,70,71,72,73,74,75,76,77,78,79,80,81,82,83,84,85,86,87,88,89,90,91,92,93,94,95,96,97,98,99,100,101,102,103,104,105,106,107,108,109,110,111,112,113,114,115,116]. The activity of such genes involved in defense is mediated by specific phytohormones like ABA, SA, JA, and Ethylene. For example, the activity of the *BOTRYTIS SUSCEPTIBLE1* (*BOS1*) gene is mediated by both ABA and JA and induces resistance against osmotic stress and necrotrophic pathogens [117], and *bos1* mutant plants are more susceptible to both stresses [117]. In Arabidopsis, the transcription factor *MYB96* plays an important role in plant protection under pathogen infection by mediating the molecular link between both ABA induced by drought stress and SA expressed following pathogen infection [118]. *SlAIM1* in tomato responds positively to the combination of abiotic stress and infection with *Botrytis cinerea* [13] and *OsMAPK5,* which has kinase activity, is a positive regulator of the rice response to drought, salt, and cold tolerance and disease resistance [86].

Interestingly, many *PR* genes are also induced upon exposure of a plant to abiotic stress ensuring disease resistance [118]. PR proteins are crucial for plant resistance against pathogens, and their expression is strongly up-regulated when plants are attacked [118]. Over-expression of certain transcription factors in plants confronted with cold stress and infection activates cold-responsive *PR* genes, thereby conferring protection against both stressors [119]. The up-regulation of some transcription factors after exposure to abiotic stress leads to an accumulation of PR proteins. The transcription factors C-repeat Binding Factors (*CBF*), Dehydration-Responsive Element-Binding proteins (*DREB*) and No Apical meristem ATAF and Cup-Shaped Cotyledon (*NAC*) have been extensively studied as players of the primary abiotic stress signaling pathways ensuring tolerance under stress [120,121,122]. *CBF* is induced under cold stress together with a group of PR proteins [123]. Transgenic Arabidopsis overproducing the *NAC* transcription factor *NTL6*, which is induced by cold stress, enhance their defense response against pathogen attack by promoting an up-regulation of the *PR1* gene [118,119,120,121,122,123,124]. Tsutsui *et al.* [125] showed that the transcription factor *DREB* could regulate the response of cross-tolerance between abiotic and biotic stress insuring the resistance of Arabidopsis response to cold and pathogen (Figure 1).

Recently, it has been proposed that the WHIRLY1 protein and *REDOX-RESPONSIVE TRANSCRIPTION FACTOR1* (*RRTF1*) could participate in the traffic of communication between plastids and the nucleus [126]. WHIRLY1 perceives the redox changes in the plastid and carries the information to the nucleus in an NPR1-independent manner. The authors propose this protein as an ideal component in retrograde signaling that will lead to acclimation and adaption to new stresses. In the same way, *RRTF1,* which is induced by biotic and abiotic stresses, could be priming distant leaves to defend themselves against further stresses.

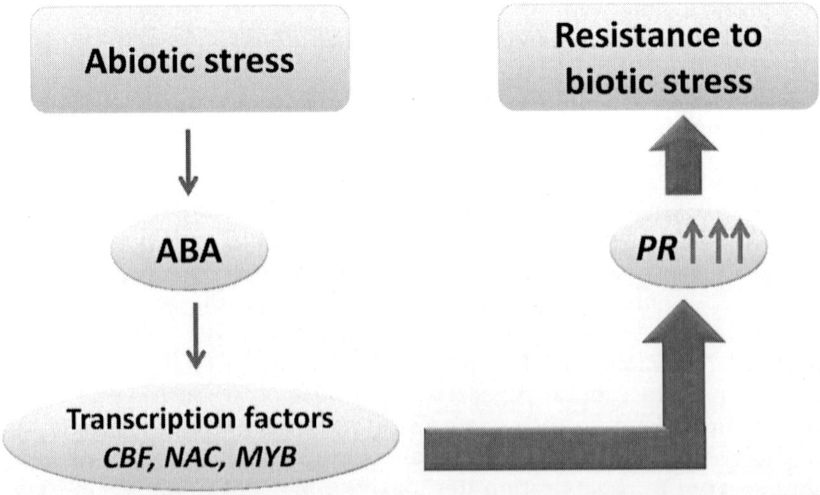

Figure 1. Abiotic stress can enhance the expression of specific transcription factors (TFs) like C-repeat Binding Factors (*CBF*), No Apical meristem ATAF and Cup-Shaped Cotyledon (*NAC*), *MYB* mediated by abscisic acid (ABA). Although the exact role of ABA in plant pathogen interactions is still a matter of debate, in some specific cases it has been shown to promote resistance against biotic stress following abiotic stress. This is attributed to the over-expression of TFs inducing the up-regulation of *PR*genes.

4. CONCLUSIONS AND OUTLOOK

A plant's response following exposure to abiotic/biotic stress strongly depends on its developmental stage [127] and the environmental conditions to which it is subjected [99]. Many stress combinations lead to phenotypic damage and, as mentioned above, the expression of defense is affected according to the type of abiotic stress and the pathogens involved. Overall, the complex response of the plant stems from the interplay of specific signaling pathways involved in abiotic and biotic stress. The combination of both stress types leads to an increased accumulation of a large number of signaling compounds that, in an ideal case, will be expressed as cross-tolerance (Figure 2).

Plants perceive the information signal of each stress and consequently activate specific molecules. Only some of them, which are common to both stressors, will participate in the defense response to the specific stress combination and thus contribute to protect the plant and enhance its resistance.

Various novel approaches can help plants to resist under combinatorial stress. The "Omics" technology is one of these approaches. Transcriptomics, proteomics, and metabolomics have revealed plant responses under stress and their underlying mechanisms and point to potential target genes, proteins or metabolites for inducing tolerance and improve plant responses. Little is known about the "Omics" characterization of abiotic and biotic stress combinations, but recently, several reports have addressed this question [16,51,70,128,129]. Although complete genome sequences are available for an increasing number of crop and model plants, in comparison, protein and metabolite databases are still rather incomplete, hence complicating the task of integrating all observations. Additionally, different plant species or even cultivars may behave differently, plant responses are also often organ-dependent, and results obtained with whole plants may be misleading.

Another approach might consist of molecular engineering of specific genes and their introduction into crop plants. By modifying a gene coding for a small antimicrobial peptide and introducing it into potato, the resistance of potato to biotic and abiotic stress was increased [130].

The manipulation of common regulators is also a promising approach. Boosting the accumulation of flavonoid biosynthesis mitigates the negative effects of abiotic and biotic stress [131,132]. Polyamines are another example. These substances have long been known to mediate resistance to pathogens [133] but they are also involved in abiotic stress resistance [134]. Genetic manipulation of polyamine accumulation could lead to multi stress tolerance [135].

A further possibility to promote cross-tolerance is the exploitation of priming. Some chemicals have been shown to prime plants for both biotic and abiotic stresses under laboratory conditions [136], and their application might allow a better management of multiple stresses under field conditions. The ultimate goal in every case is to maintain or even enhance plant performance, yield, and productivity under adverse conditions.

The interaction of SA, JA, and ET signaling results in increased resistance to pathogens. Hormones, secondary metabolites, priming agents, and further chemicals located in the cytoplasm finally up-regulate transcription factors (TF), pathogenesis related (PR) and defense genes, heat shock protein (HSP) genes, and further genes involved in protection against stress and thus lead to the phenotypic expression known as cross-tolerance. Arrows: induction; flat-ended lines: repression.

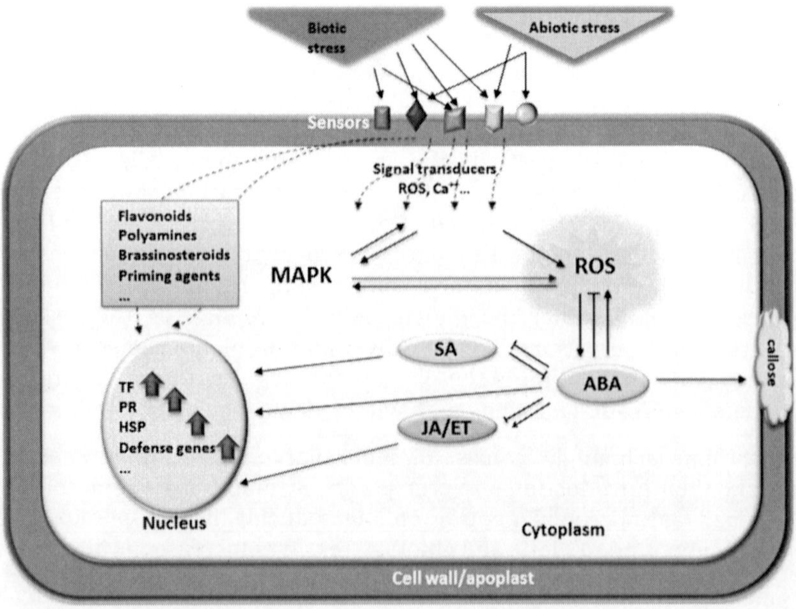

Figure 2. Elements possibly involved in cross-tolerance between biotic and abiotic stress. Both biotic and abiotic stress have to be first sensed by the plant cell, and then the information is transduced to appropriate downstream-located pathway(s). Sensors as well as signal transducers might be shared by both types of stressors. Reactive oxygen species (ROS) and Ca^{++} are known among others to play a prominent role as transducers (messengers) and mitogen-activated protein kinases (MAPK) cascades have been shown to be used by both types of stresses. MAPKs are centrally positioned in Ca^{2+}-ROS crosstalk as well as in the signal output after exposure to a specific stress. The importance of ROS has repeatedly been described for both types of stresses too, and, therefore, ROS might represent crucial elements in the integration of both stresses during cross-tolerance. Plant hormone signaling is of utter importance for stress adaptation. While abscisic acid (ABA) is predominantly involved in abiotic stress adaptation, salicylic acid (SA) and jasmonate/ethylene (JA/ET) are more responsible for the plant's reaction to biotic stress. However, there is a tremendous amount of crosstalk taking place between the various hormonal pathways, and the exact nature of this crosstalk during simultaneous biotic and abiotic stress remains to be investigated. ABA signaling contributes positively to pre-invasion defense and is responsible for enhancing callose deposition. ABA presents a positive interaction with JA/ET signaling. The activation of SA signaling by pathogen challenge can attenuate ABA responses. ABA signaling negatively affects signals that trigger systemic acquired resistance, enhancing pathogen spread from the initial site of infection.

ACKNOWLEDGMENTS

We apologize to all the colleagues whose work has not been cited in this short review. The financial support of the Swiss National Science Foundation, grant 31003A_140593, is gratefully acknowledged. We also thank the Swiss Confederation for attributing a Swiss Government Excellence Research Scholarship for foreign students to I. Ben Rejeb.

AUTHOR CONTRIBUTIONS

All authors contributed equally to this review article.

CONFLICTS OF INTEREST

The authors declare no conflicts of interest.

REFERENCES

1. Heil, M.; Bostock, R.M. Induced systemic resistance (ISR) against pathogens in the context of induced plant defences. Ann. Bot. 2002, 89, 503–512.
2. Swarbrick, P.J.; Schulze-Lefert, P.; Scholes, J.D. Metabolic consequences of susceptibility and resistance in barley leaves challenged with powdery mildew. Plant Cell Environ. 2006, 29, 1061–1076.
3. Bolton, M.V. Primary metabolism and plant defense-Fuel for the fire. Mol. Plant Microbe Interact. 2009, 22, 487–497.
4. Massad, T.J.; Dyer, L.A.; Vega, C.G. Cost of defense and a test of the carbon-nutrient balance and growth-differentation balance hypotheses for two co-occurring classes of plant defense. PLoS One 2012, 7, e7554.
5. Shao, H.B.; Chu, L.Y.; Jaleel, C.A.; Zhao, C.X. Water-deficit stress—Induced anatomical changes in higher plants. C. R. Biol. 2008, 331, 215–225.
6. Wang, W.; Vinocur, B.; Altman, A. Plant responses to drought; salinity and extreme temperatures: Towards genetic engineering for stress tolerance. Planta 2003, 218, 1–14.
7. Mordecai, E.A. Pathogen impacts on plant communities: Unifying theory, concepts, and empirical work. Ecol. Monogr.2011, 81, 429–441.
8. Maron, J.L.; Crone, E. Herbivory: Effects on plant abundance, distribution and population growth. Proc. R. Soc. B2006, 273, 2575–2584.
9. Maron, J.L.; Kauffman, M. Habitat-specific consumer impacts on plant population dynamics. Ecology 2006, 87, 113–124.
10. Strauss, S.Y.; Zangerl, A.R. Plant-insect interactions in terrestrial ecosystems. In Plant-Animal Interactions. An Evolutionary Approach; Herrera, C.M., Pellmyr, O., Eds.; Blackwell Science: Oxford, UK, 2002; pp. 77–106.

11. Brown, J.K.M.; Hovmoller, M.S. Aerial dispersal of pathogens on the global and continental scales and its impact on plant disease. Science 2002, 297, 537–541.
12. Andreasson, E.; Ellis, B. Convergence and specificity in the Arabidopsis MAPK nexus. Trends Plant Sci. 2010, 15, 106–113.
13. Abou Qamar, S.; Luo, H.; Laluk, K.; Mickelbart, V.M.; Mengiste, T. Crosstalk between biotic and abiotic stress responses in tomato is mediated by AIM1 transcription factor. Plant J. 2009, 58, 1–13.
14. Chinnusamy, V.; Schumaker, K.; Zhu, J.K. Molecular genetics perspectives on cross-talk and specifcity in abiotic stress signalling in plants. J. Exp. Bot. 2004, 55, 225–236.
15. Fraire-Velázquez, S.; Rodríguez-Guerra, R.; Sánchez-Calderón, L. Abiotic and Biotic Stress Response Crosstalk in Plants-Physiological, Biochemical and Genetic Perspectives; Shanker, A., Ed.; InTech Open Access Company: Rijeka, Croatia, 2011; pp. 1–26.
16. Laloi, C.; Appel, K.; Danon, A. Reactive oxygen signalling: The latest news. Curr. Opin. Plant Biol. 2004, 7, 323–328.
17. Spoel, S.H.; Dong, X. Making sense of hormone crosstalk during plant immune response. Cell Host Microbe 2008, 3, 348–351.
18. Fujita, M.; Fijita, Y.; Noutoshi, Y.; Takahashi, F.; Narusaka, Y.; Yamaguchi-Shinozaki, K.; Shinozaki, K. Crosstalk between abiotic and biotic stress responses: A current view from the points of convergence in the stress signaling networks. Curr. Opin. Plant Biol. 2006, 9, 436–442.
19. Todaka, D.; Nakashima, K.; Shinozaki, K.; Yamaguchi-Shinozaki, K. Toward understanding transcriptional regulatory networks in abiotic stress responses and tolerance in rice. Rice J. 2012, 5, 1–9.
20. Stotz, H.U.; Mitrousia, G.K.; de Wit, P.J.G.M.; Fitt, B.D.L. Effector-triggered defence against apoplastic fungal pathogens. Trends Plant Sci. 2014, 19, 491–500.
21. Thakur, M.; Sohal, B.S. Role of elicitors in inducing resistance in plants against pathogen infection: A review. ISRN Biochem. 2013.
22. Qin, F.; Shinozaki, K.; Yamaguchi-Shinozaki, K. Achievements and challenges in understanding plant abiotic stress responses and tolerance. Plant Cell Physiol. 2011, 52, 1569–1582.
23. Atkinson, N.; Urwin, P.-E. The interaction of plant biotic and abiotic stresses: From genes to the field. J. Exp. Bot.2012, 63, 3523–3544.
24. Mittler, R.; Blumwald, E. Genetic engineering for modern agriculture: Challenges and perspectives. Ann. Rev. Plant Biol. 2010, 61, 443–462.
25. Rasmussen, S.; Barah, P.; Suarez-Rodriguez, M.C.; Bressendorff, S.; Friis, P.; Costantino, P.; Bones, A.M.; Nielsen, H.B.; Mundy, J. Transcriptome responses to combinations of stresses on Arabidopsis. Plant Physiol. 2013, 161, 1783–1794.
26. Ramegowda, V.; Senthil-Kumar, M.; Ishiga, Y.; Kaundal, A.; Udayakumar, M.; Mysore, K.S. Drought stress acclimation impacts tolerance to Sclerotina sclerotiorum and Pseudomonas syringae in Nicotiana benthamiana. Int. J. Mol. Sci. 2013, 14, 9497–9513.
27. Anderson, J.P.; Badruzsaufari, E.; Schenk, P.M.; Manners, J.; Desmond, O.J.; Ehlert, C.; Maclean, D.J.; Ebert, P.R.; Kazan, K. Antagonistic interaction between abscisic acid and jasmonate-ethylene signaling pathways modulates defense gene expression and disease resistance in Arabidopsis. Plant Cell 2004, 16, 3460–3479.
28. Asselbergh, B.; Achuo, A.E.; Hofte, M.; van Gijegem, F. Abscisic acid deficiency leads to rapid activation of tomato defence responses upon infection with Erwinia chrysanthemi. Mol. Plant Pathol. 2008, 9, 11–24.
29. Suleman, P.; Al-Musallam, A.; Menezes, C.A. The effect of solute potential and water stress on black scorch caused by Chalara paradoxa and Chalara radicicola on date palms. Plant Dis. 2001, 85, 80–83.
30. Ton, J.; Ent, V.D.S.; Hulten, V.M.; Pozo, M.; Oosten, V.V.; Loon, L.C.V; Mauch-Mani, B.; Turlings, T.C.J.; Pieterse, C.M.J. Priming as a mechanism behind induced resistance against pathogens; insects and abiotic stress. IOBC/wprs Bull. 2009, 44, 3–13.

31. Yasuda, M.; Ishikawa, A.; Jikumaru, Y.; Seki, M.; Umezawa, T.; Asami, T.; Maruyama-Nakashita, A.; Kudo, T.; Shinozaki, K.; Yoshida, S.; et al. Antagonistic interaction between systemic acquired resistance and the abscisic acid-mediated abiotic stress response in Arabidopsis. Plant Cell 2008, 20, 1678–1692.
32. Audenaert, K.; de Meyer, G.B.; Hofte, M. Abscisic acid determines basal susceptibility of tomato to Botrytis cinerea and suppresses salicylic acid-dependent signaling mechanisms. Plant Physiol. 2002, 128, 491–501.
33. Bowler, C.; Fluhr, R. The role of calcium and activated oxygens as signals for controlling cross-tolerance. Trends Plant Sci. 2000, 5, 241–246.
34. Capiati, D.A.; Pais, S.M.; Tellez-Iñon, M.T. Wounding increases salt tolerance in tomato plants: Evidence on the participation of calmodulin-like activities in cross-tolerance signaling. J. Exp. Bot. 2006, 57, 2391–2400.
35. Suzuki, N.; Koussevitzky, S.; Mittler, R.; Miller, G. ROS and redox signalling in the response of plants to abiotic stress. Plant Cell Environ. 2012, 35, 259–270.
36. Stout, M.J.; Fidantsef, A.L.; Duffey, S.S.; Bostock, R.M. Signal interactions in pathogen and insect attack: Systemic plant-mediated interactions between pathogens and herbivores of the tomato. Lycopersicum esculentum. Physiol. Mol. Plant Pathol. 1999, 54, 115–130.
37. Achuo, E.A.; Prinsen, E.; Hofle, M. Influence of drought; salt stress and abscisic acid on the resistance of tomato to Botrytis cinerea and Oidium neolycopersici. Plant Pathol. 2006, 55, 178–186.
38. Sharma, Y.; Leon, J.; Raskin, I.; Davis, K.R. Ozone-induced responses in Arabidopsis thaliana: The role of salicylic acid in the accumulation of defense-related transcripts and induced resistance. Plant Biol. 1996, 93, 5099–5104.
39. Goel, A.K.; Lundberg, D.; Torres, M.A.; Matthews, R.; Akimoto-Tomiyama, C.; Farmer, L.; Dangl, J.L.; Grant, S.R. The Pseudomonas syringae type III effector HopAM1 enhances virulence on water-stressed plants. Mol. Plant Microbe Interact. 2008, 21, 361–370.
40. Xu, P.; Chen, F.; Mannas, J.P.; Feldman, T.; Sumner, L.W.; Roossinck, M.J. Virus infection improves drought tolerance. New Phytol. 2008, 180, 911–921.
41. Reusche, M.; Thole, K.; Janz, D.; Truskina, J.; Rindfleisch, S.; Drübert, C.; Polle, A.; Lipka, V.; Teichmann, T. Verticillium infection triggers VASCULAR-RELATED NAC DOMAIN7-dependent de novo xylem formation and enhances drought tolerance in arabidopsis. Plant Cell 2012, 24, 3823–3837.
42. Koornneef, A.; Pieterse, C.M.J. Cross talk in defense signaling. Plant Physiol. 2014, 146, 839–844.
43. Jakab, G.; Ton, J.; Flors, V; Zimmerli, L; Métraux, J.P.; Mauch-Mani, B. Enhancing Arabidopsis salt and drought stress tolerance by chemical priming for its abscisic acid responses. Plant Physiol. 2005, 139, 267–274.
44. Ton, J.; Mauch-Mani, B. Beta-amino-butyric acid-induced resistance against necrotrophic pathogens is based on ABA-dependent priming for callose. Plant J. 2004, 38, 119–130.
45. Umezawa, T.; Fujita, M.; Fujita, Y.; Yamaguchi-Shinozaki, K.; Shinozaki, K. Engineering drought tolerance in plants: Discovering and tailoring genes to unlock the future. Curr. Opin. Biotechnol. 2006, 17, 113–122.
46. Chalfoun, N.R.; Castagnaro, A.P.; Díaz Ricci, J.C. Induced resistance activated by a culture filtrate derived from an avirulent pathogen as a mechanism of biological control of anthracnose in strawberry. Biol. Control 2001, 58, 319–329.
47. Pastori, G.M.; Foyer, C.H. Common components, networks, and oathways of cress-tolerance to stress. The central role of "redox" and abscisic acid-mediated controls. Plant Physiol. 2002, 129, 460–468.
48. Weise, J.; Kranz, T.; Schbert, S. Induction of pathogen resistance in barley by abiotic stress. Plant Biol. 2004, 6, 529–536.
49. Ramirez, V.; Coego, A.; Lopez, A.; Agorio, A.; Flors, V.; Vera, P. Drought tolerance in Arabidopsis is controlled by the OCP3 disease resistance regulator. Plant J. 2009, 58, 578–591.

50. Tippmann, H.F.; Schluter, U.; Collinge, D.B. Common themes in Biotic and abiotic stress signaling in plants. Floric. Ornam. Plant Biotechnol. 2006, 3, 52–67.
51. Suzuki, N.; Rivero, R.M.; Shulaev, V.; Blumwald, E.; Mittler, R. Abiotic and biotic stress combinations. New Phytol.2014, 203, 32–43.
52. Besser, K.; Jarosch, B.; Langen, G.; Kogel, K.-H. Expression analysis of genes induced in barley after chemical activation reveals distinct disease resistance pathways. Mol. Plant Pathol. 2000, 5, 277–286.
53. Takahashi, S.; Seki, M.; Ishida, G.; Satou, M.; Sakurai, T.; Narusaka, M.; Kamiya, A.; Nakajima, M.; Enju, A.; Akiyama, K.; et al. Monitoring the expression profiles of genes induced by hyperosmotic; high salinity and oxidative stress and abscisic acid treatment in Arabidopsis cell culture using a full-length cDNA microarray. Mol. Biol. 2004, 56, 29–55.
54. Mittler, R. Abiotic stress, the field environment and stress combination. Trends Plant Sci. 2006, 11, 15–19.
55. Nakashima, K.; Ito, Y.; Yamaguchi-Shinozaki, K. Transcriptional regulatory networks in response to abiotic stresses in arabidopsis and grasses. Plant Physiol. 2009, 149, 88–95.
56. Bartoli, C.G.; Casalongué, C.A.; Simontacchi, M.; Marquez-Garcia, B.; Foyer, C.H. Interactions between hormone and redox signaling pathways in the control of growth and cross-tolerance to stress. Environ. Exp. Bot. 2013, 94, 73–88.
57. Wojtaszek, P. Oxidative burst: An early plant response to pathogen infection. Biochem. J. 1997, 322, 681–692.
58. Foyer, C.; Noctor, G. Redox homeostasis and antioxidant signaling: A metabolic interface between stress perception and physiological responses. Plant Cell 2005, 17, 1866–1875.
59. Apel, K.; Hirt, H. Reactive oxygen species: Metabolism; oxidative stress; and signal transduction. Ann. Rev. Plant Biol.2004, 55, 373–399.
60. Hancock, J.; Desikan, R.; Harrison, J.; Bright, J.; Hooley, R.; Neill, S. Doing the unexpected: Proteins involved in hydrogen peroxide perception. J. Exp. Bot. 2006, 57, 1711–1718.
61. Spoel, S.H.; Loake, G.J. Redox-based protein modifications: The missing link in plant immune signalling. Curr. Opin. Plant Biol. 2011, 14, 358–364.
62. Meng, L.; Wong, J.H.; Feldman, L.J.; Lemaux, P.G.; Buchanan, B.B. A membrane-associated thioredoxin required for plant growth moves from cell to cell; suggestive of a role in intercellular communication. Proc. Natl. Acad. Sci. USA2010, 107, 3900–3905.
63. Mittler, R.; Vanderauwera, S.; Suzuki, N.N.; Miller, G.; Tognetti, V.B.; Vandepoele, K.; Gollery, M.; Shulaev, V.; van Breusegem, F. ROS signaling: The new wave? Trends Plant Sci. 2011, 16, 1360–1385.
64. Vinocur, B.; Altman, A. Recent advances in engineering plant tolerance to abiotic stress: Achievement and limitations. Curr. Opin. Biotech. 2005, 16, 123–132.
65. Bhattacharjee, S. The language of reactive oxygen species signaling in plants. J. Bot. 2012, 2012, 1–22.
66. Foyer, C.; Noctor, G. Ascorbate and glutathione: The heart of the redox hub. Plant Physiol. 2011, 155, 1–18.
67. Choudhury, S.; Panda, P.; Sahoo, L.; Panda, S.K. Reactive oxygen species signaling in plants under abiotic stress. Plant Signal. Behav. 2013, 8, e23681.
68. Xia, X.-J.; Wang, Y.-J.; Zhou, Y.-H.; Tao, Y.; Mao, W.-H.; Shi, K.; Asami, T.; Chen, Z.; Yu, J.-Q. Reactive oxygen species are involved in brassinosteroid-induced stress tolerance in cucumber. Plant Physiol. 2012, 158, 1034–1045.
69. Mori, I.C.; Schroeder, J.I. Reactive oxygen species activation of plant Ca2+ channels. A signaling mechanism in polar growth, hormone transduction, stress signaling, and hypothetically mechanotransduction. Plant Physiol. 2004, 135, 702–708.
70. Kissoudis, C.; van de Wiel, C.; Visser, R.G.F.; van der Linden, G. Enhancing crop resilience to combined abiotic and biotic stress through the dissection of physiological and molecular crosstalk. Front. Plant Sci. 2014, 5, e207.

71. Pastor, V.; Luna, E.; Ton, J.; Cerezo, M.; García-Agustín, P.; Flors, V. Fine tunning of reactive oxygen species homeostasis regulates primed immune responses in Arabidopsis. Mol. Plant Microbe Interact. 2013, 11, 1334–1344.
72. Miller, G.; Suzuki, N.; Ciftci-Yilmaz, S.; Mittler, R. Reactive oxygen species homeostasis and signalling during drought and salinity stresses. Plant Cell Environ. 2010, 33, 453–467.
73. Spoel, S.H.; Tad, Y.; Loake, G.J. Post-translational protein modification as a tool for transcription reprogramming. New Phytol. 2010, 186, 333–339.
74. Davletova, S.; Schlauch, K.; Coutu, J.; Mittler, R. The Zinc-Finger protein Zat12 plays a central role in reactive oxygen and abiotic stress signaling in Arabidopsis. Plant Physiol. 2005, 139, 847–856.
75. Desikan, R.; Mackerness, S.A.H.; Hancock, J.T.; Neill, S. Regulation of the Arabidopsis transcriptome by oxidative stress. Plant Physiol. 2001, 127, 159–172.
76. Miller, G.; Shulaev, V.; Mittler, R. Reactive oxygen signaling and abiotic stress. Physiol. Plant. 2008, 133, 481–489.
77. Gechev, T.S.; van Breusegem, F.; Stone, J.M.; Denev, I.; Laloi, C. Reactive oxygen species as signals that modulate plant stress responses and programmed cell death. Bioessays 2006, 28, 1091–1101.
78. Tanou, G.; Molassiotis, A.; Diamantidis, G. Hydrogen peroxide and nitric oxide-induced systemic antioxidant prime-like activity under NaCl-stress and stress-free conditions in citrus plants. J. Plant Physiol. 2009, 166, 1904–1913.
79. Zhang, D.; Martyniuk, C.J.; Trudeau, V.L. SANTA domain: A novel conserved protein module in Eukaryote with potential involvement in chromatin regulation. Bioinformatics 2006, 22, 2459–2462.
80. Wurzinger, B.; Mair, A.; Pfister, B.; Teige, M. Cross-talk of calcium-dependent protein kinase and MAP kinase signaling. Plant Signal. Behav. 2011, 6, 8–12.
81. Teige, M.; Scheikl, E.; Eulgem, T.; Roczi, F.; Ichimura, K.; Shinozaki, K.; Dangl, J.L.; Hirt, H. The MKK2 pathway mediates cold and salt stress signaling in Arabidopsis. Mol. Cell 2004, 15, 141–152.
82. Brader, G.; Djamei, A.; Teige, M.; Palva, E.T.; Hirt, H. The MAP kinase kinase MKK2 affects disease resistance in Arabidopsis. Mol. Plant Microbe Interact. 2007, 20, 589–596.
83. Zhang, S.; Liu, Y. Activation of salicylic acid-induced protein kinase; a mitogen-activated protein kinase; induces multiple defense responses in tobacco. Plant Cell 2001, 13, 1877–1889.
84. amajová, O.; Plíhal, O.; Al-Yousif, M.; Hirt, H.; Šamaj, J. Improvement of stress tolerance in plants by genetic manipulation of mitogen-activated protein kinases. Biotechnol. Adv. 2013, 31, 118–128.
85. Li, Y.; Zhang, L.; Wang, X.; Hao, L.; Chu, X.; Guo, X. Cotton GhMPK6a negatively regulates osmotis tolerance and bacterial infection in transgenic Nicotiana benthaminana and plays a pivotal role in development. FEBS J. 2013, 280, 5128–5144.
86. Xiong, L.; Yang, Y. Disease resistance and abiotic stress tolerance in rice are inversely modulated by an abscisic acid-inducible mitogen-activated protein kinase. Plant Cell 2003, 15, 745–759.
87. Pitzschke, A.; Djamel, A.; Teige, M.; Hirt, H. VIP1 response elements mediate mitogen-activated protein kinase 3-induced stress gene expression. Proc. Natl. Acad. Sci. USA 2009, 106, 18414–18419.
88. Chinchilla, D.; Zipfel, C.; Robatzek, S.; Kemmerling, B.; Nurnberger, T.; Jones, J.D.G.; Felix, G.; Boller, T. A flagellin-induced complex of the receptor FLS2 and BAK1 initiates plant defence. Nature 2007, 448, 497–500.
89. Ichimura, K.; Mizoguchi, T.; Yoshida, R.; Yuasa, T.; Shinozaki, K. Various abiotic stresses rapidly activate Arabidopsis MAP kinases ATMPK4 and ATMPK6. Plant J. 2000, 24, 655–665.

90. Gudesblat, G.E.; Iusem, N.D.; Morris, P.C. Guard cell-specific inhibition of Arabidopsis MPK3 expression causes abnormal stomatal responses to abscisic acid and hydrogen peroxide. New Phytol. 2007, 173, 713–721.
91. Beckers, G.J.M.; Jaskiewicz, M.; Conrath, U. Mitogen-Activates Protein Kinases 3 and 6 are required for full priming of stress responses in Arabidopsis thaliana. Plant Cell 2009, 21, 944–953.
92. Sharma, R.; de Vleesschauwer, D.; Sharma, M.K.; Ronald, P.C. Recent advances in dissecting stress-regulatory crosstalk in rice. Mol. Plant 2013, 6, 250–260.
93. Holley, S.R.; Yalamanchili, R.D.; Moura, D.S.; Ryan, C.A.; Stratmann, J.W. Convergence of signaling pathways induced by systemin; oligosaccharide elicitors; and Ultraviolet-B radiation at the level of mitogen-activated Protein Kinases in Lycopersicon peruvianum suspension-cultured cells. Plant Physiol. 2003, 132, 1728–1738.
94. Lu, C.; Han, M.-H.; Guevara-Garcia, A.; Fedoroff, N.V. Mitogen-activated protein kinase signaling in postgermination arrest of development by abscisic acid. Proc. Natl. Acad. Sci. USA 2002, 99, 15812–15817.
95. Miura, K.; Tada, Y. Regulation of water; salinity and cold stress responses by salicylic acid. Front. Plant Sci. 2014, 5, e4.
96. Zhou, J.; Xia, X.-J.; Zhou, Y.-H.; Shi, K.; Chen, Z.; Yu, J.-Q. RBOH1-dependent H2O2 production and subsequent activation of MPK1/2 play an important role in acclimation-induced cross-tolerance in tomato. J. Exp. Bot. 2014, 65, 595–607.
97. Cramer, G.R.; Urano, K.; Delrot, S.; Pezzotti, M.; Shinozaki, K. Effects of abiotic stress on plants: A systems biology perspective. BMC Plant Biol. 2011, 11, 163–177.
98. Xiong, L.; Schumaker, K.S.; Zhu, J.K. Cell signaling during cold; drought; and salt stress. Plant Cell. 2002, 14, 165–183.
99. Liu, C.; Ruan, Y.; Lin, Z.; Wei, R.; Peng, Q.; Guan, C.; Ishii, H. Antagonism between acibenzolar-S-methyl-induced systemic acquired resistance and jasmonic acid-induced systemic acquired susceptibility to Colletotrichum orbiculare infection in cucumber. Physiol. Mol. Plant Pathol. 2008, 72, 141–145.
100. Flors, V.; Ton, J.; van Doorn, R.; Jakab, G.; García-Agustín, P.; Mauch-Mani, B. Interplay between JA, SA and ABA signalling during basal and induced resistance against Pseudomonas syringae and Alternaria brassicicola. Plant J. 2008, 54, 81–92.
101. Ton, J.; Flors, V.; Mauch-Mani, B. The multifaceted role of ABA in disease resistance. Trends Plant Sci. 2009, 14, 310–317.
102. Truman, W.; Torres de Zabala, M.; Grant, M. Type III effectors orchestrate a complex interplay between transcriptional networks to modify basal defence responses during pathogenesis and resistance. Plant J. 2006, 46, 14–33.
103. De Torres-Zabala, M.; Truman, W.; Bennett, M.H.; Lafforgue, G.; Mansfield, J.W.; Egea, P.R.; Bogre, L.; Grant, M. Pseudomonas syringae pv. Tomato hijicks the Arabidopsis abscisic acid signaling pathway to cause disease. EMBO J.2007, 26, 1434–1443.
104. García-Andrade, J.; Ramirez, V.; Flors, V.; Vera, P. Arabidopsis ocp3 mutant reveals a mechanism linking ABA and JA to pathogen-induced callose deposition. Plant J. 2011, 67, 783–794.
105. Luna, E.; Pastor, V.; Robert, J.; Flors, V.; Mauch-Mani, B.; Ton, J. Callose deposition: A multifaceted plant defense response. Mol. Plant Microbe Interact. 2011, 24, 183–193.
106. Mauch-Mani, B.; Mauch, F. The role of abscisic acid in plant-pathogen interactions. Curr. Opin. Plant Biol. 2005, 8, 409–414.
107. Robert-Seilaniantz, A.; Navarro, L.; Bari, R.; Lones, J.D.G. Pathological hormone imbalances. Curr. Opin. Plant Biol.2007, 10, 372–379.
108. Melotto, M.; Underwood, W.; Koczan, J.; Nomura, K.; He, S.Y. Plant stomata function in innate immunity against bacterial invasion. Cell 2006, 126, 969–980.
109. Chul Lee, S.; Luan, S. ABA signal transduction at the crossroad of biotic and abiotic stress responses. Plant Cell Environ. 2012, 35, 53–60.
110. Kim, T-H.; Hauser, F.; Ha, T.; Xue, S.; Bohmer, M.; Nishimura, N.; Munemasa, S.; Hubbard, K.; Peine, N.; Lee, B.; et al. Chemical genetics reveals negative regulation of

abscisic acid signaling by a plant immune response pathway. Curr. Biol. 2011, 21, 990–997.
111. Garcia-Andrade, J.; Ramirez, V.; Lopez, A.; Vera, P. Mediated plastid RNA editing in plant immunity. PLoS Pathog.2013, 9, e1003713.
112. Ton, J.; Jakab, G.; Toquin, V.; Flors, V.; Iavicoli, A.; Maeder, M.N.; Metraux, J.P.; Mauch-Mani, B. Dissecting the β-aminobutyric acid induced priming phenomenon in Arabidopsis. Plant Cell 2005, 17, 987–999.
113. Nishimura, M.T.; Stein, M.; Hou, B.H.; Vogel, J.P.; Edwards, H.; Somerville, S.C. Loss of a callose synthase results in salicylic acid-dependent disease resistance. Science 2003, 301, 969–972.
114. Jensen, M.K.; Hagedorn, P.H.; de Torres-Zabala, M.; Grant, M.R.; Rung, J.H.; Collinge, D.B.; Lyngkjaer, M.F. Transcriptional regulation by an NAC (NAM-ATAF1;2-CUC2) transcription factor attenuates ABA signalling for efficient basal defence towards Blumeria graminis f. sp. hordei in Arabidopsis. Plant J. 2008, 56, 867–880.
115. Bandurska, H.; Stroiński, A. The effect of salicylic acid on barley response to water deficit. Acta Physiol. Plant. 2005, 27, 379–386.
116. Swindell, W.R. The association among gene expression responses to nine abiotic stress treatments in Arabidopsis thaliana. Genet. Soc. Am. 2006, 174, 1811–1824.
117. Mengiste, T.; Chen, X.; Salmeron, J.; Dietrich, R. The BORTYTIS SUSCEPTIBLE1 gene encodes an R2R3MYB transcription factor protrin that is required for biotic and abiotic stress responses in Arabidopsis. Plant Cell 2003, 15, 2551–2565.
118. Seo, P.J.; Kim, M.J.; Park, J.Y.; Kim, S.Y.; Jeon, J.; Lee, Y.H.; Kim, J.; Park, C.M. Cold activation of a plasma membrane-tethered NAC transcription factor induces a pathogen resistance response in Arabidopsis. Plant J. 2010, 61, 661–671.
119. Seo, P.J.; Park, C.-M. MYB96-mediated abscisic acid signals induce pathogen resistance response by promoting salicylic acid biosynthesis in Arabidopsis. New Phythol. 2010, 186, 471–483.
120. Oh, S.J.; Song, S.I.; Kim, Y.S.; Jang, H.J.; Kim, S.Y.; Kim, M.; Kim, Y.K.; Nahm, B.H.; Kim, J.K. Arabidopsis CBF3/DREB1A and ABF3 in transgenic rice increased tolerance to abiotic stress without stunting growth. Plant Physiol. 2005, 138, 341–351.
121. Tran, L.S.P.; Nishiyama, R.; Yamaguchi-Shinozaki, K.; Shinozaki, K. Potential utilization of NAC transcription factors to enhance abiotic stress tolerance in plants by biotechnological approach. GM Crops 2010, 1, 32–39.
122. Shinozaki, K.; Yamaguchi-Shinozaki, K. Gene networks involved in drought stress response and tolerance. J. Exp. Bot.2007, 58, 221–227.
123. Snider, C.S.; Hsiang, T.; Zhao, G.; Griffith, M. Role of ice nucleation and antifreeze activities in pathogenesis and growth of snow molds. Phytopathology 2000, 90, 354–361.
124. Kim, S.-Y.; Kim, S.-G.; Kim, Y.-S.; Seo, P.-J.; Bae, M.; Yoon, H.-K.; Park, C.-M. Exploring membrane-associated NAC transcription factors in Arabidopsis: Implications for membrane biology in genome regulation. Nucleic Acids Res.2007, 35, 203–213.
125. Tsutsui, T.; Kato, W.; Asada, Y.; Sako, K.; Sato, T.; Sonoda, Y.; Kidokoro, S.; Yamaguchi-Shinozaki, K.; Tamaoki, M.; Arakawa, K.; et al. DEAR1; a transcriptional repressor of DREB protein that mediates plant defense and freezing stress responses in Arabidopsis. J. Plant Res. 2009, 122, 633–643.
126. Foyer, C.H.; Karpinska, B.; Krupinska, K. The functions of WHIRLY1 and REDOX-RESPONSIVE TRANSCRIPTION FACTOR 1 in crosstolerance responses in plants: A hypothesis. Philos. Trans. R. Soc. Lond. B Biol. Sci. 2014, 369, 20130226.
127. Zhang, Y.; Butelli, E.; de Stefano, R.; Schoonbeek, H.J.; Magusin, A.; Pagliarani, C.; Wellner, N.; Hill, L.; Orzaez, D.; Granell, A.; et al. Anthocyanins double the shelf life of tomatoes by delaying over ripening and reducing susceptibility to graymold. Curr. Biol. 2013, 23, 1094–1100.
128. Marco, F.; Altabella, T.; Alcázar, R.; Cuevas, J.C.; Bortolotti, C.; González, M.E.; Carrasco, P. Transcriptome analysis of polyamine overproducers reveals activation of plant stress responses and related signalling pathways tolerance in plants. Omics Plant Abiotic Stress Toler. 2011, 1, 82–90.

129. Atkinson, N.J.; Lilley, C.J.; Urwin, P.E. Identification of genes involved in the response of Arabidopsis to simultaneous biotic and abiotic stresses. Plant Physiol. 2013, 162, 2028–2041.
130. Goyal, R.K.; Hancock, R.E.W.; Mattoo, A.K.; Misra, S. Expression of an engineered heterologous antimicrobial peptide in potato alters plant development and mitigates normal abiotic and biotic responses. PLoS One 2013, 8, e77505.
131. Treutter, D. Significance of flavonoids in plant resistance: A review. Environ. Chem. Lett. 2006, 4, 147–157.
132. Nakabayashi, R.; Yonekura-Sakakibara, K.; Urano, K.; Suzuki, M.; Yamada, Y.; Nishizawa, T.; Matsuda, F.; Kojima, M.; Sakakibara, H.; Shinozaki, K.; et al. Enhancement of oxidative and drought tolerance in Arabidopsis by over accumulation of antioxidant flavonoids. Plant J. 2014, 77, 367–379.
133. Walters, D. Resistance to plant pathogens: Possible roles for free polyamines and polyamine catabolism. New Phytol.2003, 159, 109–115.
134. Bitrián, M.; Zarza, X.; Altabella, T.; Tiburcio, A.F.; Alcázar, R. Polyamines under abiotic stress: Metabolic crossroads and hormonal crosstalks in plants. Metabolites 2012, 2, 516–528.
135. Hussain, S.S.; Ali, M.; Ahmad, M.; Siddique, K.H. Polyamines: Natural and engineered abiotic and biotic stress tolerance in plants. Biotechnol. Adv. 2011, 29, 300–311.
136. Prime-A-Plant Group; Conrath, U.; Beckers, G.J.; Flors, V.; García-Agustín, P.; Jakab, G.; Mauch, F.; Newman, M.A.; Pieterse, C.M.; Poinssot, B.; et al. Priming: Getting ready for battle. Mol. Plant Microbe Interact. 2006, 19, 1062–1071.

CHAPTER 2

Metabolomic Profiling of Soybeans (*Glycine max* L.) Reveals the Importance of Sugar and Nitrogen Metabolism under Drought and Heat Stress

Aayudh Das [1,2], Paul J. Rushton [1,3] and Jai S. Rohila [1,4,*]

[1]*Department of Biology and Microbiology, South Dakota State University, Brookings, SD 57007, USA*

[2]*Department of Plant Biology, University of Vermont, Burlington, VT 05405, USA*

[3]*22nd Century Group Inc., Clarence, NY 14031, USA*

[4]*Dale Bumpers National Rice Research Center, USDA-ARS, Stuttgart, AR 72160, USA*

ABSTRACT

Soybean is an important crop that is continually threatened by abiotic stresses, especially drought and heat stress. At molecular levels, reduced yields due to drought and heat stress can be seen as a result of alterations in metabolic homeostasis of vegetative tissues. At present an incomplete understanding of abiotic stress-associated metabolism and identification of associated metabolites remains a major gap in soybean stress research. A study with a goal to profile leaf metabolites under control conditions (28/24 °C), drought [28/24 °C, 10% volumetric water content (VWC)], and heat stress (43/35 °C) was conducted in a controlled environment. Analyses of non-targeted metabolomic data showed that in response to drought and heat stress, key metabolites (carbohydrates, amino acids, lipids, cofactors, nucleotides, peptides and secondary metabolites) were differentially accumulated in soybean leaves. The metabolites for various cellular processes, such as glycolysis, the tricarboxylic acid (TCA) cycle, the pentose phosphate pathway, and starch biosynthesis, that regulate carbohydrate metabolism, amino acid metabolism, peptide metabolism, and purine and pyrimidine biosynthesis, were found to be affected by drought as well as heat stress. Computationally based regulatory networks predicted additional compounds that address the possibility of other metabolites and metabolic pathways that could also be important for soybean under drought and heat stress conditions. Metabolomic profiling demonstrated that in soybeans, keeping up with sugar and nitrogen metabolism is of prime significance, along with phytochemical metabolism under drought and heat stress conditions.

Keywords: crop yield; drought stress; heat stress; metabolomic profiling; phytochemicals; soybean

1. INTRODUCTION

Environmental stresses, such as drought and heat, are serious threats to soybean cultivation in the US and in other soybean growing countries, causing multi-billion dollar losses every year. Stress in plants can arise as a reduction in growth and development that could be the result of disrupted metabolic homeostasis which requires the temporal readjustment of several metabolic pathways [1]. Under abiotic stress it is a general trend that growth and yield of crops are reduced below optimum levels [2]. Deryng et al. (2014) showed that by the year 2080, soybean yields at global level may reduce by a quarter ($\Delta Y = 15.3 \pm 26.5\%$ versus $20.4 \pm 22.1\%$ without heat stress) as a result of heat stress [3]. The growth, development and yields of soybean are significantly affected by drought and heat stress; thus, in mid-south United States, an alternative production method, i.e., the Early Soybean Production System, is in practice to help farmers to shift the planting and harvest times by incorporating maturity group III–V in soybean cultivation to minimize late season drought stress effects on yield [4]. Further, it has also been shown that abiotic stresses are directly related to changes in various cellular metabolic pathways [4,5,6,7,8], including carbohydrate, amino acid, and peptide metabolism [9]. Hence, abiotic stresses that alter the metabolomic profiles of soybeans leaf tissues (source) could reduce its yield (sink) potential. In the past, genetic and agronomic advancements have enhanced the drought tolerance of crop plants, and have resulted in stabilizing yields over time [10], but now with advancements in mass spectrometry (MS) technology, it is imperative to study the global metabolomic responses of soybeans to drought and heat stresses for developing metabolomic markers, utilizing metabolic pathways, and assisting soybean breeding programs. Non-targeted metabolomics aided by the advancements of MS technologies is capable of profiling diverse classes of metabolites, and thus has become a platform of choice to study metabolites perturbed by environmental cues [5], and are continually adding significant understanding to our knowledge of molecular mechanisms that could lead to predictive biology in the future by taking a systems biology approach [6]. Recently, Rabara et al. [7] compared two metabolomic profiles of soybean and tobacco to elucidate unique signature metabolites during progressive dehydration treatment, and showed that 4-hydroxy-2-oxoglutaric acid in tobacco roots and coumestrol in soybean roots show the highest tissue-specific accumulations. Previously, in a similar progressive drought stress study in *Medicago truncatula*, Zhang et al. [11] meticulously combined transcriptomic and metabolomic data of roots and shoots from control, mild-, moderate-, severely-stressed, and re-watered plants, and provided insights into the regulation of metabolic pathways operating under different levels of drought stress in plants. The study concluded that myo-inositol and proline had striking regulatory profiles indicating involvement in the *Medicago* drought tolerance mechanism.

Previously, our group reported differential proteome expression of soybean leaves under drought and heat stress conditions [12]. The present study was a logical extension of the proteomic evaluations, to the metabolomic level. Genetic variability within a crop species is strongly expressed at the metabolome level. In other words, the genotypic effect on the metabolome could be much greater than the analytical noise or within-group biological variations. Likewise, the magnitude of metabolic or proteomic variations for a single variety under constant conditions is typically so low that it can be used to easily detect many types of stress responses including drought and heat.

Metabolomics is being used to measure differential abundance of metabolites at a global level; and has successfully been used in many crops such as soybean, rice, corn, and barley [13,14,15,16,17]. In Arabidopsis, metabolic profiling has elucidated the temporal dynamics associated with heat stress [18]. Earlier integrated metabolomic-proteomics studies have shown comprehensive exploration of soybean roots and hypocotyls under abiotic [19] and wheat biotic stresses [20]. Similar metabolomic studies were performed in grapes for drought and salinity stresses to define alterations in the metabolic pathways related to gluconeogenesis, energy metabolism and nitrogen assimilation [21]. At present an incomplete understanding of abiotic stress-associated metabolism and identification of associated metabolites remains a major gap in soybean stress research. In addition, metabolomic studies of economically important crops, such as soybean that is being used for food as well as feed, are of special interest, and are advantageous for genomics-assisted selection for plant breeders to develop improved germplasm and release novel climate resilient varieties in the future [22].

A previous report of metabolomic profiling of soybean leaves has been performed using gas chromatography-mass spectrometry (GC/MS) analysis [23]. GC/MS is a very powerful when focusing on primary metabolites whereas liquid chromatography-mass spectrometry (LC/MS) is suitable for the detection of secondary metabolites [17]. A combination of GC/MS- and LC/MS-based metabolite profiling study in *Medicago truncatula* showed a significant alteration in amino acid and carbohydrate metabolism in response to biotic and abiotic elicitors [24]. In our soybean metabolomic study of leaves using GC- and LC-coupled with MS, we determined differential abundances of various primary and secondary metabolites, such as carbohydrates, amino acids, lipids, cofactors, nucleotides, peptides, and secondary metabolites, in response to drought and heat stresses. We demonstrated that in response to drought and heat stress, key metabolites were differentially regulated in various metabolic processes that included: (1) glycolysis, the TCA cycle, the pentose phosphate pathway, and starch biosynthesis that regulates carbohydrate metabolism; (2) amino acid metabolism; (3) peptide metabolism; (4) purine and pyrimidine biosynthesis that regulates nucleotide metabolism as a whole; and (5) metabolism involving secondary metabolites, including essential phytochemicals. The results clearly reveal various metabolic regulations and changes in metabolic pathways that correlate with plant growth and development.

2. RESULTS AND DISCUSSION

2.1. Differential Expression of Metabolites in Response to Heat and Drought Stress

Metabolites are the end products of the cellular regulatory processes, and their levels are regulated as biological system responses to

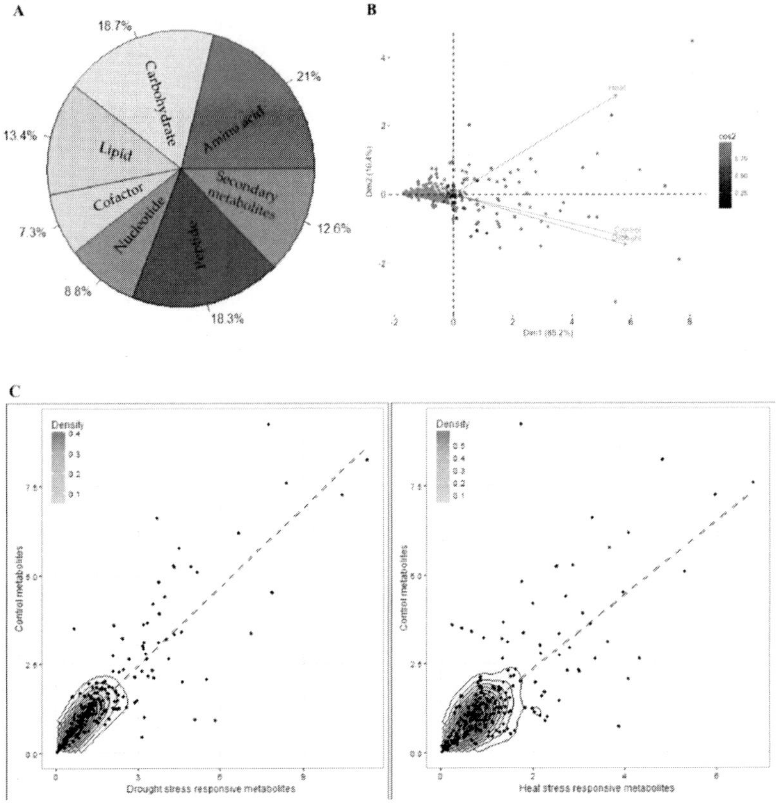

Figure 1. Metabolomics data of soybean leaves under drought and heat stress. (**A**) Classification of differentially expressed metabolites into chemical groups, in response to both drought and heat stress, based on their metabolic designations. Percentage values symbolize abundance of metabolites in each functional class. (**B**) Principal component analysis (PCA) biplot of all the variable metabolites originating from three different treatments i.e., control, drought and heat is shown here, (cos2 = the quality of the individuals on the factor map displayed as color intensity). (**C**) Scatter plot distribution of all groups of metabolites revealed that there is a linear positive correlation between the expression levels of control metabolites, and drought stress and heat stress-responsive metabolites. The best scatter of the metabolites was observed at the base of both the plot, which is visualized with an intensity level gradient ranging from 0.1 to 0.5. (Plot created by using R 3.3.1 software with the ggplot package).

environmental stress conditions. This analysis demonstrates that in response to drought and heat stresses the global metabolite present at a given time in plant cell "i.e., metabolome", changes and show differential regulation when compared between the treatments. Figure 1A, Figure S1, and Table S1 show the differential abundance of metabolites detected by GC-MS and LC-MS. Our metabolomics profiling using GC-MS and LC-MS revealed the concurrent detection of 266 putative metabolites, including primary and secondary metabolites. Interestingly, this study demonstrated that under heat stress, the number of metabolites (carbohydrates, amino acids, lipids, cofactors, nucleotides, peptides and secondary metabolites) decreased to a greater extent than under drought stress.

A principal component analysis (PCA) was performed on metabolite data to identify a pattern between the samples. Figure 1B (biplot of variables) shows that metabolites have been clustered, as each vector represents the accumulated effect of all the metabolites in that particular treatment, differently in response to heat stress compared to the control samples. Drought stress was found to have less of an effect compared to heat stress, based on PCA analysis. A scatter plot analysis (Figure 1C) revealed that there was a linear positive relationship for metabolite expression among the drought and heat stresses when compared with the control condition. We fitted a linear model to ascertain whether it was statistically rational to fit a linear correlation to those variables and to compare their differential expression. Based on the statistical inference that was derived by comparing the control versus drought, and control versus heat stress metabolites, we determined that there was a positive linear relationship among the fold changes of metabolites between stress and the control conditions.

2.2. Effect of Drought and Heat Stress on Carbohydrate Metabolism

The rate of glucose utilization by cells is known to be influenced by various environmental factors [25]. One of the outcomes of the current study was to observe the effects of heat stress on carbohydrate metabolism, specifically in glycolysis and the pentose phosphate pathway (PPP). Oxidative PPP is a major source of reducing power and metabolic intermediates for several biosynthetic processes in the cell, and apparent change in the cellular abundance of certain PPP metabolites often severely affects several other metabolism pathways [26]. Analysis of this study revealed that during heat stress, the amount of major glycolysis and PPP metabolites, such as glucose, dihydroxy acetone phosphate and pyruvate, decreased drastically compared to the control treatment

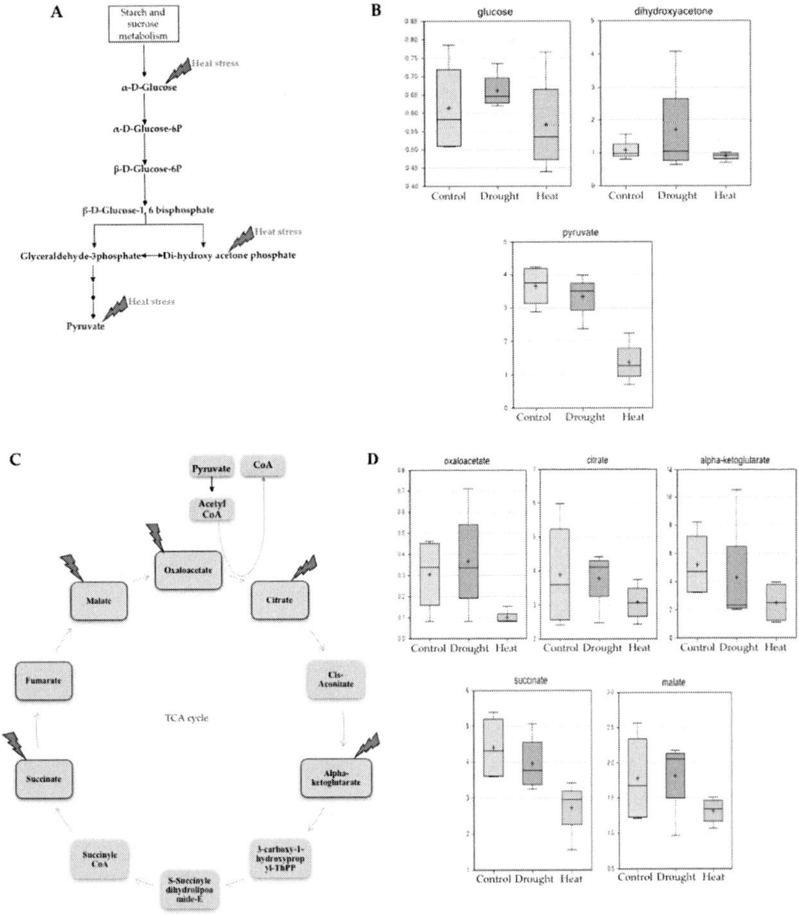

Figure 2. Effects of drought and heat stresses on the metabolites of glycolysis and the tricarboxylic acid (TCA) cycle. (**A**) Glycolysis pathway showing the metabolites that had abundances significantly affected by stresses in the study; (**B**) Box plots showing the relative abundances of glucose, dihydroxyacetone and pyruvate under the control, drought, and heat stress conditions. (**C**) TCA cycle showing metabolites that were found to have significant effect on their abundance of stress (**D**) Box plots showing relative abundances of metabolites of TCA cycle under various stressed conditions [the *Y* axis defines the relative abundances of specific metabolite and *X* axis defines the treatment group; the red lightning sign in panels (**A,C**) indicates heat stress; green box plot: control, orange box plot: drought and pink box plot: heat]. Inside the box plots, "+" indicates mean value, and "—" indicates median value.

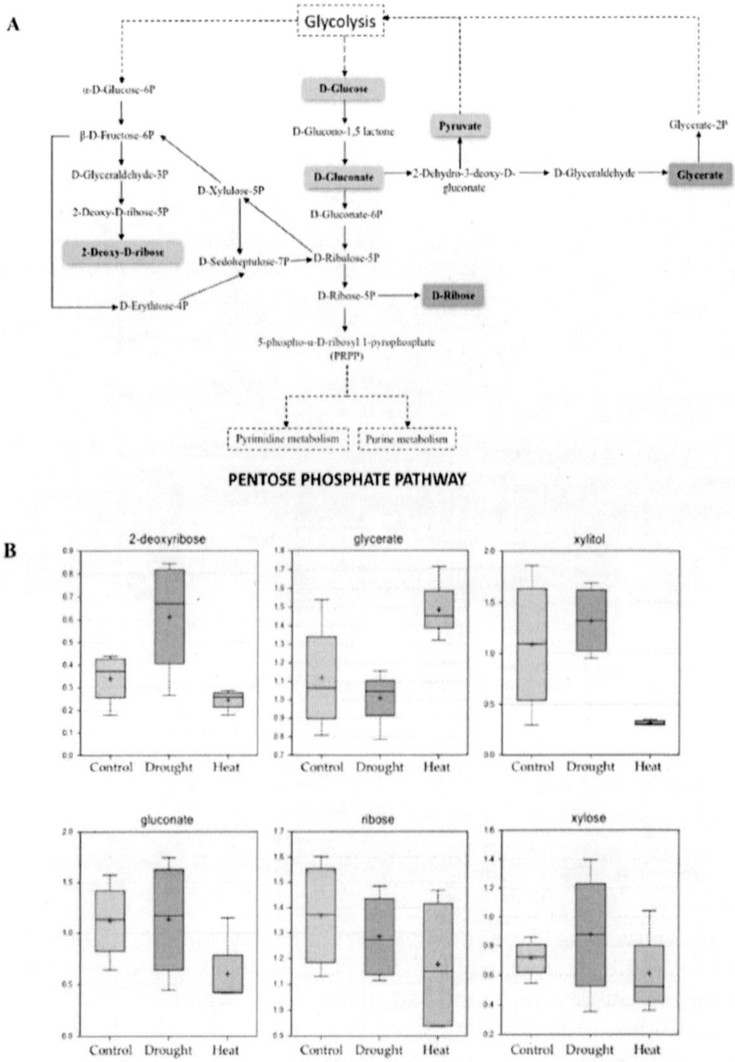

Figure 3. Effect of drought and heat stresses on metabolites related to the pentose phosphate pathway. (**A**) The pink colored metabolites indicate downregulation in response to heat stress, orange colored metabolites indicate downregulation in response to drought stress. (**B**) Box plots showing relative abundances of PPP metabolites and sugar compounds under control, drought, and heat stress conditions (pink highlights indicate heat stress-affected; box plot description is the same as Figure 2).

(Figure 2 and Figure 3, and Table S2). Respiratory pathways, such as glycolysis and the tricarboxylic acid (TCA) cycle, are essential for energy delivery to different organelles, and for maintaining various physiological functions [27]. The study indicates that during heat stress, the level of several TCA cycle metabolites, such as oxaloacetate, citrate, alpha-ketoglutarate, succinate and malate decrease significantly (Figure 2, and Table S2). This would be expected to cause a reduction in adenosine triphosphate (ATP), diminishing plant photosynthesis and carbon fixation, which may have correlating effects on soybean yields. Furthermore, proteomic studies [12] have also indicated that in response to drought and heat stresses several major enzymes of the Calvin cycle and PPP were highly affected. The metabolomic findings of the current study support such proteomic results, and illustrate how drought and heat stress may affect carbohydrate metabolism.

It has been reported that constant exposure of plants to high temperature stress reduces the consistent increase in starch and soluble sugar (e.g., maltose, sucrose) concentration, and decreases the soluble sugar concentrations in maturing pollen grains [28]. The results of this study also show that due to high air temperature (heat) stress, maltose, galacititol, sucrose, and mannitol concentrations decrease in leaves by more than two-fold (Figure S2). Primarily, due to a low sugar concentration, the transport of sugars to the reproductive cells may be greatly affected, resulting in poor development of the gametophyte [29].

The metabolic fate of maltose due to heat stress will ultimately be seen in glycolysis, as maltose is broken down into glucose followed by glucose-6-phosphate, which regulates glycolysis, 2-keto-3-deoxy-6-phosphogluconate (KDPG) and the pentose phosphate pathways [30]. Plants use mannitol, one of the crucial sugar alcohols that is synthesized simultaneously with sucrose, as an important photo-assimilate, and almost half the fixed CO_2 is converted to mannitol, which is used during growth of non-photosynthetic [31] and sink tissues [32]. Interestingly, our results indicate that mannitol is reduced due to heat stress, which may affect phloem translocation in soybeans. Furthermore, galactitol, one of the major sugar alcohols produced by the reduction of galactose, was reduced during heat stress, which indicates a decrease in galactose that may directly affect glycolysis and other major metabolic pathways that use galactose. One study shows that under stress conditions, when galactose cannot be utilized routinely in a conventional way, galactitol is formed in tissues to prevent galactose toxicity [33]. In response to drought stress, we found that most of the major carbohydrates such as glucose, fructose, sucrose and raffinose were accumulated at a higher level as compared to the control, which indicates that in response to drought, soybean growth is restricted and more carbohydrates are stored [34]. Additionally, we found that sugar compounds such as ribose, deoxyribose, gluconate, xylose and xylitol are also increased during drought stress; however, the increase is relatively lower during heat stress, which indicates that heat stress will have a greater impact on amino acid metabolism and nucleic acid biosynthesis (Figure 3 and Figure S3, and Table S2). A comparatively recent study in soybean also indicated that under drought conditions, levels of the individual sugars varied considerably among the genotypes and among tissue types; the effect was found to be more pronounced in the leaves than in nodules [8]. Similarly, a transcriptomic profiling of soybean (cv. BR16) during water deficit is also in agreement with the finding of significantly increased transcripts for sugars during water deficit conditions [35].

2.3. Effect of Drought and Heat Stress on Amino Acid Metabolism

The metabolism of amino acids includes various biosynthetic reactions by which they are assembled as precursors of polypeptides and broken down to recuperate metabolic energy [36]. Under control conditions, during the daytime, photosynthesis provides the majority of the energy to the plants, and during the night, certain metabolites, primarily starch and amino acids, drive the plant's energy generation processes via glycolysis, the TCA cycle, and amino acid metabolism [37,38].

Exposure of plants to drought and heat stresses reduces the photosynthesis rate, resulting in reduced growth and several physiological abnormalities. Convergent signal transduction associated with energy production regulates the response of the plants' metabolome to various stresses [39]. Phosphoenolpyruvate (PEP) and erythrose-4-phosphate, which are main precursors of the aromatic amino acids, form 2-keto-3-deoxy-D-arabinoheptulosonate-7-phosphate by condensation reaction, eventually producing chorismate. Chorismate is the branch point for the synthesis of all aromatic amino acids (tryptophan, tyrosine and phenylalanine) [40]. The current study indicated that heat stress significantly affected aromatic amino acid metabolism. One of the precursor compounds of the aromatic amino acid biosynthetic pathway, shikimate, was greatly reduced during heat stress, which ultimately affected tryptophan, tyrosine and phenylalanine biosynthesis (Figure 4, and Table S2). The aspartic acid (Asp) family (oxaloacetic acid and pyruvate derived) of biosynthetic pathways leads to the biosynthesis of essential amino acids. i.e., lysine, alanine, threonine, methionine and isoleucine [37].

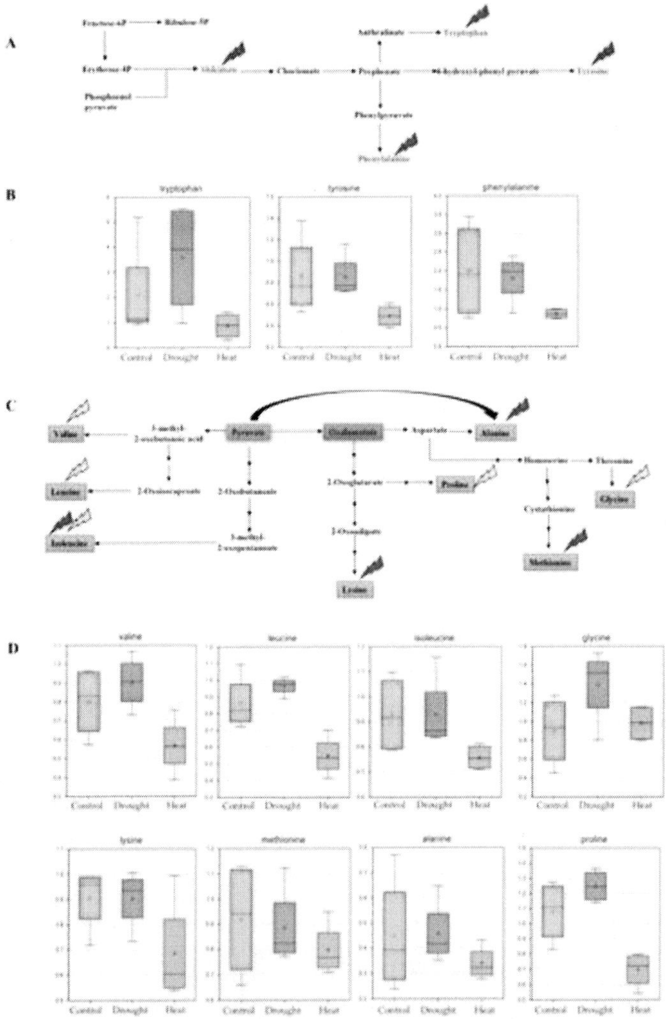

Figure 4. Effect of drought and heat stresses on metabolites of aromatic amino acid metabolism. (**A**) Pathway for aromatic amino acid metabolism showing the metabolites that were found to have significant effect of stresses on their abundance in the study (**B**) Box plots showing the relative abundances of aromatic amino acids in control, drought, and heat stress conditions (**C**) Effect of drought and heat stress on pyruvate and oxaloacetate-derived amino acid biosynthesis (**D**) Box plots showing relative levels of pyruvate and oxaloacetate-derived amino acids in control, drought and heat stress conditions. (Box plot description is the same as in Figure 2; red and yellow lightning signs in (**A**) and (**C**) indicate the effects of heat and drought stress, respectively).

Metabolic regulation between the Asp-family pathway and the TCA cycle has a key impact on the physiological response of plants to drought and heat stresses. In our analysis, we also observed that all aspartate family-derived amino acids such as lysine, alanine, methionine and isoleucine, were reduced in abundance in response to heat stress (Figure 4, and Table S2). Recent studies also show that lysine is a precursor for a metabolic pathway that functions in plant development, and in response to various stresses [41].

It has been reported that alanine, one of the essential amino acids, is activated in response to flooding stress in *Lotus japonicus*, and is involved in regulating the TCA cycle, including anaplerotic and cataplerotic fluxes [42,43]. Increases in valine and leucine upon drought stress can also result from the consistent increase in the substrate pyruvate, which also featured in our GC-MS-based results [44]. The relative increase of isoleucine and leucine (Figure 4, and Table S2) in drought-stressed soybean leaves were in agreement with previous findings in Arabidopsis [45]. Glycine and proline are two major organic osmolytes that were highly accumulated (Figure 4) in response to drought stress [46,47]. Very high accumulation of cellular proline and glycine under drought stress was also observed in our study, which clearly shows the potential for an advanced tolerance response during adverse environmental conditions, such as drought, in soybeans. Thus, proline accumulation under drought stress might function as a sink for excess products that are necessary for the maintenance of photosynthetic and respiratory processes [47]. Similar findings specifically for amino acid metabolism, were presented by Tripathi et al. (2016) where metabolomic alterations during drought stress were monitored in hydroponics-grown soybeans [48].

2.4. Effect of Abiotic Stress on Nitrogen Metabolism

In agreement with proteomic findings (Das et al. [49]), a direct correlation for key amino acids such as glutamine and glutamate was reflected in this metabolomic investigation. The proteomics study [49] indicated that cytosolic glutamine synthetase, which assimilates ammonium to glutamine using glutamate as a substrate, is reduced significantly during drought, and alleviates nitric oxide (active radicals) levels to activate drought tolerance mechanisms in soybeans. Our metabolic data supports proteomic findings, as glutamate was found to be reduced during drought stress. By combining this metabolomics data with published proteomics data, we can make an inference that with lower levels of glutamine synthetase during drought stress it is difficult for soybean plants to assimilate ammonia, which in turn results in an abundance of glutamine (Figure S4). Interestingly, based on proteomics data [49] during heat stress, when glutamine synthetase is abundant, it helps the plant to use up the glutamate rapidly, in order to generate ample amounts of glutamine.

2.5. Differential Response in Peptide Metabolism

Metabolites are the structural intermediates between amino acids and proteins during biological inter-conversions between these two groups of molecules [50]. It has been reported that severe oxidative stress induces protein mistranslation through the impairment of aminoacyl-tRNA biosynthesis [51]. One of the inevitable consequences of drought stress in plants is enhanced reactive oxygen species (ROS) generation and ROS-mediated degradation of the photosynthetic machinery during moderate heat stress [52,53]. Our analysis of amino acid metabolism suggested severe defects in the level of accumulation of cellular amino acids in response to drought and heat stress, which may directly reflect the rate of peptide synthesis via amino acyl tRNA biosynthesis.

We have found a relatively lower accumulation of major gamma-glutamyl peptides, namely gamma-glutamylalanine, gamma-glutamylglutamate, gamma-glutamylphenylalanine and gamma-glutamyltryptophan, upon drought and heat stress (Table S1). The results indicated a direct correlation between amino acid synthesis and dipeptide synthesis, especially during heat stress when we found that almost 81% of the detectable dipeptides were at a relatively low level compared to the controls (Table S1). Furthermore, during drought stress, only 47% of detectable dipeptides were lowered compared to controls (Table S1). Others were at a similar level or were higher than the control, which suggests that the plant protects against the adverse effects of drought stress by activating a tolerance mechanism to alleviate translation without any impairment of aminoacyl-tRNA biosynthesis. The heat map analysis shows the differential expression of all the identified peptides upon drought and heat stress (Figure 5).

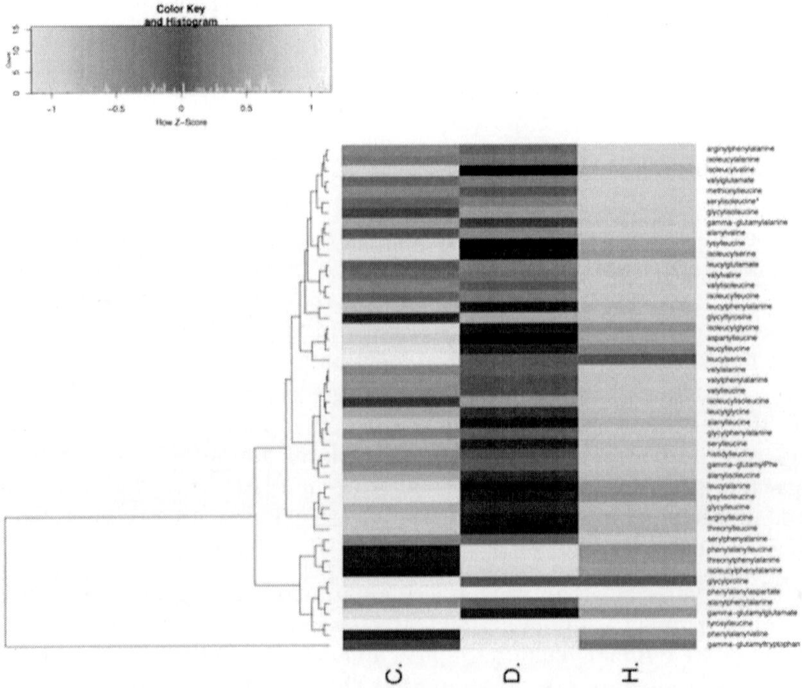

Figure 5. Differential abundance of peptides under drought and heat stresses. Heat map showing the differences in the abundance of the various peptides in soybean leaves under control (C.), drought (D.), and heat stress (H.) conditions.

2.6. Effect of Drought and Heat Stress on Nucleotide Metabolism and Protein Synthesis

Nucleotide metabolism is one of the essential components of plant metabolism and development, which affects many metabolic pathways [54].

Specifically, purine and pyrimidine nucleotides directly participate in nucleic acid synthesis, which provides the ultimate energy source for the synthesis of carbohydrates, lipids, peptides and secondary metabolites [55]. The de novo pyrimidine biosynthetic pathway (orotate pathway) is defined as the formation of uridine monophosphate (UMP)from carbamoyl phosphate (CP), orotate, 5-phosphoribosyl-1-pyrophosphate (PRPP), and a few other phosphate metabolites [56]. Our metabolomics profiling result indicates that the major building blocks, which are pyrimidine bases such as thymine, cytosine and uracil, are reduced significantly upon heat stress, which affects overall pyrimidine biosynthesis and nucleic acid biosynthesis (Figure 6).

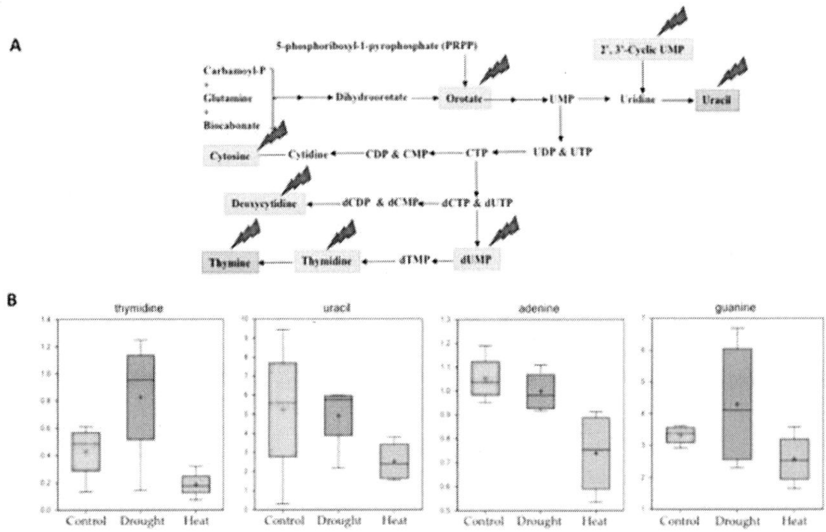

Figure 6. Effect of heat stress on regulation of pyrimidine metabolism. (**A**) Pathway for pyrimidine metabolism showing the metabolites that were found to have a significant effect of stresses on their abundance in the study. (**B**) Box plot showing major purine and pyrimidine bases and their differential abundances under control, drought and heat stress (Box plot description is the same as Figure 2; red lightning sign in (**A**) indicates heat stress).

Additionally, other precursor molecules of the pyrimidine biosynthetic pathway, such as thymidine, deoxycytidine, orotate, 2′,3′-cyclic UMP and dUMP are also reduced due to heat stress. Conversely, during heat stress, pyrimidine bases and precursor molecules are increased compared to the control, indicating the activation of a tolerance mechanism to protect nucleic acids and ultimately protein synthesis. The de novo purine biosynthetic pathway consists primarily of key amino acids, such as glutamine, aspartate, and glycine, activated ribose precursor PRPP, 10-formyl tetrahydrofolate, and CO_2 [54]. The results from the current investigation show that upon heat stress, major purine bases such as adenine and guanine are downregulated compared to the control, along with small precursor molecules, such as 2′-deoxyadenosine, 2′-deoxyguanosine, and guanosine-2′,3′-cyclic monophosphate (Figure 6). Similarly, upon drought stress, the relative levels of all metabolites in purine biosynthesis increased, clearly indicating that the drought tolerance machinery is activated during stress conditions. The 2′,3′-cNMPs were more variable in soybeans, especially in the drought stressed samples. However, under heat stress, they appeared to respond in a similar manner to that observed for wheat, i.e., lower relative to the control. As for 3′,5′-cGMP, it was not detected in the drought stressed samples, and may have been lower in heat stress than in the control, although the variation in the control was high.

2.7. Differential Regulation of Secondary Metabolites in Response to Drought and Heat Stress

Studies on plant secondary metabolites indicate that they play a major role in the adaptation of plants to the changing environment and in overcoming various stress constraints [57,58]. It has been reported that heat stress greatly reduces the leaf area, which is an immediate response to heat. This effect is alleviated by the synthesis of vital amino acids and secondary metabolites [59]. Furthermore, studies specify that various polyphenols, particularly flavonoids and hydroxycinnamates accumulate at a higher level under drought-stress conditions to normalize the CO_2 assimilation rate in the leaves [60].

In this metabolomic profiling, we found that the alkaloid caffeate accumulated at a comparatively higher level (1.5-fold) in response to drought stress than control plants, thus assuaging growth-related adversities and lignin synthesis [61]. One important benzenoid, vanillate, which is transformed to guaiacol and facilitates lignin biosynthesis, has a two-fold higher level during drought stress, but is at lower levels in heat stressed plants compared to the control plants. Flavonoid biosynthesis has been revealed to be upregulated by drought-induced oxidative stress, whereas photosynthesis is downregulated concomitantly [62]. This metabolic profiling also indicated that 73% of detected flavonoids and phenylpropanoids are upregulated upon drought stress, supporting the drought responsive role of flavonoids in soybeans (Figure 7A). Conversely, in response to heat stress, 84% of the detected flavonoids and phenylpropanoids accumulated at a lower level compared to the control, which indicates that during heat stress soybeans, especially cv. Williams-82, struggle to activate defense mechanisms effectively to relieve stress.

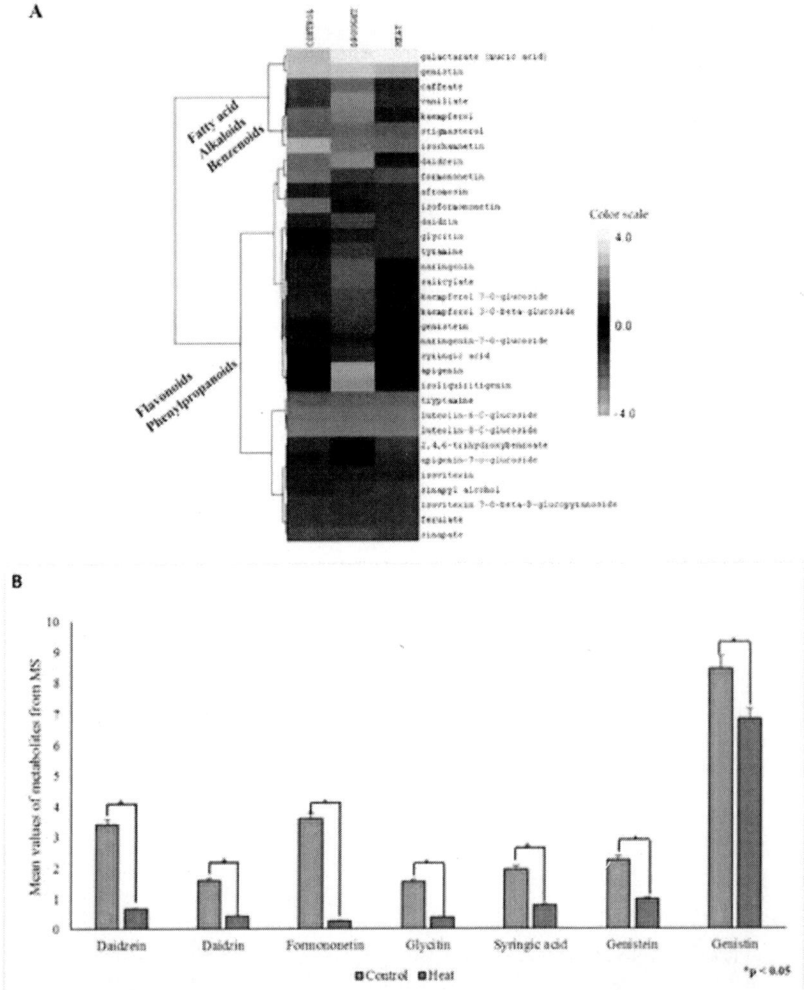

Figure 7. Effect of drought and heat stress on secondary metabolites. (**A**) Heat map showing differential expressions of various secondary metabolites during drought and heat stress compared to the control treatment. (**B**) Bar diagram showing relative expression of various phytochemicals in response to heat stress (blue) compared to the control (green). Various phytochemicals are shown on the *X*-axis, and the Y-axis has the scaled intensity of the respective phytochemical. Each value represents the mean ± S.E., and the asterisks designate the significance of changes from their corresponding control ($p < 0.05$).

2.8. Differential Expression of Soybean Phytochemicals

Soybeans are considered as one of the plant that produces a vast amount of isoflavones, which have profound effects as antioxidants [63]. Hosseinian et al. reported that in purple wheat, the anthocyanin content is increased due to heat stress, suggesting that abiotic stress may affect plant phytochemicals [64]. Conversely, studies with lettuce showed that they respond to heat stress by inducing the synthesis of various phytochemicals that act as antioxidants to activate defense mechanisms [65].

Interestingly, our results showed a significant reduction for some of the major soybean phytochemicals (daidzein, daidzin, formononetin, glycitin, syringic acid, genistein and genistin) in response to heat stress (Figure 7B). Thus, it can be implicated that heat stress affects the synthesis of these essential soy phytochemicals, and therefore diminishes the antioxidant capability of the plant. Conversely, our results showed that in response to drought stress, the same phytochemicals increased compared to the control (Figure S5), suggesting that plants might be activating their tolerance mechanisms and antioxidant activity to alleviate the stress [66]. The findings, especially the abundance of various phytochemicals (daidzein, daidzin, formononetin, syringic acid, genistein and genistin) in response to drought stress, are also supported by recent studies [48].

2.9. Metscape Analysis of Metabolomics Data Derived from LC- and GC-MS

For understanding of biological functions via usage of the computational approach, it is desirable to elucidate all functional metabolic interactions undergoing in the cell at a given time. There were many cellular compounds which were not detected in the metabolomics experimental pool for various reasons, for example, poor extraction or extremely low abundance, but they may have been be functionally associated with the detected compounds during drought and heat stress. MapMan and Metscape have the capability of illustrating this using a chart approach, to show an overall possible scenario for all of the related cellular compounds. Here we displayed a variety of functional metabolic interactions between the differentially expressed metabolites in soybeans in response to drought and heat stress, to obtain an improved understanding of metabolic pathways and metabolic regulation (Figure S6). Figure S7 has 27 input metabolites (red color) demonstrating how various carbohydrates are involved in glycolysis, TCA cycle, pentose phosphate pathway and other significant biological processes, and functionally interact with other related metabolites, and thus eventually may have an effect on biological and cellular processes when the soybean is exposed to stress. Figure S7 also shows related metabolites (yellow color), which could also be present in the same pathway as that of the input metabolites, such as pyruvate and oxaloacetic acid (OAA) of the TCA cycle. Thus, this network predicted that if a stress affected an input metabolite abundance, the abundance of corresponding related metabolites that were in the same pathway could also be affected as a result. Similarly, Figure S8 with 30 input amino acids shows related metabolites for amino acid biosynthesis metabolism. In sections above, we have discussed how drought and heat stress affects amino acid metabolism at various levels; in extension to that, this network predicts how the same pathway-derived undetected metabolites can also be affected by the same stress.

3. MATERIALS AND METHODS

3.1. Plant Growth Conditions

Soybean (*Glycine max* L. cultivar: Williams-82 [67]) seeds were surface sterilized using 50% Clorox and planted in pots filled with Metro mix-360 (Sun Gro Horticulture Agawam, MA, USA). A total of nine pots (14 L volume; C1600, NurserySupplies.com) consisting of three groups (control, drought, and heat stress) with three replications of each in complete randomization were planted in a growth chamber (Conviron, Winnipeg, MB, Canada) illuminated with white fluorescent and incandescent bulbs (500 µmoles m^{-2} s^{-1}, 16 h of light), at 28/24 °C and 60% relative humidity. Each pot had four individual soybean plants, and the stress treatments were performed as explained in Das et al. [49]. Briefly, watering was done daily, and as needed, with tap water until the second trifoliate leaves emerged. At this stage, the nine pots were separated into three groups where each group contained three treatments as follows: control, drought, and heat stress. In the first 'control' group, plants were maintained under water-saturated soil conditions at 28/24 °C. In the second 'drought' group, the plants were not watered, and gradually reached 10% volumetric water content (VWC) on the seventh day of the experiment under 28/24 °C. A separate growth chamber contained pots that were exposed to high temperatures (43/35 °C with 50% humidity) with water saturated soils; these were the third 'heat' group. After the seventh day of treatments, young trifoliate leaves from developmentally matched plants (number of plants >3) were harvested at noon in bulk from one pot as one replication unit, based on previously published studies in Soybean [49], and were immediately snap-frozen in liquid nitrogen and stored at −80 °C.

3.2. Metabolomic Profiling

Metabolomic profiles were analyzed using harvested soybean leaves. Leaf samples ground in liquid nitrogen were submitted to Metabolon, Inc. (Durham, NC, USA). The sample preparation and analysis process was performed as previously described [68,69,70]. Briefly, samples were extracted in 400 μL of methanol using an automated liquid handling system; the resultant samples were split into three aliquots for analysis on three MS instrument platforms. These included two UPLC/MS platforms, one optimized for positive ionization and a second optimized for negative ionization. The third aliquot was derivatized and analyzed by GC/MS. The UPLC-MS/MS platform included a Waters ACQUITY ultra-performance liquid chromatography (UPLC) system, a Thermo Scientific Q-Exactive high resolution/accurate mass spectrometer with heated electrospray ionization (HESI-II) source, and an Orbitrap mass analyzer operated at 35,000 mass resolution. The MS scan range was from 80 to 1000 *m/z*. GC samples were analyzed on a Thermo-Finnigan Trace DSQ fast-scanning single-quadrupole mass spectrometer using electron impact ionization (EI). The MS scan range was from 50 to 750 *m/z*. The mass spectrometry peaks were identified using Metabolon's proprietary peak integration software. Peaks were quantified using the area-under-the-curve method. The identification of known chemical entities was based on comparisons to metabolomic library entries of purified standards [71]. Because the analysis spanned multiple days, a data normalization step was performed to correct any variation resulting from instrument inter-day tuning differences. Essentially, each compound was corrected in run-day blocks by setting the medians equal to one (1.00) and normalizing each data point proportionately. Since all the treatments were comprised of three replicates (n = 3 for all groups), the statistical significance of the results was evaluated using Welch's two sample t-test and a level of significance of $p \leq 0.05$ for two group comparisons.

3.3. Data Analyses

The log2-transformed values of the metabolites mean abundances were used to cluster the metabolomic data. The heat maps and hierarchical clustering of the metabolites were performed using Gene Cluster 3.0 software [72]. The metabolite clusters were visualized using JAVA TREEVIEW software [73]. The statistical significance of the results was evaluated using Welch's two sample *t*-test. A value of $p \leq 0.05$ was considered statistically significant for two-group comparisons. To visualize the relationships of metabolites expression levels in control versus stress (drought or heat) a scatter plot was constructed using R 3.3.1 software [74]. MapMan application software (MapMan Version 3.5.1R2) was used for understanding the metabolic distributions and metabolic regulation in response to the stresses [75]. Selected metabolomic interactions were portrayed using MetScape 3.1 plugin for Cytoscape; we also displayed several protein networks using Cytoscape software, Version 3.0.1, and the KEGG database was used for ID referencing [76,77,78].

4. CONCLUSIONS

Metabolomic analysis of soybean leaves revealed a dynamic alteration in metabolites in response to drought and heat stress. This analysis clearly demonstrates the changes in metabolic pathways that may affect soybean growth and development. However, in response to heat stress, the soybean plants were severely impacted. From a metabolomic standpoint, we found that several metabolites were downregulated during heat stress, complicating the homeostasis of the general metabolism of the soybean plant. A detailed combined proteomic-transcriptomic study using soybean roots, showed that proteins involved in amino acids, and lipid biosynthesis were severely affected under heat stress [79]. This study corroborates our metabolomics findings in response to heat stress. However, an opposite scenario was observed during drought stress, where most of the metabolites significantly increased in abundance or remained the same as the control plants in our investigation. Interestingly enough, a similar reflection was observed during a previous transcriptome-based study on soybean (cv. BR16, and cv. Williams 82), which revealed that several genes, involved in nucleotide metabolism, amino acid biosynthesis, lipid metabolism secondary metabolism, were significantly increased during drought stress compared to the control [35,80]. Another study using drought-tolerant (cv. Embrapa 48) and drought-sensitive (cv. BR16) soybeans indicated that under drought stress, more than 50% genes related to various metabolic processes increased in expression independent of the tolerance factor [81].

Based on metabolomic profiling and using a MapMan-derived metabolic pathway overview (Figure S6), we proposed a model (Figure 8), which explains the heat stress-related changes in soybeans. The present study showed that in response to drought stress, soybeans experience reduced growths to alleviate the stress effects. In addition, we suggested that during drought, soybeans try to activate a tolerance pathways to ensure an adjusted metabolism, which is likely one of the soybeans defense strategies, similar to other plants, for combating drought stress [82,83]. By using compounds that were detected by GC- and LC-MS as input data, we also built regulatory networks, addressing related compounds that were not detected but might be affected in response to abiotic stress (Figures S6 and S7). Likewise, we also found that heat stress affects the synthesis of phytochemicals, such as daidzein, daidzin, formononetin, glycitin, syringic acid, genistein and, genistin.

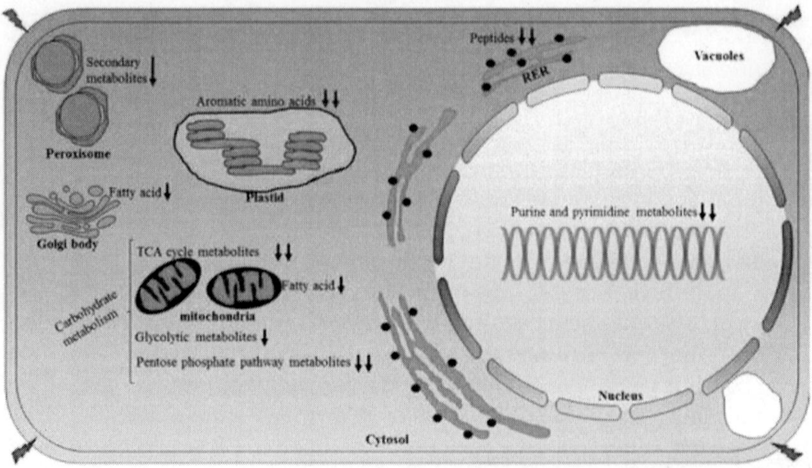

Figure 8. A proposed model shows the metabolomic responses of soybean leaves during heat stress. The figure shows the effects of heat stress on key metabolites in key cell organelles. Single and double downward-pointing arrows (based on scaled quantity of metabolites) indicate metabolites reduced due to heat stress. There was no metabolite found in this investigation that showed increased abundance during heat stress. The red lightning arrows on the corners of the cell represent heat stress.

In addition to revealing the dynamic alteration in metabolic pathways, metabolomic approaches also provide a parallel insight into metabolite composition that can be used for crop improvement projects [22]. Recent studies in tomatoes using comprehensive metabolic profiling uncovered traits for improved crop quality [84]. Metabolite profiling using NMR and GC-MS studies also revealed potentially unintended effects in genetically modified tomatoes [85]. Studies have also reported the emerging importance of using metabolites as selection markers for crop breeding, as metabolite biomarkers are associated with relevant traits under environmental stresses [12,86,87,88].

This study provides information on several metabolites that include major carbohydrates, amino acids, nucleotides and soybean phytochemicals, which can be useful for the development of better models to establish the connection between yield-associated traits and various metabolic pathways. The present study also has potential to support metabolomics-assisted breeding for the development of more stress-tolerant soybean varieties. The report, combined with high-throughput reverse genetics studies, can also assist in the rapid development of elite soybean lines. This study identifies a detailed mechanism of metabolic alteration in soybeans in response to drought and heat stresses.

SUPPLEMENTARY MATERIALS

The following are available online at www.mdpi.com/2223-7747/6/2/21/s1, Figure S1: Categorization of identified metabolites in seven broad groups. *Y* axis has number of metabolites; *X* axis has 7 different broad categories of identified metabolites. Green color bar represents number of metabolites that were found increased in abundance; golden color bar represents number of metabolites decreased in abundance compared to the control condition, Figure S2: Effect of stress on metabolites involved in starch and carbohydrate metabolism (A) Box plots showing the effect of drought and heat stress on the expression of maltose, galactitol, sucrose, and mannitol. (B) Heat map showing the effect of drought and heat stress on the abundance of metabolites that are involved in starch and carbohydrate metabolism, Figure S3: Metscape network shows the effect of abiotic stress on sugar metabolism. Red colored nodes indicate input metabolites from GC/LC-MS results, green squares indicate enzymes, grey squares indicate pathway number from KEGG database and other hexagons indicate other predicted metabolites that might be associated with the reaction, Figure S4: A proposed model by combining proteomic and metabolomic studies in the realm of nitrogen metabolism. Figure shows how glutamate and glutamine metabolite level is regulated in response to drought and heat stress based on differential expression of glutamine synthetase. (Based on Das et al., 2016 [49], and this study), Figure S5: Differential expression of phytochemicals in response to drought stress, Figure S6: Metabolic pathways regulation in response to heat stress derived from MapMan. 93 data points were mapped and 72 data points are visible in this figure. Arabidopsis ATH1_TAIR9 was used as mapping template,

Figure S7: A fully connected network of all carbohydrate derived metabolites detected in response to heat stress. It is a compound-gene based network. Red color nodes indicate input data and yellow color nodes indicate related compounds with input compounds. This network comprises of 110 nodes and 217 edges, Figure S8: Network elucidating all amino acid derived metabolites detected in response to drought and heat stress. It is a compound-gene based network. Red color nodes indicate input data and yellow color nodes indicate related compounds with input compounds. This network comprises of 244 nodes and 502 edges, Table S1: Ratio for differential expression, fold change and mean values of metabolites in response to drought and heat stress compared to the control, Table S2: Mean and p-values of differentially expressed carbohydrates and amino acids. (A) Summary of differentially expressed carbohydrates during drought and heat stress compared to the control, along with its mean and p-values. A value of $p \leq 0.05$ was considered statistically significant (red font) for the two group comparisons. (B) Summary of differentially expressed amino acids during drought and heat stress compared to the control, along with its mean and p-values. A value of $p \leq 0.05$ was considered statistically significant (red font) for the two group comparisons.

ACKNOWLEDGMENTS

Authors sincerely acknowledge SD Soybean Research & Promotion Council for financial support, and Anna McClung, Research Leader USDA-ARS for discussions, encouragements, and constructive comments during the writing of this manuscript. Mention of a trademark or proprietary product does not constitute a guarantee or warranty of the product by the U.S. Department of Agriculture and does not imply its approval to the exclusion of other products that also can be suitable. USDA is an equal opportunity provider and employer. All experiments complied with the current laws of the United States, the country in which they were performed.

AUTHOR CONTRIBUTIONS

Paul J. Rushton and Jai S. Rohila conceived and designed the experiments; Aayudh Das performed the experiments; Aayudh Das, Paul J. Rushton and Jai S. Rohila analyzed the data; Jai S. Rohila contributed reagents/materials/analysis tools; Aayudh Das, Paul J. Rushton and Jai S.

Rohila wrote the paper. All authors have read and approved the final version of the manuscript.

CONFLICTS OF INTEREST

The authors declare no conflict of interest.

REFERENCES

1. Shulaev, V.; Cortes, D.; Miller, G.; Mittler, R. Metabolomics for plant stress response. Physiol. Plant. 2008, 132, 199–208.
2. Cramer, G.R.; Urano, K.; Delrot, S.; Pezzotti, M.; Shinozaki, K. Effects of abiotic stress on plants: A systems biology perspective. BMC Plant Biol. 2011, 11, 163.
3. Deryng, D.; Conway, D.; Ramankutty, N.; Price, J.; Warren, R. Global crop yield response to extreme heat stress under multiple climate change futures. Environ. Res. Lett. 2014, 9, 3.
4. Chebrolu, K.K.; Fritschi, F.B.; Ye, S.; Krishnan, H.B.; Smith, J.R.; Gillman, J.D. Impact of heat stress during seed development on soybean seed metabolome. Metabolomics 2016, 12, 28.
5. Clarke, J.D.; Alexander, D.C.; Ward, D.P.; Ryals, J.A.; Mitchell, M.W.; Wulff, J.E.; Guo, L. Assessment of genetically modified soybean in relation to natural variation in the soybean seed metabolome. Sci. Rep. 2013, 3, 3082.
6. Li, L.; Hur, M.; Lee, J.Y.; Zhou, W.; Song, Z.; Ransom, N.; Demirkale, C.Y.; Nettleton, D.; Westgate, M.; Arendsee, Z.; et al. A systems biology approach toward understanding seed composition in soybean. BMC Genom. 2015, 16, S9.
7. Rabara, R.C.; Tripathi, P.; Rushton, P.J. Comparative metabolome profile between tobacco and soybean grown under water-stressed conditions. BioMed Res. Int. 2017, 2017, 3065251.
8. Silvente, S.; Sobolev, A.P.; Lara, M. Metabolite adjustments in drought tolerant and sensitive soybean genotypes in response to water stress. PLoS ONE 2012, 7, e38554.
9. Huang, B.; Gao, H. Growth and carbohydrate metabolism of creeping bentgrass cultivars in response to increasing temperatures. Crop Sci. 2000, 40, 1115–1120.
10. Lobell, D.B.; Roberts, M.J.; Schlenker, W.; Braun, N.; Little, B.B.; Rejesus, R.M.; Hammer, G.L. Greater sensitivity to drought accompanies maize yield increase in the us midwest. Science 2014, 344, 516–519.
11. Zhang, J.Y.; de Carvalho, M.H.C.; Torres-Jerez, I.; Kang, Y.U.N.; Allen, S.N.; Huhman, D.V.; Tang, Y.; Murray, J.; Sumner, L.W.; Udvardi, M.K. Global reprogramming of transcription and metabolism in medicago truncatula during progressive drought and after rewatering. Plant Cell Environ. 2014, 37, 2553–2576.
12. Das, A.; Paudel, B.; Rohila, J.S. Potentials of Proteomics in Crop Breeding. In Advances in Plant Breeding Strategies: Breeding, Biotechnology and Molecular Tools; Al-Khayri, M.J., Jain, M.S., Johnson, V.D., Eds.; Springer International Publishing: Cham, Switzerland, 2015; pp. 513–537.
13. Frank, T.; Röhlig, R.M.; Davies, H.V.; Barros, E.; Engel, K.H. Metabolite profiling of maize kernels–genetic modification versus environmental influence. J. Agric. Food Chem. 2012, 60, 3005–3012.
14. Jung, Y.; Ha, M.; Lee, J.; Ahn, Y.G.; Kwak, J.H.; Ryu, D.H.; Hwang, G.S. Metabolite profiling of the response of burdock roots to copper stress. J. Agric. Food Chem. 2015, 63, 1309–1317.

15. Saito, K. Phytochemical genomics—A new trend. Curr. Opin. Plant Biol. 2013, 16, 373–380.
16. Shu, X.L.; Frank, T.; Shu, Q.Y.; Engel, K.H. Metabolite profiling of germinating rice seeds. J. Agric. Food Chem. 2008, 56, 11612–11620.
17. Takuji, N.; Keiki, O.; Noureddine, B.; Jun, W.; Toshihiro, W.; Hideyuki, M.; Hirofumi, U.; Setsuko, K.; Takuro, S. Metabolomics Approach in Soybean; Science Publishers: Enfield, NH, USA, 2010; pp. 307–323.
18. Kaplan, F.; Kopka, J.; Haskell, D.W.; Zhao, W.; Schiller, K.C.; Gatzke, N.; Sung, D.Y.; Guy, C.L. Exploring the temperature-stress metabolome of arabidopsis. Plant Physiol. 2004, 136, 4159–4168.
19. Komatsu, S.; Yamamoto, A.; Nakamura, T.; Nouri, M.Z.; Nanjo, Y.; Nishizawa, K.; Furukawa, K. Comprehensive analysis of mitochondria in roots and hypocotyls of soybean under flooding stress using proteomics and metabolomics techniques. J. Proteom. Res. 2011, 10, 3993–4004.
20. Gunnaiah, R.; Kushalappa, A.C.; Duggavathi, R.; Fox, S.; Somers, D.J. Integrated metabolo-proteomic approach to decipher the mechanisms by which wheat qtl (fhb1) contributes to resistance against fusarium graminearum. PLoS ONE2012, 7, e40695.
21. Cramer, G.R.; Ergül, A.; Grimplet, J.; Tillett, R.L.; Tattersall, E.A.; Bohlman, M.C.; Vincent, D.; Sonderegger, J.; Evans, J.; Osborne, C. Water and salinity stress in grapevines: Early and late changes in transcript and metabolite profiles. Funct. Integr. Genom. 2007, 7, 111–134.
22. Fernie, A.R.; Schauer, N. Metabolomics-assisted breeding: A viable option for crop improvement? Trends Genet. 2009, 25, 39–48.
23. Benkeblia, N.; Shinano, T.; Osaki, M. Metabolite profiling and assessment of metabolome compartmentation of soybean leaves using non-aqueous fractionation and gc-ms analysis. Metabolomics 2007, 3, 297–305.
24. Broeckling, C.D.; Huhman, D.V.; Farag, M.A.; Smith, J.T.; May, G.D.; Mendes, P.; Dixon, R.A.; Sumner, L.W. Metabolic profiling of medicago truncatula cell cultures reveals the effects of biotic and abiotic elicitors on metabolism. J. Exp. Bot. 2005, 56, 323–336.
25. Wu, R.; Racker, E. Regulatory mechanisms in carbohydrate metabolism iii. Limiting factors in glycolysis of ascites tumor cells. J. Biol. Chem. 1959, 234, 1029–1035. [PubMed]
26. Kruger, N.J.; von Schaewen, A. The oxidative pentose phosphate pathway: Structure and organisation. Curr. Opin. Plant Biol. 2003, 6, 236–246.
27. Fernie, A.R.; Carrari, F.; Sweetlove, L.J. Respiratory metabolism: Glycolysis, the TCA cycle and mitochondrial electron transport. Curr. Opin. Plant Biol. 2004, 7, 254–261.
28. Pressman, E.; Peet, M.M.; Pharr, D.M. The effect of heat stress on tomato pollen characteristics is associated with changes in carbohydrate concentration in the developing anthers. Anna. Bot. 2002, 90, 631–636.
29. Saini, H.S. Effects of water stress on male gametophyte development in plants. Sex. Plant Reprod. 1997, 10, 67–73.
30. Seibold, G.M.; Wurst, M.; Eikmanns, B.J. Roles of maltodextrin and glycogen phosphorylases in maltose utilization and glycogen metabolism in corynebacterium glutamicum. Microbiology 2009, 155, 347–358.
31. Stoop, J.M.; Williamson, J.D.; Mason Pharr, D. Mannitol metabolism in plants: A method for coping with stress. Trends Plant Sci. 1996, 1, 139–144.
32. Pharr, D.M.; Stoop, J.M.H.; Williamson, J.D.; Feusi, M.E.S.; Massel, M.O.; Conkling, M.A. The dual role of mannitol as osmoprotectant and photoassimilate in celery. HortScience 1995, 30, 1182–1188.
33. Weinstein, A.N.; Segal, S. The metabolic fate of 14C galactitol in mammalian tissue. Biochim. Biophys. Acta Gen. Subj.1968, 156, 9–16.
34. Myers, J.A.; Kitajima, K. Carbohydrate storage enhances seedling shade and stress tolerance in a neotropical forest. J. Ecol. 2007, 95, 383–395.

35. Rodrigues, F.A.; Fuganti-Pagliarini, R.; Marcolino-Gomes, J.; Nakayama, T.J.; Molinari, H.B.C.; Lobo, F.P.; Harmon, F.G.; Nepomuceno, A.L. Daytime soybean transcriptome fluctuations during water deficit stress. BMC Genom. 2015, 16, 505.
36. Voet, D.; Voet, J.; Pratt, C. Fundamentals of Biochemistry: Life at the Molecular Level, 4th ed.; John Wiley & Sons: Hoboken, NJ, USA, 2013.
37. Galili, G. The aspartate-family pathway of plants. Plant Signal. Behav. 2011, 6, 192–195.
38. Zhang, J.; Yang, D.; Li, M.; Shi, L. Metabolic profiles reveal changes in wild and cultivated soybean seedling leaves under salt stress. PLoS ONE 2016, 11, e0159622.
39. Baena-González, E.; Sheen, J. Convergent energy and stress signaling. Trends Plant Sci. 2008, 13, 474–482.
40. Salter, M.; Knowles, R.G.; Pogson, C.I. Quantification of the importance of individual steps in the control of aromatic amino acid metabolism. Biochem. J. 1986, 234, 635–647.
41. Galili, G. New insights into the regulation and functional significance of lysine metabolism in plants. Annu. Rev. Plant Biol. 2002, 53, 27–43.
42. Rocha, M.; Licausi, F.; Araújo, W.L.; Nunes-Nesi, A.; Sodek, L.; Fernie, A.R.; van Dongen, J.T. Glycolysis and the tricarboxylic acid cycle are linked by alanine aminotransferase during hypoxia induced by waterlogging of lotus japonicus. Plant Physiol. 2010, 152, 1501–1513.
43. Sweetlove, L.J.; Beard, K.F.; Nunes-Nesi, A.; Fernie, A.R.; Ratcliffe, R.G. Not just a circle: Flux modes in the plant tca cycle. Trends Plant Sci. 2010, 15, 462–470.
44. Good, A.G.; Zaplachinski, S.T. The effects of drought stress on free amino acid accumulation and protein synthesis in brassica napus. Physiol. Plant. 1994, 90, 9–14.
45. Nambara, E.; Kawaide, H.; Kamiya, Y.; Naito, S. Characterization of an arabidopsis thaliana mutant that has a defect in ABA accumulation: ABA-dependent and aba-independent accumulation of free amino acids during dehydration. Plant Cell Physiol. 1998, 39, 853–858.
46. Ashraf, M.; Foolad, M. Roles of glycine betaine and proline in improving plant abiotic stress resistance. Environ. Exp. Bot. 2007, 59, 206–216.
47. Kishor, P.B.K.; Sangam, S.; Amrutha, R.N.; Laxmi, P.S.; Naidu, K.R.; Rao, K.R.S.S.; Rao, S.; Reddy, K.J.; Theriappan, P.; Sreenivasulu, N. Regulation of proline biosynthesis, degradation, uptake and transport in higher plants: Its implications in plant growth and abiotic stress tolerance. Curr. Sci. 2005, 88, 424–438.
48. Tripathi, P.; Rabara, R.C.; Reese, R.N.; Miller, M.A.; Rohila, J.S.; Subramanian, S.; Shen, Q.J.; Morandi, D.; Bücking, H.; Shulaev, V.; et al. A toolbox of genes, proteins, metabolites and promoters for improving drought tolerance in soybean includes the metabolite coumestrol and stomatal development genes. BMC Genom. 2016, 17, 1–22.
49. Das, A.; Eldakak, M.; Paudel, B.; Kim, D.-W.; Hemmati, H.; Basu, C.; Rohila, J.S. Leaf proteome analysis reveals prospective drought and heat stress response mechanisms in soybean. BioMed Res. Int. 2016, 2016, 23.
50. Fruton, J.S.; Simmonds, S. Cold Spring Harbor Symposia on Quantitative Biology; Cold Spring Harbor Laboratory Press: New York, NY, USA, 1950; pp. 55–64.
51. Ling, J.; Söll, D. Severe oxidative stress induces protein mistranslation through impairment of an aminoacyl-trna synthetase editing site. Proc. Natl. Acad. Sci. USA 2010, 107, 4028–4033.
52. Allakhverdiev, S.I.; Kreslavski, V.D.; Klimov, V.V.; Los, D.A.; Carpentier, R.; Mohanty, P. Heat stress: An overview of molecular responses in photosynthesis. Photosynth. Res. 2008, 98, 541–550.
53. De Carvalho, M.H.C. Drought stress and reactive oxygen species. Plant Signal. Behav. 2008, 3, 156–165.
54. Zrenner, R.; Stitt, M.; Sonnewald, U.; Boldt, R. Pyrimidine and purine biosynthesis and degradation in plants. Annu. Rev. Plant Biol. 2006, 57, 805–836.
55. Stasolla, C.; Katahira, R.; Thorpe, T.A.; Ashihara, H. Purine and pyrimidine nucleotide metabolism in higher plants. J. Plant Physiol. 2003, 160, 1271–1295.

56. Boldt, R.; Zrenner, R. Purine and pyrimidine biosynthesis in higher plants. Physiol. Plant. 2003, 117, 297–304.
57. Bourgaud, F.; Gravot, A.; Milesi, S.; Gontier, E. Production of plant secondary metabolites: A historical perspective. Plant Sci. 2001, 161, 839–851.
58. Edreva, A.; Velikova, V.; Tsonev, T.; Dagnon, S.; Gürel, A.; Aktaş, L.; Gesheva, E. Stress-protective role of secondary metabolites: Diversity of functions and mechanisms. Gen. Appl. Plant Physiol. 2008, 34, 67–78.
59. Wahid, A. Physiological implications of metabolite biosynthesis for net assimilation and heat-stress tolerance of sugarcane (saccharum officinarum) sprouts. J. Plant Res. 2007, 120, 219–228.
60. Tattini, M.; Galardi, C.; Pinelli, P.; Massai, R.; Remorini, D.; Agati, G. Differential accumulation of flavonoids and hydroxycinnamates in leaves of ligustrum vulgare under excess light and drought stress. New Phytol. 2004, 163, 547–561.
61. Zhong, R.; Morrison, W.H.; Himmelsbach, D.S.; Poole, F.L.; Ye, Z.H. Essential role of caffeoyl coenzyme a O-methyltransferase in lignin biosynthesis in woody poplar plants. Plant Physiol. 2000, 124, 563–578.
62. Smirnoff, N. Tansley review no. 52. The role of active oxygen in the response of plants to water deficit and desiccation. New Phytol. 1993, 125, 27–58.
63. Liu, R.H. Health benefits of fruit and vegetables are from additive and synergistic combinations of phytochemicals. Am. J. Clin. Nutr. 2003, 78, 517S–520S. [PubMed]
64. Hosseinian, F.S.; Li, W.; Beta, T. Measurement of anthocyanins and other phytochemicals in purple wheat. Food Chem.2008, 109, 916–924.
65. Oh, M.-M.; Carey, E.E.; Rajashekar, C. Environmental stresses induce health-promoting phytochemicals in lettuce. Plant Physiol. Biochem. 2009, 47, 578–583.
66. Gill, S.S.; Tuteja, N. Reactive oxygen species and antioxidant machinery in abiotic stress tolerance in crop plants. Plant Physiol. Biochem. 2010, 48, 909–930.
67. Schmutz, J.; Cannon, S.B.; Schlueter, J.; Ma, J.; Mitros, T.; Nelson, W.; Hyten, D.L.; Song, Q.; Thelen, J.J.; Cheng, J.; et al. Genome sequence of the palaeopolyploid soybean. Nature 2010, 463, 178–183.
68. Evans, A.M.; Dehaven, C.D.; Barrett, T.; Mitchell, M.; Milgram, E. Integrated, nontargeted ultrahigh performance liquid chromatography/electrospray ionization tandem mass spectrometry platform for the identification and relative quantification of the small-molecule complement of biological systems. Anal. Chem. 2009, 81, 6656–6667.
69. Lawton, K.A.; Berger, A.; Mitchell, M.; Milgram, K.E.; Evans, A.M.; Guo, L.; Hanson, R.W.; Kalhan, S.C.; Ryals, J.A.; Milburn, M.V. Analysis of the adult human plasma metabolome. Pharmacogenomics 2008, 9, 383–397.
70. Oliver, M.J.; Guo, L.; Alexander, D.C.; Ryals, J.A.; Wone, B.W.; Cushman, J.C. A sister group contrast using untargeted global metabolomic analysis delineates the biochemical regulation underlying desiccation tolerance in sporobolus stapfianus. Plant Cell 2011, 23, 1231–1248.
71. DeHaven, C.; AM, E.; Dai, H.; Lawton, K. Organization of GC/MS and LC/MS metabolomics data in to chemical libraries. J. Cheminform. 2010, 2, 9.
72. De Hoon, M.J.; Imoto, S.; Nolan, J.; Miyano, S. Open source clustering software. Bioinformatics 2004, 20, 1453–1454.
73. Saldanha, A.J. Java treeview–extensible visualization of microarray data. Bioinformatics 2004, 20, 3246–3248.
74. Wiklund, S.; Johansson, E.; Sjöström, L.; Mellerowicz, E.J.; Edlund, U.; Shockcor, J.P.; Gottfries, J.; Moritz, T.; Trygg, J. Visualization of GC/TOF-MS-based metabolomics data for identification of biochemically interesting compounds using opls class models. Anal. Chem. 2008, 80, 115–122.
75. Akula, R.; Ravishankar, G.A. Influence of abiotic stress signals on secondary metabolites in plants. Plant Signal. Behav.2011, 6, 1720–1731.
76. Kanehisa, M.; Goto, S. Kegg: Kyoto encyclopedia of genes and genomes. Nucleic Acids Res. 2000, 28, 27–30.

77. Karnovsky, A.; Weymouth, T.; Hull, T.; Tarcea, V.G.; Scardoni, G.; Laudanna, C.; Sartor, M.A.; Stringer, K.A.; Jagadish, H.; Burant, C. Metscape 2 bioinformatics tool for the analysis and visualization of metabolomics and gene expression data. Bioinformatics 2012, 28, 373–380.
78. Shannon, P.; Markiel, A.; Ozier, O.; Baliga, N.S.; Wang, J.T.; Ramage, D.; Amin, N.; Schwikowski, B.; Ideker, T. Cytoscape: A software environment for integrated models of biomolecular interaction networks. Genom. Res. 2003, 13, 2498–2504.
79. Valdés-López, O.; Batek, J.; Gomez-Hernandez, N.; Nguyen, C.T.; Isidra-Arellano, M.C.; Zhang, N.; Joshi, T.; Xu, D.; Hixson, K.K.; Weitz, K.K.; et al. Soybean roots grown under heat stress show global changes in their transcriptional and proteomic profiles. Front. Plant Sci. 2016, 7, 517.
80. Le, D.T.; Nishiyama, R.; Watanabe, Y.; Tanaka, M.; Seki, M.; Ham, L.H.; Yamaguchi-Shinozaki, K.; Shinozaki, K.; Tran, L.-S.P. Differential gene expression in soybean leaf tissues at late developmental stages under drought stress revealed by genome-wide transcriptome analysis. PLoS ONE 2012, 7, e49522.
81. Neto, J.R.C.F.; Pandolfi, V.; Guimaraes, F.C.M.; Benko-Iseppon, A.M.; Romero, C.; Silva, R.L.O.; Rodrigues, F.A.; Abdelnoor, R.V.; Nepomuceno, A.L.; Kido, E.A. Early transcriptional response of soybean contrasting accessions to root dehydration. PLoS ONE 2013, 8, e83466.
82. Chen, W.; Yao, Q.; Patil, G.B.; Agarwal, G.; Deshmukh, R.K.; Lin, L.; Wang, B.; Wang, Y.; Prince, S.J.; Song, L.; et al. Identification and comparative analysis of differential gene expression in soybean leaf tissue under drought and flooding stress revealed by rna-seq. Front. Plant Sci. 2016, 7, 1044.
83. Xoconostle-Cázares, B.; Ramirez-Ortega, F.A.; Flores-Elenes, L.; Ruiz-Medrano, R. Drought tolerance in crop plants. Am. J. Plant Physiol. 2010, 5, 241–256.
84. Schauer, N.; Semel, Y.; Roessner, U.; Gur, A.; Balbo, I.; Carrari, F.; Pleban, T.; Perez-Melis, A.; Bruedigam, C.; Kopka, J.; et al. Comprehensive metabolic profiling and phenotyping of interspecific introgression lines for tomato improvement. Nat. Biotechnol. 2006, 24, 447–454.
85. Le Gall, G.; Colquhoun, I.J.; Davis, A.L.; Collins, G.J.; Verhoeyen, M.E. Metabolite profiling of tomato (lycopersicon esculentum) using 1 h nmr spectroscopy as a tool to detect potential unintended effects following a genetic modification. J. Agric. Food Chem. 2003, 51, 2447–2456.
86. Eldakak, M.; Milad, S.I.; Nawar, A.I.; Rohila, J.S. Proteomics: A biotechnology tool for crop improvement. Front. Plant Sci. 2013, 4, 35.
87. Ramalingam, A.; Kudapa, H.; Pazhamala, L.T.; Weckwerth, W.; Varshney, R.K. Proteomics and metabolomics: Two emerging areas for legume improvement. Front. Plant Sci. 2015, 6, 1116.
88. Steinfath, M.; Strehmel, N.; Peters, R.; Schauer, N.; Groth, D.; Hummel, J.; Steup, M.; Selbig, J.; Kopka, J.; Geigenberger, P.; et al. Discovering plant metabolic biomarkers for phenotype prediction using an untargeted approach. Plant Biotechnol. J. 2010, 8, 900–911.

CHAPTER 3

Attenuation of Drought Stress in *Brassica* Seedlings with Exogenous Application of Ca^{2+} and H_2O_2

Akram Khan [1,2], Yasir Anwar [3], Md. Mahadi Hasan [3], Aqib Iqbal [1], Muhammad Ali [2], Hesham F. Alharby [3], Khalid Rehman Hakeem [3,*] and Mirza Hasanuzzaman [4]

[1]*Institute of Biotechnology and Genetic Engineering, University of Agriculture, Peshawar 25120, Pakistan*

[2]*Department of Biotechnology, Bacha Khan University Charsadda, Charsadda 24420, Pakistan*

[3]*Department of Biological Sciences, Faculty of Science, King Abdulaziz University, Jeddah 21589, Saudi Arabia*

[4]*Department of Agronomy, Faculty of Agriculture, Sher-e-Bangla Agricultural University, Sher-e-Bangla Nagar, Dhaka 1207, Bangladesh*

ABSTRACT

Drought is one of the most common abiotic stresses, affecting the growth and productivity of crop plants globally, particularly in arid and semi-arid regions. Different strategies are used to mitigate the impact of drought among crop plants. Exogenous application of different substances are known to decrease the effects of various abiotic stresses, including drought stress. The aim of this study was to evaluate the effect of Ca^{2+} and H_2O_2 in developing drought stress tolerance in *Brassica napus* "Bulbul-98" seedlings. *Brassica napus* "Bulbul-98" seedlings were exposed to 5, 10 and 15 mM Ca^{2+} and 2, 5 and 10 μM H_2O_2 concentrations twice at an interval of two days for up to 20 days after germination. Drought stress decreased relative water content (RWC), chlorophyll content and increased proline, H_2O_2, soluble protein and electrolyte leakage in *Brassica* seedlings. Exogenous Ca^{2+} (5, 10,15 mM) and H_2O_2 (2, 5, 10 μM) supplementations, during drought stress induction, showed a significant increase in RWC by 5.4%, 18.06%, 26.2% and 6.87%, 13.9%, 18.3% respectively. Similarly, with the exogenous application of Ca^{2+} (5, 10, 15 mM) and H_2O_2 (2, 5, 10 μM), chlorophyll content was increased by 15.03%, 22.2%, and 28.4%, and 9.6%, 23.3%, and 27.5% respectively. It was confirmed that the seedlings under drought stress that were supplemented with Ca^{2+} and H_2O_2 recovered from water content reduction and chlorosis, and were able to grow normally.

Keywords: drought stress; relative water content; electrolyte leakage; proline; SDS-PAGE

1. INTRODUCTION

A number of abiotic stresses like drought, temperature, and salinity, usually reduce crop yield [1]. It has been approximated that crops reachonly 25% of their likely yield, because of the damaging effects of environmental stresses [2]. These stresses can take place at any phase of plant growth, thus illustrating the dynamic nature of crop plants and their yield. Drought is one of the main abiotic stresses, and significantly affects yield and growth of plants, and plays a vital role in their geographical division [3,4,5]. According to the Food and Agriculture Organization (FAO), 45% of the agricultural land on earth is exposed to drought stress [6]. Water deficiency induces a set of physiological and biochemical reactions in plants and is one of the most composite unfavorable conditions, since it not only depends on the severity and period of the stress occurrence, but also on the plant developmental period and its morphology [7,8]. As an adaptive and protection mechanism, plant hormonal and signaling networks are involved in various ways to manage stress under various abiotic stress conditions [9]. Even though an assortment of genotypes with improved yield in drought conditions has been a vital feature of crop reproduction, the biological basis for drought tolerance is still poorly understood. High photosynthesis rate maintenance [10], osmotic modification to decrease water loss [11], high instantaneous water effectiveness maintenance (defined as the ratio of transpiration to leaf photosynthesis) [12], waxy coatings on the plant exterior, and deeper root morphologies, are some of the traits found in drought tolerant genotypes. The inhibition of development leading to the production of a range of modifications in plant physiological, biochemical and molecular features is generally caused by drought stress [3,4,5].

A general occurrence in plants subjected to various abiotic and biotic stresses, is the production of reactive oxygen species (ROS). By the commencement of an antioxidant defense system consisting of enzymatic and non-enzymatic components, the cells usually retain a stable-state ROS level [13]. ROS are greatly reactiveto DNA, membrane lipids, and protein, and they are key causative factors for stress-induced cellular injures. High antioxidant ability or high levels of antioxidants can avoid cell death and is associated with stress tolerance [14,15]. Several studies have shown that H_2O_2, one of the mobile forms of ROS, is a major signal molecule, mediating a seriesof reactions [16]. Exogenous Ca^{2+} can improve plant stress resistance, guard the structure of cellular plasma membranes, slow down the synthesis of activating oxides, control the metabolism of plant hormones, and sustain normal photosynthesis [17,18,19]. Besides this, cellular Ca^{2+} also transmits drought signals, therefore modifying physiological reactions introduced by drought stress [20,21]. Related results of improved stress tolerance have been observed subsequent to pre-treatment with H_2O_2 [22,23].

Among the oilseeds crops, *Brassica* is one of the most important crops, due to its edible oil production [24]. *Brassica* has been developed in high rainfall areas, and does not grow well in low rainfall areas [25]. Reduction of the yield of *Brassica* due to drought stress have been reported by many authors [26,27]. However, the influence of various exogenous elements in the reduction of drought stress is still in its infancy.

Hence, the current study was undertaken to find out the consequences of the exogenous application of Ca^{2+} and H_2O_2pre-treatment on the drought stress tolerance of *Brassica napus* "Bulbul-98" at early growth stages. This study has also elucidated the physiological and biochemical changes under drought stress conditions associated with the pre-treatment of these chemicals and determination of differentially expressed proteins with these pre-treatments under normal (irrigated) and drought stress conditions.

2. RESULTS

2.1. Rate of Water Loss (RWL)

The highest rate of water loss was 320.45 mg·g^{-1}h^{-1} (DM) found in the non-supplemented seedlings (Figure 1A). During the 1 h measuring period, the rate of water loss of the seedlings supplemented with 2, 5 and 10 µM H_2O_2 decreased by 18.5%, 31.1% and 37.18% respectively. The water loss rate from leaf discs of the seedlings sprayed with 5, 10 or 15 mM $CaCl_2$ was 252.45, 221.4 and 202.3 mg·g^{-1}h^{-1} (DM) respectively.

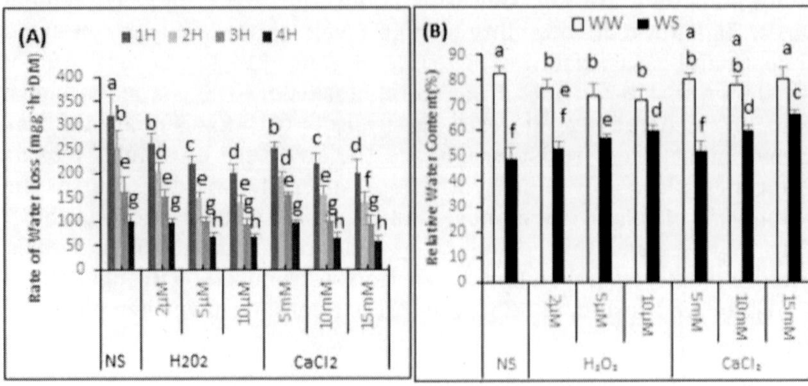

Figure 1. H_2O_2 and $CaCl_2$ pre-treatment effect on the of the rate of water loss (**A**) over time (h, 1H, 2H, 3H and 4H); Relative water content (**B**) of *Brassica napus* "Bulbul-98" seedlings under irrigated (WW) and drought stress (WS) conditions (**B**). In accordance with Least Significant Difference (LSD) test, the bars with at least one common alphabet are not significantly different at $p \leq 0.05$.

2.2. Relative Water Content (RWC)

The data showed that the RWC of the non-supplemented (NS) seedlings was 82.39% ± 3.13% under irrigated conditions and 49.02% ± 4.20% after exposure to drought stress. Thus there was 40.5% decrease in the RWC of the non-supplemented seedlings under drought stress conditions. It can be further observed from the data that H_2O_2 application significantly reduced RWC under irrigated conditions (Figure 1B). The RWC of the seedlings supplemented with 2, 5 and 10 µM $H_2O_2$was 76.49% ± 3.53%, 73.99% ± 4.27% and 72.19% ± 3.67%, respectively. Thus, compared with seedling that were non-supplemented in irrigated conditions, the RWC of the seedlings supplemented with 2, 5 and 10 µM

H_2O_2 decreased by 7%, 10% and 12%, respectively. In contrast, when exposed to drought stress situations, there was a gradual increase in the RWC of the seedlings supplemented with H_2O_2, compared with non-supplemented seedlings. The RWC of the seedlings exposed to drought stress and supplemented with 2, 5, and 10 μM H_2O_2, was 52.64% ± 2.79%, 56.95% ± 1.38% and 60.01% ± 1.68%. Thus, compared with non-supplemented seedlings, there was an increase of 6.87%, 13.9% and 18.3% in the RWC of seedlings under drought stress supplemented with 2, 5, and 10 μM H_2O_2, respectively. There was no major effect on the RWC of seedlings under irrigated conditions after supplementation with $CaCl_2$. The "Bulbul-98" seedlings maintained a RWC of 80.06% ± 2.67%, 78.04% ± 3.29% and 80.25% ± 4.68%, respectively under irrigated conditions after supplementation with 5, 10 and 15 mM of $CaCl_2$. Spraying the seedlings with 5, 10 and 15 mM $CaCl_2$ prior to exposure to drought gradually increased the RWC to 51.85% ± 3.79%, 59.83% ± 1.95% and 66.37% ± 1.63%.

2.3. Chlorophyll Content

Mean chlorophyll content of the non-supplemented (NS) seedlings was 768.20 ± 19.58 $\mu g \cdot g^{-1}$ fresh weight (FW) under irrigated conditions and 364.96 ± 14.53 $\mu g \cdot g^{-1}$ FW after exposure to drought stress. Thus, the exposure to drought stress decreased the chlorophyll content by 53%, signifying enhanced modification to chloroplast (Figure 2A). The content of chlorophyll was 743.09 ± 25.51, 704.48 ± 15.54 and 674.03 ± 48.82 $\mu g \cdot g^{-1}$ FW in the seedlings treated with 2, 5 and 10μM H_2O_2. Thus, H_2O_2 application upon irrigated situations has enhanced alteration to chlorophyll in the seedlings. After 2, 5 and 10 μM H_2O_2 application, chlorophyll content was 403.01 ± 15.56, 475.90 ± 29.21 and 503.67 ± 13.98 $\mu g \cdot g^{-1}$ FW under drought stress conditions. This data indicated that there was 10%, 25% and 29% less modification to chlorophyll respectively, in the supplemented seedlings compared with non-supplemented seedlings under similar conditions. Before applying the drought, the chlorophyll content was 429.53 ± 49.10, 469.13 ± 24.27 and 519.94 ± 10.47 $\mu g \cdot g^{-1}$ FW under 5, 10 and 15 mM $CaCl_2$ treatment. Thus the seedlings supplemented with $CaCl_2$ before drought imposition incurred 13%, 24% and 33% less modification to chlorophyll compared with non-supplemented seedlings.

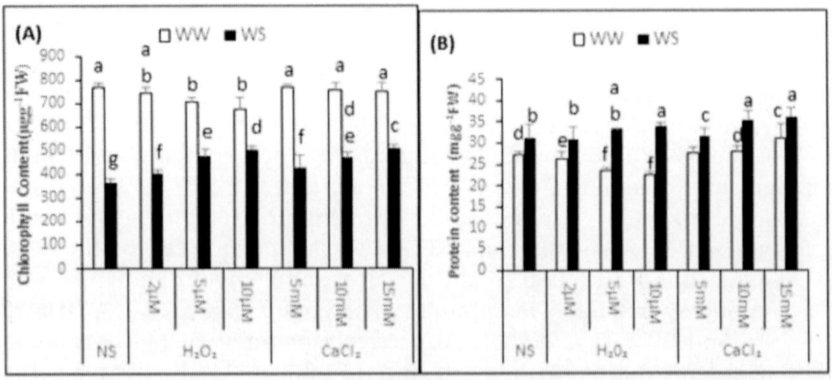

Figure 2. H_2O_2 and $CaCl_2$ supplementation effect on the content of chlorophyll (**A**) and protein content (**B**) of the *Brassica napus* "Bulbul-98" seedlings under irrigated (WW) and drought stress (WS) conditions. In accordance to Least Significant Difference (LSD) test, the bars with at least one common alphabet are not significantly different at $p \leq 0.05$.

2.4. Soluble Protein Content

In the non-supplemented seedlings, the protein content was 26.42 ± 1.40 and 31.30 ± 2.29 mg·g^{-1} FW in irrigated and drought stress conditions, respectively (Figure 2B). The protein content was 26.17 ± 1.80, 25.66 ± 1.36 and 25.33 ± 0.38 mg·g^{-1} FW in irrigated and 30.70 ± 2.09, 32.82 ± 0.45 and 32.34 ± 2.78 mg·g^{-1} FW in drought stress conditions after pre-treatment of the seedlings with 2, 5 and 10 µM H_2O_2, respectively. Thus, compared with non-supplemented seedlings, there was a 0.94%, 2.90% and 4.25% decrease in the soluble protein content under irrigated conditions and an initial 1.92% decrease, then a 4.95% and 3.17% increase in protein content under drought stress conditions after 2, 5 and 10 µM H_2O_2pre-treatment, respectively. Though statistically non-significant, pre-treatment of the "Bulbul-98" seedlings with $CaCl_2$resulted in induction of soluble protein accumulation under both irrigated and drought stress conditions. The protein content was 27.55 ± 1.38, 27.84 ± 1.31 and 30.45 ± 2.50 mg·g^{-1} FW in irrigated and 33.06 ± 2.79, 34.36 ± 0.54 and 35.97 ± 1.50 mg·g^{-1} FW under drought stress conditions after pre-treatment of the seedlings with 5, 10 and 15 mM $CaCl_2$, respectively. Thus, compared with the non-supplemented seedlings under respective conditions, pre-treatment with 5, 10 and 15 mM $CaCl_2$ increased the protein content by 4.28%, 5.15% and 14.48% under irrigation, and 5.62%, 9.26% and 13.59% underdrought stress conditions, respectively.

2.5. Electrolyte Leakage

The data showed that electrolyte leakage of the non-supplemented (NS) seedlings was 20.74% ± 1.65% under irrigated conditions and 66.60% ± 4.35% after exposure to drought stress. Thus, exposure to drought stress increased the electrolyte leakage by 2.21-fold, signifying an increase in membrane damage (Figure 3). In irrigated conditions, increasing H_2O_2concentration supplementation gradually increased electrolyte leakage. Electrolyte leakage was 24.64% ± 2.90%, 29.64% ± 4.15% and 30.94% ± 2.16% in seedlings treated with 2, 5 and 10 μM H_2O_2, respectively. Thus, application of H_2O_2 under irrigated situations enhances damage to cellular membranes, resulting in increased water loss and lower relative seedling water content. $CaCl_2$ supplementation, on the other hand, was not significantly affected for electrolyte leakage in irrigated conditions. It can be noted from the data that application of H_2O_2 or $CaCl_2$ upon drought stress conditions partly decreased electrolyte leakage, signifying lesser damage to cellular membranes. After 2, 5 and 10 μM application of H_2O_2, the electrolyte leakage was 61.37% ± 3.06%, 49.27% ± 2.59% and 44.18% ± 2.70% under drought stress conditions respectively. This indicated that there was 8%, 26% and 34% less electrolyte leakage from the supplemented seedlings compared with non-supplemented seedlings under similar conditions. Similarly, the electrolyte leakage in seedlings applied 5, 10 and 15 mM $CaCl_2$ before drought was 57.54% ± 3.28%, 48.33% ± 2.91% and 45.26% ± 3.04%, respectively. The seedlings supplemented with $CaCl_2$ before drought treatment in curred 14%, 27% and 32% less membrane damage compared with non-supplemented seedlings.

2.6. Proline Content

Variance analysis exhibited a significant difference ($p < 0.05$) in the proline content of "Bulbul-98" seedlings following H_2O_2 or $CaCl_2$ supplementation under both irrigated and drought stress conditions (Figure 4A). In the non-supplemented seedlings, the proline content increased to 8.99 ± 0.89 from 2.27 ± 0.28 $\mu mol \cdot g^{-1}$ DW. In irrigated conditions, a major increase in proline content was noted with increasing quantity of supplemented H_2O_2. Thus, the proline content was 2.79 ± 0.37, 2.98 ± 0.18 and 3.55 ± 0.55 $\mu mol \cdot g^{-1}$ DW in the seedlings regularly irrigated and with 2, 5 and 10 μM H_2O_2respectively. $CaCl_2$, in contrast, did not significantly affect the proline content in irrigated

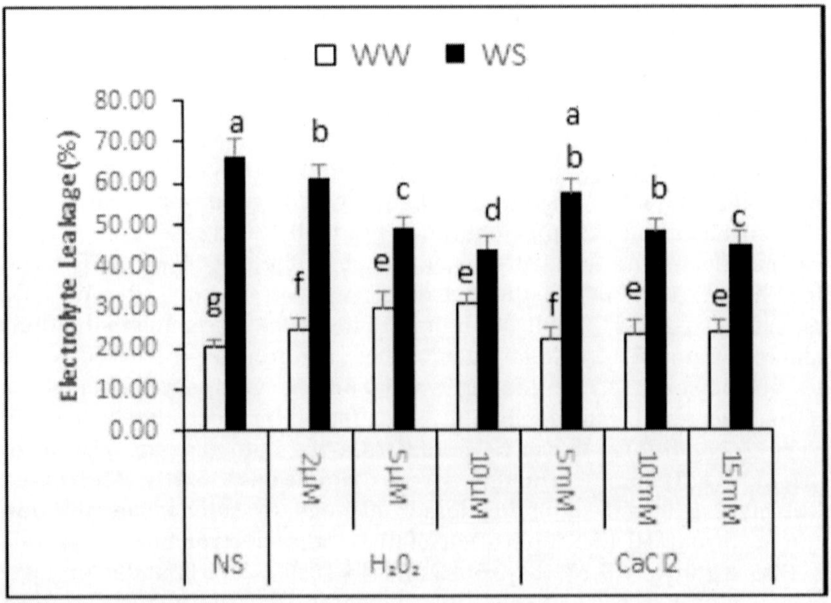

Figure 3. Effect of H_2O_2 and $CaCl_2$ supplementation on percent electrolyte leakage (%) from *Brassica napus* "Bulbul-98" seedlings under irrigated (WW) and drought stress (WS) conditions. In accordance with the Least Significant Difference (LSD) test, the bars with at least one common alphabet are not significantly different at $p \leq 0.05$.

conditions. A significant induction of proline was noted with H_2O_2 or $CaCl_2$ application under drought stress conditions. The application of 2, 5 and 10 μM H_2O_2 improved the proline content to 10.35 ± 0.58, 12.09 ± 0.79 and 14.31 ± 0.88 $\mu mol \cdot g^{-1}$ DW respectively, and 5, 10 and 15 mM $CaCl_2$ application before drought increased the proline content to 8.75 ± 0.62, 10.74 ± 0.58 and 13.38 ± 0.90 $\mu mol \cdot g^{-1}$ DW, respectively. Thus, application of 2, 5 and 10 μM H_2O_2 improved the proline content of seedlings by 15%, 34% and 59%, and 10 and 15 mM $CaCl_2$ increased the proline content by 20% and 49% respectively, over non-supplemented seedlings, under similar conditions.

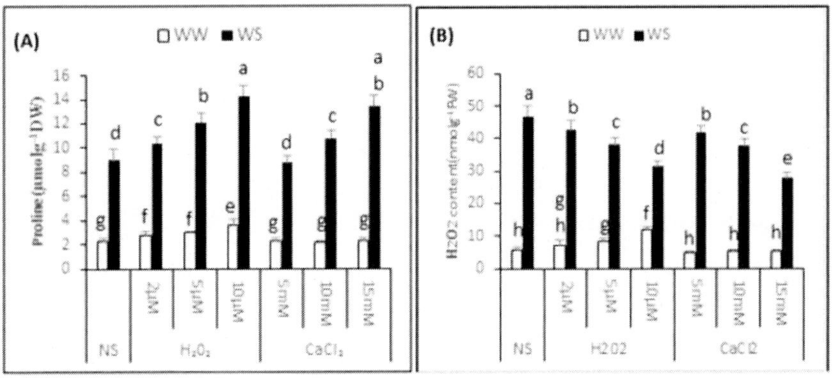

Figure 4. H_2O_2 and $CaCl_2$ supplementation effect on the proline (**A**) and H_2O_2 content (**B**) of the *Brassica napus* "Bulbul-98" seedlings under irrigated (WW) and drought stress (WS) conditions. In accordance to Least Significant Difference (LSD) test, the bars with at least one common alphabet are not significantly different at $p \leq 0.05$.

2.7. H_2O_2 Content

The analysis of variance exhibited a significant difference ($p \leq 0.05$) in endogenous H_2O_2 content of "Bulbul-98" seedlings after H_2O_2 or $CaCl_2$ supplementation in both irrigated and drought stress conditions (Figure 4B). In the non-supplemented seedlings, drought stress conditions improved the content of H_2O_2 to 46.46 ± 3.68 from 6.29 ± 0.55 nmol·g^{-1} FW. In irrigated conditions, a major increase in endogenous content of H_2O_2 was noted with an increasing amount of supplemented H_2O_2, and 8.35 ± 0.84, 9.82 ± 0.76 and 11.44 ± 0.52 nmol·g^{-1} FW H_2O_2 was recorded in the seedlings that were regularly irrigated and treated with 2, 5 and 10 µM H_2O_2. Seedlings pre-treated with $CaCl_2$, in contrast, were not considerably affected in endogenous H_2O_2 content under irrigated conditions. A significant interactive effect of pre-treatment of seedlings with H_2O_2 or $CaCl_2$ application and drought stress was noted with regards to the endogenous content of H_2O_2. Compared with non-supplemented seedlings under drought stress conditions, a decrease in content of endogenous H_2O_2 was observed with increased amounts of H_2O_2 or $CaCl_2$. The content of endogenous H_2O_2 under drought stress was 41.77 ± 3.68, 35.08 ± 1.46 and 28.68 ± 1.24 nmol·g^{-1} FW, respectively after pre-treatment of seedlings with 2, 5 and 10 µM H_2O_2 before exposure to drought. Similarly, pre-treatment of the seedlings with 5, 10 and 15 mM $CaCl_2$ application before drought decreased the content of endogenous H_2O_2 to 40.66 ± 0.37, 37.26 ± 2.22 and 27.04 ± 3.97

nmol·g^{-1} FW, respectively. Thus, application of 2, 5 and 10 μM H_2O_2 decreased the content of endogenous H_2O_2 in the seedlings by 10%, 24% and 38%respectively, and 5, 10 and 15 mM $CaCl_2$ decreased the content of endogenous H_2O_2 by 12%, 20% and 42% respectively, over the non-supplemented seedlings under similar conditions.

2.8. SDS-PAGE Analysis

Total soluble proteins from the non-supplemented and the seedlings pre-treated with different concentrations of $H_2O_2$and $CaCl_2$ under irrigated and drought stress conditions were separated through a one dimensional 15% sodium dodecyl sulfate polyacrylamide gel electrophoresis (SDS-PAGE) (Figure 5). It was observed from the intensity of the bands, that approximately equal amounts of protein were loaded in each well. The protein bands obtained through the SDS-PAGE gels from seedlings exposed to different treatments were quantified through computer software BandLeader®, and this was also visually confirmed. The data clearly indicated that both qualitative and quantitative changes occurred in the seedlings as a result of water availability and different treatments. A total of 20 protein bands were identified by the Band Leader®software in the non-supplemented seedlings under irrigated conditions. Quantitative changes were noted in the SDS-PAGE banding pattern of the seedlings regularly irrigated after pre-treatment with H_2O_2 or $CaCl_2$ did not affect the protein content, as predicted from the band intensities; however, an increase in the protein concentration was noted after 10 and 15 mM $CaCl_2$ application. Rubisco large and small sub-units (RbcL and RbcS, respectively) were the most abundant proteins in the gel under all conditions (Bands No. 5 and 12). Band 5 and Band 12 had intensities of approximately 150 pixels in the non-supplemented seedlings, and those pre-treated with 2, 5 and 10 μM H_2O_2 and 5 mM $CaCl_2$under irrigated conditions. Related to the increase in protein content, Band 5 and 12 intensities have increased to 220 pixels after pre-treating the seedlings with 10 and 15 mM $CaCl_2$, respectively. Similarly, there was a slight decrease in the intensity of band 7 after H_2O_2but this increased after $CaCl_2$ pre-treatment under irrigated conditions. The intensity of this band was 135 pixels in the non-supplemented seedlings, and decreased to 124 pixels after H_2O_2 pre-treatment, and increased to 190 and 210 pixels after 10 and 15 mM $CaCl_2$ pre-treatment. The intensity of band 18 also increased after both H_2O_2 and $CaCl_2$ pre-treatments. The disappearance of band 17 after H_2O_2 pre-treatment was the only qualitative difference in the SDS-PAGE. When exposed to drought stress, a total of 23 bands were observed in the non-supplemented seedlings (Figure 5B). Among these, 19 bands were those also expressed in the non-supplemented seedlings under irrigated conditions. Band 17 in the irrigated conditions was not expressed under the drought stress conditions. However, an increase was observed in the intensity of most of the bands. Band 12 had intensities of approximately 130 pixels in the non-supplemented seedlings, and those pre-treated with 2, 5 and 10 μM H_2O_2, and 5 mM $CaCl_2$ under drought stress conditions.

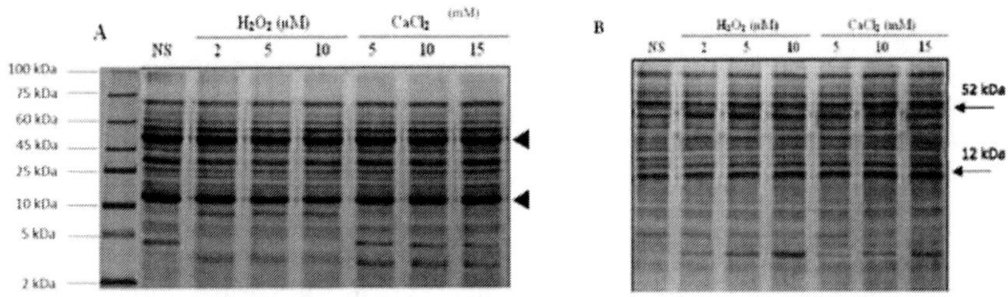

Figure 5. SDS-PAGE electrophoregram of total soluble proteins of "Bulbul-98" seedlings under irrigated (**A**) and drought stress (**B**) conditions after H_2O_2 and $CaCl_2$pre-treatment. Upper and lower arrowheads in each electrophoregram shows the large and small subunits of Rubisco. NS, non-supplemented.

Band 5 and 12 intensities increased to 190 pixels after pre-treating the seedlings with 10 and 15 mM $CaCl_2$, respectively. However, there was a slight increase in the intensity of several bands after $CaCl_2$ pre-treatment under drought stress conditions. The intensity of these band were between 100–110 pixels in the non-supplemented seedlings, as well as those pre-treated with 2, 5 and 10 μM H_2O_2 and 5 mM $CaCl_2$ which increase to 170 pixels after 10 mM $CaCl_2$, and gradually increased to 190 pixels after 15 mM $CaCl_2$ pre-treatment under drought stress conditions.

3. DISCUSSION

Evidence showed that H_2O_2 influences the activation or inhibition of various cellular processes in a dose-dependent manner. H_2O_2atlow concentration enhances plant tolerance to a range of abiotic and biotic stresses [28]. Similarly, Ca^{2+} acts as a secondary messenger to couple a large variety of extra-cellular stimuli with intracellular responses in plant cells. It also has a stabilizing effect on cell wall and membranes, and improves the drought tolerance of plant cells [19,29]. Though $CaCl_2$ is lethal to plants in higher concentrations, in low quantities it may enhance stress tolerance by the provision of $Ca^{2+}$for cell stabilization and signaling, thus inducing the production of different stress peptides. Furthermore, Ca^{2+} and Cl^- are also essential cofactors for photosynthetic water oxidation. This experiment was designed to study whether pre-treatment with H_2O_2 or $CaCl_2$activates plant signaling mechanisms and acclimatizes the seedlings under drought stress conditions. The initial reaction of plants to a diminishing water supply is reduction of water loss, which is attained by either the stomata closing, or reducing the potential of water by accumulation of different solutes. In this experiment, the decrease in water loss rate from excised leaves was first measured to ascertain the beneficial effect of H_2O_2 and $CaCl_2$ pre-treatment before drought imposition. RWL has been suggested as a screening technique to identify genotypes under drought stress [30]. The data suggest that H_2O_2 and $CaCl_2$ pre-treatments both resulted in significantly reduced water loss from leaf disks in a dose dependent manner (Figure 1A). The excised leaves from the non-supplemented "Bulbul-98" had a significantly higher rate of water loss compared with water loss from leaves of H_2O_2 and $CaCl_2$-treated plants. A similar difference in the rate of water loss from untreated plants and those exposed to periodic drought has been observed in tobacco [31]. Different factors including the opening of stomata, accumulation of compatible proteins and solutes, protection of cellular membranes from lipid peroxidation, and deposition of cuticular waxes could affect the water loss rate from a plant. The physiological adaptation of "Bulbul-98" pre-treated with H_2O_2 or $CaCl_2$ before drought imposition was further probed by determining the relative water content of the leaves from seedling under each treatment. The relative water content changes also reflected the ameliorative effect of H_2O_2 and $CaCl_2$ pre-treatment before drought stress (Figure 1B). It could also be noted that H_2O_2 application under irrigated conditions resulted in a decrease in RWC, but there was no adverse effect of $CaCl_2$ pre-treatment under similar conditions. Furthermore, the ameliorative effect of $CaCl_2$ treatment on the RWC was more pronounced under drought stress conditions, compared with pre-treatment of H_2O_2. Similar to our results, the improvement of water relations after H_2O_2 pre-treatment in maize [32] and *Cistus albidus* [33] under drought and soybean [34] under salinity stress has been reported. Furthermore, improved water relations have been reported after pre-treatment with $CaCl_2$ under salinity and flooding stress in wheat, rice, and barley. It is known that under drought stress conditions, a positive turgor pressure is maintained by stomatal closure [35] or osmotic adjustment, through the accumulation of compatible solutes [36]. The results of RWL and RWC in drought- stressed seedlings after H_2O_2 or $CaCl_2$ pre-treatment indicated an improvement in water relations. However, if this improvement is only due to stomatal closure, it will typically induce the limitation of gas exchange and alter the rate of photosynthesis and metabolism [37]. It is known that H_2O_2triggers proline accumulation in maize seedlings, a compatible solute [38]. Similarly, elevated content of proline and glycine betaine, improved the water status and resulted in minimum damage to cellular membranes, and Ca^{2+} in the medium appeared to reduce damaging effect of stress [39]. To properly understand the physiological mechanism of tolerance after H_2O_2 and $CaCl_2$ pre-treatment, the concentration of osmoprotectant proline was determined under control and drought stress conditions. A minor increase in proline content was noted in seedlings pre-treated with H_2O_2 in the irrigated conditions, showing a minor alteration of cellular metabolism and induction of stress (Figure 4A). However, the data showed a significant increase in proline content upon conditions of drought stress. Furthermore, the application of H_2O_2 or $CaCl_2$ upon conditions of drought stress strongly induced proline production. This increase in the proline content could be due to the induction of a proline-producing enzyme and the inhibition of the catabolic enzyme proline oxidase. The proline content increase under drought helps with osmotic adjustment. Though the data at the last of drought stage indicated a lower rate of water loss with pre-treatments, it appears that the pre-treated seedlings maintained a steady state of transpiration compared with non-supplemented seedlings. The non-supplemented seedlings maintained a higher transpiration, resulting in depletion of water and enhanced damage to cells. Thus, this data provided a further proof of the hypothesis that H_2O_2 or $CaCl_2$applications induces proline production under drought stress conditions. Stressful conditions induce complex and highly regulated ROS accumulation through plasma membrane-bound NADPH oxidase and NADPH peroxidase of cell walls. This ROS accumulation, especially H_2O_2, stimulates or down-regulate differently located enzymes, some of which are involved in H_2O_2 generation & degradation [40]. Furthermore, the increased production of the hydroxyl radical (·OH), induces lipid peroxidation, resulting in damage to cellular membranes [41]. The damage to the membranes results in the uncontrolled loss of water and nutrients, and entry of extracellular hydrolases, thus adversely affecting cellular metabolism [42]. The physiological response of the "Bulbul-98"; seedlings to drought imposition after each pre-treatment was further elucidated with the determination of damage to cellular membranes and concentration of endogenous H_2O_2. The data showed an enhancement in electrolyte leakage from the membranes, and endogenous H_2O_2 concentration after exposure of "Bulbul-98" seedlings to drought stress conditions (Figure 3). The pre-treatment of seedlings with $CaCl_2$ enhanced damage to membranes, and endogenous H_2O_2 concentration under irrigation conditions, but the damage to membranes was more severe and endogenous H_2O_2 concentration was greater after the pre-treatment of H_2O_2 under similar conditions. However, H_2O_2 and $CaCl_2$ combined pre-treatment protected the cellular membranes and reduced H_2O_2 accumulation upon conditions of drought stress. It could be further interpreted from the data that the endogenous H_2O_2 concentration was lowest in seedlings supplemented with 15 mM $CaCl_2$. The lower electrolyte leakage and accumulation of endogenous H_2O_2 in the seedlings pre-treated by H_2O_2 or $CaCl_2$ could be due to the activation of ROS scavenging enzymes. Previous studies have indicated an increase in the activities of peroxidase, catalase and the enzymes of water-water cycle after pre-treatment with $CaCl_2$. In case of SDS-PAGE procedure, the size of Band 5 was 52 kDa, and the size of Band 12 was found to be almost 12 kDa. Previous studies have suggested that these band sizes correspond to the larger and smaller subunits of rubisco protein respectively [43]. Plants under abiotic stress seem to cause overexpression of this protein, which may play a possible role in plant growth.

4. MATERIAL AND METHODS

4.1. Plant Material

The greenhouse experiment was carried out at the Institute of Biotechnology and Genetic Engineering (IBGE, Khyber Pakhtunkhwa Agricultural University, and Peshawar in November 2015. Every pot was filled with well-rottedfarm yard manure and silt (1:1). During this experiment, *Brassica napus* "Bulbul-98" was used. Plants were arranged in a completely randomized design with three replications for precision, and allowed to grow for 35 days after germination. The seedlings were allowed to grow under controlled conditions (light, 100 μmol photon $m^{-2}\cdot s^{-1}$; temperature, 25 ± 2 °C; Relative humidity, 65–70%). The plants were sprayed separately two times at two day intervals regularly up to 20 days after germination with the following concentrations of 5, 10 and 15 mM Ca^{2+} in the form of $CaCl_2\cdot 2H_2O$) and 2, 5, and 10 μM H_2O_2. After 20 days of germination, half of the pots were sufficiently watered and maintained at 100% field capacity (as well-watered), and remaining pots were subjected to drought stress by withholding the water supply at 30% field capacity (as drought stressed). Field capacity was maintained by weighing the pots every day. We included plants with no spray treatment and with irrigation maintained throughout the experiment as extra control treatments.

4.2. Rate of Water Loss (RWL)

The rate of water loss was calculated according to the modified Ristic and Jenks (2002) method [44]. Leaf blades were excised from the pots and brought to relative water content (100%) by placing in de-ionized water for 2 h. Excess water was removed, and the leaf blades were weighed by an electric balance. Leaf blades were exposed to air circulation under darkness produced by an electric fan for 500 min. When the leaf blades were measured, the data was taken at different times [Tx (min); x = 0 min]. During exposure to circulating air, leaf blades were weighed at four times and recorded as time Tx where x = 1, 2, 3, and 4. Leaf blades were dried for 48 h at 80 °C and dry mass (DM) was determined. The rate of water loss was calculated as:

Leaf water loss ($mg \cdot g^{-1}h^{-1}$ DM) = [(FMTx − FMTx + 1) × 60]/[DM × (Tx + 1 − Tx)]

where FM is fresh mass, DM is dry mass, Tx and Tx + 1 is measuring time.

4.3. Relative Water Content

Fresh weight (FW) was obtained by weighing the leaf disc at harvest time. Leaf discs were fully immersed in double distilled water at 4 °C for 24 h. The samples were blotted dry on filter paper after 24 h to determine the turgid weight (TW) by another weighing. Finally, the leaf disc was oven dried at 70 °C for 48 h and dry weight (DW) was obtained. Relative water content was calculated by using the following formula:

Relative Water Content (RWC) = (FW − DW)/(TW − DW) × 100

4.4. Chlorophyll Content

The total chlorophyll content was measured according to the method by Arnon [45] by homogenizing leaf samples (100 mg) with 3 mL of 80% acetone. The homogenate were centrifuged at 15,000 rpm and the supernatant was collected. The absorbance were taken at 470, 645 and 663 nm by a UV spectrophotometer (UV 1900, Rayleigh, Beijing Beifen-Ruili Analytical Instrument (Group) Co. Ltd, Beijing, China) according to Lichtenthaler and Wellburn (1983) [46].

4.5. Soluble Protein Content

Leaves (100 mg) were homogenized in 2 mL of potassium phosphate buffer (0.05 M, pH 7.4) containing 1 mM PMSF, 2mMdithiothreitol, 0.1

mM (ethylene diamine tetra acetic acid) EDTA, and 20% polyvinyl polypyrrolidone (PVP) using a homogenizer. The sample was then centrifuged at 15,000 rpm. Supernatant was collected and the soluble protein content was quantified using bovine serum albumin as a standard (Bradford 1976) [47].

4.6. Electrolyte Leakage

A Consort C-931 conductivity meter was used to measuring the electrolyte leakage in the leaf (5 cm^2). In 5 mL double distilled water, the leaf discs were incubated at 25 °C for 3 h with shaking and initial conductivity of the solution were determined. After autoclaving the samples, the final conductivity of the solution were determined (100% electrolyte leakage). The quantity of electrolytes leakage were estimated as a percentage (%) of initial to final conductivity.

4.7. Proline Content

Proline was measured according to the method by Bates et al. [48]. The plant materials (500 mg) were homogenized in 3% sulphosalicylic acid. The sample was separated by centrifugation at 5000 rpm. At 100 °C for 1 h, 100 μL of the extract was reacted with acid ninhydrin, and then the reaction was terminated in an ice bath. The optical density was measured at 520 nm by mixing the reaction mixture. From a standard curve in the range of 0–20 μg/mL of L-proline, the amount of proline was determined.

4.8. H_2O_2 Content

Plant materials (100 mg FW) were homogenized with 0.5 mL of trichloroacetic acid (TCA) in an ice bath. The homogenate was centrifuged at 12,000 rpm for 15 min. 100 mM potassium phosphate buffer and 1 M KI were added to each supernatant. Absorbance was measured at 390 nm. H_2O_2 was quantified based on a standard curve.

4.9. SDS-PAGE

SDS-PAGE gel electrophoresis was done according to the method by Hames et al. [49] by using a 3% stacking and a 15% running gel. Stacking gel: 35.4% (*w/v*) acrylamide, 0.62% (*w/v*) bis-acrylamide, 10% *w/v* SDS, 1M Tris (pH = 6.8), 5 μL of tetra methylene diamine (TEMED), 10% (*w/v*) ammonium persulfate (APS) solution, 3.71 mL of distilled water (H_2O). Running gel: 35.4% (*w/v_* acrylamide, 0.62% (*w/v*) bis-acrylamide,

10% *w/v* SDS, 1 M Tris (pH = 8.8), 3 mL of distilled water (H_2O), 6 μL TEMED, and 10% (*w/v*) APS solution. The gels were stained with $AgNO_3$ solution and rocked at normal room temperature for 30 min. These gels were destained in 10% (*v/v*) acetic acid, 3% (*v/v*) glycerol, 40% (*v/v*) methanol, at normal room temperature.

4.10. Statistical Analysis

Analysis of variance (ANOVA) was done by applying the Fisher LSD test with Minitab (17) statistical software. Means with different letters are regarded as statistically significant at $p \leq 0.05$.

5. CONCLUSIONS

Drought stress is known to cause disruptions in almost all physiological parameters. However, the exogenous application of various components could help in recovery of the damage caused by this stress. In the current study, we have found that *Brassica* seedlings with drought stress-induced physiological damage recovered with the application of Ca^{2+} and H_2O_2supplementations. Hence, exogenous application of these components could be a suitable strategy to improve crop production under abiotic stress.

ACKNOWLEDGMENTS

Authors highly acknowledge all the colleagues and staff members who helped while performing this work. We are also thankful to Deanship of Scientific Research (DSR), King Abdulaziz University, Saudi Arabia for providing the financial support.

AUTHOR CONTRIBUTIONS

Akram Khan, Md. Mahadi Hasan, Muhammad Ali and Yasir Anwar wrote the manuscript, perform data analysis and prepare figures, Aqib Iqbal designed the experiment, helps in planning and conducting the experiments, Khalid Rehman Hakeem and Hesham F. Alharby helps in the data analysis, writing the manuscript and provide critical revision, Mirza Hasanuzzaman helped in the data analysis.

CONFLICTS OF INTEREST

The authors declare no conflicts of interest.

REFERENCES

1. Hakeem, K.R.; Chandna, R.; Ahmad, P.; Ozturk, M.; Iqbal, M. Relevance of proteomic investigations in plant stress physiology. OMICS J. Integr. Biol. 2012, 16, 621–635.
2. Boyer, J.S. Plant productivity and environment. Science 1992, 218, 443–448.
3. Bajji, M.; Kinet, J.M.; Lutts, S. The use of the electrolyte leakage method for assessing cell membrane stability as a water stress tolerance test in durum wheat. J. Plant Growth Reg. 2002, 36, 61–70.
4. Hsu, S.Y.; Hsu, Y.T.; Kao, C.H. The effect of polyethylene glycol on proline accumulation in rice leaves. Biol. Plant.2003, 46, 73–78.
5. Ratnayaka, H.H.; Molin, W.T.; Sterling, T.M. Physiological and antioxidant responses of cotton and spurred anoda under interference and mild drought. J. Exp. Bot. 2003, 54, 2293–2305.
6. Bot, A.J.; Nachtergaele, F.O.; Young, A. Land Resource Potential and Constraints at Regional and Country Levels; World Soil Resources Reports 90; Land and Water Development Division, FAO: Rome, Italy, 2000.
7. Riznsky, L.H.; Liang, H.; Hittler, R. The combined effect of drought stress and heat shock on gene expression in tobacco plant. Plant Physiol. 2002, 130, 1143–1151.
8. Bartels, D.; Sunkar, R. Drought and salt tolerance in plants. Crit. Rev. Sci. 2005, 24, 23–58.
9. Pastori, G.M.; Foyer, C.H. Common components, networks, and pathways of cross-tolerance to stress. The central role of “redox” and abscisic acid-mediated controls. Plant Physiol. 2002, 129, 460–468.
10. Gummuluru, S.; Hobbs, S.L.A.; Jana, S. Genotypic variability in physiological characters and its relationship to drought tolerance in Durum wheat. Can. J. Plant Sci. 1999, 69, 703–711.
11. Blum, A. Osmotic adjustment and growth of barley genotypes under stress. Crop Sci. 1999, 29, 230–233.
12. Morgan, J.A.; LeCain, D.R. Leaf gas exchange and related leaf traits among 15 winter wheat genotypes. Crop Sci.1991, 31, 443–448.
13. Shigeoka, S.T.; Ishikawa, M.; Tamoi, Y.; Miyagawa, Y.; Takeda, T.; Yabuta, Y.; Yoshimura, K. Regulation and function of ascorbate peroxidase isoenzymes. J. Exp. Bot. 2002, 53, 1305–1319.
14. Smirnoff, N. The role of active oxygen in the response of plants to water deficit and desiccation. New Phytol. 1993, 125, 27–58.
15. Ahmad, P.; Hakeem, K.R.; Kumar, A.; Ashraf, M.; Akram, N.A. Salt induced changes in photosynthetic activity and oxidative defense system of three cultivars of mustard (Brassica juncea L.). Afr. J. Biotechnol. 2012, 11, 2694–2703.
16. Mittler, R.; Herr, E.H.; Orvar, B.L.; Van Camp, W.; Willekens, H.; Inze, D.; Ellis, B.E.E. Transgenic tobacco plants with reduced capability to detoxify reactive oxygen intermediates are hyper responsive to pathogen infection. Proc. Natl. Acad. Sci. USA 1999, 96, 14165–14170.
17. Mlot, C. Plant biology in the genome era. Science 1998, 281, 331–332.
18. Shao, H.B.; Liang, Z.S.; Shao, M.A. Changes of some physiological and biochemical indices for soil water deficits among 10 wheat genotypes at seedling stage. Colloids Surf. B Biointerfaces 2005, 42, 107–113.

19. Shao, H.B.; Chu, L.Y.; Shao, M.A. Calcium as a versatile plant signal transducer under soil water stress. BioEssays2008, 30, 634–641.
20. Qureshi, M.I.; Qadir, S.; Zolla, L.J. Proteomics based dissection of stress-responsive pathways in plants. Plant Physiol.2007, 164, 1239–1260.
21. Bush, D.S. Calcium regulation in plant cells and its role in signaling. Annu. Rev. Plant Physiol. 1995, 46, 95–122.
22. Yu, C.W.; Murphy, T.M.; Sung, W.W.; Lin, C.H. H2O2 treatment induces glutathione accumulation and chilling tolerance in mung bean. Funct. Plant Biol. 2002, 29, 1081–1087.
23. Yu, C.W.; Murphy, T.M.; Lin, C.H. Hydrogen peroxide-induced chilling tolerance in mung beans mediated through ABA-independent glutathione accumulation. Funct. Plant Biol. 2003, 30, 955–963.
24. Sadaqat, H.A.; Tahir, M.H.N.; Hussain, M.T. Physiogenetic aspects of drought tolerance in canola (Brassica napus). Int. J. Agric. Biol. 2003, 5, 611–614.
25. Resketo, P.; Szabo, L. The effect of drought on development and yield components of soybean. Int. J. Trop. Agric. 1992, 8, 347–354.
26. Maliwal, G.L.; Thakkar, K.R.; Sonani, V.V.; Patel, P.H.; Trivedi, S.N. Response of mustard (Brassica juncea L.) to irrigation and fertilization. Ann. Agric. Res. 1998, 19, 353–359.
27. Patel, J.R. Effect of irrigation and nitrogen on mustard. J. Maharashtra Agric. Univ. 1999, 23, 259–261.
28. Delledonne, M.; Zeier, J.; Marocco, A.; Lamb, C. Signal interactions between nitric oxide and reactive oxygen intermediates in the plant hypersensitive disease resistance response. Proc. Natl. Acad. Sci. USA 2001, 98, 13454–13459.
29. Ma, Y.Y.; Song, W.Y.; Liu, Z.H.; Zhang, H.M.; Guo, X.L.; Shao, H.B.; Ni, F.T. The dynamic changing of Ca2+ cellular localization in maize leaflets under drought stress. Cell Biol. 2009, 332, 350–362.
30. Clarke, J.M.; Romagosa, I.; Jana, S.; Srivastava, J.P.; Caig, M.C. Relationship of excised—Leaf water loss rate and yield of durum wheat in diverse environments. Can. J. Plant Sci. 1988, 69, 1075–1081.
31. Cameron, K.D.; Teece, M.A.; Smart, L.B. Increased accumulation of cuticular wax and expression of lipid transfer protein in response to periodic drying events in leaves of tree tobacco. Plant Physiol. 2006, 140, 176–183.
32. Neto, A.D.A.; Prisco, J.T.; Filho, J.E.; Medeivo, J.V.R.; Filho, E.G. Hydrogen peroxide pre-treatment induces salt stress acclimation in maize plants. J. Plant Physiol. 2005, 162, 1114–1122.
33. Jubany-Mari, T.; Munné-Bosch, S.; López-Carbonell, M.; Alegre, L. Hydrogen peroxide is involved in the acclimation of the Mediterranean shrub, Cistusalbidus L., to summer drought. J. Exp. Bot. 2009, 60, 107–120.
34. Ishibashi, Y.; Zheng, S.H.; Arima, S. Effect of seed pre-treatment with hydrogen peroxide in soybean seed germination and seedling growth. Coast. Bioenv. 2010, 15, 55–60.
35. Bosabalidis, A.M.; Kofidis, G. Comparative effects of drought stress on leaf anatomy of two olive cultivars. Plant Sci.2002, 163, 375–379.
36. Chaves, M.M.; Flexas, J.; Pinheiro, C. Photosynthesis under drought and salt stress: Regulation mechanisms from whole plant to cell. Anal. Bot. 2009, 103, 551–560.
37. Yordanov, I.; Velikova, V.; Tsone, V. Plant response to drought, acclimation and stress tolerance. Photosynthetica 2000, 30, 171–186.
38. Yang, S.L.; Llan, S.S.; Gong, M. Hydrogen peroxide-induced proline and metabolic pathway of its accumulation in maize seedlings. J. Plant Physiol. 2009, 166, 1694–1699.
39. Nayyar, H. Variation in osmoregulation in differentially drought-sensitive wheat genotypes involves calcium. Biologia Plant. 2003, 47, 541–547.
40. Ranieri, A.; Castagna, A.; Pacini, J.; Baldan, B.; Mensuali, S.A.; Soldatini, G.F. Early production and scavenging of hydrogen peroxide in the apoplast of sunflowers plants exposed to ozone. J. Exp. Bot. 2003, 54, 2529–2540.

41. Ferrari-Iliou, R.; d'Arcy-Lameta, A.; Pham-Thi, A.T.; Zuily-Fodil, Y.; Mazliak, P. Effect of drought on photodynamic peroxidation of leaf total lipophilic extracts. Phytochemistry 1994, 37, 1237–1243.
42. Morse, M.; Rafudeen, M.S.; Farrant, J.M. An overview of the current understanding of desiccation tolerance in the vegetative tissues of higher plants. In Plant Responses to Drought and Salinity Stress: Developments in a Post-Genomic Era, 1st ed.; Advances in Botanical Research 2011; Elsevier Ltd.: Burlington County, NJ, USA, 2011; pp. 319–347.
43. Chen, Y.; Wang, B.; Chen, J.; Wang, X.; Wang, R.; Peng, S.; Chen, L.; Ma, L.; Luo, J. Identification of Rubiscorbc L. and rbcS in Camellia oleifera and their potentials as molecular markers for selection of high tea oil cultivars. Front. Plant Sci. 2015, 6, 189.
44. Ristic, Z.; Jenks, M.A. Leaf cuticle and water loss in maize lines differing in dehydration avoidance. J. Plant Physiol.2002, 159, 645–651.
45. Arnon, D.I. Copper enzymes in isolated chloroplasts. Polyphenoloxidase in Beta vulgaris. Plant Physiol. 1949, 24, 1–15.
46. Lichtenthaler, H.K.; Wellburn, A.R. Determination of Total Carotenoids and Chlorophylls a and b of Leaf Extracts in Different Solvents. Biochem. Soc. Trans. 1983, 11, 591–603.
47. Bradford, M.M. A rapid and sensitive for the quantization of microgram quantities of protein utilizing the principle of protein-dye binding. Anal. Biochem. 1976, 72, 248–254.
48. Bates, L.S.; Waldren, R.P.; Teare, I.D. Rapid determination of free proline for water stress studies. Plant Soil 1973, 39, 205–207.
49. Hames, B.D.; Rickwood, D. Gel Electrophoresis of Proteins: A Practical Approach, 2nd ed.; IRL Press Limited: Oxford, UK, 1999; p. 383.

CHAPTER 4

Plants Coping Abiotic and Biotic Stresses: A Tale of Diligent Management

Hatem Rouached,[1] Sikander Pal,[2] Shimon Rachmilevitch,[3] Marc Libault,[4] and Lam-Son Phan Tran[5]

[1]*Biochimie et Physiologie Moléculaire des Plantes, Unité Mixte de Recherche CNRS-INRA, Université Montpellier 2, Montpellier SupAgro, Place Viala, 34060 Montpellier, France*
[2]*Department of Botany, University of Jammu, Jammu, Jammu and Kashmir 180-006, India*
[3]*Jacob Blaustein Institute for Desert Research, BGU, 84990 Sede-Boqer Campus, Israel*
[4]*Department of Microbiology and Plant Biology, University of Oklahoma, 770 Van Vleet Oval, Norman, OK 73019, USA*
[5]*Signaling Pathway Research Unit, RIKEN Center for Sustainable Resource Science, 1-7-22 Suehiro-cho, Tsurumi-ku, Yokohama 230-0045, Japan*

Plants unlike other living forms are sessile thereby facing severe biotic and abiotic stresses. Plants have evolved different efficient defence responses which thrive upon a number of intrinsic factors, such as genotypic and phenotypic constitutions and developmental circumstances, and extrinsic factors like severity and duration of the stresses. Stress management uses molecular and biochemical level controls, the competence, and speed, at which a stress signal is perceived and transmitted to generate stress signal molecules and activate stress-protective mechanisms. A well-concerted action of the plants' competence at morphological, physiological, biochemical, and molecular strata regulates numerous adaptive responses to biotic and abiotic stresses. Genetic manipulations of signalling networks have been widely used to improve plant productivity under stressful conditions. Advanced biotechnological application will enable maintaining agriculture in a sustainable manner. In this special issue, we present two reviews and four research papers which address genomic, molecular, and physiological regulations as well as signalling networks dealing with plant responses to abiotic and biotic factors.

Climate change, desertification, and the rise in human population have put a severe load on agriculture and are deteriorating crop productivity. In recent years, numerous molecular and metabolic pathways involved in plant responses and adaptation to various types of environmental stresses have been identified and reported. Among hundreds of metabolic pathways identified, the role of polyamines in stress management to enhance plant acclimation and adaptation is emerging rapidly. In this special issue, a timely review by P. Rangan et al. (2014) summarizes our knowledge on biosynthesis and catabolism of polyamines and highlights recent progress in elucidating the functions of polyamines in regulation of plant responses to abiotic stresses. Given a huge genetic variation among plant species, the authors also discuss a systematic approach based on polyamine-mediated enhancement of stress tolerance which might be used as a potential strategy for screening and identification of natural variants within existing crop species. The identified genotypes that possess compatible allelic variants could be then used for the improvement of stress tolerance.

Plant-microbe interactions are at the core of symbiotic, parasitic, or mutualistic plant-microbe relationships. These interactions have displayed a unique way of mutualistic communications for a resource sharing. To shed light on the topic, a detailed review of M. Libault (2014) explains the unique mechanisms in using legumes to interact with bacteria. Leguminous plants have developed a mutualistic symbiotic relationship with rhizobium (a type of soil bacteria). Upon bacterial infection, a new root organ called nodule is developed that enables the leguminous plants to access a steady source of nitrogen through the fixation and assimilation of the atmospheric N_2 by the symbiotic bacteria. In return, the bacteria also get benefit from the symbiotic plants that provide photosynthesis product to bacteria as source of carbon. Environmental stress or climate change, which influences the concentration of the atmospheric carbon dioxide (CO_2), will have a significant impact on plant photosynthesis. As a consequence, this will affect the nitrogen and carbon metabolism, leading to altered nitrogen fixation efficiency. The key regulatory mechanisms controlling carbon/nitrogen balances with particular attention to legume nodulation are reviewed by coeditor M. Libault in his review article. In addition, readers can also get an overview about the effect of the change in CO_2 level on nitrogen fixation efficiency through this review, giving rise to idea as to how we could mitigate the impact of the change in atmospheric CO_2 concentration.

Besides the change in atmospheric CO_2 level, water deficit is one of the major constraints for nodulation, and, as a consequence for plant productivity. This topical issue is presented here by a research article of S. Sulieman et al. (2014). In their study, the authors assessed the growth and nodulation attributes of two soybean varieties DT2008 and Williams 82 (W82), which have contrasting drought-tolerant capacity, in a symbiotic association with Bradyrhizobium japonicum under drought and subsequent rehydration. The authors aimed to understand the correlation between N_2 fixation efficiency and differential drought-responsive phenotypes of DT2008 and W82. Their results also provide genetic resources and basis foundation for further genomic research that would lead to better understanding of mechanisms involved in regulation of N_2 fixation in soybean under drought at molecular level.

Heavy metal pollution has been a matter of grave concern. Until recently, efforts have been mainly restricted to phytoremediation of soils using plant species with high metal uptake capacity such as Brassica species. Chromium (Cr) is a highly phytotoxic heavy metal affecting crop productivity and human health via entering the food chain. Phytoremediation of Cr-polluted soils has been mostly demonstrated in using herbaceous plants, whereas use of cotton cultivars in Cr phytoremediation is least explored. In the present issue, M. K. Daud et al. (2014) have shown the potentials of two transgenic cotton cultivars (J208 and Z905) and their hybrid line (ZD14) in Cr phytoremediation using a multiple biomarker approach. Their work showed that these cotton cultivars and hybrid line could effectively uptake and sequestrate Cr in dead parts of the plants, such as vacuole and cell wall, besides having a more highly accelerated antioxidant system. This study thus proposes a new role of cotton cultivars in phytoremediation of Cr-polluted soils.

The interactions between macro- and micronutrient homeostases have been reported in many physiological and nutritional situations. N. Bouain et al. (2014) studied the interaction between phosphate (Pi) and zinc (Zn) homeostasis in two lettuce varieties. The results revealed that the variation in Pi and Zn supplies affects the biomass and photosynthesis as well as the dynamics of Pi transport in a contrasting manner between the two varieties, indicating a genetic basis for such physiological responses. On the basis of their results, the authors proposed a working model of how Pi and Zn signalling pathways are integrated into functional networks to control plant growth.

Salinity is a major abiotic stress worldwide claiming agriculture lands and affecting productivity. Research paper by A. Matsui et al. (2014) investigated salt stress in Arabidopsis. The authors reported that the tasiRNA-ARF pathway is involved in controlling floral architecture in plants under drought and high salinity. The ta-siRNA biosynthesis mutants and mutated ARF3-overexpressing plants showed short stamen under drought and salt stress, which hampered self-pollination. This study reports for the first time that a type of ta-siRNAs (tasiRNA-ARF) plays an important role in maintaining normal stamen growth and fertilization under drought and high salinity.

ACKNOWLEDGMENTS

The editors would like to thank the authors for contributing their manuscripts to this special issue.

CHAPTER 5

Combined Effects of Ozone and Drought on the Physiology and Membrane Lipids of Two Cowpea (*Vigna unguiculata* (L.) Walp) Cultivars

Deborah Moura Rebouças [1], Yuri Maia De Sousa [2], Matthieu Bagard [1,*], Jose Helio Costa [2], Yves Jolivet [3,4], Dirce Fernandes De Melo [2] and Anne Repellin [1]

[1]*Institut d'Ecologie et des Sciences de l'Environnement de Paris, Faculté des Sciences et Technologie, Université Paris-Est Créteil, 61 Avenue du Général De Gaulle, 94010 Créteil, France*

[2]*Departamento de Bioquímica e Biologia Molecular, Universidade Federal do Ceará, Fortaleza, P.O. Box 6029, 60455-760 Fortaleza, Ceará, Brazil*

[3]*Unité Mixte de Recherche Ecologie et Ecophysiologie Forestières, Université de Lorraine, BP239, F-54506 Vandœuvre-lès-Nancy, France*

[4]*Institut National de la Recherche Agronomique, Unité Mixte de Recherche Ecologie et Ecophysiologie Forestières, BP239, F-54280 Champenoux, France*

ABSTRACT

The interactive effects of drought and ozone on the physiology and leaf membrane lipid content, composition and metabolism of cowpea (*Vigna unguiculata* (L.) Walp.) were investigated in two cultivars (EPACE-1 and IT83-D) grown under controlled conditions. The drought treatment (three-week water deprivation) did not cause leaf injury but restricted growth through stomatal closure. In contrast, the short-term ozone treatment (130 ppb 12 h daily during 14 day) had a limited impact at the whole-plant level but caused leaf injury, hydrogen peroxide accumulation and galactolipid degradation. These effects were stronger in the IT83-D cultivar, which also showed specific ozone responses such as a higher digalactosyl-diacylglycerol (DGDG):monogalactosyl-diacylglycerol (MGDG) ratio and the coordinated up-regulation of DGDG synthase (*VuDGD2*) and ω-3 fatty acid desaturase 8 (*VuFAD8*) genes, suggesting that membrane remodeling occurred under ozone stress in the sensitive cultivar. When stresses were combined, ozone did not modify the stomatal response to drought and the observed effects on whole-plant physiology were essentially the same as when drought was applied alone. Conversely, the drought-induced stomatal closure appeared to alleviate ozone effects through the reduction of ozone uptake.

Keywords: drought; ozone; stress combination; membrane lipids; cowpea

1. INTRODUCTION

Plants, as sessile organisms, are permanently subjected to changing environmental conditions that might compromise homeostasis, growth and even survival. As a consequence, plants have developed elaborate mechanisms to sense and respond to changes in their environment through appropriate adjustments. Water availability constitutes a major limiting factor for plant productivity and drought is considered one of the most influent abiotic constraint to crop yield [1]. Rising temperatures and altered precipitation patterns are two main components of climate change and both contribute to increase the frequency and intensity of drought episodes regionally. Air pollution is another aspect of the anthropogenic alteration of the environment. Among air pollutants, tropospheric ozone is considered the most detrimental to plants [2]. Ozone is a secondary product of photochemical reactions that involve nitrogen oxides and volatile organic compounds. Due to the increased emissions of these precursors from vehicle and industrial sources, tropospheric ozone has increased since the 1950s [3]. Although emission reduction policies have successfully limited the frequency and extent of high ozone episodes and despite recent evidence that tropospheric ozone has leveled off in the last decade, at least in the Northern hemisphere [4], background concentrations have now reached levels potentially damaging to crop production in several regions of the world [5].

As high temperature and irradiance promote both ozone production and water deficit, the occurrence of high ozone exposure during drought episodes is common in semi-arid regions. It is clear that the response of plants to a combination of stress conditions is unique and cannot be predicted from the response to the stresses applied individually [6]. The interacting effects of ozone and drought are still unclear as contradictory findings are reported in the literature [7]. Meta-analyses of the effects of ozone on trees [8] and crops [9] showed that drought tended to mitigate the adverse effects of the pollutant. It is assumed that the drought-induced stomatal closure limits ozone uptake, thereby protecting leaf tissues from the oxidative stress caused by the pollutant. Reciprocal cross-protection was observed in *Medicago truncatula* [10] but ozone can also in some cases exacerbate the effects of water stress [11]. For instance, in oak seedlings, ozone combined with drought caused a stronger malondialdehyde accumulation and larger biomass losses than drought applied alone [12]. Ozone can also cause stomatal sluggishness [13]: recent results suggest that ozone can disrupt the ABA-induced signal transduction pathway for stomatal control, which could in turn impair the ability of plants to respond to water stress [14]. Overall, the interactions between ozone and drought depend largely on the temporal distribution of ozone and drought episodes [7].

At the cellular level, drought promotes the production of reactive oxygen species (ROS), essentially in the chloroplasts where the drought-induced over-reduction of the electron transport chain can lead to the formation of singlet oxygen ($^1O_2$) and superoxide ($O_2^{-\bullet}$) [15]. Ozone is itself a strong oxidant and its decomposition in the apoplast generates a range of ROS. In all cases, if the oxidative load exceeds the apoplastic antioxidant capacity, ROS can spread within the cell and add to the ROS produced endogenously [16]. ROS, being reactive molecules, can oxidize all types of cellular components [15]. In addition, ROS play a role in signaling and induce defense responses [17]. Potential interacting effects between drought and ozone may thus arise from the stimulation of ROS production as a common feature of the two environmental stresses.

Membranes are among the first cellular components to perceive drought stress and the role of phospholipases and lipid-derived messengers in triggering signaling cascades has been largely demonstrated [18]. Likewise, the ozone-triggered intracellular signal transduction is initiated at the plasma membrane through mechanisms that involve lipid peroxidation and subsequent formation of signal molecules such as jasmonates, although the primary site of ozone reaction within leaf tissues is the extracellular matrix [19]. Besides their role in cell signaling, membrane lipids may also contribute to stress tolerance through the adjusting of membrane fluidity and the reorganization of cell components [20]. From this perspective, the effects of drought and ozone on leaf membrane lipids share many similarities. Both stresses induce a decrease in membrane lipid content [21,22], an inhibition of lipid biosynthesis [23,24], a stimulation of lipolytic activities [25,26] and a decrease in linolenic acid (18:3) content [27,28]. It appears that changes in membrane lipid metabolism play an essential part in plant responses to the combination of ozone and drought stresses.

In the present work, we examined the effects of drought, ozone and the combination of these stresses on cowpea (*Vigna unguiculata* (L.) Walp.). Cowpea is a staple legume crop in semi-arid regions of the tropics and subtropics, where meteorological conditions promote the co-occurrence of drought and ozone stresses. While the impact of water stress on cowpea has been extensively studied [21,29,30,31], the response of this crop to ozone is less documented [32]. In recent studies, the responses of cowpea to ozone have been investigated in Asian and African cultivars with respect to biomass production, yield, photosynthesis, nitrogen fixation and ROS detoxification [33,34,35,36,37]. Here, we investigated the response of a Southamerican (EPACE-1) and an Asian cultivar (IT83-D) to the combination of ozone and drought stresses. In addition to the characterization of the physiological responses of the two cultivars to ozone and drought, a specific attention was drawn to plastidial membrane lipid content and fatty acid composition as well as to the expression of genes encoding enzymes involved in membrane lipid biosynthesis and degradation. The results show that ozone and drought stresses induced contrasting responses in cowpea plants and bring new insights into the interplay between general and specific multi-stress responses.

2. RESULTS

2.1. Physiological Responses of Cowpea Plants to Ozone and Drought Stresses

The individual and interactive effects of ozone and drought stresses on the two cowpea cultivars were first assessed at the whole-plant level on aerial parts by evaluating shoot biomass production (shoot dry weight, SDW), leaf water content (relative water content, RWC), stomatal conductance (g_s) and chlorophyll fluorescence (effective quantum yield of photosystem II (Φ_{PSII}) (Figure 1). Damage at the leaf level were evaluated by monitoring visible symptoms of damage, by detecting hydrogen peroxide (H_2O_2) (Figure A1) and by measuring plastidial membrane lipid contents (Figure 2). From now on, treatments will be referred to as C for control, O for ozone, D for drought and OD for the combination of ozone and drought. During the 14 d of experiment, plants of the O and OD treatments were exposed to an average of 130.4 ppb of ozone 12 h daily (Table 1). At the end of the experiment, the concentration-based index of ozone exposure AOT40 reached 16.7 ppm·h on average. The stomatal uptake of ozone (or Phytotoxic Ozone Dose, POD_0) was 3.8 and 4.6 mmol·m^{-2} in EPACE-1 and IT83-D, respectively. Visible symptoms of leaf injury (dark/brown necrosis) and H_2O_2 accumulation were observed in both cultivars in response to the O treatment, to a greater extent in IT83-D than in EPACE-1 (Figure A1). However, no significant effect on biomass production, leaf water content, stomatal conductance or chlorophyll fluorescence could be observed in response to the O treatment, except a 40% decrease in shoot dry weight in IT83-D as compared to the control after 7 day (Figure 1a).

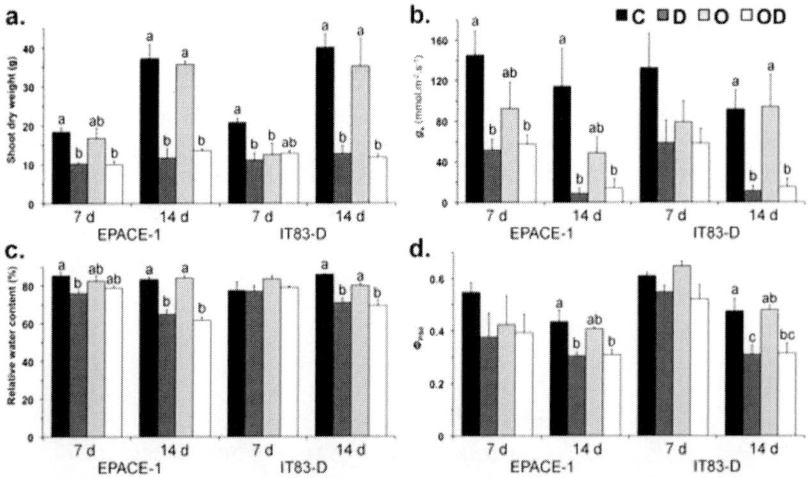

Figure 1. (**a**) Shoot dry weight; (**b**) stomatal conductance (g_s); (**c**) relative water content and (**d**) quantum yield of PSII (Φ_{PSII}) of *Vigna unguiculata* plants subjected to the following treatments: Control (C, black bars), Drought (D, dark grey bars), Ozone (O, light grey bars), Ozone + Drought (OD, white bars). Measurements were performed 7 and 14 days after the onset of the stress treatments. Means ± SEM are shown (n = 3–4). When significant differences were found, letters indicate homogenous subsets at each time point for a given cultivar (one-way ANOVA and Tukey post-hoc tests, α = 0.05).

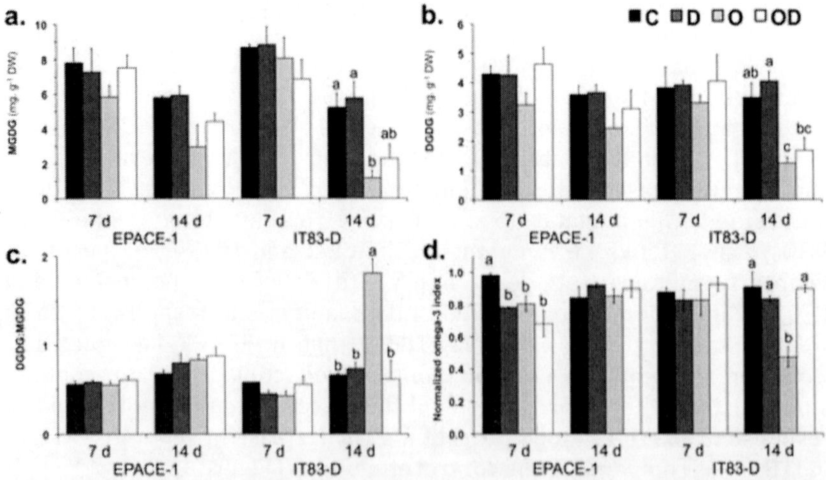

Figure 2. (**a**) MGDG contents; (**b**) DGDG contents; (**c**) DGDG:MGDG ratio and (**d**) normalized omega-3 index in the leaves of *Vigna unguiculata* plants subjected to the following treatments: Control (C, black bars), Drought (D, dark grey bars), Ozone (O, light grey bars), Ozone + Drought (OD, white bars). Measurements were performed 7 and 14 days after the onset of the stress treatments. Means ± SEM are shown (n = 3). When significant differences were found, letters indicate homogenous subsets at each time point for a given cultivar (one-way ANOVA and Tukey post-hoc tests, α = 0.05). MGDG, Monogalactosyl-diacylglycerol; DGDG, digalactosyl-diacylglycerol, omega-3 index 18:3/(18:0 + 18:1 + 18:2).

In contrast to the O treatment, the drought treatment alone (D) did not cause leaf injury or H_2O_2 accumulation (Figure A1) but had strong effects at the whole-plant level. Most of these effects were of the same degree in the two cowpea cultivars. First, the drought treatment induced stomatal closure as compared to the control, already after 7 day in EPACE-1 (−50%) and to a larger degree after 14 day in both cultivars (−90%) (Figure 1b). This response limited water loss as RWC was only 20% lower than in control plants after 14 day of drought stress (Figure 1c), but correlated with a severe restriction of plant growth: the aerial biomass of the drought-treated plants was 50% and 70% lower than that of the control plants after 7 and 14 day of experiment, respectively (Figure 1a). This strong restriction of above-ground biomass production was associated to a 35% decrease in the quantum yield of PSII (Φ_{PSII}) in drought-stressed plants as compared to control plants for both cultivars (Figure 1d).

Table 1. Twelve-hour mean ozone concentration and indices of ozone exposure (AOT40, ozone concentrations accumulated over a threshold of 40 ppb) and stomatal uptake (POD_0, phytotoxic ozone dose, expressed as the accumulated stomatal flux of ozone) to which cowpea plants were subjected in the following treatments: Control (C), Drought (D), Ozone (O), Ozone + Drought (OD). Mean ± SD are shown (n = 4). Ozone concentration and AOT40 were averaged from the values recorded in 4 identical chambers. Ozone stomatal uptake (POD_0) was averaged from 4 biological replicates. Control (C) and drought-treated (D) plants were supplied with activated charcoal-filtered air and ozone concentration was close to zero (nd, not detected).

Cultivar	Treatment	12-h Mean (O_3) (ppb)	AOT40 (ppm·h)	POD_0 (mmol·m^{-2})
EPACE-1	C	nd	nd	nd
	D			nd
	O	130.4 ± 5.2	16.7 ± 1.0	3.8 ± 0.7
	OD			2.9 ± 0.5
IT83-D	C	nd	nd	nd
	D			nd
	O	130.4 ± 5.2	16.7 ± 1.0	4.6 ± 1.2
	OD			3.1 ± 1.4

When the plants were exposed to drought in combination with ozone, the POD_0 was 24% (2.9 mmol·m^{-2}) and 33% (3.1 mmol·m^{-2}) lower in EPACE-1 and IT83-D, respectively, as compared to plants treated with ozone only (Table 1). Consistently, the extent of leaf injury was less important in OD-treated than in O-treated plants and no H_2O_2 could be detected in leaves of the in OD-treated plants (Figure A1). At the whole-plant level, the effects of the combination of ozone and drought stresses were essentially the same as those of the drought treatment alone (Figure 1).

2.2. Galactolipid Content and Fatty Acid Composition

To examine further the interactive and individual effects of ozone and drought, we investigated the impact of the two stresses applied alone or in combination on pastidial galactolipids, which are known cellular targets of environmental stresses. We analyzed plastidial galactolipid content and fatty acid composition (Figure 2) as well as the transcript abundance of genes encoding enzymes involved in galactolipid synthesis, desaturation and degradation (Figure 3).

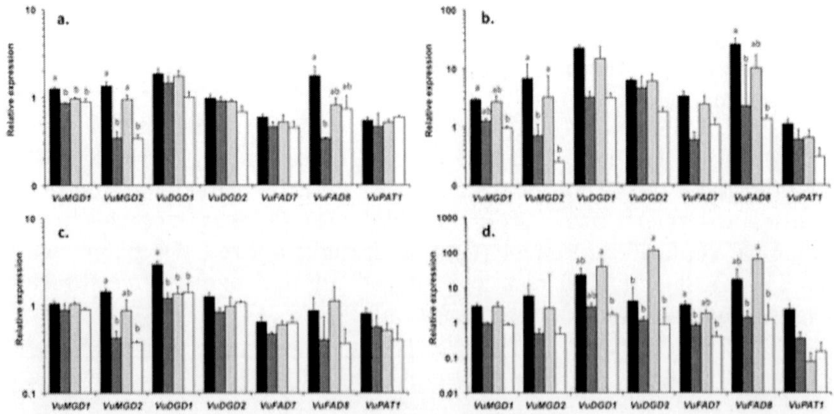

Figure 3. Relative expression of genes involved in lipid metabolism in leaves of (**a**,**b**) EPACE-1 and (**c**,**d**) IT83-D plants exposed to the following treatments: Control (C, black bars), Drought (D, dark grey bars), Ozone (O, light grey bars), Ozone + Drought (OD, white bars). Measurements were performed 7 and 14 days after the onset of the stress treatments. Means ± SEM are shown (n = 3). Each value is the average of two technical replicates. Expression of target genes was normalized to the expression of the reference gene *VuEF-1α*. When significant differences were found, letters indicate homogenous subsets at each time point for a given cultivar (one-way ANOVA and Tukey post-hoc tests, α = 0.05). *VuMGD1*, type 1 monogalactosyl-diacylglycerol synthase; *VuMGD2*, type 2 monogalactosyl-diacylglycerol synthase; *VuDGD1*, type 1 digalactosyl-diacylglycerol synthase; *VuDGD2*, type 2 digalactosyl-diacylglycerol synthase; *VuFAD7*, ω-3 fatty acid desaturase 7; *VuFAD8*, ω-3 fatty acid desaturase 8; *VuPAT1*, patatin-like lipid acyl hydrolase; *VuEF-1α,* elongation factor 1 alpha.

Galactolipids (monogalactosyl-diacylglycerol, MGDG and digalactosyl-diacylglycerol, DGDG) were the dominant lipid classes in cowpea leaf tissues, where their presence in plastidial membranes supports the bioenergetic and metabolic functions of the chloroplasts. The D treatment had no effect either on MGDG and DGDG contents (Figure 2a,b) or on their fatty acid composition and insaturation level (Table A1 and Table A3). In contrast, the O treatment affected plastidial galactolipids, in a significant manner in IT-83D (Figure 2a,b). In this cultivar, after 14 day of O treatment, MGDG and DGDG contents were reduced by 77% and 64%, respectively, as compared to the controls (Figure 2a,b), and these effects correlated with a 3-fold increase in the DGDG:MGDG ratio (Figure 2c). In the EPACE-1 cultivar, plastidial galactolipids were also affected by ozone (after 14 day MGDG and DGDG contents were reduced by 48% and 32%, respectively, as compared to the control), but these effects were not significant (Figure 2a,b) and the DGDG:MGDG ratio remained unchanged (Figure 2c). The polyunsaturated fatty acid 18:3 was predominant in the fatty acid composition of MGDG (ca. 95%) and DGDG (ca. 88%) in the leaves of both cultivars (Table A1). In EPACE-1, the FA composition of galactolipids was not significantly modified by the various treatments (Table A1 and Table A3). In IT83-D, the proportion of 18:3 in MGDG decreased from 95% in the control to 81% after 14 day of ozone treatment alone. This ozone-induced decrease in 18:3% was balanced by increases in the other FA % in MGDG, mainly the saturated FA 16:0 and 18:0 (Table A1) and translated into a decrease in the omega-3 index (Figure 2d). No apparent modification in FA % was observed in DGDG (Table A1).

2.3. Expression of Genes Coding Enzymes Involved in Lipid Metabolism

To better understand the observed modifications in plastidial galactolipid, the abundance of transcripts encoding enzymes involved in their biosynthesis, desaturation and degradation were analyzed using quantitative real-time PCR (Figure 3). The genes studied encoded two enzymes of the galactolipid biosynthesis pathway (MGDG synthase, EC 2.4.1.46: *VuMGD1* and *VuMGD2*; DGDG synthase, EC 2.4.1.241: *VuDGD1* and *VuDGD2*), two chloroplastic ω-3 fatty acid desaturases (EC 1.14.19.3, *VuFAD7* and *VuFAD8*) and the patatin-like lipid acyl hydrolase (EC 3.1.1.26: *VuPAT1*). Values obtained for the target genes were normalized with respect to the expression of the reference gene *VuEF1-α*.

In the control leaves of both cultivars, the expression level of most genes was higher at 14 day than at 7 day, by a factor of 2 to 10 (Figure 3). The O treatment had limited effects on the expression of the tested genes, except a down regulation of *VuMGD1* in EPACE-1 and *VuDGD1* in IT83-D at 7 day (Figure 3a,c), and a 27-fold up-regulation of *VuDGD2* in IT83-D at 14 day (Figure 3d). In contrast, the genes tested were globally down regulated in response to the D treatment, although a relatively high variability in the data limited the significance of the observed effects (Figure 3). Among the genes tested, *VuMGD2*, *VuDGD1* and *VuFAD8* showed the earliest and strongest repression under the drought treatment. As seen with most parameters that were analyzed in this study, the effects on transcipt levels of the analyzed genes of the OD treatment were essentially the same as those of the D treatment. In both cultivars, the expression of *VuPAT1* was unchanged, or slightly repressed, in response to the various treatments as compared to the control (Figure 3).

3. DISCUSSION

3.1. The Drought Treament Restricted Plant Biomass Production but Did Not Cause Cellular Damage

The drought treatment caused an early and progressive stomatal closure in both cowpea cultivars (Figure 1b). As described in [30], the two cultivars showed a drought avoidance strategy through the early regulation of stomatal aperture, which allowed them to limit water loss as shown by the moderate reductions in RWC (Figure 1c). This, in turn, most likely restricted CO_2 assimilation and explain the dramatic reduction in shoot biomass production after the first week of treatment (Figure 1a). Despite this strong restriction on plant growth and although the water-deprived plants showed signs of wilting, the results obtained on the first mature trifoliate leaves indicate that both cultivars were able to maintain cellular integrity under drought stress. First, leaves of drought-stressed plants did not display visible symptoms of injury or H_2O_2accumulation (Figure A1); Second, membrane lipid content and composition were not modified in leaf tissues of plants submitted to the drought treatment alone (Figure 2 and Figure 3; Table A1 and Table A3). Contrary to the results of previous studies [29,30,38], IT83-D did not appear more sensitive to water stress than EPACE-1. It is likely that, in our experimental conditions, the severity of the drought treatment was not sufficient to reveal differences in drought tolerance between the two cultivars. Indeed, when compared to previous data obtained with the same cowpea cultivars submitted to drought, the RWC of 60%–70% measured at 14 day corresponds to a leaf water potential of −1.5 MPa [29], which is equivalent to a water stress defined as moderate in [31,39]. Consistent with this, several features of the responses of cowpea to severe drought, such as the decrease in leaf membrane lipid content [21,31] and the induction of *VuPAT1* expression [26], were not observed in the present work. Both cultivars maintained the FA unsaturation level of membrane lipids under drought stress (Table A1 and Table A3), even though *VuFAD7* and *VuFAD8* were repressed (Figure 3), which might indicate a capacity for stress acclimation [20].

3.2. The Ozone Treatment Caused Leaf Injury and Decreased Plastidial Galactolipid Content but Had Limited Effect at the Whole-Plant Level

In contrast with drought, the 14-day ozone treatment had a limited impact on photosynthesis and biomass production. These results are consistent with references [34,35]), where these parameters were strongly reduced but only after 40–50 days of treatment. Umponstira et al. [36] found strong ozone effects on cowpea biomass production after only 7 days of treatment, but subsequent results showed that cowpea plants at the vegetative growth stage, as in the present study, were less sensitive to ozone than plants at later growth stages [37]. In the present paper, significant effects of ozone were observed at the cellular level. Some of these effects were found in both cowpea cultivars. First, ozone exposure provoked symptoms of leaf injury and H_2O_2 accumulation (Figure A1). Ozone-induced cell death has been reported in various plant species and is associated to ROS accumulation [40]; Second, the ozone treatment induced decreases in the plastidial galactolipids MGDG and DGDG in both cultivars (Figure 2a,b). Reductions in galactolipid contents have also been reported in leaves of spinach [41] and snapbean [42] plants exposed to acute ozone exposure, and in pea and wheat plants subjected to moderate ozone concentrations [22,43]. Studies on the effects of short-term, acute exposure showed that the decrease in plastidial galactolipids did not result from the direct oxidative damage of ozone but rather from secondary effects on lipid metabolism [44,45]. The putative pathways for the degradation of galactolipids in ozone-treated leaves involves a number of enzymes, including one or several deacylating enzymes capable of hydrolyzing galactolipids, presumably lipid acyl hydrolases (LAHs) [46]. The expression of *VuPAT1*, a drought-inducible LAH purified from cowpea leaves with high substrate specificity for galactolipids [26], was very low in response to ozone in the present study, suggesting that other deacylating enzyme(s) are involved in galactolipid degradation in ozone-treated cowpea leaves.

3.3. Inter-Varietal Differences in Ozone Sensitivity

Most of the observed effects of ozone on cowpea leaves were more pronounced in IT83-D than in EPACE-1, including the extent of leaf injury and the decrease in galactolipid content. Intervarietal differences in ozone sensitivty were found in African cowpea cultivars, but could not be explained by differential radical scavenging capacities [34,35]. Here, this discrepancy between the two cultivars could be explained by a higher oxidative load in leaf tissues of IT83-D than in those of EPACE-1 as indicated by a higher ozone uptake in IT83-D (+21% in POD_0, Table 1), even though this difference was not significant. Nevertheless, IT83-D showed responses to ozone that were not found in EPACE-1, such as the increase in the DGDG:MGDG ratio, as observed in snapbean cultivars subjected to acute ozone exposure [45,42]. The relative larger decline in MGDG is consistent with its extreme sensitivity to degradation processes activated by environmental stresses, as reported in drought-treated *Arabidopsis* [47] and cowpea plants [31]. Furthermore, the ozone treatment reduced the 18:3 in MGDG in IT83-D (Table A1), as found in the leaves of pea [22] and wheat [43] plants exposed to ozone. A LAH purified from cowpea leaves had the highest substrate specificity for (18:3/18:3)-MGDG [48] and LAH activities in cowpea leaf extracts were shown to selectively reduce 18:3 content in MGDG [49]. Taken together, these results suggest that the reduction in 18:3 content in MGDG reflects the preferential degradation of MGDG by LAHs upon ozone exposure in the sensitive cowpea cultivar IT83-D. Constraints such as wounding, low temperature and drought have been shown to increase the expression of the plastidial ω-3 fatty acid desaturase *FAD8* [39,50]. Furthermore, the over expression of *FAD8* increased 18:3 content and tolerance to salt and drought stresses in transgenic tobacco plants [51]. In the present work, the stimulation of *VuFAD8* expression found in IT83-D in response to ozone exposure might have mitigated the ozone-induced loss of 18:3 in membrane lipids but was not sufficient to maintain the unsaturation level of MGDG (Table A2). Beside their role in adjusting FA unsaturation, ω-3 FADs are involved in the regulation of plant defense responses through the production of trienoic FAs that serve as precursors for the synthesis of oxylipins such as jasmonic acid (JA) [52]. JA accumulates upon ozone exposure and acts in lesion containment during the process of ozone-induced cell death [17]. Since *VuFAD7* was repressed by the ozone treatment (Figure 3d), *VuFAD8* could have a specific role in the JA-mediated induction of defense responses to ozone in the leaves of the sensitive cowpea cultivar. Iyer et al. [10] showed that JA signaling was involved in the responses of *Medicago truncatula* to combined ozone and water stresses specifically. The potential role of *VuFAD8* in drought-ozone interactions could be further investigated with a longer water deprivation treatment.

Although the omega-3 index was determined from fatty acid contents in total membrane lipids, the lower 18:3 content in MGDG in IT83-D in response to ozone translated into a significant decrease of this biomarker. This indicates that the omega-3 index, which has been validated as an indicator of metal bioavailability in polluted soils and of the associated adverse effects on plants [53], can be used also as a relevant biomarker of ozone impact. However, the fact that the omega-3 index remained unchanged despite substantial ozone exposures in EPACE-1 in the O and OD treatments and in IT83-D in the OD treatment suggests that the biomarker's response is not linearly related to the ozone dose and shows intraspecific variations.

3.4. Drought Alleviated the Effects of Ozone by Reducing Its Stomatal Uptake

Ozone is known to alter stomatal responses to a variety of environmental stimuli [54,55]. By inducing stomatal sluggishness [13] or by disrupting the ABA-induced signal transduction pathway for stomatal control [14], ozone can impair the ability of plants to respond to water stress. However, in our study, the rate and intensity of the stomatal closure caused by the water deficit was not modified by ozone in the OD treatment as compared to the D treatment (Figure 1b). Reciprocally, however, the stomatal closure induced by the water deficit decreased the uptake of ozone in EPACE-1 and IT83-D when drought and ozone were applied in combination (Table 1). Accordingly, several effects of ozone observed in the O treatment were less pronounced in the OD treatment, including leaf injury, H_2O_2 accumulation and galactolipid depletion (Figure A1 and Figure 2a,b). Furthermore, some effects of ozone in IT83-D were not found in the combined OD treatment, such as the decrease in the omega-3 index, the increase in DGDG:MGDG ratio and the coordinated up regulation of *VuDGD2* and *VuFAD8* (Figure 2c and Figure 3d). Taken together, these data suggest that the drought-induced stomatal closure mitigated the detrimental impacts of ozone on cowpea leaf tissues, which is consistent with the findings of meta-analyses of the effects of ozone on trees [56] and crops [9].

4. MATERIALS AND METHODS

4.1. Plant Materials and Growth Conditions

Two cultivars of *Vigna unguiculata* (L.) Walp. were used in these experiments: EPACE-1 originating from the semi-arid Northeastern part of Brazil and IT83-D from humid regions in Southern China. Based on electrolyte leakage tests, these cultivars have been classified as drought-tolerant and drought-susceptible, respectively [38]. Seeds were germinated in 5-L pots (one seed per pot) containing a mixture of 0.4 kg of compost (N/P/K 14/16/18, 1.2 $kg \cdot m^{-3}$, Gramoflor Repiquage, France) and 15 g of fertilizing granules (Nutricote T-100, N/P/K/MgO 13/13/13/2; Fertil, Boulogne-Billancourt, France), under controlled conditions (photosynthetic photon flux density of 250 $\mu mol \cdot m^{-2} \cdot s^{-1}$, 14 h daylight, 25 °C/20 °C day/night temperatures and 80% relative humidity) for two weeks. All 48 pots (24 pots per cultivar) were dispatched in eight identical phytotronic chambers (ground surface 1.44 m^2, height: 2 m) where the seedlings were allowed to acclimate for one week before stress treatments were applied as described in the following paragraph. At the onset of the treatments, i.e., three weeks after sowing, plants had two mature trifoliate leaves.

4.2. Stress Treatments

For 14 day, three-week old cowpea seedlings were exposed to the following treatments: control (C), drought (D), ozone (O) and combined ozone and drought (O + D). Control plants were exposed to ambient air and received 400 mL of water daily. Drought was applied by withholding watering from the beginning of the acclimation period to the end of the experiment. Pots were weighed daily to make sure that water loss and soil drying progressed at the same rate for all droughted plants. At the end of the experiment (14 day), droughted plants had been submitted to a three-week water deprivation period and all showed signs of wilting. Ozone stress was imposed by exposing plants to 120 ± 10 ppb of ozone through 12-h daily fumigations (11 a.m. to 11 p.m.). Ozone was produced from pure O_2 by two ozone generators (OZ500; Fischer, Bonn, Germany and CMG3-3; Innovatec II, Rheinbach, Germany) and continuously monitored by an ozone analyzer (O341M; Environment S.A., Paris, France). For molecular and biochemical assays, the first fully developed trifoliolate leaves were collected after 7 and 14 day of treatments, immediately frozen in liquid N_2 and stored at −80 °C for further analysis.

4.3. Shoot Biomass Production, Relative Water Content, Chlorophyll Fluorescence and Stomatal Conductance Measurements

Shoots were collected and weighed after 7 and 14 day of treatments. Fresh tissues were then dehydrated at 60 °C during 48 h before determination of shoot dry weight (SDW). Relative water content (RWC) of leaf tissues was assessed by determining fresh weight (FW), turgid weight (TW) and dry weight (DW) of 2 cm diameter leaf discs according to the following formula [57]:

$$RWC = ((FW - DW)/(TW - DW)) \times 100 \quad (1)$$

In vivo chlorophyll fluorescence was measured with a Fluorescence Monitoring System (FMS 1, Hansatech). Leaves were dark-adapted for 30 min before the minimal fluorescence (F_0) was recorded. A saturating flash (7200 $\mu mol \cdot m^{-2} \cdot s^{-1}$ for 1 s) was applied to obtain the maximal fluorescence of dark-adapted leaves (F_m). Actinic light (400 $\mu mol \cdot m^{-2} \cdot s^{-1}$) was then turned on to drive photosynthesis. When the leaves reached steady-state conditions, the steady-state fluorescence (F_s) was recorded. A saturating flash (7200 μmol·m^{-2}·s^{-1} for 1 s) and a dark pulse were then applied to obtain the maximal (F_m') and minimal (F_0') fluorescence of light-adapted leaves, respectively. The photochemical efficiency of PSII (F_v/F_m) and the quantum yield of PSII (Φ_{PSII}) were calculated as follows:

$$F_v/F_m = (F_m - F_o)/F_m \quad (2)$$

$$\Phi_{PSII} = (F_m' - F_0')/F_m' \quad (3)$$

Stomatal conductance to water vapor (g_s) was measured after 7 and 14 day of experiment with a SC-1 portable leaf porometer (Decagon Devices, Inc., Pullman, WA, USA). Measurements were performed inside the phytotronic chambers, before the beginning of the ozone fumigation period, on one of the leaflets of the first fully developed trifoliolate leaf.

4.4. Ozone Exposure and Dose Indices

The AOT40 (ozone concentration accumulated over a threshold concentration of 40 ppb, under a minimum irradiance of 50 W·m^{-2}, in ppb·h) index is defined as the sum of the difference between the hourly mean ozone concentration at the top of the canopy and 40 ppb, for all daylight hours within a specified time period [58]. Instantaneous ozone uptake was calculated from the monitoring of ozone concentration and stomatal conductance to water vapor (g_s) as described in [59]. The accumulated stomatal flux or Phytotoxic Ozone Dose (POD, in mmol·m^{-2}) was determined by summing hourly ozone uptakes. Since no dose-response relationship is available for *Vigna* or related species, the POD was calculated without any threshold of instantaneous ozone flux [58]. Measurements of g_s were performed at 1, 3, 5, 8, 10, 12, 14 days of experiment for the calculation of each hourly ozone uptake of the day. For days when g_s was not measured, an average value was calculated from the values recorded the flanking days.

4.5. DAB Staining

Hydrogen peroxide (H_2O_2) accumulation was investigated using 3,3′-diaminobenzidine (DAB) staining. DAB solution (1 mg·mL^{-1}) was vacuum-infiltrated into 1-cm diameter leaf discs and the disks were left to impregnate for 24 h in the dark. Leaf discs were then discolored in three successive 95% ethanol baths and stored in glycerol 30%. Pictures were taken with a stereoscopic microscope (Nikon SMZ1000) under 0.8× magnification, coupled with a digital camera (Nikon D70S).

4.6. Leaf Lipid Extraction and Separation

Leaf samples were ground in liquid nitrogen and approximately 1 g was boiled in distilled water for 2 min to stop lipolytic activities. Lipophilic compounds were extracted in a chloroform:methanol:water (2:1:1, *v*:*v*:*v*) mixture [60], dried under nitrogen stream and immediately resuspended in 1 mL of an ethanol:toluene mixture (1:4, *v*:*v*) for storage at 4 °C. All extraction solutions were supplemented with 0.01% (*v*:*v*) butylated hydroxytoluene (BHT) to prevent lipid oxidation. Total lipids were separated by thin layer chromatography (TLC) on silica gel plates (G60; Merck) with the solvent system developed by [61]. Bands corresponding to lipid classes were visualized with primuline (0.01% in 80% acetone, *m*:*v*) under UV light and scraped. Lipids were then saponified and the obtained fatty acids were methylated with boron trifluoride [62]. Fatty acid methyl esters (FAMEs) were quantified relative to heptadecanoic acid (17:0), which was added as an internal standard before methylation.

4.7. GC-MS Analysis

FAMEs were separated using a gas chromatograph (Clarus 680, Perkin Elmer, Waltham, MA, USA) fitted with a fused silica capillary column (60 m × 0.25 mm i.d., 0.25 μm film thickness, Elite-WAX ETR, Perkin Elmer). Samples and standards were introduced by a 1 μL splitless injection system at 250 °C. Helium was used as the carrier gas at a constant flow of 1 mL·min^{-1}. The oven temperature was programmed to rise continuously from 75 °C to 200 °C for a total run time of 60 min. Separated FAMEs were analyzed through electronic ionization (70 eV) with a single quadripole mass spectrometer (Clarus 600, Perkin Elmer). FAMEs were identified by comparing the obtained spectral data to a NIST database (National Institute of Standards Technology, Gaithersburg, MD, USA). Calibration standards were prepared from a commercially available mixture of standards (F.A.M.E. Mix, C8-C22 unsaturates, Supelco, Bellefonte, PA, USA), using heptadecanoic acid (17:0) as an internal standard. Calibration standards were analyzed in triplicate and linear standard curves were established for the methyl esters of the five main fatty acids found in membrane lipids of *Vigna* leaves (16:0, 18:0, 18:1; 18:2, 18:3) [21]. To maximize sensitivity and specificity, FAMEs were quantified in single-ion-recording (SIR) mode. Mass spectra of eluting FAME compounds were identified using the commercial mass spectral library supplied with the manufacturer's software (TurboMass, Perkin Elmer). For each sample of total lipids, the responses (peak areas) of FAMEs were normalized separately to the response of the internal standard 17:0 and quantified using the corresponding calibration relationship. To determine the lipid content in a given leaf, the amounts of fatty acids were summed and the result expressed in relation to the leaf tissue dry weight. The omega-3 index as defined as the 18:3/(18:0 + 18:1 + 18:2) fatty acid ratio was determined as described in [53]. Omega-3 index values were normalized to the highest value measured at a given time point for a given cultivar.

4.8. RNA Extraction and cDNA Synthesis

Leaf samples were ground in liquid N_2 with pestle and mortar and approximately 70 mg of powder were used for total RNA extraction, using the Qiagen RNeasy Plant Mini kit following the manufacturer's instructions. Total RNA (50 µg) was treated with TURBO DNase (Invitrogen, Carlsbad, CA, USA). RNA concentration was measured using a Nanodrop ND-1000 spectrophotometer (NanoDrop Technologies, Wilmington, DE, USA) and RNA integrity was controlled by electrophoresis on 1.5% agarose gels. cDNA synthesis was performed on 800 ng of total RNA extracts by reverse transcription with the SuperScript III RT kit (Invitrogen) according manufacturer's instructions. cDNA samples were kept at −20 °C before analysis.

4.9. Real-Time PCR Analysis

Leaf cDNAs were used for the detection of transcript accumulation for the following genes involved in membrane lipid biosynthesis and degradation: *VuMGD1, VuMGD2, VuDGD1, VuDGD2, VuFAD7, VuFAD8, VuPAT1*. Additional information on real-time PCR analyses is given in Table A5. Real-time PCR reactions were performed using the Power SYBR Green PCR Master Mix kit (Applied Biosystems, Carlsbad, CA, USA) in a StepOne Plus termocycler (Applied Biosystems). The 20 µL reaction mixtures contained 1 µL of cDNA (from 4 ng of reverse-transcribed RNA), 5 pmol of both sense and antisense primers and 10 µL of 2× SYBR Green DNA Polymerase mix. All samples were amplified under the following conditions: one denaturing step at 95 °C for 10 min, 40 cycles: 95 °C for 15 s (denaturing), 62 °C for 30 s (annealing), 72 °C for 30 s (elongation). Efficiencies of real-time PCR reactions were determined using dilution series of calibrator cDNA samples. Reactions were run in two replicates and the relative gene expression levels were normalized to the expression levels of a selected reference gene (elongation factor 1 alpha *VuEF-1α*). The reference gene *VuEF1α* was selected among a number of potential housekeeping genes using the GeNorm method [63].

4.10. Statistical Analysis

As described in the appropriate figure captions, parameters were measured on 3 or 4 plants per treatment, considered as independent biological replicates. At each time point and for a given cultivar, data from the different treatments (control, C; drought, D; ozone, O; ozone + drought OD) were submitted to a one-way analysis of variance (ANOVA) followed by Tukey's post-hoc tests ($\alpha = 0.05$). Homogenous subsets were indicated by appropriate letters. In addition, the effects of experimental factors (time, cultivar, drought and ozone) and their interactions were assessed through the multivariate ANOVA of the whole data set and the between-subject effects are included as supplementary information (Supporting Information Table A2, Table A3 and Table A4. Statistical analysis were conducted using the R software [64].

5. CONCLUSIONS

Both cowpea cultivars showed acclimatory responses to drought such as stomatal closure that led to growth impairment, but the water stress was not severe enough to cause leaf and membrane injury nor to reveal differences in drought tolerance between IT83-D and EPACE-1 cultivars. In contrast, the ozone treatment had a limited impact at whole-plant level but provoked leaf injury and altered membrane lipids. These effects were more pronounced in IT83-D, revealing intervarietal differences in ozone tolerance. When stresses were combined, ozone did not modify the stomatal response to drought and the observed effects on whole-plant physiology were essentially the same as when drought was applied alone. Conversely, the drought-induced stomatal closure appeared to alleviate ozone effects through the reduction of ozone uptake. Although the impact of the combined stresses on yield could not be evaluated in this study, the interactive effects of combined ozone and drought may play a major role for cowpea and others crops exposed in the field to a combination of the two stresses, and should be taken into account in environmental risk assessment for vegetation.

ACKNOWLEDGMENTS

Research was supported by the Brazilian *Coordenação de Aperfeiçoamento de Pessoal de Nível Superior* (CAPES) to Deborah Moura Reboucas and Yuri Maia De Sousa in the form of a double-diploma scholarship between Brazilian and French Laboratories. The authors would like to acknowledge Stéphane Martin for his technical help with the phytotronic chambers.

AUTHOR CONTRIBUTIONS

Deborah Moura Reboucas, plant growth and treatment, physiological measurements, gene expression and lipid analyses, data analysis and manuscript editing; Yuri Maia De Sousa, plant growth and treatment, physiological measurements; Matthieu Bagard, experimental design, physiological measurements, gene expression and lipid analyses, data analysis and manuscript writing; Jose Helio Costa, bioinformatics and manuscript proofreading; Yves Jolivet, experimental design and manuscript writing; Dirce Fernandes De Melo, experimental design and manuscript writing; Anne Repellin, experimental design and manuscript writing.

CONFLICTS OF INTEREST

The authors declare no conflict of interest.

REFERENCES

1. Araus, J.L.; Slafer, G.A.; Reynolds, M.P.; Royo, C. Plant breeding and drought in C3 cereals: What should we breed for? Ann. Bot. 2002, 89, 925–940.
2. Ainsworth, E.A.; Yendrek, C.R.; Sitch, S.; Collins, W.J.; Emberson, L.D. The effects of tropospheric ozone on net primary productivity and implications for climate change. Annu. Rev. Plant Biol. 2012, 63, 637–661.
3. Vingarzan, R. A review of surface ozone background levels and trends. Atmos. Environ. 2004, 38, 3431–3442.
4. Oltmans, S.J.; Lefohn, A.S.; Shadwick, D.; Harris, J.M.; Scheel, H.E.; Galbally, I.; Tarasick, D.W.; Johnson, B.J.; Brunke, E.G.; Claude, H.; et al. Recent tropospheric ozone changes—A pattern dominated by slow or no growth. Atmos. Environ. 2013, 67, 331–351.

5. Mills, G.; Pleijel, H.; Braun, S.; Buker, P.; Bermejo, V.; Calvo, E.; Danielsson, H.; Emberson, L.; Fernandez, I.G.; Grunhage, L.; et al. New stomatal flux-based critical levels for ozone effects on vegetation. Atmos. Environ. 2011, 45, 5064–5068.
6. Suzuki, N.; Rivero, R.M.; Shulaev, V.; Blumwald, E.; Mittler, R. Abiotic and biotic stress combinations. New Phytol.2014, 203, 32–43.
7. Matyssek, R.; Le Thiec, D.; Löw, M.; Dizengremel, P.; Nunn, A.J.; Häberle, K.-H. Interactions between drought and O3stress in forest trees. Plant Biol. 2006, 1, 11–17.
8. Wittig, V.E.; Ainsworth, E.A.; Naidu, S.L.; Karnosky, D.F.; Long, S.P. Quantifying the impact of current and future tropospheric ozone on tree biomass, growth, physiology and biochemistry: A quantitative meta-analysis. Glob. Chang. Biol. 2009, 15, 396–424.
9. Feng, Z.; Kobayashi, K. Assessing the impacts of current and future concentrations of surface ozone on crop yield with meta-analysis. Atmos. Environ. 2009, 43, 1510–1519.
10. Iyer, N.J.; Tang, Y.; Mahalingam, R. Physiological, biochemical and molecular responses to a combination of drought and ozone in Medicago truncatula. Plant Cell Environ. 2013, 36, 706–720.
11. Maier-Maercker, U. Predisposition of trees to drought stress by ozone. Tree Physiol. 1998, 19, 71–78.
12. Cotrozzi, L.; Remorini, D.; Pellegrini, E.; Landi, M.; Massai, R.; Nali, C.; Guidi, L.; Lorenzini, G. Variations in physiological and biochemical traits of oak seedlings grown under drought and ozone stress. Physiol. Plant. 2016, 157, 69–84.
13. Paoletti, E.; Grulke, N.E. Ozone exposure and stomatal sluggishness in different plant physiognomic classes. Environ. Pollut. 2010, 158, 2664–2671.
14. Wilkinson, S.; Davies, W.J. Drought, ozone, ABA and ethylene: New insights from cell to plant to community. Plant Cell Environ. 2010, 33, 510–525.
15. Møller, I.M.; Jensen, P.E.; Hansson, A. Oxidative modifications to cellular components in plants. Annu. Rev. Plant Biol.2007, 58, 459–481.
16. Sharma, P.; Jha, A.B.; Dubey, R.S.; Pessarakli, M. Reactive oxygen species, oxidative damage, and antioxidative defense mechanism in plants under stressful conditions. Am. J. Bot. 2012, 2012, 217037.
17. Kangasjärvi, J.; Jaspers, P.; Kollist, H. Signalling and cell death in ozone-exposed plants. Plant Cell Environ. 2005, 28, 1021–1036.
18. Meijer, H.J.; Munnik, T. Phospholipid-based signaling in plants. Annu. Rev. Plant Biol. 2003, 54, 265–306.
19. Baier, M.; Kandlbinder, A.; Golldack, D.; Dietz, K.-J. Oxidative stress and ozone: Perception, signalling and response. Plant Cell Environ. 2005, 28, 1012–1020.
20. Upchurch, R.G. Fatty acid unsaturation, mobilization, and regulation in the response of plants to stress. Biotechnol. Lett. 2008, 30, 967–977.
21. De Paula, F.M.; Pham Thi, A.T.; da Silva, J.V.; Justin, A.M.; Demandre, C.; Mazliak, P. Effects of water stress on the molecular species composition of polar lipids from Vigna unguiculata L. leaves. Plant Sci. 1990, 66, 185–193.
22. Carlsson, A.S.; Hellgren, L.I.;Selldén, G.; Sandelius, S. Effects of moderately enhanced levels of ozone on the acyl lipid composition of the leaves of garden pea (Pisum sativum). Physiol. Plant. 1994, 91, 754–762.
23. Mudd, J.B.; McManus, T.T.; Ongun, A.; McCullogh, T.E. Inhibition of glycolipid biosynthesis in chloroplasts by ozone and sulfhydryl reagents. Plant Physiol. 1971, 48, 335–339.
24. De Paula, F.M.; Pham Thi, A.T.; Zuily-Fodil, Y.; Ferrari-Iliou, R.; da Silva, J.V.; Mazliak, P. Effects of water stress on the biosynthesis and degradation of polyunsaturated lipid molecular species in leaves of Vigna unguiculata. Plant Physiol. Biochem. 1993, 31, 707–715.
25. Hellgren, L.I.; Carlsson, A.S.; Sellden, G.; Sandelius, A.S. In situ leaf lipid-metabolism in garden pea (Pisum sativum L.) exposed to moderately enhanced levels of ozone. J. Exp. Bot. 1995, 46, 221–230.

26. Matos, A.R.; d'Arcy-Lameta, A.; Franca, M.; Petres, S.; Edelman, L.; Kader, J.; Zuily-Fodil, Y.; Pham-Thi, A.T. A novel patatin-like gene stimulated by drought stress encodes a galactolipid acyl hydrolase. FEBS Lett. 2001, 491, 188–192.
27. Dakhma, W.S.; Zarrouk, M.; Cherif, A. Effects of drought-stress on lipids in rape leaves. Phytochemistry 1995, 40, 1383–1386.
28. Carlsson, A.S.; Wallin, G.; Sandelius, S. Species- and age-dependent sensitivity to ozone in young plants of pea, wheat and spinach: Effects on acyl lipid and pigment content and metabolism. Physiol. Plant. 1996, 98, 271–280.
29. Campos, P.S.; Ramalho, J.C.; Lauriano, J.A.; Silva, M.J.; do Céu Matos, M. Effects of Drought on Photosynthetic Performance and Water Relations of Four Vigna Genotypes. Photosynthetica 1999, 36, 79–87.
30. De Carvalho, M.H.C.; Laffray, D.; Louguet, P. Comparison of the physiological responses of Phaseolus vulgaris and Vigna unguiculata cultivars when submitted to drought conditions. Environ. Exp. Bot. 1998, 40, 197–207.
31. Torres-Franklin, M.L.; Gigon, A.; de Melo, D.F.; Zuily-Fodil, Y.; Pham-Thi, A.T. Drought stress and rehydration affect the balance between MGDG and DGDG synthesis in cowpea leaves. Physiol. Plant. 2007, 131, 201–210.
32. Hewitt, D.K.L.; Mills, G.; Hayes, F.; Norris, D.; Coyle, M.; Wilkinson, S.; Davies, W. N-fixation in legumes—An assessment of the potential threat posed by ozone pollution. Environ. Pollut. 2016, 208, 909–918.
33. Malaiyandia, M.; Natarajana, M. Impact of Ozone on Morphological, Physiological, and Biochemical Changes in Cow Pea (Vigna unguiculata (L.) Walp.). Ozone Sci. Eng. 2014, 36, 36–42.
34. Tetteh, R.; Yamaguchi, M.; Wada, Y.; Funada, R.; Izuta, T. Effects of ozone on growth, net photosynthesis and yield of two African varieties of Vigna unguiculata. Environ. Pollut. 2015, 196, 230–238.
35. Tetteh, R.; Yamaguchi, M.; Izuta, T. Effect of ambient levels of ozone on photosynthetic components and radical scavenging system in the leaves of African cowpea varieties. Afr. Crop Sci. J. 2016, 24, 127–142.
36. Umponstira, C.; Pimpa, W.; Nanegrungsun, S. Physiological and biochemical responses of cowpea (Vigna unguiculata (L.) Walp) to ozone. Songklanakarin J. Sci. Technol. 2006, 28, 861–869.
37. Umponstira, C.; Kawayaskul, S.; Chuchaung, S.; Homhaul, W. Effect of Ozone on Nitrogen Fixation, Nitrogenase Activity and Rhizobium of Cowpea (Vigna unguiculata (L.) Walp.). Naresuan Univ. J. 2009, 17, 213–220.
38. Vasquez-Tello, A.; Zuily-Fodil, Y.; Pham Thi, A.T.; da Silva, J.V. Electrolyte and Pi leakages and soluble sugar contents as physiological tests for screening resistance to water stress in Phaseolus and Vigna species. J. Exp. Bot. 1990, 228, 827–832.
39. Torres-Franklin, M.L.; Repellin, A.; Huynh, V.-B.; d'Arcy-Lameta, A.; Zuily-Fodil, Y.; Pham-Thi, A.-T. Omega-3 fatty acid desaturase (FAD3, FAD7, FAD8) gene expression and linolenic acid content in cowpea leaves submitted to drought and after rehydration. Environ. Exp. Bot. 2009, 65, 162–169.
40. Langebartels, C.; Wohlgemuth, H.; Kschieschan, S.; Grun, S.; Sandermann, H. Oxidative burst and cell death in ozone-exposed plants. Plant Physiol. Biochem. 2002, 40, 567–575.
41. Sakaki, T.; Saito, K.; Kawaguchi, A.; Kondo, N.; Yamada, M. Conversion of monogalactosyldiacylglycerols to triacylglycerols in ozone-fumigated spinach leaves. Plant Physiol. 1990, 94, 766–772.
42. Whitaker, B.D.; Lee, E.H.; Rowland, R.A. EDU and ozone protection: Foliar glycerolipids and steryl lipids in snapbean exposed to O3. Physiol. Plant. 1990, 80, 286–293.
43. Sandelius, A.S.; Näslund, K.; Carlsson, A.S.; Pleijel, H.; Selldén, G. Exposure of spring wheat (Triticum aestivum) to ozone in open–top chambers. Effects on acyl lipid composition and chlorophyll content of flag leaves. New Phytol.1995, 131, 231–239.
44. Fong, F.; Heath, R.L. Lipid-content in the primary leaf of bean (Phaseolus vulgaris) after ozone fumigation. Z. Pflanzenphysiol. 1981, 104, 109–115.

45. Nouchi, I.; Toyama, S. Effects of ozone and peroxyacetyl nitrate on polar lipids and fatty acids in leaves of morning glory and kidney bean. Plant Physiol. 1988, 87, 638–646.
46. Matos, A.R.; Pham-Thi, A.T. Lipid deacylating enzymes in plants: Old activities, new genes. Plant Physiol. Biochem.2009, 47, 491–503.
47. Gigon, A.; Matos, A.-R.; Laffray, D.; Zuily-Fodil, Y.; Pham-Thi, A.-T. Effect of drought stress on lipid metabolism in the leaves of Arabidopsis thaliana (ecotype Columbia). Ann. Bot. 2004, 94, 345–351.
48. Sahsah, Y.; Pham Thi, A.T.; Roy-Macauley, H.; d'Arcy-Lameta, A.; Repellin, A.; Zuily-Fodil, Y. Purification and characterization of a soluble lipolytic acylhydrolase from cowpea (Vigna unguiculata L.) leaves. Biochim. Biophys. Acta1994, 1215, 66–73.
49. Sahsah, Y.; Campos, P.; Gareil, M.; Zuily-Fodil, Y.; Pham-Thi, A.T. Enzymatic degradation of polar lipids in Vigna unguiculata leaves and influence of drought stress. Physiol. Plant. 1998, 104, 577–586.
50. Nishiuchi, T.; Iba, K. Roles of plastid omega-3 fatty acid desaturases in defense response of higher plants. J. Plant Res.1998, 111, 481–486.
51. Zhang, M.; Barg, R.; Yin, M.; Gueta-Dahan, Y.; Leikin-Frenkel, A.; Salts, Y.; Shabtai, S.; Ben-Hayyim, G. Modulated fatty acid desaturation via overexpression of two distinct omega-3 desaturases differentially alters tolerance to various abiotic stresses in transgenic tobacco cells and plants. Plant J. 2005, 44, 361–371.
52. Yaeno, T.; Matsuda, O.; Iba, K. Role of chloroplast trienoic fatty acids in plant disease defense responses. Plant J. 2004, 40, 931–941.
53. Le Guédard, M.; Faure, O.; Bessoule, J.J. Early changes in the fatty acid composition of photosynthetic membrane lipids from Populus nigra grown on a metallurgical landfill. Chemosphere 2012, 88, 693–698.
54. Paoletti, E. Ozone slows stomatal response to light and leaf wounding in a Mediterranean evergreen broadleaf, Arbutus unedo. Environ. Pollut. 2005, 134, 439–445.
55. Dumont, J.; Spicher, F.; Montpied, P.; Dizengremel, P.; Jolivet, Y.; Le Thiec, D. Effects of ozone on stomatal responses to environmental parameters (blue light, red light, CO2 and vapour pressure deficit) in three Populus deltoides × Populus nigra genotypes. Environ. Pollut. 2013, 173, 85–96.
56. Wittig, V.E.; Ainsworth, E.A.; Long, S.P. To what extent do current and projected increases in surface ozone affect photosynthesis and stomatal conductance of trees? A meta-analytic review of the last 3 decades of experiments. Plant Cell Environ. 2007, 30, 1150–1162.
57. Weatherley, P. Studies in the water relations of the cotton plant: I—The field measurement of water deficit in leaves. New Phytol. 1950, 49, 81–97.
58. United Nations Economic Commission for Europe. Convention on Long-Range Transboundary Air Pollution 2004 Modelling and Mapping Manual of the LRTAP Convention; United Nations Economic Commission for Europe: Geneva, Switzerland, 2004.
59. Bagard, M.; Le Thiec, D.; Delacote, E.; Hasenfratz-Sauder, M.-P.; Banvoy, J.; Gérard, J.; Dizengremel, P.; Jolivet, Y. Ozone-induced changes in photosynthesis and photorespiration of hybrid poplar in relation to the developmental stage of the leaves. Physiol. Plant. 2008, 134, 559–574.
60. Bligh, E.G.; Dyer, W.J. A rapid method of total lipid extraction and purification. Can. J. Biochem. Physiol. 1959, 235, 8–17.
61. Lepage, M. Identification and composition of turnip root lipids. Lipids 1967, 2, 244–250.
62. Metcalfe, L.D.; Schmitz, A.A. The rapid preparation of fatty acid esters for gas chromatographic analysis. Anal. Chem.1961, 33, 363–364.
63. Vandesompele, J.; de Preter, K.; Pattyn, F.; Poppe, B.; van Roy, N.; de Paepe, A.; Speleman, F. Accurate normalization of real-time quantitative RT-PCR data by geometric averaging of multiple internal control genes. Genome Biol. 2002, 3.

64. R Core Team. R: A Language and Environment for Statistical Computing; R Foundation for Statistical Computing: Vienna, Austria, 2013.

CHAPTER 6

Impact of Combined Abiotic and Biotic Stresses on Plant Growth and Avenues for Crop Improvement by Exploiting Physio-morphological Traits

Prachi Pandey[1], Vadivelmurugan Irulappan[1], Muthukumar V. Bagavathiannan[2] and Muthappa Senthil-Kumar[1] *

[1] *National Institute of Plant Genome Research, New Delhi, India,*

[2] *Department of Soil and Crop Sciences, Texas A&M University, College Station, TX, USA*

ABSTRACT

Global warming leads to the concurrence of a number of abiotic and biotic stresses, thus affecting agricultural productivity. Occurrence of abiotic stresses can alter plant–pest interactions by enhancing host plant susceptibility to pathogenic organisms, insects, and by reducing competitive ability with weeds. On the contrary, some pests may alter plant response to abiotic stress factors. Therefore, systematic studies are pivotal to understand the effect of concurrent abiotic and biotic stress conditions on crop productivity. However, to date, a collective database on the occurrence of various stress combinations in agriculturally prominent areas is not available. This review attempts to assemble published information on this topic, with a particular focus on the impact of combined drought and pathogen stresses on crop productivity. In doing so, this review highlights some agriculturally important morpho-physiological traits that can be utilized to identify genotypes with combined stress tolerance. In addition, this review outlines potential role of recent genomic tools in deciphering combined stress tolerance in plants. This review will, therefore, be helpful for agronomists and field pathologists in assessing the impact of the interactions between drought and plant-pathogens on crop performance. Further, the review will be helpful for physiologists and molecular biologists to design agronomically relevant strategies for the development of broad spectrum stress tolerant crops.

Keywords: stress interaction, stress combinations, morpho-physiological traits, drought and pathogen infection, crop production, productivity

INTRODUCTION

Due to global warming, and potential climate abnormalities associated with it, crops typically encounter an increased number of abiotic and biotic stress combinations, which severely affect their growth and yield (Mittler, 2006; Prasad et al., 2011; Atkinson et al., 2013; Narsai et al., 2013; Prasch and Sonnewald, 2013; Suzuki et al., 2014; Mahalingam, 2015; Pandey et al., 2015a; Ramegowda and Senthil-Kumar, 2015). Concurrent occurrence of abiotic stresses such as drought and heat has been shown to be more destructive to crop production than these stresses occurring separately at different crop growth stages (Mittler, 2006; Prasad et al., 2011). Abiotic stress conditions such as drought, high and low temperature and salinity are known to influence the occurrence and spread of pathogens, insects, and weeds (Coakley et al., 1999; Scherm and Coakley, 2003; McDonald et al., 2009; Ziska et al., 2010; Peters et al., 2014). They can also result in minor pests to become potential threats in future (Duveiller et al., 2007). These stress conditions also directly affect plant–pest interactions by altering plant physiology and defense responses (Scherm and Coakley, 2003). Additionally, abiotic stress conditions such as drought enhance competitive interactions of weeds on crops as several weeds exhibit enhanced water use efficiency than crops (Patterson, 1995; Ziska et al., 2010; Valerio et al., 2013).

The effect of combined stress factors on crops is not always additive, because the outcome is typically dictated by the nature of interactions between the stress factors (Atkinson et al., 2013; Prasch and Sonnewald, 2013; Pandey et al., 2015a,b; Choudhary et al., 2016; Ramu et al., 2016). Plants tailor their responses to combined stress factors and exhibit several unique responses, along with other common responses. Therefore, to fully recognize the impact of combined abiotic and biotic stresses on plants, it is important to understand the nature of such interactions. Mittler and colleagues developed a "stress matrix" to compile the interactions among various abiotic and biotic stresses on plant growth and productivity (Mittler, 2006; Suzuki et al., 2014). This matrix illustrates that the stress combinations can have negative as well as positive effects on plants. Therefore, development of plants with enhanced tolerance to combined abiotic and biotic stresses involves identification of physio-morphological traits that are affected by combined stresses.

Based on the currently available studies on the effect of concurrent stresses on plants, this review attempts to improve and amend the current understanding of stress combinations by explaining some fundamental concepts pertaining to them, highlighting their global occurrence and assessing their influence on crop growth. In this review, we provide a general overview of different stress combinations and their impact on agriculture and discuss in detail the effect of combined drought and pathogen infection on some important crops. The importance of undertaking simulation studies for assessing the impact of combined stresses on plants is also highlighted. Taking leads from some important studies on individual stresses, we have also presented some of the potential traits which can be utilized for crop improvement under combined drought and pathogen infection.

EXAMPLES OF DIFFERENT STRESS COMBINATIONS OCCURRING IN NATURE

Based on the number of interacting factors, stresses can be grouped into three categories: single, multiple individual, and combined stresses (Supplementary Figure S1). A single stress represents only one stress factor affecting plant growth and development, whereas multiple stress represents the impact of two or more stresses occurring at different time periods without any overlap (multiple individual) or occurring concurrently with at least some degree of overlap between them (combined). The co-occurrence of drought and heat stresses during summer is an example of a combined abiotic stress, whereas a bacterial and fungal pathogen attacking a plant at the same time represents a case of combined biotic stress. For example, brown apical necrosis of *Juglans regia* (walnut) is caused by combinations of fungal pathogens *Fusarium* spp., *Alternaria* spp., *Cladosporium* spp., *Colletotrichum* spp., and *Phomopsis* spp., and a bacterium, *Xanthomonas arboricola* (Belisario et al., 2002). A first stress factor preceded by another stress factor in sequence may either "endure" (due to priming) or "predispose" the plants to the subsequent stress. For example, drought predisposes *Sorghum bicolor* (sorghum) to *Macrophomina phaseolina* (causal agent of charcoal root rot) (Goudarzi et al., 2011). There are also scenarios where plants are exposed to "repetitive" stresses, where a single or multiple stresses are intervened by short or long recovery periods. For instance, incidences of multiple spells of hot days or multiple occurrences of drought and high temperature at different phenological stages of plants represent repetitive stresses.

Some examples of different stress combinations that are expected to arise due to climate change and their impact on plants is given in Supplementary Table S1. Simultaneously occurring drought and heat stress stands as the most evident stress combination (Prasad et al., 2011; Jedmowski et al., 2015). Likewise, plants growing in arid and semi-arid regions often face a combination of salinity and heat stress. High light stress also often accompanies heat stress. *Vitis vinifera* (grapes) growing in regions characterized by a continental climate, such as North China, face a combination of drought and cold stress which affects their productivity (Su et al., 2015). Plants growing in the Mediterranean region encounter combined cold and high light stress (Loreto and Bongi, 1989). *Triticum aestivum* (winter wheat) is also known to experience a combination of ozone and cold stress which reduces its frost hardiness (Barnes and Davison, 1988). Likewise, salinity combined with ozone stress reduces yields of *Cicer arietinum* (chickpea) and *Oryza sativa* (rice) (Welfare et al., 2002).

Similar to the different abiotic stress combinations, plants also encounter more than one biotic stresses simultaneously or sequentially. Infection by a combination of fungi, bacteria, and viruses are common and are known to cause severe disease symptoms, compared to infections by individual pathogens. Various biotic stress combinations and their impact on plants have been discussed by Lamichhane and Venturi (2015), and are also tabulated in Supplementary Table S1.

Plants also encounter biotic stressors simultaneously with abiotic stressors (Supplementary Table S2). The impact of environmental factors on plant diseases popularly known as the "disease triangle" has always been an important consideration for plant pathologists. Reports have documented the effect of drought or salinity leading to resistance or susceptibility of plants to *Puccinia* spp. (causal agent of rust), *Verticillium* spp. (causal agent of verticillium wilt), *Fusarium* spp. (causal agent of Fusarium wilt), *Pythium* spp. (causal agent of root rot), and *Erysiphe* spp. (causal agent of powdery mildew) (Supplementary Tables S1, S2). The influence of co-occurring drought (Valerio et al., 2013), high temperature (Cordes and

Bauman, 1984), or cold (Patterson and Flint, 1979) stress on increased competitiveness of weeds over crops has also been documented.

Stress Interactions as an Important Aspect Governing the Impact of Stress Combinations on Plants

Different types of stress interactions can have a range of effects on plants depending on the nature, severity, and duration of the stresses (**FIGURE Figure11**). In case of some abiotic–abiotic and majority of abiotic–biotic stress combinations, interactions not only occur between the plant and the stressors at the plant interface, but also directly between the stressors at or outside the plant interface (Supplementary Figure S2). In fact, the nature of such interactions between the stressors governs the magnitude of their impact on crop response. For example, a concurrent heat wave during a drought period may lead to more soil water evaporation resulting in aggravated drought conditions and increased crop yield loss. In addition to this, drought and heat stresses have synergistic effects on plant physiology, resulting in greater negative net impact manifested as drastic yield reductions (Mittler, 2006). Likewise, concurrent drought and weed stress further reduces water availability to crops and subsequently increases the competitiveness of weeds on them (Stuart et al., 1984).

In case of stress combinations involving heat and pathogen stress, high temperatures not only affect plants but also pathogens. Temperature is, in fact, one of the most important factors affecting the occurrence of bacterial diseases such as those caused by *Ralstonia solanacearum* (causal agent of wilt in tomato), *Acidovorax avenae* (causal agent of seedling blight and bacterial fruit blotch of cucurbits) and *Burkholderia glumae* (causal agent of bacterial panicle blight in rice) (Kudela, 2009). An increase in temperature modifies the growth rate and reproduction of pathogens (Ladanyi and Horvath, 2010). Temperature also affects the incidence of vector-borne diseases by altering the population development and spread of vectors. Similarly, the effect of salt stress on plant diseases might be the outcome of its modulation on the pathogen virulence, the host physiology and microbial activity in soils (Triky-Dotan et al., 2005). For example, increased incidence of *Fusarium* wilt in *Solanum lycopersicum* (tomato) under salt stress was found to be caused by more sporulation of the fungi under saline conditions (Daami-Remadi et al., 2009).

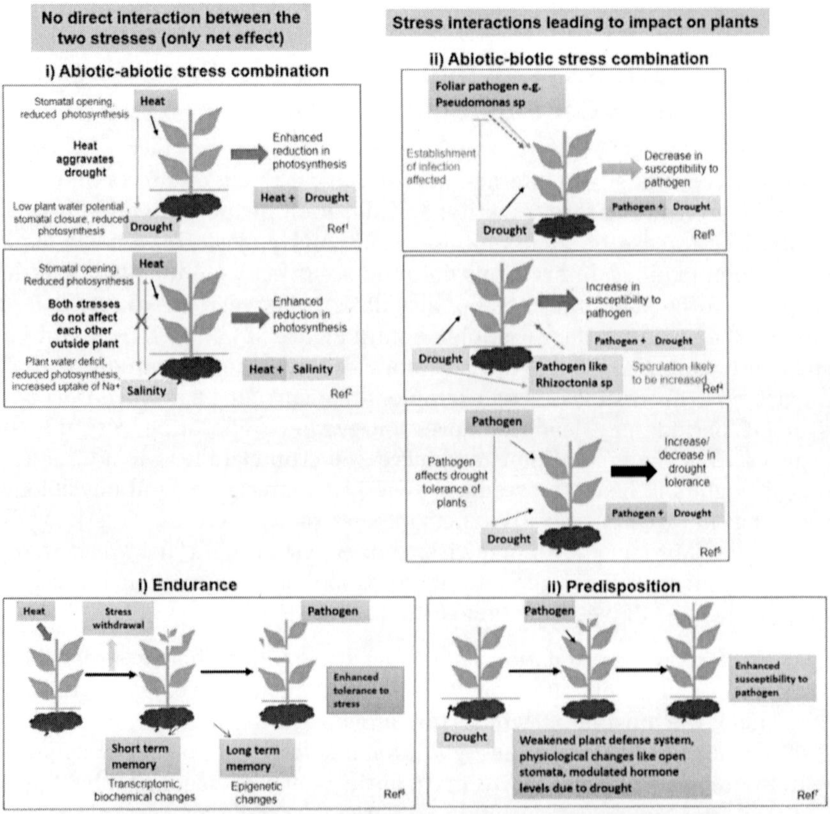

FIGURE 1. SCHEMATIC REPRESENTATION OF EFFECT OF STRESS COMBINATION ON PLANTS. **(A)**Effect of combined stresses on plants is explained by representative examples of heat and drought (abiotic–abiotic stress) and drought and pathogen stress (abiotic–biotic ...

The combination of two stresses (abiotic–abiotic or abiotic–biotic) does not always lead to negative impact on plants. Some stress combinations negate the effect of each other, leading to a net neutral or positive impact on plants. One stress may also provide endurance to plants against another stress and hence yield is not always negatively impacted. For example, individual drought and ozone stresses are detrimental to the growth of *Medicago truncatula* (alfalfa), but the combination of drought and ozone results in increased tolerance of plants to the stress combination (Puckette et al., 2007). High CO_2 has been shown to ameliorate the effect of drought stress in *T. aestivum* (Kaddour and Fuller, 2004) and *Poa pratensis* (bluegrass) (Song et al., 2014). Likewise, an increase in CO_2level from 350 to 675 ppm favored the competitiveness of the C_3 crop *Glycine max* (soybean) over the C_4 weed *Sorghum halepense* (johnsongrass) (Patterson, 1995). *S. lycopersicum* exposed to combined salinity and heat stress performs better than plants subjected to these stresses separately (Rivero et al., 2014). Ozone treatment also provides enhanced resistance to *Puccinia* spp. in *T. aestivum, Pseudomonas glycinea* (causal agent of bacterial blight) in *G. max* and *Erysiphe polygoni* in *Pisum sativum* (pea) (Supplementary Table S1).

Some stress combinations exhibit far more complex interactions and their effect on plants are variable. Heat–pathogen and drought–pathogen stress combinations are examples of such complex interactions. For example, with increased temperature, *T. aestivum* and *Avena sativa* (oats) become more susceptible to *Puccinia* spp., but some forage species such as *Cynodon dactylon* (Bermuda grass) become more resistant to rust disease (Coakley et al., 1999). Heat–pathogen and drought–pathogen interactions can be regarded as two agriculturally important stress combinations. The impact of combined heat and pathogen interaction on plants has been discussed by Pautasso et al. (2012) and Garrett et al. (2006). In the present review, we specifically focus on drought and pathogen stress combination as a case study and discuss it as a model for understanding the impact of abiotic and biotic stress combinations on plants.

Drought–Pathogen Stress Combination: A Model for Understanding Combined Abiotic–Biotic Stresses

Drought stress interacts with pathogen infection both additively and antagonistically. On the basis of the number of reports of plant diseases being affected by drought stress and the frequency of occurrence of drought stress, this combination can be considered as one of the most important stress combinations affecting crop yields worldwide (**FIGURE Figure22**). Drought stress is reported to enhance the susceptibility of *S. bicolor, T. aestivum, Senecio vulgaris* (groundsel), *Hordeum vulgare* (barley), *Gossypium* spp. (cotton), and *C. arietinum* to *M. phaseolina, Puccinia* sp., *Erysiphe graminis* f. sp. *hordei, Fusarium oxysporum* f. sp. *vasinfectum*, and *Rhizoctonia bataticola*, respectively (Supplementary Table S1). On the other hand, drought stress is reported to provide endurance to tomato, *Medicago sativa* and *Arabidopsis thaliana* against *Botrytis cinerea* (causal agent of gray mold), *Oidium neolycopersici* (causal agent of powdery mildew), *Verticillium albo-atrum* (causal agent of verticillium wilt), and *Pseudomonas syringae* (causal agent of bacterial speck disease), respectively (Achuo et al., 2006; Gupta et al., 2016). In some cases, concurrent pathogen infection helps plants to endure drought stress, resulting in increased yield (Supplementary Figure S3) (Davis et al., 2014).

For example, infection with *Cucumber mosaic virus* (CMV) led to improved drought tolerance of *Capsicum annum* (pepper), *S. lycopersicum*, and *Nicotiana tabacum* (tobacco) (Xu et al., 2008).

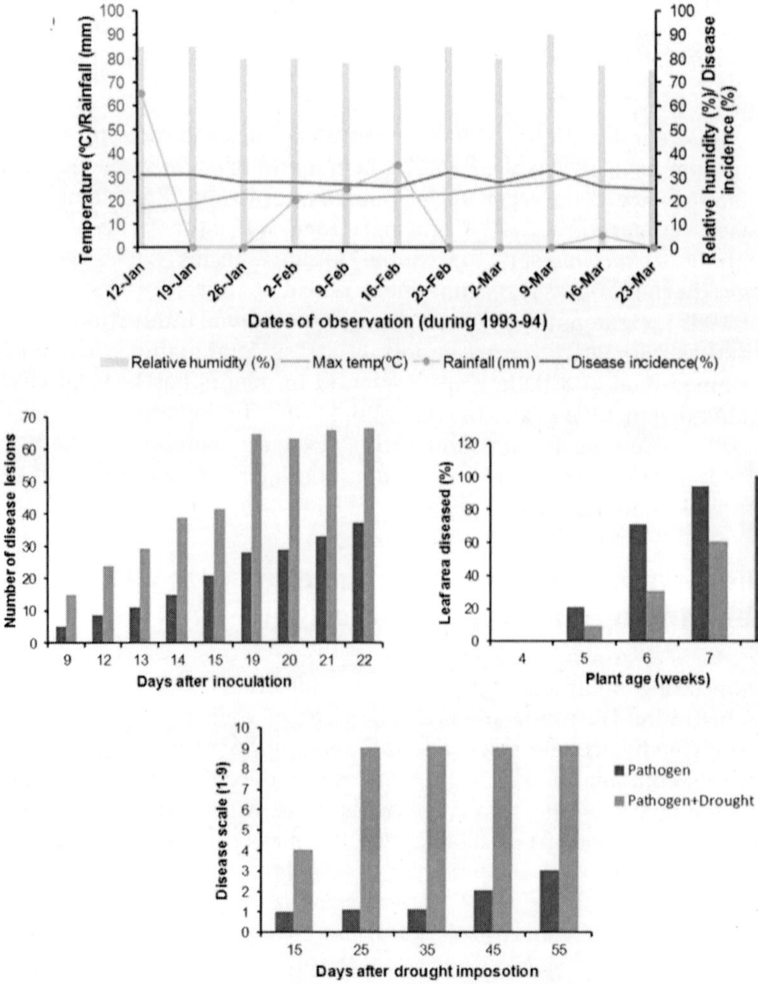

FIGURE 2. IMPACT OF COMBINED ABIOTIC STRESS AND PATHOGEN INFECTION ON PLANTS. The impact of combined abiotic stresses (mainly drought) and pathogen infection has been shown by taking examples from a few representative studies. **(A)** Impact of weather variables like **...**

The effect of combined drought and pathogen infection at physiological and molecular levels has been discussed in a number of recent reports (Pandey et al., 2015a; Ramegowda and Senthil-Kumar, 2015; Choudhary et al., 2016; Gupta et al., 2016) and also summarized in Supplementary Figures S2, S3. In this review, we focus on some important plant diseases favored by drought stress.

One of the important diseases known to be aggravated by high temperature and water deficit conditions is dry root rot (DRR), caused by a necrotrophic fungus *R. bataticola*, Sharma and Pande (2013) have shown the interaction between *R. bataticola* and drought stress in laboratory conditions by infecting *C. arietinum* plants grown at different soil moisture contents with this fungi. This study showed that the disease incidence was the highest at 40% soil moisture content (**FIGURE Figure2D2D**). Less disease incidence at high soil moisture content was attributed to the inability of the fungal sclerotia to survive under wet soil conditions (Olaya and Abawi, 1996; Umamaheswari et al., 2000).

Long periods of drought accompanied with warm days and cool nights generally favor powdery mildew in *Beta vulgaris* (sugar beet) caused by the fungus *Erysiphe betae*. Increased occurrence of powdery mildew infection was observed in several parts of United States in the drought year of 1988 (Lamey, 1988). Occurrence of powdery mildew infections also coincided with extended periods of drought in Germany (Lamey, 1988). In contrast to the above report, drought stress delayed powdery mildew disease development in *Alliaria petiolata* (garlic mustard; Enright and Cipollini, 2007), which could have been due to osmotic stress mediated stomatal closure that typically reduces the pathogen's ability to enter through the leaf (Thaler and Bostock, 2004). However, the exact reason for the same is not yet known. Enright and Cipollini (2007) showed that drought stress reduced plant growth, resulting in the powdery mildew fungi infecting all available leaf area by the end of the experiment, though there was a delay in disease development (**FIGURE Figure2C2C**). In cases such as this, although drought did not aggravate disease development, the net impact of the two stresses resulted in loss of plant performance.

Drought stress accompanied by high soil temperature has been correlated with increased charcoal stalk rot development, caused by *M. phaseolina*, in *S. bicolor* (Odvody and Dunkle, 1979; Mihail, 1989). This disease has also recently emerged as a threat in regions with warmer summers and low rainfall (Smith et al., 2015). Soil moisture content affects microsclerotia survival, root infection, and disease development. It has been found that microsclerotia can survive in dry soils for prolonged periods, but is unable to survive in saturated soils for more than a week (Mayek-Perez et al., 2002). Such interaction between drought and charcoal root rot has also been shown in *Phaseolus vulgaris* (common bean) under laboratory conditions (Mayek-Perez et al., 2002).

It has been reported that drought conditions in England and Wales have resulted in higher incidences of common scab caused by *Streptomyces scabiei* in *Solanum tuberosum* (potato) (Potato Council News, 2011). Infection occurs for 6 weeks after the start of tuber initiation and dry soils facilitate rapid infection of the fungus on developing tubers. The amount of scab on a tuber's surface is directly related to the length of time that the plants are deprived of irrigation (Lapwood and Hering, 1968). The timing of drought occurrence also affects the severity of scabs on surface of tubers and it was found that drought during early stages of tuber development resulted in more scabs (Lapwood and Hering, 1968). Research by Davis et al. (1974) showed that irrigating fields to as high as 90% field capacity effectively suppresses common scab.

Given that a number of drought–pathogen stress combinations have a net negative influence on crop yields, it is important to devise strategies for improving crop performance under these stresses. A promising way of doing so is to identify measurable parameters or traits that are affected by combined stress conditions, which can be modified favorably to improve crop productivity under combined stress conditions. In the section below, we highlight some key traits that can be used for crop improvement under combined drought and pathogen infection.

Potential Traits for Screening Genotypes for Tolerance to Combined Drought and Pathogen Infection

Root System Architecture

Root system architecture (RSA) acts as a major interface between the plants and several biotic as well as abiotic factors and enables the plants to circumvent the environmental challenges by sensing and responding to them. The length and density of primary as well as lateral roots play a crucial role in drought stress tolerance. Development of high root length density (RLD) along with increased root diameter in response to drought stress confers drought tolerance in rice. For example, rice lines with low RLD show reduced drought tolerance (Allah et al., 2010). High RLD favors improved plant growth under drought conditions as it provides access to moisture present at deeper soil depths (Lynch et al., 2014; Zhan et al., 2015). Likewise, under drought stress, *Zea mays* (maize) with high RLD and few lateral roots had high plant water status, increased leaf photosynthesis, stomatal conductance, and increased overall growth, compared to plants with low RLD and more lateral roots. The presence of fewer but longer lateral roots results in enhanced rooting depth thereby increasing water acquisition from deeper layers of soil which helps in improved plant performance under drought (Lynch et al., 2014; Zhan et al., 2015).

Interestingly, RSA also plays a significant role under pathogen infection in plants. In an evaluation of virulence of *Pythium debaryanum* and *Pythium ultimum* (causal agents of root rot) on *T. aestivum*, plants with high root length had less fungal infection (Higginbotham et al., 2004). In contrast, infection of *Rhizoctonia solani* (causal agent of root rot) on *S. lycopersicum* caused reduction in total root length, number of root tips, and magnitude of root branching, which compromised water exploration from deep soil layers and consequently the shoot growth (Berta et al., 2005; Simonetta et al., 2007). Thus there seems to be a correlation between RLD and the extent of pathogen infection by root infecting fungi. Thus we speculate that increasing the RLD of plants might help in reducing pathogen infection.

Combined drought and root infecting pathogens cause greater damage to plants as both stresses appear to additively disrupt the RSA. For example, under drought conditions, *Fusarium solani* f. sp. *phaseoli* (causal agent of root rot in beans) infects the roots of *P. vulgaris* in deep layers of soil and affects water absorption. As a result of infection, accessibility to water present at deeper soil profiles is compromised under drought conditions, leading to severe reductions in plant growth (Dryden and Van Alfen, 1984).

Under combined stress, the time of occurrence of pathogen infection or drought stress has a significant effect on the net impact. *Phytophthora cryptogea* (causal agent of root and crown rot) infection on drought stressed *Carthamus tinctorius* (safflower) resulted in severe root rot disease development and a marked reduction in fresh weight of roots, compared to the conditions where drought stress followed the infection (Duniway, 1977). Similarly, *Phytophthora parasitica* (causal agent of root rot) infection on drought stressed *S. lycopersicum* resulted in greater disease severity, represented by an increase in the number of brown roots, reduced root length, and low fresh weight, compared to pathogen infection followed by drought stress (Ristaino and Duniway, 1989). Drought stress induced increase in root growth and exudation of amino acids (such as alanine and proline) and carbohydrates (such as pentose and glucose) are known to be responsible for enhanced root rot disease development on drought stressed plants (Schroth and Hildebrand, 1964; Duniway, 1977). Root exudates serve as nutrients for the growth of soil borne pathogens. These drought induced changes in the host physiology enhance pathogen infection by directly attracting more pathogens as well as intensifying the existing infection on plant roots. Additionally, pathogen infection modulates the composition of root exudates. Tomato roots infected with *F. oxysporum* f. sp. *radicis-lycopersici* exhibited decreased exudation of citric acid, but increased secretion of succinic acid as compared to the non-infected roots.

Moreover, co-infection with bio-control bacterium *Pseudomonas fluorescens* WCS365 resulted in less disease and more secretion of succinic acid (Kamilova et al., 2006). Identification and analysis of exudates commonly secreted under drought and pathogen stress, and attempts toward manipulating the secretion of these exudates by inhibiting or over-expressing the secretory pathways may help in conferring tolerance to drought and pathogen infection. However, studies in this direction needs to be done to prove the suitability of this approach.

A number of studies have demonstrated no influence of drought stress on pathogen infection induced root damage in plants. For example, in an assessment of the effect of drought stress on infection by *Gaeumannomyces graminis* (Sacc.) var. *tritici* (causal agent of root rot) in wheat under low and severe drought stress conditions, Balota et al. (2005) found that infection under both the drought levels caused similar reduction in root dry mass. In addition, carbon assimilation rate and root decay were also found to be reduced similarly under both drought intensities, indicating that increasing drought intensities had little effect on disease development. Furthermore, *Pythium irregulare* and *R. solani* infection on *T. aestivum* cultivars under drought stress did not result in any significant change in root lesions inflicted by pathogen infection compared to infection on well-watered plants (Aldahadha, 2012).

Taken together, in most cases plant survival under concurrent drought and pathogen infection is compromised if RLD is affected as it influences the acquisition of water. Plants with the ability to maintain high RLD may perform better under combined drought and pathogen infection. Considering the role of RLD in both drought tolerance and pathogen infection, this trait can be utilized as a potential morpho-physiological trait for selecting cultivars with resistance to combined drought and pathogen stress. Additionally, root phenotyping tools can be exploited for screening plants with combined stress tolerance. Several studies have reported quantitative trait loci (QTLs) associated with RSA under drought stress (Comas et al., 2013). For example, a constitutive QTL, designated as root-abscisic acid 1 (ABA1), associated with root traits like branching, diameter, angle, and total dry mass has been identified in maize (Giuliani et al., 2005). Similarly, a QTL for ABA-induced reduction in lateral root growth and size has been identified in *A. thaliana* (Fitz Geral et al., 2006; Xiong et al., 2006). Moreover QTL (ARR2.1) for root rot resistance and tap root diameter (TD2-1) are correlated and increase in tap root diameter was related to enhanced resistance (Hagerty et al., 2015). Hence, we expect that some QTLs associated with efficient RSA may also be used in breeding programs to develop combined drought and pathogen resistant crops.

LEAF PUBESCENCE

Trichomes (leaf hairs) are modified epidermal cells found in uni- or multi-cellular, branched or unbranched, and glandular or non-glandular forms all over the surface of a plant. Though number and types of trichomes are genetically controlled, the environmental conditions also determine their pattern of occurrence. Plants grown in semi-arid environments maintain water levels by foliar absorption of water with the help of trichomes. Trichomes entrap water droplets and it has been shown that *Phlomis fruticosa* (Jerusalem sage) leaves with trichomes in mesophyll cells absorb dew deposits, which results in decreased water potential of drought stressed leaves, compared to the leaves of *Hedera helix* without trichomes. In addition, photosynthetic performance of hairy leaves is greater than that of non-hairy leaves under water stress conditions (Grammatikopoulos and Manetas, 1994). In some cases, drought conditions also increase trichome production in plants as a means of adaptation. For instance, drought stressed *Sinapis arvensis* (wild mustard) plants had more trichomes compared to control plants of the same line (Roy et al., 1999).

Studies have found that trichomes can serve as a barrier to infection by foliar pathogens (Lai et al., 2000). For example, *Phytophthora infestans* (causal agent of late blight) infection in *S. tuberosum* is negatively correlated with the presence of glandular trichomes. Presence of trichomes can reduce the relative humidity at the leaf surface, which is unfavorable for the germination of fungal spores (Lai et al., 2000). Trichomes may also secrete exudates that possess anti-fungal activities (Armstrong-Cho and Gossen, 2005; Nonomura et al., 2009). For example, exudates secreted by glandular trichomes present all over the plant surface of chickpea are shown to decrease infection by *Ascochyta rabiei* (causal agent of ascochyta blight) due to the anti-fungal properties of the exudates (Armstrong-Cho and Gossen, 2005). It was found that increased concentrations of exudates inhibited the conidial germination of *A. rabiei* while low concentrations promoted it. Identification of the pathways and genetic elements behind the glandular secretions from the trichomes under pathogen stress and their careful manipulation can enhance the resistance of plants.

In contrast to the above reports, trichomes in some plants may favor pathogen growth. For example, trichomes present on the leaf surface of common beans were reported to favor the growth of *P. syringae*. As trichomes retain water, exudates released from the broken cuticle at the base of trichomes might favor microbial growth (Monier and Lindow, 2003). Similarly, *A. thaliana* mutant *gl1* (GLABROUS1) plants with less trichome density were found to be tolerant to infection by necrotrophic fungus *B. cinerea* but *try* (TRYPTYCHON) mutants with more trichome density were found susceptible to infection (Calo et al., 2006). Not many studies have been done to probe into the role of trichomes in pathogen infection. A closer understanding of plant–pathogen interaction at this interface may help in further unraveling the role of trichomes in enhancing pathogen infection. It is evident that plants produce more trichomes under drought to minimize transpiration. In many cases, the presence of glandular trichomes has been shown to provide tolerance against pathogen invasion. However, as mentioned above, there are exceptions to this rule. Moreover, role of glandular trichomes and their secretory products under drought stress needs to be studied. Taken together, an extensive understanding of the nature of plant–pathogen interaction under drought stress would be needed in cases where trichomes enhance pathogen growth. Mapping for leaf pubescence related QTLs have been done for many plants like *Gossypium hirsutum* and *A. thaliana* (Lacape and Nguyen, 2005; Bloomer et al., 2014). It can be hypothesized that enhancement of trichome production impart protection against combined drought and pathogen infection in many cases and trichomes can be considered as a potential morpho-physiological trait conferring tolerance to the stress combination. Moreover, identification of QTLs related to leaf trichome density and secretion under drought and pathogen infection can also help in breeding genotypes better adapted to the combined stress. It can also be utilized for exploring the genes and pathways regulating trichome production and secretion which can be suitably modified to confer enhanced resistance under combined stress scenarios.

LEAF WATER POTENTIAL REGULATION

A change in plant water potential is directly correlated to soil moisture level and is also affected by fungal and bacterial pathogens that disrupt the function of the plant vascular system. However, some traits related to maintenance of plant water potential are negatively affected by drought and pathogen stress. For example, plants close stomata under drought stress in order to reduce the transpirational loss of water. In contrast, infection by *Uromyces phaseoli* (causal agent of leaf rust) inhibits stomatal closure on *P. vulgaris* due to the toxins produced by the pathogen (Duniway and Durbin, 1971) which indicates that pathogen infection in cases like this can compromise drought tolerance.

Some pathogens may reduce plant water content even under sufficient soil moisture conditions. For example, *U. phaseoli* infection in *P. vulgaris* results in wilting at high soil water potential due to xylem damage, whereas uninfected plants experience wilting only under drought. Inhibition of stomatal closure by toxins secreted by *U. phaseoli*, disruption of cuticle layer and impaired stomatal resistance account for the increased water loss, which further reduce leaf water potential of plants under drought stress (Duniway and Durbin, 1971). Similarly, Burman and Lodha (1996) demonstrated a marked reduction in shoot water potential, leaf turgidity and transpiration in *Vigna unguiculata* (cowpea) plants subjected to concurrent drought and *M. phaseolina* (causal agent of charcoal rot and stem blight) infection. McElrone et al. (2003) showed that *V. vinifera* subjected to combined drought and *Xylella fastidiosa* (causal agent of leaf scorch) infection experience a significant reduction in leaf water potential and stomatal conductance, which aggravates the scorch symptoms more in drought stressed plants, compared to well-watered plants.

When *P. vulgaris* was exposed to simultaneous drought and *M. phaseolina* infection, high transpiration rate, decreased water potential and low stomatal resistance was observed in the stressed plants (Mayek-Perez et al., 2002). Drought stress caused plants to produce carbohydrates which facilitated the growth and infection of *M. phaseolina* (Mayek-Perez et al., 2002). In addition, it was found that varieties resistant to infection maintained high leaf water potential compared to susceptible varieties (Mayek-Perez et al., 2002). In case of charcoal rot due to infection by *M. phaseolina* in *G. max*, it has been found that maturation of the sclerotia was induced only by the reduced leaf water potential due to drought stress. It was also found that symptoms appeared only after imposition of drought stress. Likewise, Diourte et al. (1995) found that post flowering drought stress caused a reduction in leaf water potential in *S. bicolor*. Plants with reduced water potential had longer *M. phaseolina* lesions, which directly resulted in a reduction of grain yield. Pastor-Corrales and Abawi (1988) had demonstrated that drought-resistant bean varieties showed resistance to *M. phaseolina* infection as well.

Taken together, leaf water potential can be influenced by both drought and vascular pathogens and improved water status of plants under drought conditions might correspond to improved pathogen as well as drought resistance. One of the factors defining plants response to vascular pathogen infection is the xylem vessel dimension; *V. vinifera* genotypes with smaller xylem diameter are known to be less susceptible to infection by fungal vascular wilt pathogens (Pouzoulet et al., 2014). Identification of QTLs related to xylem diameter and xylem pit anatomy can be helpful to identify mechanisms for tolerance against combined drought and pathogen infection. Thus, plant water potential can be used as an important morpho-physiological trait to screen plants resistant to combined drought and pathogen infection.

CUTICULAR WAX

Cuticle plays a vital role in protecting plants from drought stress and pathogen invasion. When stomata are closed under drought stress, a small amount of water is lost through cuticular layer. Cuticular layer also acts as a barrier to pathogen infection as it is hydrophobic and devoid of moisture (Martin, 1964).

The significance of cuticular layer has been studied under drought stress conditions. For example, drought stress led to an increase in the concentration of cuticular wax components such as alkanes, aldehydes, and ketones in *A. thaliana*, resulting in increased wax coverage in the stressed plants (Kosma et al., 2009). Under drought stress, the drought tolerant *T. aestivum* plants exhibit enhanced thickness of the cuticular layer while the susceptible varieties do not show any change in cuticle thickness (Hameed et al., 2002).

Likewise, the importance of cuticular wax has also been studied under pathogen infection. Marcell and Beattie (2002) exposed wild type and glossy mutants of *Z. mays* (*gl4*) to *Clavibacter michiganensis* (causal agent of leaf blight and Goss's wilt of maize). Compared to the wild-type, more bacterial colonies were observed on the *gl4* mutants, which had a thin cuticular layer due to an alteration in the wax biosynthesis pathway (Marcell and Beattie, 2002). Nutrient and water exudation through the weak cuticular layer might have encouraged the colonization of bacteria, leading to more pathogen growth in the *gl4* mutants. Jenks et al. (1994) showed that bloomless (*bm*) mutants of *S. bicolor*, deficient in the synthesis of epicuticular wax and having a thin cuticular layer, were highly susceptible to infection by *Exserohilum turcicum* (causal agent of leaf blight) compared to the wild type plants. The rate of water loss was found to be high in the *bm* mutant compared to wild type plants. This apparently suggests that the thickness of cuticular wax can be used as a trait to identify plants tolerant to *E. turcicum*. Plants without stomata and deficient in cuticular wax have been used to study the significance of cuticular wax under pathogen infection. Isaacson et al. (2009) showed that penetration of pathogen through stomata was more with astomatous fruits (cutin deficient, *cd* mutant) of *S. lycopersicum*. Only the *cd* fruits were found to be infected with *B. cinerea* (causal agent of gray mold) depicting a role of cutin in pathogen resistance. Along with the cutin content, the composition and architecture of the wax layer also determines their role in defense. In the above study, it was also found that among the three cd mutants, cd1 which showed lack of microfissures, elevated level of amyrins and decreased levels of alkanes of chain length >30 showed maximum water loss and minimum susceptibility to *B. cinerea*.

Although, there are no studies showing the direct role of cuticular wax under combined drought and pathogen infection, the above evidence suggests its probable role in combined stress. Thus, plants produce more complex and thick cuticular wax layer in response to drought stress, which in turn might impart tolerance to pathogen infection. Additionally, the composition and architecture of the wax layer is equally important in defining the role of cuticle in defense mechanism. Detailed investigation of the pathways that determine the composition and structure of cuticle layer may help in identifying targets which can be manipulated to impart improved resistance to plants against combined drought and pathogen infection. Srinivasan et al. (2008) have found that QTLs for epicuticular wax, rate of water loss from excised leaves and harvest index co-located with QTLs associated with shoot and root-related drought resistance traits in rice. One example of such QTL is a region on chromosome 8 of rice. Considering the importance of cuticular wax in providing resistance against many pathogen infection, identification of QTLs linked to wax content and disease resistance should also be done. Cuticular wax may be considered as a trait that can be used to screen plants tolerant to combined drought and pathogen infection. The measurement of wax content can be made by simple weight analysis by immersing leaves in chloroform and determining the wax content after chloroform evaporation (Zhou et al., 2013). Thus, the trait can be efficiently utilized for large scale screenings of plants better adapted to combined drought and pathogen infection.

CANOPY TEMPERATURE

Canopy temperature (Tc) has been used to measure drought stress tolerance of plants (Gonzalez-Dugo et al., 2005). Tc varies with each leaf under drought and pathogen infection, as stress induced drooping and curling of leaf reflects radiation differently (Jackson, 1986). Tc plays a significant role in plant growth under drought stress. *T. aestivum* plants under drought stress were found to have high Tc and yielded less than irrigated plants (Blum et al., 1989). Plants which maintain low Tc under drought stress conditions possess high plant water status and thus are better adapted for drought stress (Blum, 2009).

The importance of Tc under pathogen infection has also been shown by Eyal and Blum (1989). Tc of *T. aestivum* plants infected with *Mycosphaerella graminicola* (causal agent of *Septoria tritici* blotch), can be positively correlated with disease occurrence as infected plants had higher Tc. The increase in Tc can be ascribed to the damaged cuticular layer due to infection by pathogens. Furthermore, a negative correlation was observed between Tc and leaf greenness as pathogen infection progressed.

Thus, measurement of plant Tc can be used to identify both infected and un-infected areas (Eyal and Blum, 1989).

The significance of Tc under concurrent drought and pathogen infection has been shown in some studies (Pinter et al., 1979; Dow et al., 1988). When *B. vulgaris* (sugar beet) was infected with *Pythium aphanidermatum* (causal agent of root rot) under drought stress, increased Tc was observed in drought stressed plants as compared to control plants (Pinter et al., 1979). Increased Tc can be attributed to infection induced root damage, resulting in interruption of water uptake and a reduction in plant water potential. Infected plants had higher Tc as compared to the drought stressed, -uninfected plants. Similar increase in Tc was observed in *Gossypium* spp. infected with *Phymatotrichum omnivorum* (causal agent of *Phymatotrichum* root rot) under drought stress (Pinter et al., 1979). Under concurrent drought and *M. phaseolina* (causal agent of charcoal rot) infection, increased leaf temperature and decreased stomatal resistance were observed in stressed *P. vulgaris* (Mayek-Perez et al., 2002). Likewise, infection of *Sclerotinia minor* Jagger (causal agent of watery mold and soft rots) on thinned and un-thinned *Arachis hypogaea* (peanut) plants under drought stress has been studied by Dow et al. (1988). Drought stressed un-thinned plant canopy showed increased disease severity compared to the thinned treatments. Thinning lead to a modification in plant canopy size, which affected the microclimate as well as Tc. High relative humidity and microclimate associated with the un-thinned canopy favored disease infection, whereas thinned canopy exhibited lower relative humidity and lesser disease infection. The effects were further supported by a reduction in soil moisture content in the thinned fields, compared to the non-thinned ones; thinned canopy might have reduced canopy level humidity and increased transpiration rate as well as water uptake from soil, resulting in lower Tc and consequently lesser infections. Thus increasing space between the plants can be utilized as an agronomic practice in areas with less soil moisture availability for providing resistance against combined drought and pathogen infection. It has been demonstrated that plants which regulate transpiration and gas exchange could maintain cooler canopies under drought conditions. Tc has also been found to increase with increasing number of dead leaves (Blum, 2009), further emphasizing the importance of transpiration and gas exchange in reducing Tc.

Overall, a combination of drought stress and pathogen infection shows a negative correlation with Tc. Plants which maintained normal Tc under these conditions did not compromise growth and yield. Thus, Tc, which is shown to be affected by both individual and combined drought and pathogen infection, can be considered as an important trait for assessing combined drought and pathogen tolerance of plants. A simple measurement of Tc using infra-red thermometers can be implemented as an efficient means to screen genotypes better adapted to grow under combined drought and pathogen infection. Some QTLs for Tc has been found to be related to root development. For example, QTL for Tc at chromosome 2B of *T. aestivum* is also the main QTL responsible for root developmental in wheat (Pinto and Reynolds, 2015). An attempt to find the relation between Tc and Cephalosporium stripe disease (CSD) of winter wheat caused by *Cephalosporium gramineum* has been recently made by Froese et al. (2016). Although the authors could not find any significant relation between QTLs for Tc and disease severity, they suggest that a better evaluation of Tc might be helpful in proving the correlation between Tc and CSD resistance. As Tc is a good indicator of water status of plants investigation to find correlation between Tc and vascular diseases can also reveal mechanisms and QTLs for resistance to combined drought and pathogen infection.

Development of Crops with Improved Performance Under Combined Drought and Pathogen Stress

Role of Simulation Studies in Assessing the Impact of Drought–pathogen Combination

Crop yield is determined as a net result of complex interactions among abiotic and biotic conditions, soil features and crop management practices. Several crop modeling approaches which can predict the effect of various weather conditions on crop yield can be used to devise strategies for farm planning and regional policy development. Similarly, a number of plant disease prediction models have also been developed and evaluated. For example, Garcia et al. (2008) developed and applied a geographical information system (GIS) based agro-meteorological disease model to determine the sowing dates with low climatic risk for the infection of potato late blight disease in the Andes region of Venezuela (Garcia et al., 2008).

Considering the role of biotic stress factors in determining the yield of plants, it becomes utmost important that the effect of biotic constraints are considered in addition to the abiotic factors in order to generate a more comprehensive and relevant projection of future global plant productivity under a changing climate. In order to assess the impact of combined biotic and abiotic stresses on plants, linked "climate-crop disease" models need to be developed. Few simulation studies have been attempted to link disease forecasting models to regional climatic scenarios (Oldenburg et al., 2009; Caffarraa et al., 2012). Simulation studies like these should be extended to more crops in order to assess the yield loss potential of diseases in the current scenario of climate change. This would demand intensive collaboration between climatologists, agronomists, and plant pathologists involved in disease epidemic modeling. Efforts in this direction would help in planning better strategies for improving crop productivity.

Role of Genomic Tools for Developing Combined Drought and Pathogen Stress Tolerant Crops

A few important molecular studies have recently been employed to elucidate the molecular responses of plants against combined drought and pathogen stresses (Supplementary Table S3). These studies have not only shed light on a plant's defense mechanism against combined stresses but also revealed some potential candidates for improvement of plant tolerance to combined stresses. Some of the important candidate genes identified so far are methionine homeostasis gene; methionine gamma lyase (AtMGL), rapid alkalinization factor-like 8 (AtRALFL8) involved in cell wall remodeling and azelaic acid induced 1 (AZI1) functioning in systemic plant immunity (Atkinson et al., 2013). Tolerance to combined drought and pathogen stress is also contributed by genes involved in crosstalk between the drought and pathogen infection associated signaling pathways. The roles of proline and polyamine metabolism in combined drought and pathogen stress tolerance of *A. thaliana* and *V. vinifera* have also been indicated by some studies (Hatmi et al., 2015; Gupta et al., 2016). The identified candidate genes can be suitably modulated to confer enhanced tolerance against the combined stresses. The modification can be done by genome editing using tools like CRISPR/Cas9 system. CRISPR/Cas9 system can also be used to modulate the transcription of the genes of interest by guiding catalytically inactive dead Cas9 (dCas9) or dCas9 fused with transcriptional repressors/activators to the promoter of a gene. Further research in this direction using the different functional genomic approaches can, thus, help in uncovering responses of plants to combined drought and pathogen stresses.

CONCLUSION AND FUTURE PERSPECTIVES

Plants under field conditions face a combination of different abiotic and biotic stresses. The interaction between these stresses and their impact on plants has been discussed earlier as part of the "disease triangle." The interaction between the two stress conditions may either negatively or positively affect plant growth. For example, a co-existing drought can also modulate the interaction of different pathogens and plants differently, leading to either suppression or increase in pathogen growth. Therefore, it becomes very important to study the interaction between the two stresses in order to better understand the net impact of stress combinations on plants. Several important diseases such as DRR, powdery mildew, and charcoal rot are significantly affected by co-occurring drought conditions and identification and development of superior cultivars can be done if a mechanistic understanding of the interaction between pathogen and drought stress is attained. The strategies for improving crop performance under combined drought and pathogen stress have been schematically represented in **FIGURE Figure33**. Attempts to understand the interactions have already been started in the form of transcriptomic studies (Supplementary Table S3). Well-designed experiments involving simultaneous drought and pathogen stress on plants have also been undertaken, revealing some aspects of drought–pathogen interactions (Gupta et al., 2016; Sinha et al., 2016). Plant genotypes can be screened for traits such as RSA, leaf water potential, leaf pubescence, and leaf cuticular waxes for identification of superior germplasm lines.

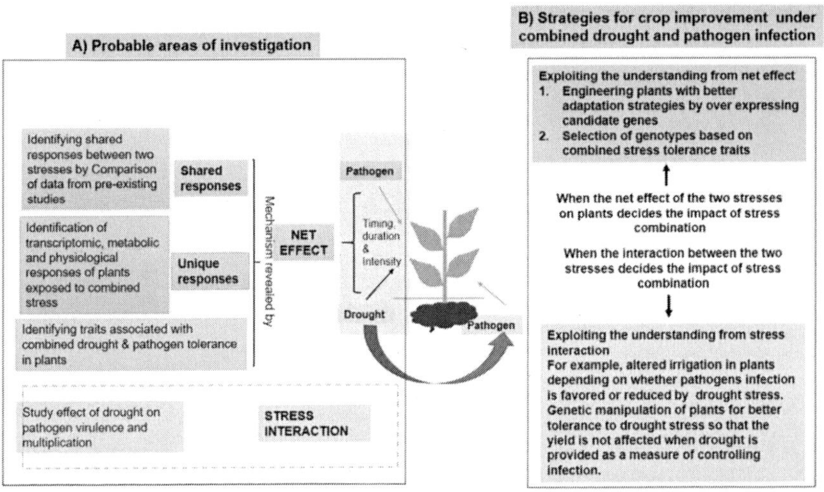

Figure 3. OUTLINE OF STRATEGIES FOR IMPROVING CROP PERFORMANCE UNDER COMBINED DROUGHT AND PATHOGEN **STRESS.** For understanding the effect of stress on plants, it is important to first understand the nature of the stress combinations, i.e., the interaction between **...**

To vividly assess the effect of different stress combinations on plants, it is imperative to design experiments that can reveal different aspects of interactions between the two stresses. A well thought about stress imposition protocol that is not very different from stresses occurring under field conditions, complemented by relevant physiological assays and the recently evolved genomic tools, can help uncover the response of plants to stress combinations. Understandings from studies on plant response to combined drought and pathogen stress can be utilized by breeders and field pathologists to better analyze the performance of the superior/tolerant genotypes. Further development of crop simulation models involving a combination of drought and pathogen stress can help in disease forecasting in places where concurrence of the two stresses is prevalent. Thus, integrative efforts from crop modeling experts, agronomists, field pathologists, breeders, physiologists, and molecular biologists can efficiently lead to development of combined stress tolerant crops that can perform well under field conditions.

AUTHOR CONTRIBUTIONS

MS-K conceived the concept and provided outline. PP drafted the manuscript. VI drafted "traits" part of the manuscript. MB contributed to "weeds/herbicides" part and also edited the manuscript. MS-K edited and finalized the manuscript.

CONFLICT OF INTEREST STATEMENT

The authors declare that the research was conducted in the absence of any commercial or financial relationships that could be construed as a potential conflict of interest.

FOOTNOTES

FUNDING. Combined stress tolerance related projects at MS-K lab were supported by DBT-innovative young biotechnologist award (BT/09/IYBA/2015/07) and DBT-Ramalingaswami re-entry fellowship grant (BT/RLF/re-entry/23/2012). PP and VI acknowledge financial support from SERB young scientist (SB/YS/LS-71/2014) and DBT-JRF (DBT/2015/NIPGR/430) schemes, respectively.

REFERENCES

1. Achuo E. A., Prinsen E., Hofte M. (2006). Influence of drought, salt stress and abscisic acid on the resistance of tomato to *Botrytis cinerea* and *Oidium neolycopersici*. *Plant Pathol.* 55 178–186.

2. Ahmed I. M., Dai H., Zheng W., Cao F., Zhang G., Sun D., et al. (2013). Genotypic differences in physiological characteristics in the tolerance to drought and salinity combined stress between Tibetan wild and cultivated barley. *Plant Physiol. Biochem.* 63 49–60.
3. Aldahadha A. M. A. (2012). *Effect of Root Diseases and Drought on Water use Efficiency of Wheat.* Doctoral thesis, University of New England; Armidale.
4. Allah A. A., Shimaa A., Zayed B., Gohary A. E. (2010). The role of root system traits in the drought tolerance of rice (*Oryza sativa* L.). *Int. J. Agric. Biol. Sci.* 183–87.
5. Armstrong-Cho C., Gossen B. D. (2005). Impact of glandular hair exudates on infection of chickpea by *Ascochyta rabiei*. *Can. J. Bot.* 83 22–27.
6. Atkinson N. J., Lilley C. J., Urwin P. E. (2013). Identification of genes involved in the response to simultaneouss biotic and abiotic stress. *Plant Physiol.* 1622028–2041.
7. Balota M., Rush C. M., Payne W. A., Lazar M. D. (2005). The effect of take-all disease on gas-exchange rates and biomass in two winter wheat lines with different drought response. *Plant Soil* 275 337–348.
8. Barnes J. D., Davison A. W. (1988). The influence of ozone on the winter hardiness of Norway Spruce [*Picea abies* (L) Karst]. *New Phytol.* 108 159–166.
9. Belisario A., Maccaroni M., Corazza L., Balmas V., Valier A. (2002). Occurrence and etiology of brown apical necrosis on Persian (English) walnut fruit. *Plant Dis.* 86 599–602.
10. Berta G., Sampo S., Gamalero E., Massa N., Lemanceau P. (2005). Suppression of Rhizoctonia root-rot of tomato by *Glomus mossae* BEG12 and *Pseudomonas fluorescens* A6RI is associated with their effect on the pathogen growth and on the root morphogenesis. *Eur. J. Plant Pathol.* 111 279–288.
11. Bethenod O., Huber L., Slimi H. (2001). Photosynthetic response of wheat to stress induced by *Puccinia recondita* and post-infection drought.*Photosynthetica* 39 581–590.
12. Bloomer R. H., Lloyd A. M., Symonds V. V. (2014). The genetic architecture of constitutive and induced trichome density in two new recombinant inbred line populations of *Arabidopsis thaliana*: phenotypic plasticity, epistasis, and bidirectional leaf damage response. *BMC Plant Biol.* 14:119
13. Blum A. (2009). Effective use of water (EUW) and not water-use efficiency (WUE) is the target of crop yield improvement under drought stress. *Field Crop Res.* 112 119–123.
14. Blum A., Shpiler L., Golan G., Mayer J. (1989). Yield stability and canopy temperature of wheat genotypes under drought-stress. *Field Crop Res.* 22 289–296.
15. Burman U., Lodha S. (1996). *Macrophomina phaseolina* induced changes in plant water relations of resistant and susceptible cowpea genotypes. *Indian Phytopathol.* 49 254–259.
16. Caffarraa A., Rinaldia M., Eccela E., Rossib V., Pertota I. (2012). Modelling the impact of climate change on the interaction between grapevine and its pests and pathogens: European grapevine moth and powdery mildew. *Agric. Ecosyst. Environ.* 148 89–101.
17. Calo L., García I., Gotor C., Romero L. C. (2006). Leaf hairs influence phytopathogenic fungus infection and confer an increased resistance when expressing a Trichoderma αα-1, 3-glucanase. *J. Exp. Bot.* 57 3911–3920.
18. Choudhary A., Pandey P., Senthil-Kumar M. (2016). "Tailored responses to simultaneous drought stress and pathogen infection in plants," in *Drought Stress Tolerance in Plants* Vol. 1 eds Hossain M. A., Wani S. H., Bhattacharjee S., Burritt D. J., Tran L.-S. P., editors. (Cham: Springer International Publishing;) 427–438.
19. Coakley S. M., Scherm H., Chakraborty S. (1999). Climate change and plant disease management. *Annu. Rev. Phytopathol.* 37 399–426.
20. Comas L. H., Becker S. R., Cruz V. M. V., Byrne P. F., Dierig D. A. (2013). Root traits contributing to plant productivity under drought. *Front. Plant Sci.*4:442
21. Cordes R. C., Bauman T. T. (1984). Field competition between ivy leaf morning glory (*Ipomoea hederacea*) and soybeans (*Glycine max*). *Weed Sci.* 32 364–370.

22. Crisp P. A., Ganguly D., Eichten S. R., Borevitz J. O., Pogson B. J. (2016). Reconsidering plant memory: intersections between stress recovery, RNA turnover, and epigenetics. *Sci. Adv.* 2:e1501340
23. Daami-Remadi M., Souissi A., Oun H. B., Mansour M., Nasraoui B. (2009). Salinity effects on fusarium wilt severity and tomato growth. *Dyn. Soil Dyn. Plant* 3 61–69.
24. Davis J. R., McMaster G. M., Callihan R. H., Garner J. G., McDole R. E. (1974). The relationship of irrigation timing and soil treatments to control potato scab. *Phytopathology* 64 1404–1410.
25. Davis S., Bosque-Perez N., Foote N., Magney T., Eigenbrode S. (2014). *REACCH Annual Report Year 4.* Available at: https://www.reacchpna.org/node/170
26. Diourte M., Starr J. L., Jeger M. J., Stack J. P., Rosenow D. T. (1995). Charcoal rot (*Macrophomina phaseolina*) resistance and the effects of water stress on disease development in sorghum. *Plant Pathol.* 44 196–202.
27. Dow R. L., Powell N. L., Porter D. M. (1988). Effects of modification of the plant canopy environment on Sclerotinia blight of peanut. *Peanut Sci.* 15 1–5.
28. Dryden P., Van Alfen N. K. (1984). Soil moisture, root system density, and infection of roots of pinto beans by *Fusarium solani* f. sp. *phaseoli* under dryland conditions. *Phytopathology* 74 132–135.
29. Duniway J. M. (1977). Predisposing effect of water stress on the severity of Phytophthora root rot in safflower. *Phytopathology* 67 884–889.
30. Duniway J. M., Durbin R. D. (1971). Detrimental effect of rust infection on the water relations of bean. *Plant Physiol.* 48 69–72.
31. Duveiller E., Singh R. P., Nicol J. M. (2007). The challenges of maintaining wheat productivity: pests, diseases and potential epidemics. *Euphytica* 157 417–430.
32. Enright S. M., Cipollini D. (2007). Infection by powdery mildew *Erysiphe cruciferarum* (Erysiphaceae) strongly affects growth and fitness of *Alliaria petiolata* (Brassicaceae). *Am. J. Bot.* 94 1813–1820.
33. Eyal Z., Blum A. (1989). Canopy temperature as a correlative measure for assessing host response to *Septoria tritici* blotch of wheat. *Plant Dis.* 73 468–471.
34. Fitz Geral J. N., Lehti-Shiu M. D., Ingram P. A., Deak K. I., Biesiada T., Malamy J. E. (2006). Identification of quantitative trait loci that regulate Arabidopsis root system size and plasticity. *Genetics* 172 485–498.
35. Froese P. S., Murray T. D., Carter A. H. (2016). Quantitative cephalosporium stripe disease resistance mapped in the wheat genome. *Crop Sci.* 56 1586–1601.
36. Garcia B. I. L., Sentelhas P. C., Tapia L. R., Sparovek G. (2008). Climatic risk for potato late blight in the Andes region of Venezuela. *Sci. Agric.* 65 32–39.
37. Garrett K. A., Dendy S. P., Frank E. E., Rouse M. N., Travers S. E. (2006). Climate change effects on plant disease: genomes to ecosystems. *Annu. Rev. Phytopathol.* 44 489–509.
38. Giuliani S., Sanguineti M. C., Tuberosa R., Bellotti M., Salvi S., Landi P. (2005). Root-ABA1 a major constitutive QTL, affects maize root architecture and leaf ABA concentration at different water regimes. *J. Exp. Bot.* 56 3061–3070.
39. Gonzalez-Dugo M. P., Moran M. S., Mateos L., Bryant R. (2005). Canopy temperature variability as an indicator of crop water stress severity. *Irrigation Sci.* 24 233–240.
40. Goudarzi S., Banihashemi Z., Maftoun M. (2011). Effect of salt and water stress on root infection by *Macrophomina phaseolina* and ion composition in shoot in sorghum. *Iran J. Plant Pathol.* 47 69–83.
41. Grammatikopoulos G., Manetas Y. (1994). Direct absorption of water by hairy leaves of *Phlomis fruticosa* and its contribution to drought avoidance. *Can. J. Bot.* 72 1805–1811.
42. Gupta A., Dixit S. K., Senthil-Kumar M. (2016). Drought stress predominantly endures *Arabidopsis thaliana* to *Pseudomonas syringae* infection. *Front. Plant Sci.* 7:808
43. Hagerty C. H., Cuesta-Marcos A., Cregan P. B., Song Q., McClean P., Noffsinger S., et al. (2015). Mapping and root rot resistance and root architecture quantitative trait loci in common bean. *Crop Sci.* 55 1969–1977.

44. Hameed M., Mansoor U., Ashraf M., Rao A. U. R. (2002). Variation in leaf anatomy in wheat germplasm from varying drought-hit habitats. *Int. J. Agric. Biol.* 4 12–16.
45. Hatmi S., Gruau C., Trotel-Aziz P., Villaume S., Rabenoelina F., Baillieul F., et al. (2015). Drought stress tolerance in grapevine involves activation of polyamine oxidation contributing to improved immune response and low susceptibility to *Botrytis cinerea*. *J. Exp. Bot.* 66 775–787.
46. Higginbotham R. W., Paulitz T. C., Kidwell K. K. (2004). Virulence of Pythium species isolated from wheat fields in eastern Washington. *Plant Dis.* 88 1021–1026.
47. Isaacson T., Kosma D. K., Matas A. J., Buda G. J., He Y., Yu B., et al. (2009). Cutin deficiency in the tomato fruit cuticle consistently affects resistance to microbial infection and biomechanical properties, but not transpirational water loss. *Plant J.* 60 363–377.
48. Jackson R. D. (1986). Remote sensing of biotic and abiotic plant stress. *Annu. Rev. Phytopathol.* 24 265–287.
49. Jedmowski C., Ashoub A., Momtaz O., Brüggemann W. (2015). Impact of drought, heat, and their combination on chlorophyll fluorescence and yield of wild barley (*Hordeum spontaneum*). *J. Bot.* 2015 9
50. Jenks M. A., Joly R. J., Peters P. J., Rich P. J., Axtell J. D., Ashworth E. N. (1994). Chemically induced cuticle mutation affecting epidermal conductance to water vapor and disease susceptibility in *Sorghum bicolor* (L.) Moench. *Plant Physiol.* 105 1239–1245.
51. Kaddour A. A., Fuller M. P. (2004). The effect of elevated CO_2 and drought on the vegetative growth and development of durum wheat (*Triticum durum* Desf.) cultivars. *Cereal Res. Commun.* 32 225–232.
52. Kamilova F., Kravchenko L. V., Shaposhnikov A. I., Makarova N., Lugtenberg B. (2006). Effects of the tomato pathogen *Fusarium oxysporum* f. sp *radicis-lycopersici* and of the biocontrol bacterium *Pseudomonas fluorescens* WCS365 on the composition of organic acids and sugars in tomato root exudate. *Mol. Plant Microbe Interact.* 19 1121–1126.
53. Kosma D. K., Bourdenx B., Bernard A., Parsons E. P., Lü S., Joubès J., et al. (2009). The impact of water deficiency on leaf cuticle lipids of Arabidopsis.*Plant Physiol.* 151 1918–1929.
54. Kudela V. (2009). Potential impact of climate change on geographic distribution of plant pathogenic bacteria in central Europe. *Plant Prot. Sci.* 45 S27–S32.
55. Lacape J.-M., Nguyen T. B. (2005). Mapping quantitative trait loci associated with leaf and stem pubescence in cotton. *J. Hered.* 96 441–444.
56. Ladanyi M., Horvath L. (2010). A review of the potential climate change impact on insect populations–general and agricultural aspects. *Appl. Ecol. Environ. Res.*8 143–152.
57. Lai A., Cianciolo V., Chiavarini S., Sonnino A. (2000). Effects of glandular trichomes on the development of *Phytophthora infestans* infection in potato (*S. tuberosum*). *Euphytica* 114 165–174.
58. Lamey H. A. (1988). *Sugarbeet Research and Extension Reports* Vol. 19. Available at: http://www.sbreb.org/brochures/Mildew/index2.htm 158–161
59. Lamichhane J. R., Venturi V. (2015). Synergisms between microbial pathogens in plant disease complexes: a growing trend. *Front. Plant Sci.* 6:385
60. Lapwood D. H., Hering T. F. (1968). Infection of potato tubers by common scab during brief periods when soil is drying. *Eur. Potato J.* 11 177–187.
61. Loreto F., Bongi G. (1989). Combined low temperature-high light effect on gas-exchange properties of jojoba leaves. *Plant Physiol.* 91 1580–1585.
62. Lynch J. P., Chimungu J. G., Brown K. M. (2014). Root anatomical phenes associated with water acquisition from drying soil: targets for crop improvement. *J Exp. Bot.* 65 6155–6166.
63. Mahalingam R. (ed.) (2015). "Consideration of combined stress: a crucial paradigm for improving multiple stress tolerance in plants," in *Combined Stresses in Plants* (Cham: Springer International Publishing;) 1–25.

64. Marcell L. M., Beattie G. A. (2002). Effect of leaf surface waxes on leaf colonization by *Pantoea agglomerans* and *Clavibacter michiganensis*. *Mol. Plant Microb. Interact.* 15 1236–1244.
65. Martin J. T. (1964). Role of cuticle in the defense against plant disease. *Annu. Rev. Phytopathol.* 2 81–100.
66. Mayek-Perez N., Garcia-Espinosa R., Lopez-Castaneda C., Acosta-Gallegos J. A., Simpson J. (2002). Water relations, histopathology and growth of common bean (*Phaseolus vulgaris* L.) during pathogenesis of *Macrophomina phaseolina* under drought stress. *Physiol. Mol. Plant Pathol.* 60 185–195.
67. McDonald A., Riha S., DiTommasob A., DeGaetanoa A. (2009). Climate change and the geography of weed damage: analysis of U.S. maize systems suggests the potential for significant range transformations. *Agric. Ecosyst. Environ.* 130 131–140.
68. McElrone A. J., Sherald J. L., Forseth I. N. (2003). Interactive effects of water stress and xylem limited bacterial infection on the water relations of a host vine.*J. Exp. Bot.* 54 419–430.
69. Mihail J. D. (1989). *Macrophomina phaseolina*: spatio-temporal dynamics of inoculum and of disease in a highly susceptible crop. *Phytopathology* 79 848–855.
70. Mittler R. (2006). Abiotic stress, the field environment and stress combination.*Trends Plant Sci.* 11 15–19.
71. Monier J. M., Lindow S. E. (2003). Differential survival of solitary and aggregated bacterial cells promotes aggregate formation on leaf surfaces. *Proc. Natl. Acad. Sci. U.S.A.* 100 15977–15982.
72. Narsai R., Wang C., Chen J., Wu J., Shou H., Whelan J. (2013). Antagonistic, overlapping and distinct responses to biotic stress in rice (*Oryza sativa*) and interactions with abiotic stress. *BMC Genomics* 14:93
73. Nonomura T., Xu L., Wada M., Kawamura S., Miyajima T., Nishitomi A., et al. (2009). Trichome exudates of *Lycopersicon pennellii* form a chemical barrier to suppress leaf-surface germination of *Oidium neolycopersici* conidia. *Plant Sci.*176 31–37.
74. Odvody G. N., Dunkle L. D. (1979). Charcoal stalk rot of sorghum: effect of environment on host-parasite relations. *Phytopathology* 69 250–254.
75. Olaya G., Abawi G. S. (1996). Effect of water potential on mycelial growth and on production and germination of sclerotia of *Macrophomina phaseolina*. *Plant Dis.* 80 1351–1354.
76. Oldenburg E., Manderscheid R., Erbs M., Weigel H. J. (2009). "Interaction of free air carbon dioxide enrichment (FACE) and controlled summer drought on fungal infections of maize," in *Crop Plant Resistance to Biotic and Abiotic Factors: Current Potential and Future Demands* eds Feldmann F., Alford D. V., Furk C., editors. (Berlin: Julius Kühn-Institut;) 75–83.
77. Pandey P., Ramegowda V., Senthil-Kumar M. (2015a). Shared and unique responses of plants to multiple individual stresses and stress combinations: physiological and molecular mechanisms. *Front. Plant Sci.* 6:723
78. Pandey P., Sinha R., Mysore K. S., Senthil-Kumar M. (2015b). "Impact of concurrent drought stress and pathogen infection on plants," in *Combined Stresses in Plants* ed. Mahalingam R., editor. (Cham: Springer International Publishing;) 203–222.
79. Pastor-Corrales M. A., Abawi G. S. (1988). Reactions of selected bean accessions to infection by *Macrophomina phaseolina*. *Plant Dis.* 72 39–41. 10.1094/PD-72-0039
80. Patterson D. T. (1995). Effects of environmental stress on weed/crop interaction. *Weed Sci.* 43 483–490.
81. Patterson D. T., Flint E. P. (1979). Effects of chilling on cotton (*Gossypium hirsutum*), velvetleaf (*Abutilon theophrasti*), and spurred anoda (*Anoda cristata*). *Weed Sci.* 27 473–479.
82. Pautasso M., Döring T. F., Garbelotto M., Pellis L., Jeger M. J. (2012). Impacts of climate change on plant diseases–opinions and trends. *Eur. J. Plant Pathol.*133 295–313.

83. Peters K., Breitsameter L., Gerowitt B. (2014). Impact of climate change on weeds in agriculture: a review. *Agric. Sustain. Dev.* 34 707–721. 10.1007/s13593-014-0245-2
84. Pinter P. J., Stanghellini M. E., Reginato R. J., Idso S. B., Jenkins A. D., Jackson R. D. (1979). Remote detection of biological stresses in plants with infrared thermometry. *Science* 205 585–586.
85. Pinto R. S., Reynolds M. P. (2015). Common genetic basis for canopy temperature depression under heat and drought stress associated with optimized root distribution in bread wheat.TAG. *Theor. Appl. Genet.* 128 575–585.
86. Potato Council News (2011). Available at. http://www.potatopro.com/news/2011/continuing-drought-increases-common-scab-threat-uk-potato-growers
87. Pouzoulet J., Pivovaroff A. L., Santiago L. S., Rolshausen P. E. (2014). Can vessel dimension explain tolerance toward fungal vascular wilt diseases in woody plants? Lessons from Dutch elm disease and esca disease in grapevine.*Front. Plant Sci.* 5:253
88. Prasad P. V. V., Pisipati S. R., Momcilovic I., Ristic Z. (2011). Independent and combined effects of high temperature and drought stress during grain filling on plant yield and chloroplast EF-Tu expression in spring wheat. *J. Agron. Crop Sci.* 197 430–441.
89. Prasch C. M., Sonnewald U. (2013). Simultaneous application of heat, drought, and virus to Arabidopsis plants reveals significant shifts in signaling networks.*Plant Physiol.* 162 1849–1866.
90. Puckette M. C., Weng H., Mahalingam R. (2007). Physiological and biochemical responses to acute ozone-induced oxidative stress in *Medicago truncatula*. *Plant Physiol. Biochem.* 45 70–79.
91. Ramegowda V., Senthil-Kumar M. (2015). The interactive effects of simultaneous biotic and abiotic stresses on plants: mechanistic understanding from drought and pathogen combination. *J. Plant Physiol.* 176 47–54.
92. Ramu V. S., Paramanantham A., Ramegowda V., Mohan-Raju B., Udayakumar M., Senthil-Kumar M. (2016). Transcriptome analysis of sunflower genotypes with contrasting oxidative stress tolerance reveals individual- and combined- biotic and abiotic stress tolerance mechanisms. *PLoS ONE* 11:e0157522
93. Ristaino J. B., Duniway J. M. (1989). Effect of pre-inoculation and post-inoculation water stress on the severity of Phytophthora root rot in processing tomatoes. *Plant Dis.* 73 349–352.
94. Rivero R. M., Mestre T. C., Mittler R., Rubio F., Garcia-Sanchez F., Martinez V. (2014). The combined effect of salinity and heat reveals a specific physiological, biochemical and molecular response in tomato plants. *Plant Cell Environ.* 37 1059–1073.
95. Roy B. A., Stanton M. L., Eppley S. M. (1999). Effects of environmental stress on leaf hair density and consequences for selection. *J. Evol. Biol.* 12 1089–1103.
96. Scherm H., Coakley S. M. (2003). Plant pathogens in a changing world.*Australas. Plant Pathol.* 32 157–165.
97. Schroth M. N., Hildebrand D. C. (1964). Influence of plant exudates on root-infecting fungi. *Annu. Rev. Phytopathol.* 2 101–132.
98. Sharma M., Pande S. (2013). Unravelling effects of temperature and soil moisture stress response on development of dry root rot [*Rhizoctonia bataticola* (Taub.)] Butler in Chickpea. *Am. J. Plant Sci.* 4 584–589.
99. Sharma S. K., Sharma P. N., Sharma B. K. (2012). Epidemiological studies on stem rot of chickpea. *Indian Phytopathol.* 54 185–187.
100. Simonetta S., Avidano L., Berta G. (2007). Morphogenetic effects induced by pathogenic and non-pathogenic *Rhizoctonia solani* Kühn strains on tomato roots. *Caryologia* 60 141–145.
101. Sinha R., Gupta A., Senthil-Kumar M. (2016). Understanding the impact of drought on foliar and xylem invading bacterial pathogen stress in chickpea.*Front. Plant Sci.* 7:902

102. Smith D., Chilvers M., Dorrance A., Hughes T., Mueller D., Niblack T., et al. (2015). *Soybean Disease Management.* Available at: http://cropprotectionnetwork.org/soybean/charcoal-rot/

103. Song Y., Yu J., Huang B. (2014). Elevated CO2-mitigation of high temperature stress associated with maintenance of positive carbon balance and carbohydrate accumulation in kentucky bluegrass. *PLoS ONE* 9:e89725

104. Srinivasan S., Gomez S. M., Kumar S. S., Ganesh S. K., Biji K. R., Senthil A., et al. (2008). QTLs linked to leaf epicuticular wax, physio-morphological and plant production traits under drought stress in rice (*Oryza sativa* L.). *Plant Growth Regul.* 56 245–256.

105. Stuart B. L., Harrison S. K., Abernathy J. R., Krieg D. R., Wendt C. W. (1984). The response of cotton (*Gossypium hirsutum*) water relations to smooth pigweed (*Amaranthus hybridus*) competition. *Weed Sci.* 32 126–132.

106. Su L., Dai Z., Li S., Xin H. (2015). A novel system for evaluating drought–cold tolerance of grapevines using chlorophyll fluorescence. *BMC Plant Biol.* 15:82

107. Suzuki N., Rivero R. M., Shulaev V., Blumwald E., Mittler R. (2014). Abiotic and biotic stress combinations. *New Phytol.* 203 32–43.

108. Thaler J. S., Bostock R. M. (2004). Interactions between abscisic-acid-mediated responses and plant resistance to pathogens and insects. *Ecology* 85 48–58.

109. Triky-Dotan S., Yermiyahu U., Katan J., Gamliel A. (2005). Development of crown and root rot disease of tomato under irrigation with saline water.*Phytopathology* 95 1438–1444.

110. Umamaheswari C., Ramakrishnan G., Nallathambi P. (2000). Role of inoculum level on disease incidence of dry root rot caused by *Macrophomina phaseolina* in groundnut. *Madras Agric. J.* 87 71–73.

111. Valerio M., Lovelli S., Perniola M., Di Tommaso T., Ziska L. (2013). The role of water availability on weed–crop interactions in processing tomato for southern Italy. *Acta Agric. Scand. Sect. B* 63 62–68.

112. Welfare K., Yeo A. R., Flowers T. J. (2002). Effects of salinity and ozone, individually and in combination, on the growth and ion contents of two chickpea (*Cicer arietinum* L.) *varieties. Environ. Pollut.* 120 397–403.

113. Xiong L., Wang R. G., Mao G., Koczan J. M. (2006). Identification of drought tolerance determinants by genetic analysis of root response to drought stress and abscisic acid. *Plant Physiol.* 142 1065–1074.

114. Xu P., Chen F., Mannas J. P., Feldman T., Sumner L. W., Roossinck M. J. (2008). Virus infection improves drought tolerance. *New Phytol.* 180 911–921.

115. Zhan A., Schneider H., Lynch J. (2015). Reduced lateral root branching density improves drought tolerance in maize. *Plant Physiol.* 168 1603–1615.

116. Zhou L., Ni E., Yang J., Zhou H., Liang H., Li J., et al. (2013). Rice OsGL1-6 is involved in leaf cuticular wax accumulation and drought resistance. *PLoS ONE*8:e65139

117. Ziska L. H., Tomecek M. B., Gealy D. R. (2010). Evaluation of competitive ability between cultivated and red weedy rice as a function of recent and projected increases in atmospheric CO2. *Agron. J.* 102 118–123.

CHAPTER 7

Quantification of Climate Warming and Crop Management Impacts on Cotton Phenology

Shakeel Ahmad [1,*], Qaiser Abbas [1], Ghulam Abbas [1], Zartash Fatima [1], Atique-ur-Rehman [1], Sahrish Naz [1], Haseeb Younis [1], Rana Jahanzeb Khan [1], Wajid Nasim [2], Muhammad Habib ur Rehman [3], Ashfaq Ahmad [4], Ghulam Rasul [5], Muhammad Azam Khan [6] and Mirza Hasanuzzaman [7]

[1]*Department of Agronomy, Bahauddin Zakariya University, Multan 60800, Pakistan*

[2]*Department of Environmental Sciences, COMSATS Institute of Information Technology, Vehari 61100, Pakistan*

[3]*Department of Agronomy, Muhammad Nawaz Shareef Agriculture University, Multan 60800, Pakistan*

[4]*US-Pakistan Center for Advanced Studies in Agriculture and Food Security (USPCAS-AFS), University of Agriculture, Faisalabad 38040, Pakistan*

[5]*Pakistan Meteorological Department, Islamabad 44000, Pakistan*

[6]*Department of Agriculture, District Chiniot 35400, Pakistan*

[7]*Department of Agronomy, Sher-e-Bangla Agricultural University, Dhaka 1207, Bangladesh*

ABSTRACT

Understanding the impact of the warming trend on phenological stages and phases of cotton (*Gossypium hirsutum* L.) in central and lower Punjab, Pakistan, may assist in optimizing crop management practices to enhance production. This study determined the influence of the thermal trend on cotton phenology from 1980–2015 in 15 selected locations. The results demonstrated that observed phenological stages including sowing (S), emergence (E), anthesis (A) and physiological maturity (M) occurred earlier by, on average, 5.35, 5.08, 2.87 and 1.12 days decade^{-1}, respectively. Phenological phases, sowing anthesis (S-A), anthesis to maturity (A-M) and sowing to maturity (S-M) were reduced by, on average, 2.45, 1.76 and 4.23 days decade^{-1}, respectively. Observed sowing, emergence, anthesis and maturity were negatively correlated with air temperature by, on average, −2.03, −1.93, −1.09 and −0.42 days °C^{-1}, respectively. Observed sowing-anthesis, anthesis to maturity and sowing-maturity were also negatively correlated with temperature by, on average, −0.94, −0.67 and −1.61 days °C^{-1}, respectively. Applying the cropping system model CSM-CROPGRO-Cotton model using a standard variety in all locations indicated that the model-predicted phenology accelerated more due to warming trends than field-observed phenology. However, 30.21% of the harmful influence of the thermal trend was compensated as a result of introducing new cotton cultivars with higher growing degree day (thermal time) requirements. Therefore, new cotton cultivars which have higher thermal times and are high temperature tolerant should be evolved.

Keywords: cotton phenology; CSM-CROPGRO-cotton model; climate warming; thermal trend

1. INTRODUCTION

Cotton (*Gossypium hirsutum* L.) is known as "White-Gold" and is a most important fiber, oilseed and cash crop in Pakistan. It accounts for 1.5% of the gross domestic product GDP and 7.1% of the agricultural economy of Pakistan. The cotton cultivated area, total production bales and yield ha^{-1} were 2.96 million ha, 13.91 million bales and 802 kg·ha^{-1}, respectively [1]. Nevertheless, a study showing the adverse effects of climate warming on the phenological stages and phases of cotton was carried out in Punjab, Pakistan. Cotton growth and development are highly impacted due to biotic as well as abiotic stresses. From these stresses, the elevated mean temperature is the major environmental factor which harmfully reduces cotton yield and quality [2,3,4,5,6,7,8].

Climate change has negatively impacted agricultural production and ecological systems of developed and developing countries, which has been demonstrated in various current worldwide climate change studies

[9]. A recent Intergovernmental Panel on Climate Change IPCC report indicated that in mostly Asian countries, the integer of cold days as well as nights has been reduced, while the integer of hot days as well as nights and the heat wave frequency has been increased due to the warming trend since about 1950. The mean temperature at the global level has increased by 0.85 °C since the industrial revolution. The decade of the 2000s was the warmest decade compared previous decades and 2014 was the warmest year compared to past years [9]. Similarly, in Punjab, Pakistan, a warming trend has been indicated for the period of the previous three decades and predominantly in the 2000s [10,11].

Phenological stages as well as phases of any crop are affected by changes in weather conditions and crop management practices, which are comprised of sowing times and variety assortments [12,13,14]. Constant variations in sowing times and the growing of new introduced varieties have made it difficult to measure the long-standing response of the phenology of a crop towards the thermal trend [15,16,17,18,19]. The crops' growth and developmental rate are accelerating in most climatic conditions due to the increasing warming trend due to climate change. Temperature is the most important factor which influences the crop development rate throughout the whole crop lifecycle and ultimately affects the biological and economical yield [20,21]. Consequently, we must be aware of the phenological response of a crop because of changes in the local mean temperature in order to be capable of developing excellent adaptation strategies, for example better agronomic management practices and improved new introduced cultivars that can compensate for the potential harmful impact of the warming trend [22,23,24]. Crop phenological phases are reduced due to the advancement in the phenological stages due to the warming trend. This negative impact on the crop phenology can be reduced by adopting newly introduced cultivars with higher growing degree day requirements and early sowing dates [24,25,26,27]. The relationship between natural changes in the crop environment, agronomical management practices, and the shifting of cultivars cannot be clarified by statistical models as a result of the perplexing collaboration between hereditary qualities, environment, and management. Nevertheless, the utilization of processed-based crop growth models, for example Decision Support System for Agrotechnology Transfer DSSAT and please define, etc., can clarify some of these multifaceted interactions [28,29,30,31,32,33] by permitting scientists to analyze either the effects of a solitary variable at once or the associations among different factors [34,35].

The purposes of this research were (1) to inspect the observed trends of the stages and phases of cotton phenology from 1980 to 2015 in central and lower Punjab, Pakistan; (2) to correlate the observed phenological stages and calculated phases with temperature trends for the same time period; and (3) to recognize the interrelated impact of warming trends and agronomical management practices on the phenology of the cotton crop.

2. RESULTS

2.1. Temperature Trend

The thermal trend of the environment was observed in the phenological phases of cotton in all 15 chosen locations in central and lower Punjab, Pakistan, for the duration of 1980 to 2015 (Figure 1). The observed air temperature increasing trend all through the phenological phases which included S-A, S-M and A-M ranged from 0.52–0.86, 0.72–1.05 and 0.56–0.99 °C decade^{-1}, respectively (Figure 2). Overall, the average enhanced temperature values were 0.73, 0.89 and 0.81 °C decade^{-1} throughout S-A, S-M and A-M, respectively, in 15 locations in upper and lower Punjab.

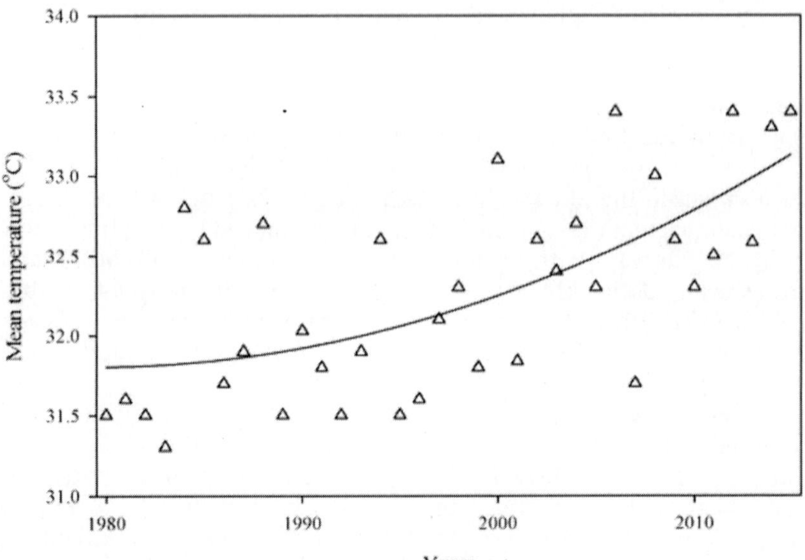

Figure 1. Mean total temperature during the cotton growing season from 1980 to 2015 at 15 sites in Punjab, Pakistan.

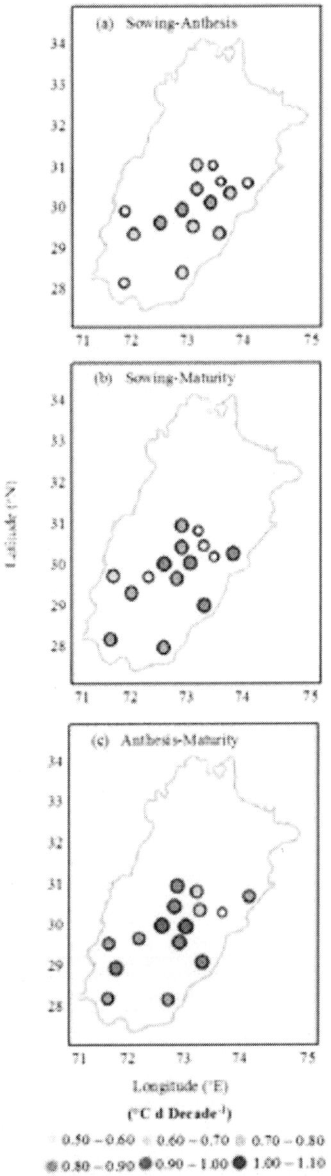

Figure 2. Observed trends in mean temperature during the phenological phases, (**a**) sowing–anthesis, (**b**) sowing–maturity and (**c**) anthesis–maturity, for cotton from 1980 to 2015 at 15 locations in Punjab, Pakistan.

2.2. Spatial and Temporal Changes in Phenological Stages of Cotton

Usually, sowing of the cotton crop is done in mid-April to mid-June in central and lower Punjab, Pakistan. The sowing dates of cotton were earlier during 1980 to 2015 in all 15 locations (Table 1). The field-observed range of earliness in the sowing dates of cotton was between 2.80 to 7.40 days decade^{-1} and, on average, the sowing date was 5.35 days decade^{-1}earlier. This earliness of the sowing dates was statistically significant ($p < 0.05$) at 13 locations and non-significant ($p > 0.05$) at two locations. Earliness in emergence dates was observed at all locations and the range of 2.66 to 7.03 days decade^{-1} for emergence dates was observed. However, on average, 5.08 days decade^{-1} were observed in the earliness of the emergence dates (Figure 3). Earliness in the emergence date was significant at 13 locations and non-significant at two locations.

Table 1. Average observed phenology of cotton in Punjab, Pakistan, during the period of 1980–2015.

Districts	Sowing DOY	EmergenceDOY	Anthesis [a] DOY	Maturity [b] DOY
Toba Tek Singh	119 ± 7.0	124 ± 6.4	190 ± 5.4	274 ± 9.3
Jhang	121 ± 6.2	126 ± 5.7	188 ± 3.5	281 ± 7.3
Okara	128 ± 5.7	133 ± 5.3	191 ± 7.2	273 ± 5.4
Sahiwal	123 ± 7.2	128 ± 6.9	182 ± 6.1	278 ± 8.3
Pakpattan	130 ± 5.2	135 ± 5.0	189 ± 5.9	270 ± 7.8
Multan	120 ± 4.3	125 ± 4.1	191 ± 7.2	281 ± 6.7
Khanewal	125 ± 5.0	130 ± 4.8	193 ± 8.4	276 ± 5.5
Vehari	128 ± 6.7	133 ± 6.2	197 ± 9.2	285 ± 7.2
Lodhran	125 ± 4.4	130 ± 4.1	189 ± 6.2	278 ± 8.1
Bahawalnagar	131 ± 3.8	136 ± 3.5	192 ± 8.1	272 ± 9.5
Bahawalpur	130 ± 5.1	135 ± 4.5	187 ± 4.9	282 ± 8.5
DG Khan	126 ± 4.2	131 ± 3.9	183 ± 5.8	276 ± 6.7
Rajanpur	122 ± 6.9	127 ± 6.3	190 ± 7.3	280 ± 7.6
Muzaffargarh	127 ± 5.2	132 ± 4.7	194 ± 6.1	276 ± 7.1
Rahim Yar Khan	123 ± 4.8	128 ± 3.9	197 ± 6.8	270 ± 8.6

[a] 50% Anthesis; [b] 90% Physiological maturity; DOY = Day of year.

Anthesis (50% flowering) of cotton in central and lower Punjab, Pakistan, usually starts in June to September (Figure 4). Anthesis dates were advanced, ranging from 1.20 to 4.10 days decade^{-1} at all locations. The average advancement in the anthesis date was −2.87 days decade^{-1}. The anthesis date was significant by statistical calculations at 11 locations and non-significant at four locations. The physiological maturity (90% open bolls) date was also advanced at all 15 locations, which ranged from 0.50 to 1.90 days decade^{-1}. On average, the cotton physiological maturity date was advanced by 1.12 days decade^{-1}. Advancement in the maturity date was significant at 14 locations and non-significant at one location.

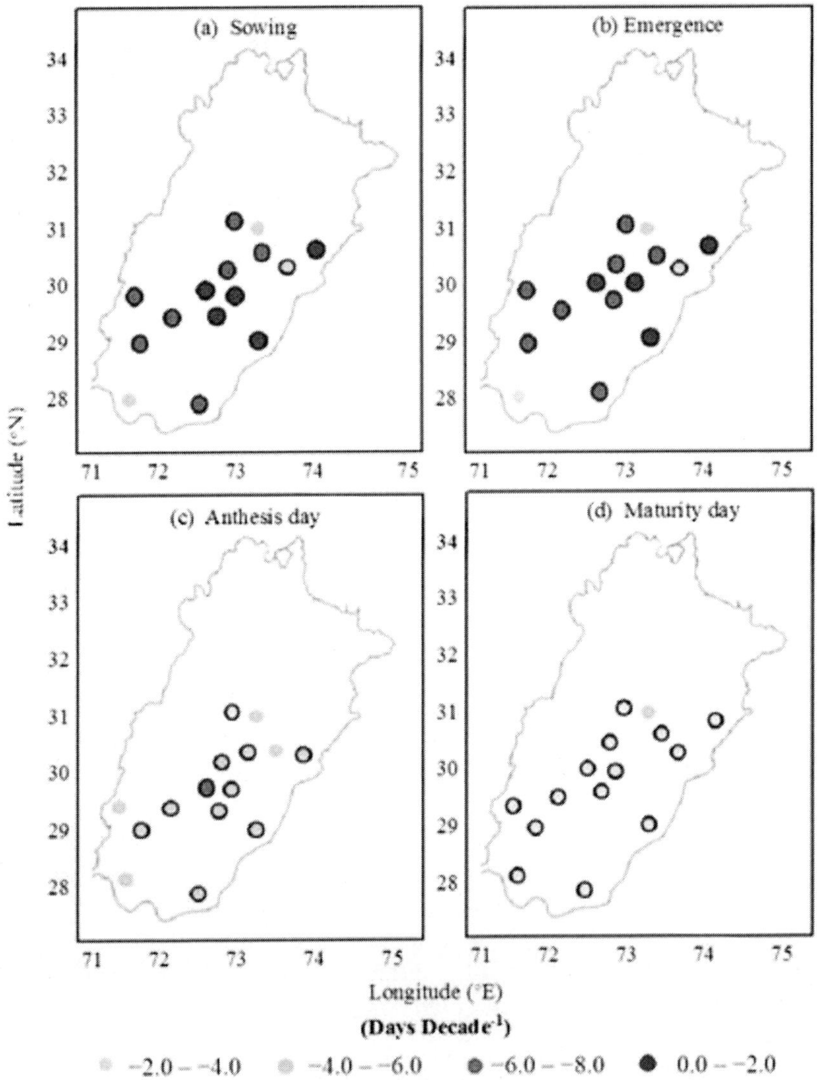

Figure 3. Observed trends in phenological stages of cotton sown from 1980 to 2015 in Punjab, Pakistan; (**a**) sowing; (**b**) emergence; (**c**) anthesis and (**d**) maturity. Circles with black border indicate statistically significant trends at $p = 0.05$ probability level.

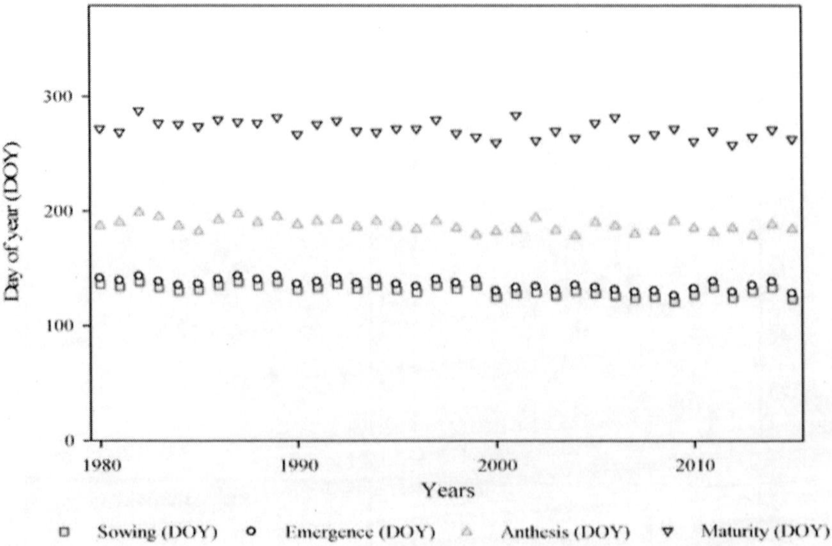

Figure 4. Time series plots of observed dates of onset of phenological stages (sowing, emergence, anthesis and maturity) for cotton from 1980 to 2015 in Punjab, Pakistan.

2.3. Spatial and Temporal Changes in Phenological Phases of Cotton

The impact of the thermal trend on the cotton crop phenological phases is represented by Figure 5. The cotton crop duration from sowing to physiological maturity was reduced, ranging from 2.30 to 5.66 days decade^{-1} due to the earliness of sowing and physiological maturity. On average, the sowing-to-maturity phenological phase was reduced by 4.23 days decade^{-1}. The reduction of the growth duration was statistically significant at 13 locations and non-significant at two locations. The sowing-to-anthesis phase was decreased from the range of 1.02 to 3.30 days decade^{-1} at the locations, which was statistically significant at 12 locations and non-significant at three locations. On average, the reduction in the sowing-to-anthesis phase was 2.45 days decade^{-1}. The anthesis-to–physiological maturity phase was decreased 1.76 days decade^{-1}on average, ranging from 0.70 to 2.36 days decade^{-1}, which was statistically significant at 11 locations and non-significant at four locations.

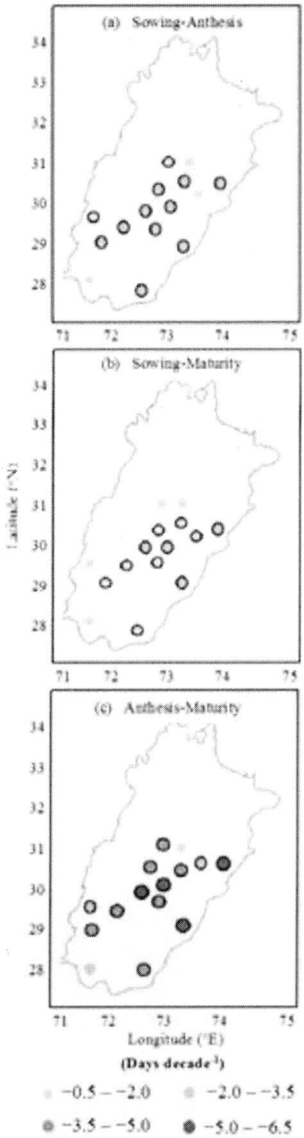

Figure 5. Observed trends in the length of phenological phases for cotton from 1980 to 2015 in Punjab, Pakistan: (**a**) sowing–anthesis; (**b**) anthesis–maturity; and (**c**) sowing–maturity. Circles with black border indicate statistically significant trend at $p = 0.05$ probability level.

2.4. Spatial and Temporal Changes in Thermal Characteristics for Cotton Cultivars

For the cotton cultivars, the total growing degree day requirement from the sowing to anthesis phase was enhanced at all 15 locations, ranging from 53 to 95 °C d decade^{-1} with an average value of 76 °C d decade^{-1}. However, this increase was significant at 10 locations and non-significant at five locations (Figure 6). Similarly, the thermal time requisite for the phenological phase of anthesis to physiological maturity was also increased, ranging from 66 to 102 °C d decade^{-1} and 85 °C d decade^{-1} on average. This total thermal time requirement enhancement was significant by statistical analysis at 12 locations and non-significant at three locations.

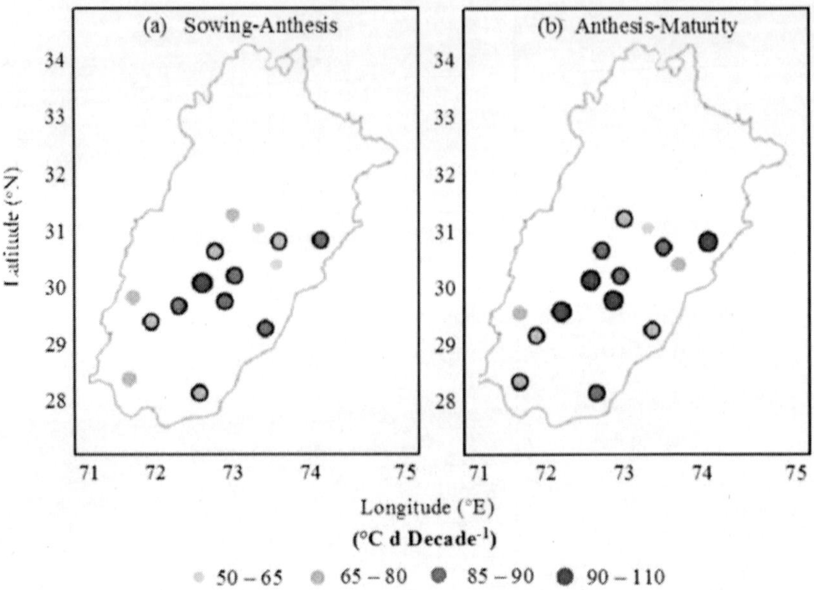

Figure 6. Observed trends in thermal time required for cotton in Punjab, Pakistan, to advance from (**a**) sowing–anthesis and (**b**) anthesis–maturity. Circles with black border indicate statistically significant trend at $p = 0.05$ probability level.

2.5. Correlation of Observed Phenology to Air Temperature

The regression coefficient of the field-observed phenological stages and phases with the temperature is shown in Table 2 and Figure 7. Negative relationships of sowing dates were obtained with the air temperature at all 15 locations, which were, on average, −2.03 days $°C^{-1}$ and ranged from −1.06 to −2.81 days $°C^{-1}$ (significant at 13 and non-significant at two locations). Emergence dates were also negatively correlated with the thermal trend, which ranged in value from −1.01 to 2.67 days $°C^{-1}$ (significant at 13 and non-significant at two locations) and were −1.93 days $°C^{-1}$ on average. The negative correlation of anthesis with the temperature was obtained, ranging from −0.46 to −1.56 days $°C^{-1}$ and, on average, −1.09 days $°C^{-1}$, which was significant and non-significant by statistical analysis at 11 and four locations, respectively. Physiological maturity was also negatively correlated with the air temperature by an average of −0.42 days $°C^{-1}$ and advanced in the range of −0.19 to −0.72 days $°C^{-1}$ (significant at 14 and non-significant at one location). A negative correlation of the sowing to anthesis phenological phase with the temperature was obtained in the range of −0.39 to −1.32 days $°C^{-1}$ (significant at 12 and non-significant at three locations) and, on average, −0.94 days $°C^{-1}$. Advancement of the sowing to the physiological maturity phase was done by, on average, −1.61 days $°C^{-1}$ and ranged from −0.87 to −2.15 days $°C^{-1}$, which was at 13 locations and non-significant at two locations. A negative correlation of the anthesis to maturity phase with the warming trend was observed, which ranged from −0.27 to −0.90 days $°C^{-1}$ (significant at 11 locations and non-significant at four locations) and was −0.67 days $°C^{-1}$ on average.

2.6. Correlation of Simulated Phenology to Air Temperature

Negative relationships among the CSM-CROPGRO-Cotton model simulated three phenological phases and the warming trend was acquired, which is represented by Table 2 and Figure 7. The sowing-to-anthesis phase was advanced by a mean of −1.42 days $°C^{-1}$ and the values ranged from −0.75 to −2.08 days $°C^{-1}$, which was statistically significant at 14 locations. Advancement of the anthesis to the physiological maturity phase ranged from −0.50 to −2.25 days $°C^{-1}$, which was significant at 13 locations and non-significant at two locations, with −1.06 days $°C^{-1}$ on average. The cotton duration (sowing-to-physiological maturity phase) was shortened by an average of −1.97 days $°C^{-1}$ and the values ranged from −1.21 to −2.50 days $°C^{-1}$ (statistically significant at 14 locations and non-significant at one location).

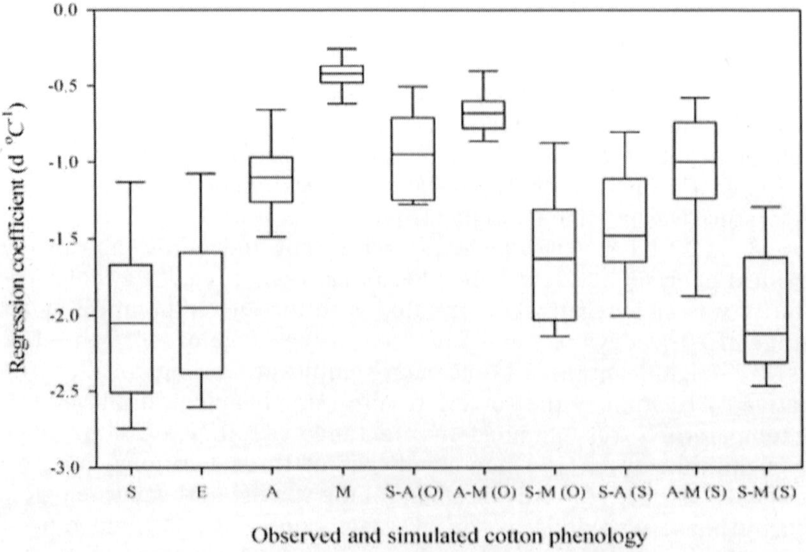

Figure 7. Observed phenology (stages and phases) and simulated phenology (phases) versus temperature trends of cotton, based on data from 15 locations in Punjab, Pakistan, from 1980 to 2015. S: sowing; E: emergence; A: anthesis; M: maturity; S-A: sowing–anthesis; A-M: anthesis–maturity; S–M: sowing–maturity; (O): observed; (S) simulated. Central horizontal line: median; box limits: 25th and 75th percentiles; whiskers: minimum and maximum values.

Table 2. Summary of observed and simulated phenology response to temperature for cotton in Punjab, Pakistan, for 1980–2015.

Phenology	No. neg. [a]	No. pos. [b]	No. sig. neg. [c]	No. sig. pos. [d]	Reg. Mean [e] (days $°C^{-1}$)
Cotton stages and phases (observed)					
Sowing	15	0	13	0	−2.03
Emergence	15	0	13	0	−1.93
Anthesis	15	0	11	0	−1.09
Maturity	15	0	14	0	−0.42
Sowing-Anthesis	15	0	12	0	−0.94
Anthesis-Maturity	15	0	11	0	−0.67
Sowing-Maturity	15	0	13	0	−1.61
Cotton phases (simulated)					
Sowing-Anthesis	15	0	14	0	−1.42
Anthesis-Maturity	15	0	13	0	−1.06
Sowing-Maturity	15	0	14	0	−1.97

[a] Number of locations with negative regression coefficients; [b] Number of locations with positive regression coefficients; [c] Number of locations with significant negative regression coefficients; [d] Number of locations with significant positive regression coefficients; [e] Mean of regression coefficients.

2.7. Model-Predicted and Observed Phenology

The relationship of the correlation of the observed and simulated cotton phenological phases with the air temperature is revealed in Table 3. Temperature sensitivity was higher for the model-predicted phenological phases in contrast to field-recorded phenological phase values. Differences among the model-simulated and field-observed recorded data for S-A, S-M and A-M were 0.48, 0.36 and 0.39 days °C^{-1}, respectively, which was statistically significant. The difference in the model-predicted and observed data regarding the cotton phenological phases indicated that the various latest approved/recommended cotton varieties grown by farmers during 1980 to 2015 have higher degree days requirements.

Table 3. Comparison of the responses of cotton phenology with average temperature using the observed and simulated data in Punjab, Pakistan, during 1980–2015.

Phenology (Phases)	Regression Coefficient [a] (days °C^{-1})		Difference between obs. and sim. Regression Correlations (Days °C^{-1})	t-Test (p-Value)
	Observed Data	Simulated Data		
Sowing-Anthesis	−0.94	−1.42	0.48	0.0012 **
Anthesis-Maturity	−0.67	−1.06	0.39	0.0053 **
Sowing-Maturity	−1.61	−1.97	0.36	0.0021 **

[a] Mean of regression coefficients; Obs. = observed; Sim. = simulated; ** Significant at the 0.01 probability level.

3. DISCUSSION

Climate warming had an enormous influence on the phenology of agricultural crops in several regions. The field-recorded changes in the cotton phenological stages and phases in central and lower Punjab, Pakistan, between 1980 and 2015 were possibly caused by the increase in temperature. Nevertheless, other integrated crop management decisions were the changeability in sowing times and the assortment of newly introduced cultivars [36,37] which had a superior total thermal time (total growing degree days) requirement. The latest introduced cultivars and sowing times are, in general, determined through the local agriculturist community, whose preferences regarding varieties and sowing times are adapted in response to the climatic warming tendency in central and lower Punjab, Pakistan. Nonetheless, the precipitation and residual soil moisture for the period of the crop-growing months could also impact the sowing time assessment along with the nutrient uptake by the crops in this region. The earliness of the emergence, anthesis, and maturity (picking) times was due to the earliness of the cotton sowing time. However, certain phenological trends were a result of the distinction in observed thermal trends. On average, the earliness in the sowing and emergence dates of cotton was 5.35 and 5.08 days decade^{-1}, respectively, while the advancement of the anthesis and maturity dates was 2.87 and 1.12 days decade^{-1}, respectively.

Plant breeding and genetics scientists continuously develop new heat-tolerant cultivars through various breeding techniques that are introduced to the local farmer community subsequent to extensive evaluation [25,38,39,40]. Frequently, these newly cultivars are adapted for local core and non-core cotton area growing conditions, including observed thermal trends. Researchers also have reported innovative phenological characteristics (heat tolerant and climate smart) in improved cotton cultivars [26,35]. In this investigation, the researchers determined the separate influence of the growth of newly introduced cultivars for the 15 selected locations from the field-observed changes through the simulation of the phenology of cotton by means of the same cultivar for the interval of the research period. This crop growth model simulation allowed for the separation of the relations of the temperature and the improved heat-tolerant varietal response [41]. The heat sensitivity of the model simulations was superior as compares to the field-observed phenological data, demonstrating that about 30.21% (Table 3) of the direct negative influence of the warming tendency was mitigated with the evolution of new cotton cultivars that require more photo-thermal time to achieve the various phenological stages in an optimal time [2]. An analogous trend of introducing the latest cultivars which are adapted to a thermal trend was established for other agricultural crops around the world, for example maize in various regions of the USA, and rice and winter wheat in China [35,42]. If the whole duration of the lifecycle of crop varieties is shorter, then as a result, there is a lessening of crop biological and economical yield as a consequence of the lesser time for total dry matter accumulation for the duration of the vegetative segment, chiefly for the highest-input agricultural crops [2,43,44]. As a consequence, the agricultural community chooses newly introduced cultivars, which have higher thermal time requirements for the compensation of the negative effect of the thermal trend on the cotton phenology [35,45].

Crop physiological growth procedures are impacted by temperature, which has a direct impact on the length of the phenological phases and, consequently, affects the ultimate biological and economic yield of cotton [4,5,6,8,46,47]. Whole dry matter accumulation is decreased with the earliness of the anthesis and physiological maturity time due to a mean temperature increasing trend, as a result reducing the crop yield [21]. In numerous regions of the globe, the yield of an assortment of crops has been improved due to a longer reproductive phase (A-M) caused by earliness in flowering and a delayed physiological maturity time [4,42,48,49].

Day and night mean temperatures are forecasted to increase in the future as compared to the precedent climatic situation [9]. At the ending of the 21st century, the average temperature is predicted to rise by 2 to 3 °C on the basis of Representative Concentration Pathways RCPs scenarios in Punjab, Pakistan [23]. Moreover, relentless events, for instance heat strokes due to extreme temperature, extreme rainfalls, floods and droughts, are forecasted to be more frequent as well, also potentially negatively impacting agricultural production [50,51]. The phenological crop phases possibly will be potentially reduced for the period of the upcoming decades if the warming trends persist, whereas, at the same time, in temperate and semi-arid regions with high mean temperatures there is also the potential for crop durations to increase if the temperatures are above the cardinal temperatures [24,34,52,53]. As a consequence, the introduction and growth of new heat-resistant cultivars with higher growing degree day (total thermal time) requirements, high temperature tolerance, as well as sowing time adjustments are imperative for mitigation of the harmful impacts of climate change in Punjab, Pakistan [19], and other states across the globe.

4. MATERIALS AND METHODS

4.1. Description of Site, Weather, Cotton Management and Phenological Data

Fifteen locations in the conventional cotton zone in Punjab (Figure 8), Pakistan were selected for this study along with 36 years observed weather data were collected from 1980 to 2015 from Pakistan Meteorological Department (PMD), Islamabad. The cotton phenological stages data comprising of sowing, emergence, anthesis (50% flowering) and maturity (90% bolls open) dates were collected from extension wing of Department of Agriculture, Government of Punjab, Pakistan for 15 selected locations. With the assist of these recorded observed phenological stages data, three phenological phases data of cotton crop including S-A, A-M and S-M were determined. Cotton crop management practices were decided by the farming community, and varieties being grown by the framers of selected locations are presented in Table 4. While, other management practices that vary among the farming community; included sowing dates, seed rates, sowing methods, fertilizer amounts, types and application dates, irrigation regimes, methods and application dates, etc. The new cotton variety was being grown by the farming community subsequent to each six to eight years (on an average nine varieties per selected location) and variety selection criteria was higher growing degree day's requirement. The newly evolved varieties by the breeders were grown by the local farmers on the advice of extension workers.

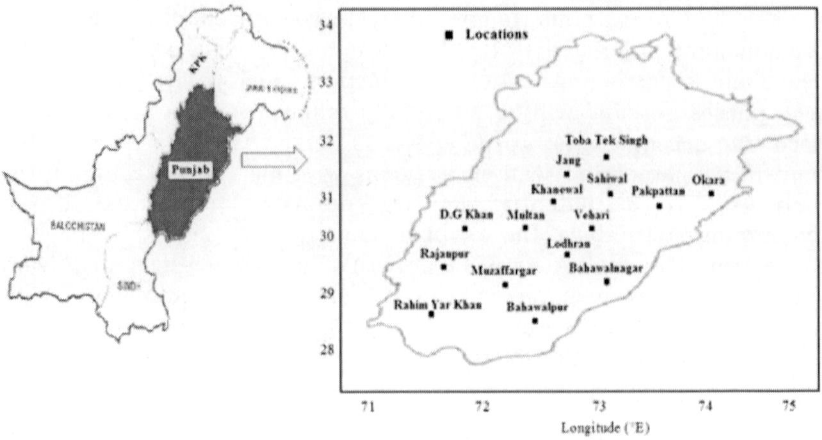

Figure 8. Map showing study locations in Punjab, Pakistan.

Table 4. Cotton cultivars in all locations during 1980–2015.

Sr. No.	Site Name	Cultivars
1	Toba Tek Singh	CIM-506a, FH-901e, FDH-228b, CYTO-124, BH-3297, B-803, MNH-552, IR-1524
2	Jhang	CIM-499a, CIM-446e, MS-240, Sitara-005, SLH-8, FH-685, BH-100a, NS-141
3	Okara	CIM-534c, CIM-554e, IR-1274, MNH-998, FH-901, CIM-110, CIM-435, CIM-602
4	Sahiwal	NIAB-111c, CIM-473d, GN-1532, CEMB-66, B-896, CIM-70, B-622, FVH-55
5	Pakpattan	BH-160e, MNH-786a, CIM-109, NIBGE-901, AGC-777, NIBGE-6, FS-631, CIM-240
6	Multan	IR-3701, CIM-240, CIM-1100, NIBGE-2d, AGC-999, CIM-109, B-820, BH-118
7	Khanewal	MG-6, NIAB-846d, CIM-707, Sitara-12, NIAB-2008e, S-12, MNH-554, FVH-57
8	Vehari	Sitara-008, NIAB-777d, CIM-600, FH-142, TCD-3, CIM-482a, MNH-93, VH-259
9	Lodhran	FH-113, CRSM-38d, TS-103, CYTO-177, FH-628b, FH-87, FH-657, MNH-516
10	Bahawalnagar	Neelam-121c, AH-151d, Tarzan-3, B-842, CIM-465c, RH-500, SLS-1, Tarzan-1
11	Bahawalpur	Ali Akbar-802, S-14 , MNH-786a, Leader-1, FDH-170b, NIAB-78, B-821, NIAB-26
12	DG Khan	IR-1524, NIBGE-1e, VH-305, FH-113, TSR-2375, Karishma, CIM-467, VH-137
13	Rajanpur	GN-2085, NIAB-846e, IUB-13, MS-240, MNH-465, BH-95, FH-649, MNH-536
14	Muzaffargarh	CIM-496c, Desi Ravib, FH-Lalazar, 149-F, S-12, FH-629, MVH-518, VS-135
15	Rahim Yar Khan	Ali Akbar-703, FDH-170b, AC-134, MM-58, CIM-448a, B-803, FVH-49, NIAB-846

Sr. No. = Serial number; a = American cultivar; b = local cultivars; c = short duration; d = medium duration; e = heat tolerant.

4.2. Analysis of Observed Data

Relationship between observed phenological stages and phases' data of cotton crop with average temperature was determined with the help of linear regression equation, in which year was used as independent variable. With the assist of maximum phenological stages at each location, windows for warming trends were gained. For instance, time window for any phenological phase like S-M was from the earlier sowing date to the most recent physiological maturity for the period of last years at each location. In this procedure, the measured warming trend was independent from the corresponded phenological stages and phases variations.

Correlation of phenological with average temperature of month for the period of month of happening of phenological stages was determined to assess that whether mean monthly temperature influence on phenological stages. Subsequent linear regression equation was used to determine the influence of temperature on cotton crop phenology:

$$OP_{nt} = a_{nt}T_{nt} + b_{nt} + \varepsilon_{nt} \quad (1)$$

In Equation (1), observed phenological phases (days) or stages (DOY; day of year) are represented by 'OP_{nt}' for corresponding nth agricultural weather station or location in respective *t* year. 'T_{nt}' represents to average daily temperature in centigrade during the relevant phenological stage or phase for particular stations in t year. Regression coefficient (days °C^{-1}) of phenological stage or phase responding to temperature is represented by 'a_{nt}' variable. Variables 'b_{nt}' and 'ϵ_{nt}' stand for intercept and error term for each particular station, respectively. Cotton phenological stages and phases responding to temperature and agronomic management variations are considered by regression coefficient (a_{nt}).

4.3. CSM-CROPGRO-Cotton Model Descriptions

CSM-CROPGRO-Cotton model was used to isolate outcome of cultivar, management and temperature from the climate warming on phenology changes in cotton phenological stages and phases in fifteen locations. The observed data were compared to simulated data sets. CSM-CROPGRO-Cotton model was calibrated and validated for fifteen locations. Thus total fifteen varieties were calibrated and evaluated. The most commonly grown variety for in a particular location was identified and used for model simulations. After that 1980 to 1982 observed data were used for model calibration and 1983 to 1985 for model validation. Then validated model was applied to simulate the A and M dates for individual location using historical weather data for 1980–2015. The CSM-CROPGRO-Cotton model needed following files for simulation runs; (1) weather file (FILE.WTH) with daily solar radiation, maximum air temperature, minimum air temperature, and precipitation; (2) soil file having soil physical and chemical properties; (3) crop management file and (4) genetic coefficients file. DSSAT version 4.6 was used for the purpose of this research study [54,55]. The detailed information can be found in earlier publications [19,56].

4.4. Phenology Simulation with CSM-CROPGRO-Cotton Model and Calculation of Growing Degree Days

The CSM-CROPGRO-Cotton model was employed for simulation of cotton phenology (stages and phases) based on cumulative degree days (GDDs) and crop variety particular GDDs demand for every developmental stage. Photoperiod sensitivity effect the phenology before anthesis. Total accumulation of growing degree days for phenological phases S-A and A-M is measured by following formula:

$$\mathrm{ATT} = \sum_{i=1}^{n} \mathrm{DTT} \quad (2)$$

In this equation, accumulated thermal time (ATT) per day is represented by daily thermal time (DTT) and n represents to number of days of crop phenology.

With the help of only one cultivar and keeping similar crop management practices during the years, crop phenological stages and phases were simulated to determine the separate influence of agronomic management practices, variety and temperature on crop phenology. Crop phenology such as anthesis, physiological maturity dates and sowing to maturity phase were predicted from 1980 to 2015 in all selected fifteen locations in upper and lower Punjab, Pakistan. Only one, mostly grown cultivar during 1980–1982 at every location was used for the purpose of calibration of CSM-CROPGRO-Cotton model. Thus, for overall calibration of CSM-CROPGRO-Cotton model at various 15 selected locations, total 15 cultivars were used. After this, crop phenological observed data from 1983 to 1985 was used for the purpose of validation of DSSAT CSM-CROPGRO-Cotton model. Then, crop model was employed to predict crop phenological stages and phases during 1980 to 2015 with the help of same variety and agronomic management practices at every year.

Influence of temperature on simulated crop phenological phases such as S-A, A-M, and S-M was determined by linear regression analysis:

$$SP_{nt} = C_{nt}T_{nt} + d_{nt} + \varepsilon_{nt} \quad (3)$$

In Equation (3), simulated phenological phases (days) or an event (day of year) is represented by 'SP_{nt}' for corresponding nth agricultural weather station or location in respective t year. 'T_{nt}' represents to average daily phenological phase temperature in centigrade during the relevant phenological phase for particular stations in t year. Regression coefficient (days °C^{-1}) of phenological phase responding to temperature is represented by 'c_{nt}' variable. Variables 'd_{nt}' and 'ϵ_{nt}' stand for intercept and error term for each particular location, respectively. Cotton predicted phenological phases responding to temperature and agronomic management variations are considered by regression coefficient (c_{nt}). Crop management data including; varieties, sowing dates, seed rates, amount and time of irrigation and method of application, amount and time of fertilizer, organic amendment, tillage and chemical applications commonly used by farming community of each location. Historical weather data and soil physico-chemical properties data were obtained from each location. All data about crop management, weather and soil data were input in CSM-CROPGRO-cotton model for calibration, evaluation and its application.

Comparison between models predicted and observed crop phenological stages such as anthesis and physiological maturity dates in years (1983–1985) of model validation at 15 locations is shown in Figure 9. The performance of CSM-CROPGRO-Cotton model was good as model simulated anthesis and physiological maturity dates well matched with observed phenology data (slope = 0.87, R^2 = 0.84, p < 0.01) and (slope = 0.85, R^2 = 0.83, p < 0.01), respectively (Figure 9).

4.5. Difference among Observing and Simulating Crop Phenological Respond to Temperature

In Equation (1), variable a_{nt} is regression coefficient which shows the respond of crop phenology to variety change, planting date and temperature. While in Equation (3), only effect of temperature on simulated crop phenology is reflected by variable c_{nt} regression coefficient. If the difference (a_{nt} – c_{nt}) between regression coefficients is showed by negative value, then it means that local farming community grown short duration varieties in last years. If difference is positive then it point out that local farmers changed to longer thermal time requirement cultivars during last decades. Paired t-test was applied to determine the significance difference between regression coefficients.

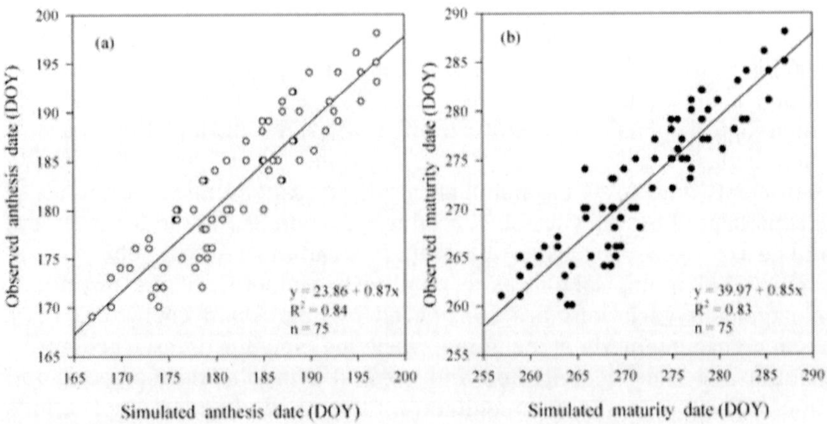

Figure 9. Observed and simulated dates of onset of phenological stages anthesis (**a**; unfilled symbol; left) and maturity (**b**; filled symbols; right) for cotton in Punjab, Pakistan, during 1983–1985; DOY; day of year.

5. CONCLUSIONS

The warming trend caused the changes in the field-observed phenological stages and phases of the cotton crop in Punjab, Pakistan, for the duration of 1980 to 2015 at 15 cotton-growing locations. Observed phenological stages S, E, A, and M occurred earlier by an average of 5.35, 5.08, 2.87 and 1.12 days decade^{-1}, respectively. Observed S, E, A, and M were negatively correlated with the air temperature by an average of −2.03, −1.93, −1.09 and −0.42 days °C^{-1}, respectively. The negative influence of the warming trend was partially diminished by the adaptation of growing new evolved cultivars with higher heat accumulation or growing degree day requirements. Approximately one-third of the negative impact of the enhanced temperature on the phenology of cotton was compensated for with the help of evolving new cultivars with higher total growing degree day requirements.

ACKNOWLEDGMENTS

The research work was funded by the Higher Education Commission (HEC), Islamabad, Pakistan through NRPU research grant 4511.

AUTHOR CONTRIBUTIONS

Shakeel Ahmad, Atique-ur-Rehman, Wajid Nasim, Ashfaq Ahmad, Ghulam Rasul, Muhammad Azam Khan, Muhammad Habib ur Rehman and Mirza Hasanuzzaman conceived and designed the study; Qaiser Abbas, Ghulam Abbas, Zartash Fatima, Sahrish Naz, Haseeb Younis, and Rana Jahanzeb Khan performed the experiments and analyzed the data, Shakeel Ahmad and Mirza Hasanuzzaman wrote the paper. However, all authors read the manuscript.

CONFLICTS OF INTEREST

The authors declare no conflict of interest.

REFERENCES

1. Government of Pakistan (GOP). Economic Survey of Pakistan, Economic Advisory Wing; Government of Pakistan: Islamabad, Pakistan, 2014–2015; pp. 29–30.
2. Reddy, K.R.; Vara Prasad, P.V.; Kakani, V.G. Crop responses to elevated carbon dioxide and interactions with temperature: Cotton. J. Crop Improv. 2005, 13, 157–191.
3. Song, Y.L.; Zhang, Q.; Dong, W.J. Impact of Climate Change on Cotton Production in Xingjiang Autonomous Region. Agric. Meteorol. 2004, 3, 15–20.
4. Wang, H.L.; Gan, Y.T.; Wang, R.Y.; Niu, J.Y.; Zhao, H.; Yang, Q.G.; Li, G.C. Phenological trends in winter wheat and spring cotton in response to climate changes in northwest China. Agric. For. Meteorol. 2008, 148, 1242–1251.
5. Yoon, S.T.; Hoogenboom, G.; Flitcroft, I.; Bannayan, M. Growth and development of cotton (Gossypium hirsutum L.) in response to CO2 enrichment under two different temperature regimes. Environ. Exp. Bot. 2009, 67, 178–187.
6. Yang, Y.; Yang, Y.; Han, S.; Macadam, I.; Liu, D.L. Prediction of cotton yield and water demand under climate change and future adaptation measures. Agric. Water Manag. 2014, 144, 42–53.
7. Luo, Q.; Bange, M.; Johnston, D.; Braunack, M. Cotton crop water use and water use efficiency in a changing climate. Agric. Ecosyst. Environ. 2015, 202, 126–134.
8. Huang, J.; Ji, F. Effects of climate change on phenological trends and seed cotton yields in oasis of arid regions. Int. J. Biometeorol. 2015, 59, 877–888.
9. Intergovernmental Panel on Climate Change (IPCC). Climate Change 2014: Impacts, Adaptation, and Vulnerability; Fifth Assessment Report on the Intergovernmental

Panel on Climate Change; Cambridge University Press: Cambridge, NY, USA, 2014; p. 688.

10. Yihui, D.; Guoyu, R.; Guangyu, S.; Peng, G.; Xunhua, Z.; Panmao, Z.; De'er, Z.; Zongci, Z.; Shaowu, W.; Huijun, W.; et al. China's National Assessment Report on Climate Change (I): Climate change in China and the future trend. Adv. Clim. Change Res. 2007, 3 (Suppl.), 1–5.
11. Wang, S.Y.; Davies, R.E.; Huang, W.R.; Gillies, R.R. Pakistan's two-stage monsoon and links with the recent climate change. J. Geophys. Res. Atmos. 2011.
12. Gordo, O.; Sanz, J.J. Phenology and climate change: A long-term study in a Mediterranean locality. Oecoloqia 2005, 146, 484–495.
13. Craufurd, P.Q.; Wheeler, T.R. Climate change and the flowering time of annual crops. J. Exp. Bot. 2009, 60, 2529–2539.
14. Beamish, A.L.; Nijland, W.; Edwards, M.; Coops, N.C.; Henry, G.H.R. Phenology and vegetation change measurements from true colour digital photography in high Arctic tundra. Arct. Sci. 2016, 2, 33–49.
15. Chmielewski, F.-M.; Muller, A.; Bruns, E. Climate changes and trends in phenology of fruit trees and field crops in Germany, 1961–2000. Agric. For. Meteorol. 2004, 121, 69–78.
16. Estrella, N.; Sparks, T.H.; Menzel, A. Trends and temperature response in the phenology of crops in Germany. Global Change Biol. 2007, 13, 1737–1747.
17. Zhenyong, D.; Heling, W.; Guochang, L.; Jiwu, X.; Yufei, Z.; Jinfang, X. Study on the cause of formation and countermeasure effect of climate warming on cotton produce in Hexi corridor. Adv. Earth Sci. 2008, 23, 160–166.
18. Cottee, N.S.; Tan, D.K.Y.; Bange, M.P.; Cothren, J.T.; Campbell, L.C. Multi-Level determination of heat tolerance in cotton (Gossypium hirsutum L.) under field conditions. Crop Sci. 2010, 50, 2553–2564.
19. Ahmad, S.; Nadeem, M.; Abbas, G.; Fatima, Z.; Khan, R.J.Z.; Ahmed, M.; Ahmad, A.; Rasul, G.; Khan, M.A. Quantification of the effects of climate warming and crop management on sugarcane phenology. Clim. Res. 2016, 71, 47–61.
20. Menzel, A.; Sparks, T.H.; Estrella, N.; Koch, E.; Aasa, A.; Ahas, R.; Alm-Kubler, K.; Bissolli, P.; Braslavska, O.; Briede, A.; et al. European phenological response to climate change matches the warming pattern. Glob. Chang. Biol. 2006, 12, 1969–1976.
21. Xiao, D.; Tao, F. Contributions of cultivars, management and climate change to winter wheat yield in the North China Plain in the past three decades. Eur. J. Agron. 2014, 52, 112–122.
22. Hoffmann, A.A.; Sgrò, C.M. Climate change and evolutionary adaptation. Nature 2011, 470, 479–485.
23. Ahmad, A.; Ashfaq, M.; Rasul, G.; Wajid, S.A.; Khaliq, T.; Rasul, F.; Saeed, U.; Habib ur Rahman, M.; Hussain, J.; Baig, I.A.; et al. Impact of climate change on the rice-wheat cropping system of Pakistan. In Handbook of Climate Change and Agro-Ecosystems; Hillel, D., Rosenzweig, C., Eds.; Imperial College Press: London, UK; The American Society of Agronomy: Madison, WI, USA, 2015; Volume 3, pp. 219–258.
24. Anwar, M.R.; Liu, D.L.; Farquharson, R.; Macadam, I.; Abadi, A.; Finlayson, J.; Wang, B.; Ramilan, T. Climate change impacts on phenology and yields of five broadacre crops at four climatologically distinct locations in Australia. Agric. Syst. 2015, 132, 133–144.
25. Liu, Z.; Yuan, Y.L.; Liu, S.Q.; Yu, X.N.; Rao, L.Q. Screening for high-temperature tolerant cotton cultivars by testing in vitro pollen germination, pollen tube growth and boll retention. J. Integrat. Plant Biol. 2006, 48, 706–714.
26. Singh, R.P.; Vara Prasad, P.V.; Sunita, K.; Giri, S.N.; Reddy, K.R. Influence of high temperature and breeding for heat tolerance in cotton: A review. Adv. Agron. 2007, 93, 313–385.
27. Wolkovich, E.M.; Cook, B.I.; Allen, J.M.; Crimmins, T.M.; Betancourt, J.L.; Travers, S.E.; Pau, S.; Regetz, J.; Davies, T.J.; Kraft, N.J.B.; et al. Warming experiments underpredict plant phenological responses to climate change. Nature 2012, 485, 494–497.

28. Dounias, I.; Aubry, C.; Capillon, A. Decision-making processes for crop management on African farms. Modelling from a case study of cotton crops in northern Cameroon. Agric. Syst. 2002, 73, 233–260.
29. Ortiz, B.V.; Hoogenboom, G.; Vellidis, G.; Boote, K.; Davis, R.F.; Perry, C. Adapting the CROPGRO-Cotton model to simulate cotton biomass and yield under southern root-knot nematode parasitism. Trans. ASAE 2009, 52, 2129–2140.
30. Boote, K.J.; Jones, J.W.; Hoogenboom, G.; White, J.W. The role of crop systems simulation in agriculture and environment. Int. J. Agric. Environ. Inf. Syst. 2010, 1, 41–54.
31. Pathak, T.B.; Jones, J.W.; Fraisse, C.W.; Wright, D.; Hoogenboom, G. Uncertainty analysis and parameter estimation for the CSM-CROPGRO-Cotton model. Agron. J. 2012, 104, 1363–1373.
32. Thorp, K.R.; Barnes, E.M.; Hunsaker, D.J.; Kimball, B.A.; White, J.W.; Nazareth, V.J.; Hoogenboom, G. Evaluation of CSM-CROPGRO-Cotton for simulating effects of management and climate change on cotton growth and evapotranspiration in an arid environment. Trans. ASAE 2014, 57, 1627–1642.
33. Modala, N.R.; Ale, S.; Rajan, N.; Munster, C.L.; DeLaune, P.B.; Thorp, K.R.; Nair, S.S.; Barnes, E.M. Evaluation of the CSM-CROPGRO-Cotton model for the Texas rolling plains region and simulation of deficit irrigation strategies for increasing water use efficiency. Trans. ASAE 2015, 58, 685–696.
34. White, J.W.; Hoogenboom, G.; Hunt, L.A. A structured procedure for assessing how crop models respond to temperature. Agron. J. 2005, 96, 426–439.
35. Liu, Y.; Wang, E.; Yang, X.; Wang, J. Contributions of climatic and crop varietal changes to crop production in the North China Plain, since 1980s. Glob. Chang. Biol. 2010, 16, 2287–2299.
36. Cleland, E.E.; Chuine, I.; Menzel, A.; Mooney, H.A.; Schwartz, M.D. Shifting plant phenology in response to global change. Trends Ecol. Evol. 2007, 22, 357–365.
37. Braunack, M.V. Cotton farming systems in Australia: Factors contributing to changed yield and fibre quality. Crop Pasture Sci. 2013, 64, 834–844.
38. Karademir, E.; Karademir, C.; Ekinci, R.; Basbag, S.; Basal, H. Screening cotton varieties (Gossypium hirsutum L.) for heat tolerance under field conditions. Afr. J. Agric. Res. 2012, 7, 6335–6342.
39. Khan, N.; Azhar, F.M.; Khan, A.A.; Ahmad, R. Measurement of canopy temperature for heat tolerance in upland cotton: Variability and its genetic basis. Pak. J. Agri. Sci. 2014, 51, 359–365.
40. Abro, S.; Rajput, M.T.; Khan, M.A.; Sial, M.A.; Tahir, S.S. Screening of cotton (Gossypium hirsutum L.) genotypes for heat tolerance. Pak. J. Bot. 2015, 47, 2085–2091.
41. Visser, M.E.; Both, C. Shifts in phenology due to global climate change: The need for a yardstick. Proc. Biol. Sci. 2005, 272, 2561–2569.
42. He, L.; Asseng, S.; Zhao, G.; Wu, D.; Yang, X.; Zhuang, W.; Jin, N.; Yu, Q. Impacts of recent climate warming, cultivar changes, and crop management on winter wheat phenology across the Loess Plateau of China. Agric. For. Meteorol.2015, 200, 135–143.
43. Zheng, J.; Ge, Q.; Hao, Z. Impacts of climate warming on plant phenophases in China for the last 40 years. Chin. Sci. Bull. 2002, 47, 1826–1831.
44. Zhang, T.; Huang, Y. Estimating the impacts of warming trends on wheat and maize in China from 1980 to 2008 based on county level data. Int. J. Climatol. 2013, 33, 699–708.
45. Qingchun, L.; Huaisui, Q.; Yuyu, R.; Tuxing, J. Temperature suitability of cotton and its changes in Henan Province. Resour. Sci. 2003, 26, 51–56.
46. Bednarz, C.W.; Nichols, R.L. Phenological and morphological components of cotton crop maturity. Crop Sci. 2005, 45, 1497–1503.
47. Luo, Q.; Bange, M.; Clancy, L. Cotton crop phenology in a new temperature regime. Ecol. Modell. 2014, 285, 22–29.

48. Tao, F.; Yokozawa, M.; Xu, Y.; Hayashi, Y.; Zhang, Z. Climate changes and trends in phenology and yields of field crops in China, 1981–2000. Agric. For. Meteorol. 2006, 138, 82–92.

49. García-Mozo, H.; Mestre, A.; Galán, C. Phenological trends in southern Spain: A response to climate change. Agric. For. Meteorol. 2010, 150, 575–580.

50. Rasul, G.; Mahmood, A.; Sadiq, A.; Khan, S.I. Vulnerability of the Indus delta to climate change in Pakistan. Pak. J. Meterol. 2012, 8, 89–107.

51. Richardson, A.D.; Keenan, T.F.; Migliavacca, M.; Ryu, Y.; Sonnentag, O.; Toomey, M. Climate change, phenology, and phenological control of vegetation feedbacks to the climate system. Agric. For. Meteorol. 2013, 169, 156–173.

52. White, J.W.; Hoogenboom, G. Crop response to climate: Ecophysiological models. In Climate Change and Food Security; Lobell, D., Burke, M., Eds.; Springer Science: Dordrecht, The Netherlands, 2010; Volume 37, pp. 59–83.

53. Soliz, L.M.A.; Oosterhuis, D.M.; Coker, D.L.; Brown, R.S. Physiological response of cotton to high night temperature. Am. J. Plant Sci. Biotechnol. 2008, 2, 63–68.

54. Jones, J.W.; Hoogenboom, G.; Porter, C.H.; Boote, K.J.; Batchelor, W.D.; Hunt, L.A.; Wilkens, P.W.; Singh, U.; Gijsman, A.J.; Ritchie, J.T. The DSSAT cropping system model. Eur. J. Agron. 2003, 18, 235–265.

55. Hoogenboom, G.; Jones, J.W.; Wilkens, P.W.; Porter, C.H.; Boote, K.J.; Hunt, L.A.; Singh, U.; Lizaso, J.I.; White, J.W.; Uryasev, O.; et al. Decision Support System for Agrotechnology Transfer (DSSAT) 2015, Version 4.6.1; DSSAT Foundation: Prosser, WA, USA, 2015; Available online: www.dssat.net (accessed on 11 March 2016).

56. Ahmad, S.; Abbas, G.; Fatima, Z.; Khan, R.J.Z.; Anjum, M.A.; Ahmed, M.; Khan, M.A.; Porter, C.H.; Hoogenboom, G. Quantification of the impacts of climate warming and crop management on canola phenology in Punjab, Pakistan. J. Agron. Crop Sci. 2017, in press.

CHAPTER 8

Heat Stress Decreases Levels of Nutrient-Uptake and -Assimilation Proteins in Tomato Roots

Anju Giri [1], Scott Heckathorn [1,*], Sasmita Mishra [1] and Charles Krause [2]

[1]*Department of Environmental Sciences, University of Toledo, Toledo, OH 43606, USA*

[2]*U.S. Department of Agriculture, University of Toledo, Toledo, OH 43606, USA*

ABSTRACT

Global warming will increase root heat stress, which is already common under certain conditions. Effects of heat stress on root nutrient uptake have rarely been examined in intact plants, but the limited results indicate that heat stress will decrease it; no studies have examined heat-stress effects on the concentration of nutrient-uptake proteins. We grew *Solanum lycopersicum* (tomato) at 25 °C/20 °C (day/night) and then transferred some plants for six days to 35 °C /30 °C (moderate heat) or 42 °C/37 °C (severe heat) (maximum root temperature = 32 °C or 39 °C, respectively); plants were then moved back to control conditions for seven days to monitor recovery. In a second experiment, plants were grown for 15 days at 28 °C/23 °C, 32 °C/27 °C, 36 °C/31 °C, and 40 °C/35 °C (day/night). Concentrations of nutrient-uptake and -assimilation proteins in roots were determined using protein-specific antibodies and ELISA (enzyme-linked immunosorbent assay). In general, (1) roots were affected by heat more than shoots, as indicated by decreased root:shoot mass ratio, shoot vs. root %N and C, and the level of nutrient metabolism proteins vs. less sensitive photosynthesis and stomatal conductance; and (2) negative effects on roots were large and slow-to-recover only with severe heat stress (40 °C–42 °C). Thus, short-term heat stress, if severe, can decrease total protein concentration and levels of nutrient-uptake and -assimilation proteins in roots. Hence, increases in heat stress with global warming may decrease crop production, as well as nutritional quality, partly via effects on root nutrient relations.

Keywords: heat stress; high temperature; nutrients; nutrient-uptake proteins; *Solanum lycopersicum*; tomato

1. INTRODUCTION

High-temperature stress, both from chronic or abrupt heating, is often a limiting factor for plant growth, development, and reproduction [1,2,3]. Both chronic and abrupt heat stress are expected to increase as a consequence of anthropogenically-driven global warming [4]. For example, in the mid-continental United States and Europe, the frequency, severity, and duration of heat waves is expected to increase in the future [5]. In many cases, it is expected that increases in extreme high-temperature events (e.g., heat waves) will affect plants more negatively than increases in average temperatures [6]. Heat stress decreases plant function in many ways, with negative effects on growth, photosynthesis, respiration, reproduction, water relations, and hormone production being especially well-studied [3,7,8,9,10].

Both shoots and roots are sensitive to heat-related damage, and roots are often as sensitive, or more, as shoots to a heat stress [7,11]. Further, roots are often subjected to high and potentially-stressful temperatures; e.g., when canopies are not closed and soil receives direct sunlight [12,13,14] or in cool-season species during hot summer months [15]. The tolerance of roots to heat stress scales with the mean temperatures of the habitats to which the species are adapted, such that optimal temperatures for root growth are lower in cool-season species, higher in warm-season species, and still higher in warm-desert plants [11]. For example, soil temperatures can exceed 30 °C in the top 5 cm of the soil profile in cool-season wheat fields in cool-temperate locations [15], 33 °C to 10 cm depth in oat fields in Texas [14], 35 °C–40 °C to 10 cm depth in a sub-tropical maize field [16], and 70 °C to 40 °C from the surface to 15 cm depth in deserts dominated by succulents [17]; in each of these cases, these soil temperatures exceed optimal temperatures for root growth for the respective species.

High-temperature stress reduces root growth, number, and mass [7], which affects the growth of aboveground tissue by restricting the supply of water and mineral nutrients, affecting production of hormones synthesized in roots and transported to shoots, and altering sink-source relationships between shoots and roots [7,8,9,18]. Relative to shoots, less research has examined the effects of heat stress on roots, and most of this past research had focused on root growth and carbon relations (especially respiration) [7]. Relatively little past research has investigated how heat stress affects plant nutrient relations [7,8,11,19], and most of this previous work has measured only heat effects on nutrient content or concentration. In addition, most of the past research on root heat stress has focused on chronic heat stress, but the responses of roots to chronic warming can differ from abrupt heat stress [11]; hence, the effects of abrupt heat stress on root nutrient-uptake is especially poorly understood.

Based on the few past studies, it is known that heat stress often decreases the concentration of nutrients in plant tissues or decreases the total content of nutrients in the plants, though effects can vary among nutrients and species [11]. Heat stress can also disrupt enzymes involved in nutrient metabolism (e.g., nitrate and ammonium assimilation) [20,21]. Decreases in nutrient acquisition with heat stress could potentially be caused by several factors, including a decrease in root mass or surface area and/or a decrease in nutrient uptake per unit root [19,22]. Decreases in nutrient uptake per unit root might be caused by depletion of labile C (total non-structural carbohydrate), and hence energy, in roots (e.g., due to a decrease in transport of shoot C to roots or an increase in root respiration) or by direct heat damage to roots [7], which might decrease the production or function of nutrient-uptake proteins. For most mineral nutrients, the bulk of their uptake is mediated by the activity of specific nutrient-uptake proteins, and uptake protein activity depends on both the concentration of uptake proteins per unit root, as well as the rate at which each protein works. However, almost nothing is known about the effect of heat stress on nutrient-uptake proteins in roots.

To investigate the effects of heat stress on nutrient uptake and nutrient-uptake proteins, we determined the effects of moderate to severe short-term heat stress in roots of tomato (*Solanum lycopersicum* L. cultivar Bigboy) on the concentration of the following key nutrient-uptake proteins: NRT1 and NRT2, the main low- and high-affinity nitrate transporters [23]; AMT1, the primary ammonium transporter [23]; PHT1, the primary root phosphorus transporter [24]; KT1, the main potassium transporter [25], FRO1, iron reductase, one of the two main Fe-uptake proteins in dicots [26]; and BOR1 and NIP5;1, the two main B transporters [27]. In addition, we also determined the effects of heat on the levels of key N assimilation enzymes: nitrate reductase (NR), glutamate dehydrogenase (GDH), glutamine synthetase (GS), and glutamate oxoglutarate amino transferase (GOGAT) [28]. Finally, we also measured the effects of heat on shoot and root growth, photosynthesis and stomatal conductance, and root total protein content, to determine how heat effects on nutrient uptake and nutrient-metabolism proteins relate to effects on root vs. shoot growth and function. Tomato was used as a model system because it is a warm-season moderately-thermotolerant species [29,30], originates from warm sub-tropical habitats that experience variable temperatures and abrupt heat stress [31], and has been used as a model in many heat stress studies [11]. In addition, given that the majority of tomato roots are in the top 10–20 cm of soil, soil temperatures can frequently exceed optimal levels in field-grown tomato (e.g., exceed 34 °C–35 °C at 10–15 cm, with optimal ca. 25 °C–27 °C) [32,33,34].

2. RESULTS

2.1. Main Experiment

Relative to control plants of the same age, both root and shoot biomass were decreased by the severe heat treatment, and root growth was more sensitive than shoot growth under severe heat stress, as indicated by the decline in root:shoot mass ratio at 42 °C (Figure 1). The biomass of moderately-stressed plants was only slightly (and non-significantly) decreased with six days of heating, and the biomass of these plants was slightly (and non-significantly) above unheated controls after the seven days recovery period under control conditions (i.e., by day 13). In contrast, relative to control plants of the same age, severely-stressed plants did not recover from heat within seven days, and their total biomass was 73% less than controls on day 13.

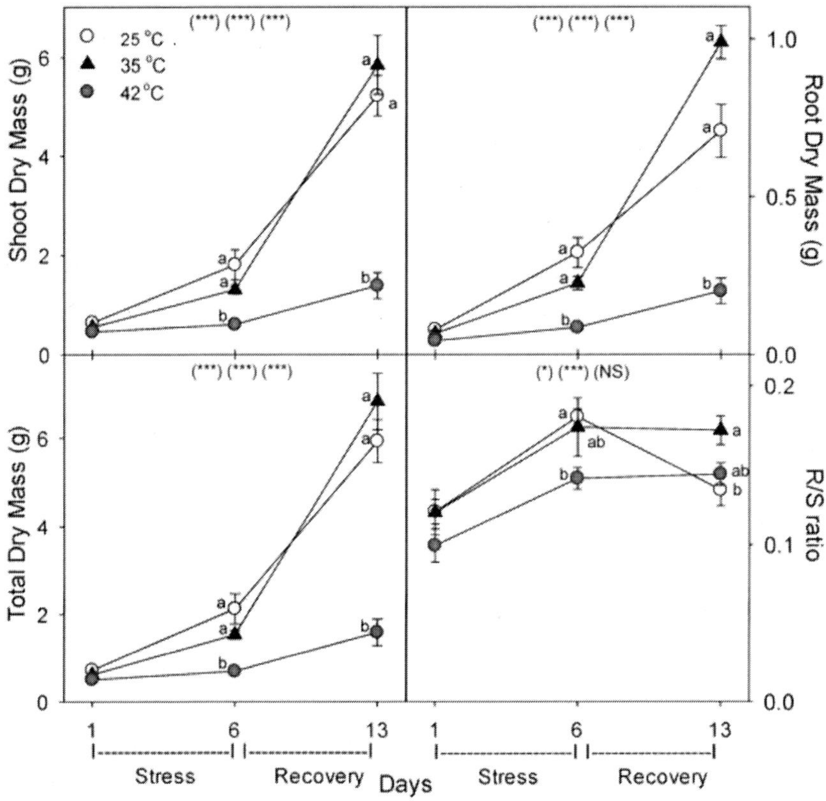

Figure 1. Effect of abrupt heat stress on biomass of tomato. Plants were grown at 25/20 °C day/night (control), and then subsets of plants were heat stressed at 35/30 or 42/37 °C day/night for one or six days (maximum root temperatures = 32 or 39 °C), and then returned to control conditions for seven days of recovery (=day 13). Values are means ± 1 SE for four independent replicates from each harvest (days 1, 6, 13). Within each variable, the significance of main treatment effects (temperature, day, temperature × day) is indicated in parentheses (* $p < 0.05$, ** $p < 0.01$, *** $p < 0.001$, NS = not significant). Different lowercase letters indicate significant difference among temperature treatments within each day.

Net photosynthesis (P_n) increased with moderate heat (35 °C), relative to control plants of the same age, but decreased with severe heat (day 6 only, with full recovery by day 13) (Figure 2). Severely-stressed plants opened their stomates more vs. controls (i.e., increased stomatal conductance, G_s), on day 1 of heat stress, which is common [29] and can

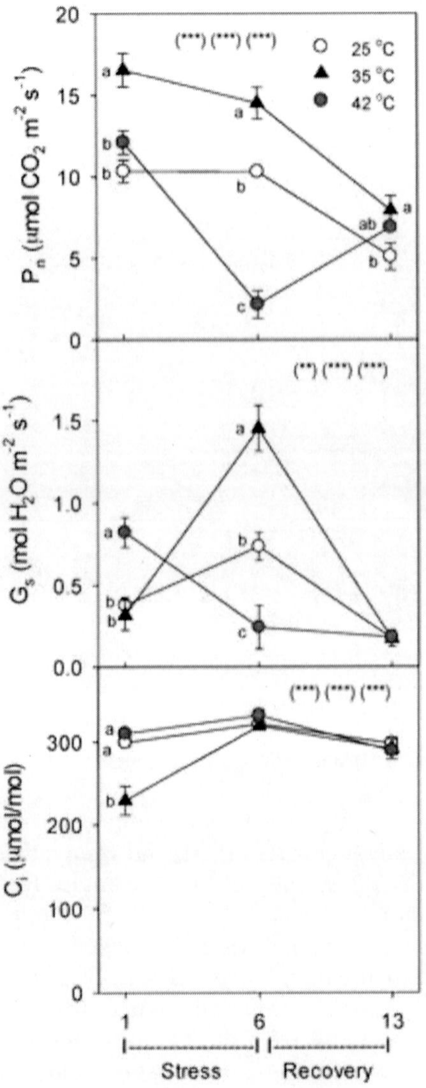

Figure 2. Effect of abrupt heat stress on net photosynthesis (P_n), stomatal conductance (G_s), and leaf internal CO_2 concentration (C_i) of tomato. Plants were grown at 25 °C/20 °C day/night (control), and then subsets of plants were heat stressed at 35/30 or 42 °C/37 °C day/night for one or six days (maximum root temperatures = 32 °C or 39 °C), and then returned to control conditions for seven days of recovery (=day 13). Values are means ± 1 SE for four independent replicates from each harvest (days 1, 6, 13). Statistics are as in Figure 1. * $p < 0.05$, ** $p < 0.01$, *** $p < 0.001$.

increase evaporative cooling, but had lower G_s on day 6 vs. controls; in moderately-stressed plants, G_s was increased on day 6, but similar to controls on day 1 (all treatments were similar on day 13). Except for day 1 (where 35 °C < controls or 42 °C), leaf internal carbon dioxide concentration (C_i) was similar among the treatments, indicating that decreases in P_n were not caused by stomatal closure in heated plants, and thus were not caused by water stress (as similar to [30,35]).

Relative to control plants of the same age, root protein content decreased only in plants heated to 42 °C, with full recovery by day 13 (Figure 3). Compared to control plants of the same age, the C concentration of roots was greater following six days of heating (especially at 42 °C), and this increase above control plants persisted after seven days of post-heat recovery, but there was no significant effect of heating on shoot %C, excluding on day 1, where heat decreased %C at 35 °C (Figure 4). In contrast to %C, %N in the roots tended to be less in heat-stressed plants relative to controls during the entire experiment (especially for 42 °C at days 1 and 6), while in shoots, heats effects on %N, compared to age controls, were significant only for 42 °C on days 6 and 13, wherein %N was lower. The uptake rate of macronutrients (NPK) and the micronutrient, Fe, by roots decreased with heat stress, compared to controls of the same age, but only at 42 °C (a non-significant decrease in K and Fe was observed at 35 °C) (Figure 5). Interestingly, B uptake rate did not decrease with heat, but rather increased at 35°C relative to other treatments (days 1–6). Notably, N and P uptake rate was still lower in severely-heat-stressed plants during recovery, whereas K and Fe uptake largely recovered by day 13, compared to age controls.

Relative to control plants of the same age, heat stress (both 35 °C and 42 °C) initially decreased the concentration (per g dry root) of all the nutrient-uptake proteins we measured (NRT1, NRT2, AMT1, KT1, PHT1, FRO1, BOR1, and NIP5;1) (Figure 6, day 1). After six days of heat stress, the levels of nutrient uptake proteins in moderately-stressed plants were similar to unheated controls, but in severely-stressed plants, levels of these proteins remained below controls in all except for NIP5;1 and AMT1. After seven days of post-heat recovery, levels of these nutrient-uptake proteins had recovered to levels observed in control plants for all but FRO1 (with smaller non-significant decreases for NRT1 and KT1). Similar patterns, in heat-stressed plants compared to controls of the same age, were observed in the relative level of the nutrient assimilation proteins examined (NR, GOGAT, GDH, GS), with decreases in their levels following one day of heat stress, recovery of their levels in moderately-stressed, but not in severely-stressed, plants by day 6, and a tendency for near-complete recovery in both heat treatments by day 13 (Figure 7). As with the nutrient-uptake proteins, heat decreased levels of the plasmalemma H^+-ATPase, compared to age controls, but decreased levels of this protein during the entire six days of heat treatment at both 35 °C and 42 °C; H^+-ATPase levels recovered to control-plant levels by the end of the seven days post-heat recovery period (Figure 8).

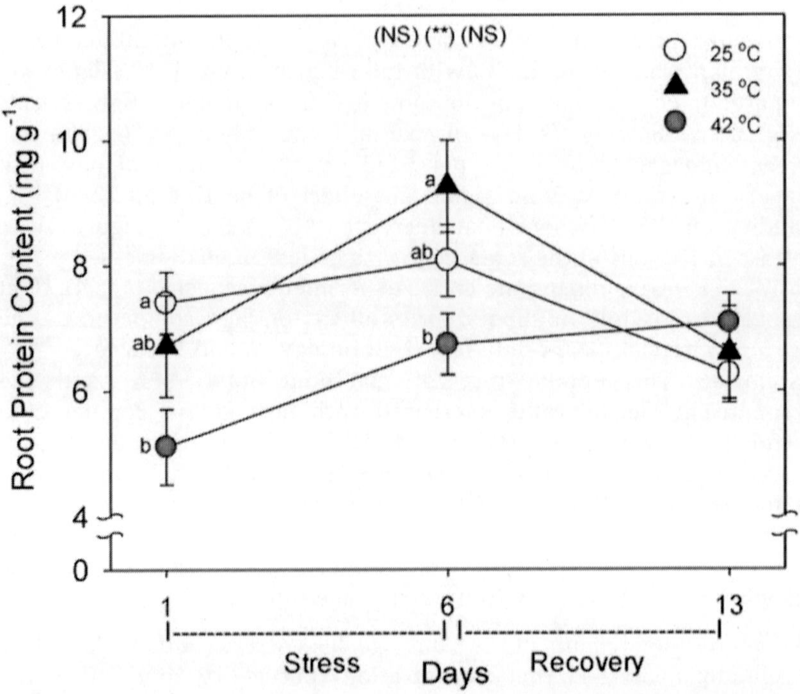

Figure 3. Effect of abrupt heat stress on protein content per gram (fresh weight) of roots. Plants were grown at 25 °C/20 °C day/night (control), and then subsets of plants were heat stressed at 35 °C/30 °C or 42 °C/37 °C day/night for one or six days (maximum root temperatures = 32 °C or 39 °C), and then returned to control conditions for seven days of recovery (=day 13). Values are means ± 1 SE for four independent replicates from each harvest (days 1, 6, 13). Statistics are as in Figure 1. ** $p < 0.01$.

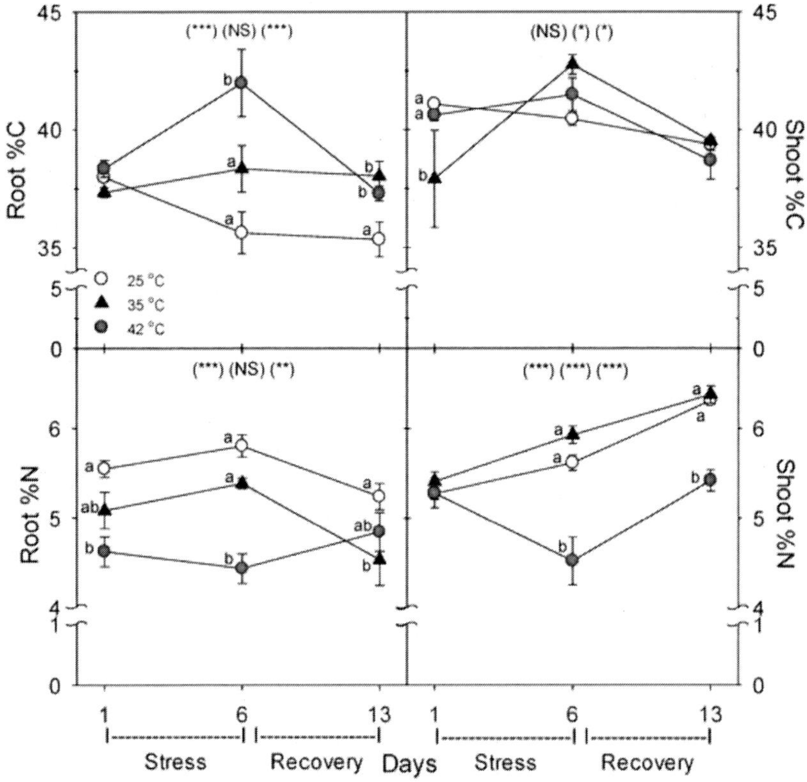

Figure 4. Effect of abrupt heat stress on the concentration of C and N in shoot (= leaves + stems) and roots of tomato (dry mass basis). Plants were grown at 25 °C/20 °C day/night (control), and then subsets of plants were heat stressed at 35 °C/30 °C or 42 °C/37 °C day/night for one or six days (maximum root temperatures = 32 °C or 39 °C), and then returned to control conditions for seven days of recovery (=day 13). Values are means ± 1 SE for four independent replicates from each harvest (days 1, 6, 13). Statistics are as in Figure 1. * $p < 0.05$, ** $p < 0.01$, *** $p < 0.001$.

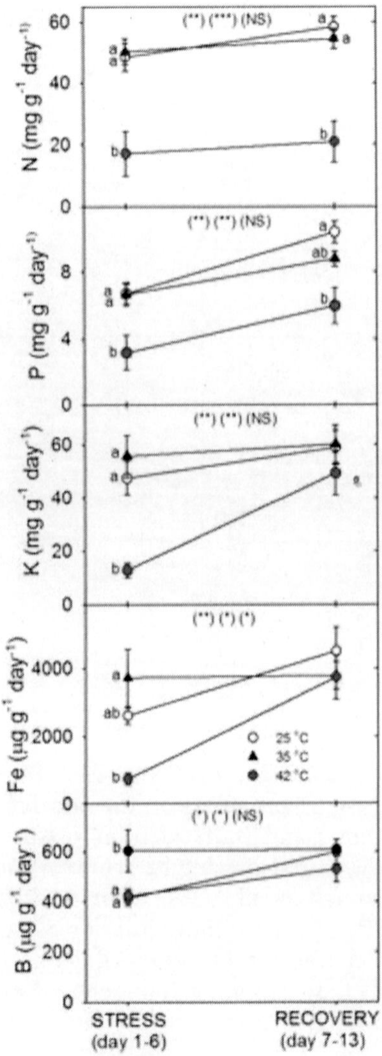

Figure 5. Effect of abrupt heat stress (days 1–6) and recovery (7–13) on the uptake rate of specific nutrients by roots of tomato (total mg for macronutrients and µg for micronutrients in the plant per g dry root per day). Plants were grown at 25 °C/20 °C day/night (control), and then subsets of plants were heat stressed at 35 °C/30 °C or 42 °C/37 °C day/night for one or six days (maximum root temperatures = 32 °C or 39 °C), and then returned to control conditions for seven days of recovery (=day 13). Values are means ± 1 SE for four independent replicates from each harvest (days 1, 6, 13). Statistics are as in Figure 1. * $p < 0.05$, ** $p < 0.01$, *** $p < 0.001$.

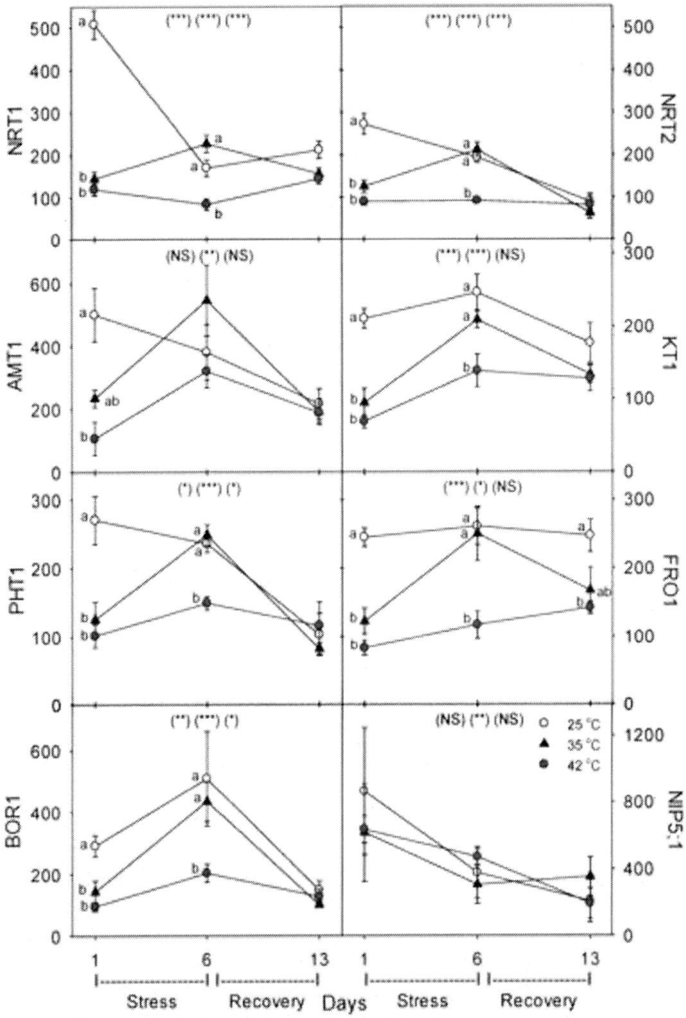

Figure 6. Effect of abrupt heat stress on relative levels of specific nutrient transport proteins (per g dry root) in roots of tomato: low-affinity NO_3, NRT1; high-affinity NO_3, NRT2; NH_4, AMT1; K, KT1; P, PHT1; Fe, FRO1; B, BOR1 and NIP5;1. Plants were grown at 25 °C/20 °C day/night (control), and then subsets of plants were heat stressed at 35 °C/30 °C or 42 °C/37 °C day/night for one or six days (maximum root temperatures = 32 °C or 39 °C), and then returned to control conditions for seven days of recovery (=day 13). Values are means ± 1 SE for four independent replicates from each harvest (days 1, 6, 13). Statistics are as in Figure 1. * $p < 0.05$, ** $p < 0.01$, *** $p < 0.001$.

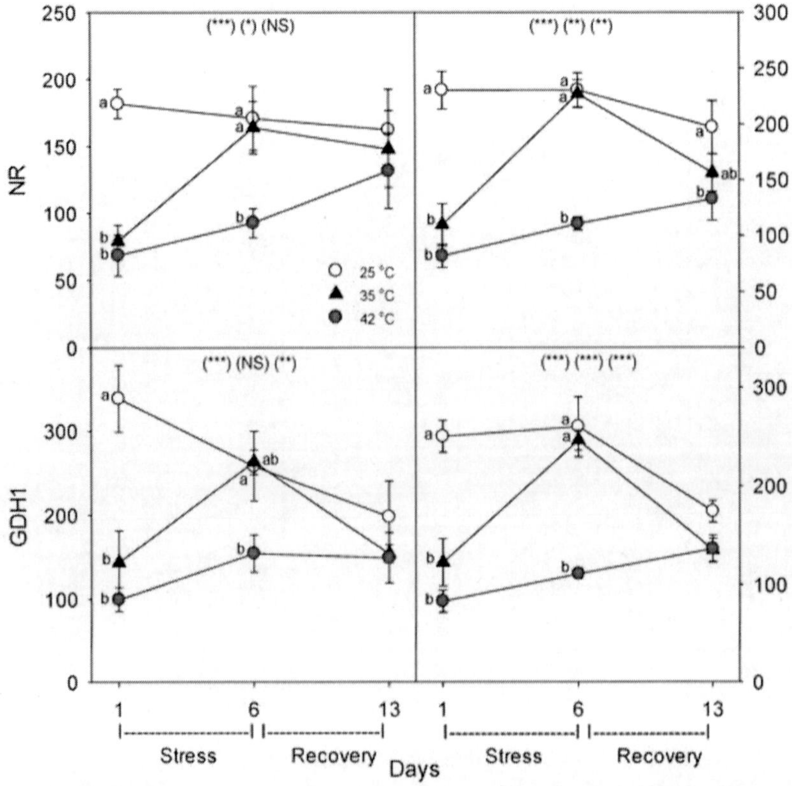

Figure 7. Effect of abrupt heat stress on relative levels of specific nutrient metabolism proteins (per g dry root) in roots of tomato: NR, nitrate reductase; GOGAT, glutamine oxoglutarate amino transferase; GDH1, glutamate dehydrogenase; GS, glutamine synthetase. Plants were grown at 25 °C/20 °C day/night (control), and then subsets of plants were heat stressed at 35 °C/30 °C or 42 °C/37 °C day/night for one or six days (maximum root temperatures = 32 °C or 39 °C), and then returned to control conditions for seven days of recovery (=day 13). Values are means ± 1 SE for four independent replicates from each harvest (days 1, 6, 13). Statistics are as in Figure 1. * $p < 0.05$, ** $p < 0.01$, *** $p < 0.001$.

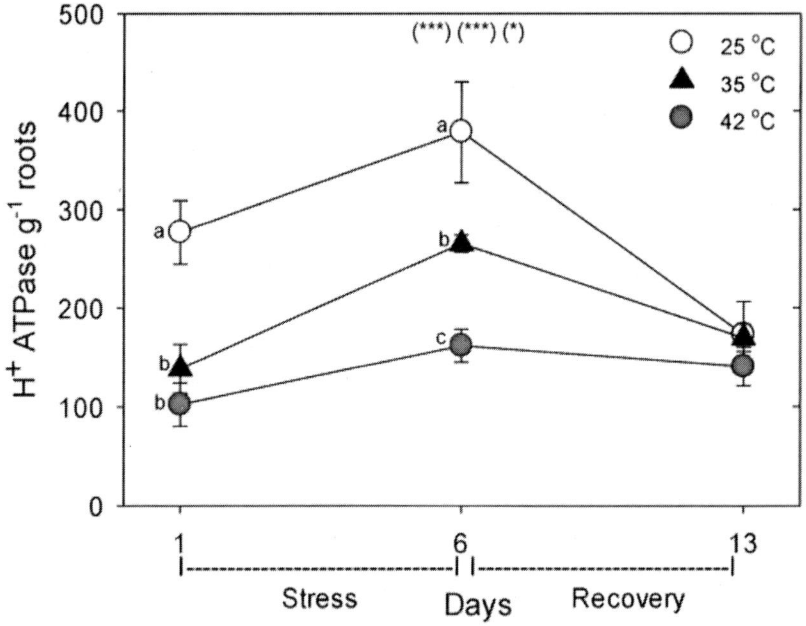

Figure 8. Effect of abrupt heat stress on relative levels of the plasmalemma H^+-ATPases (per g dry root) in roots of tomato. Plants were grown at 25 °C/20 °C day/night (control), and then subsets of plants were heat stressed at 35 °C/30 °C or 42 °C/37 °C day/night for one or six days (maximum root temperatures = 32 °C or 39 °C), and then returned to control conditions for seven days of recovery (=day 13). Values are means ± 1 SE for four independent replicates from each harvest (days 1, 6, 13). Statistics are as in Figure 1. * $p < 0.05$, ** $p < 0.01$, *** $p < 0.001$.

2.2. Second Confirmatory Experiment

As in the first main experiment (six days at 35 °C or 42 °C), root mass and root:shoot mass ratio decreased at high temperatures (15 days at 36 °C and 40 °C vs. 28 °C and 32 °C) (Figure 8). Also, as in the first experiment, the concentration of NRT1 and NRT2 decreased only at very high temperature (40 °C) (Figure 9).

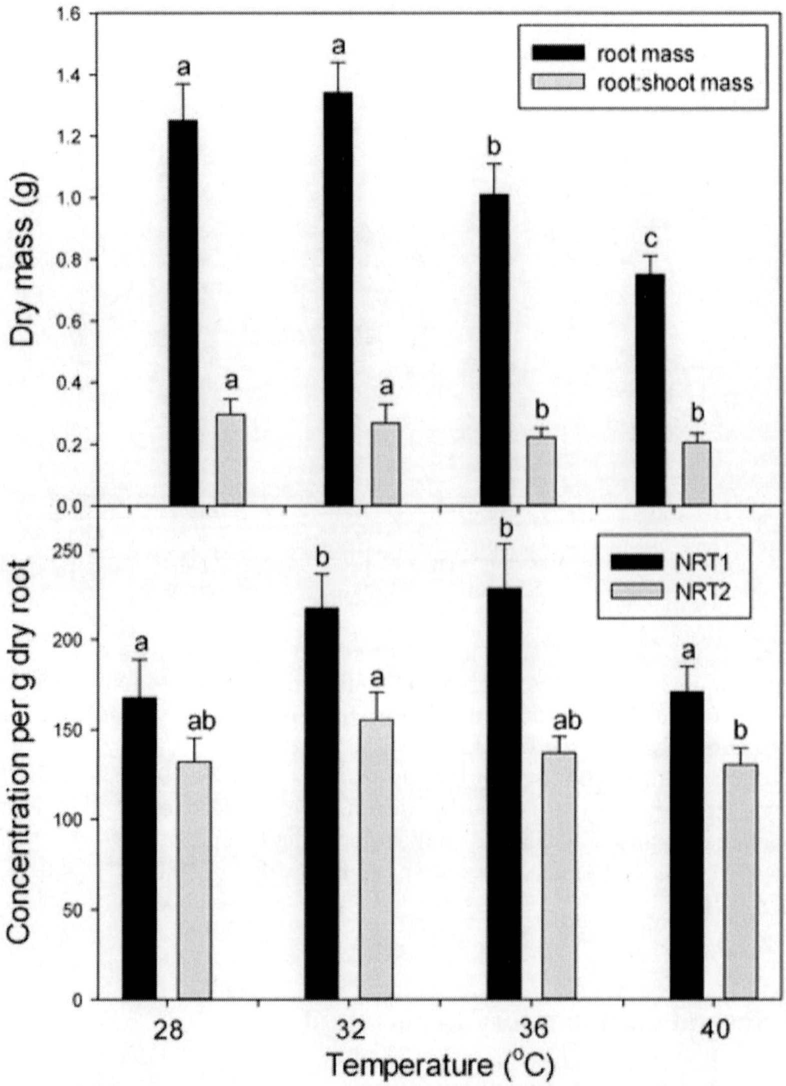

Figure 9. Effect of chronic heat stress on (top panel) plant growth and (bottom panel) relative levels of the nitrate-uptake proteins, NRT1 and NRT2 (per g dry root), in roots of tomato. Plants were grown at 28 °C/23 °C, 32 °C/27 °C, 36 °C/31 °C, or 40 °C/35 °C (day/night) for 15 days. Values are means ± 1 SE for four independent replicates. Different lowercase letters indicate significant difference among temperature treatments within each response variable.

3. DISCUSSION

With increases in temperatures due to global warming, plants are likely to experience increasingly frequent, hotter, and longer episodes of abrupt heat stress (e.g., heat waves) in the future, and this will negatively impact plant function. Based on the limited past studies, we know that heat stress can negatively affect plant nutrient relations [11], but the effect of heat stress, chronic or abrupt, on root nutrient uptake rate has been little studied, excluding a few studies mostly using root pieces in vitro or heating only soil in intact plants, and we are aware of no previous research on effects of heat stress on nutrient-uptake proteins [7,19]. Our results show that in tomato, heat stress decreased total N content per plant [not shown, but = tissue N concentration (%N) × biomass (g), which both decreased, relative to control plants of the same age], by decreasing both plant growth (root growth more than shoot growth) and decreasing uptake rate of nutrients per g of root; similar negative effects of heat on total plant P, K, Fe, and B content were also observed [36].

Decreases in root growth and plant nutrient-uptake rate were likely caused by damage, rather than by lower available C for metabolism, given that heat stress increased root membrane damage [36] and %C, while decreasing root growth and protein concentration. The uptake rate of nutrients by roots decreased for most nutrients during severe, but not moderate heat stress, which was correlated with relative decreases in the concentration of nutrient-uptake proteins. It is also possible that decreases in the passive uptake and root-to-shoot transport of nutrients by transpiration-driven mass flow occurred during heat stress, and this contributed to decreases in plant nutrient uptake during heat stress. However, decreases in stomatal conductance were evident only at 42 °C on day 6, and though stomatal conductance was decreased then, transpiration was not lower due to increases in leaf-to-air vapor pressure differences. Further, results for B, which should be affected by mass flow more than the other mineral nutrients, given that it is smallest and uncharged, are not consistent with a heat-related decrease in passive nutrient uptake.

As in other previous studies [11], in roots and shoots subjected to the same high temperatures, roots were more sensitive to heat stress than shoots. In this study, both root and shoot growth decreased with heat stress, but the effect was larger for roots, as indicated by a decrease in root-to-shoot ratio (especially for severely-stressed plants). Also, photosynthesis (P_n) was decreased by heat only on day 6 at 42 °C, while root membrane damage increased at both 35 °C and 42 °C [36]. Similarly, there was little effect of heat on shoot %C, but root %C increased with heat stress, and shoot %N decreased only at 42 °C, but decreased at both 35 °C and 42 °C in roots. The decrease in %N in shoots and roots with severe heat stress indicate that total plant N uptake was more impacted by heat than total plant growth. These results also indicate that heat likely did not impair C translocation from shoots to roots, otherwise, root %C would not increase, which is consistent with results of a study on abrupt heat stress in a heat-tolerant grass [35]. In addition, root protein content decreased with heat stress, which would result from an increase in protein degradation and/or decrease in protein synthesis, and either cause would constitute heat damage and both are known to occur during heat stress in roots [7].

Consistent with the effects of moderate vs. severe heat stress on total plant N content and %N, moderate heat stress did not affect the rate of nutrient uptake per g of root, but severe heat stress did decrease the uptake rate of four of five nutrients examined (N, P, K, Fe vs. B). Heat stress also decreased N-uptake rate per g root in the warm-season C_4 grass, *Andropogon gerardii*, as measured by both sequential harvesting and ^{15}N labeling [35]. Heat-related decreases in nutrient-uptake rates by roots can be caused by decreases in the concentration of nutrient-uptake proteins and/or by decreases in the activity (or transport or reaction rate) of individual uptake proteins. In this study, most of the eight nutrient-uptake proteins examined showed similar responses to heat stress; e.g., levels of the proteins per g root decreased compared to controls after 24 h of heat treatment at 35 °C and 42 °C (excluding NIP5;1), but after six days of heat treatment, only plants heated at 42 °C exhibited decreased levels of uptake proteins (excluding NIP5;1 and AMT1), and after seven days of post-heat recovery, levels of most uptake proteins had recovered to control levels (excluding FRO1, and non-significantly, KT1 and NRT1). Perhaps the heat tolerance of NIP5;1 vs. BOR1 and the other proteins is related to the fact that NIP5;1 is a channel protein, while BOR1 is not, and B is typically an uncharged molecule at physiological pH, but the other nutrients examined here are charged [37]. The effects of heat stress on the levels of the four N-assimilatory proteins examined were very similar to effects on nutrient-uptake proteins, though levels of all four proteins tended to remain slightly lower than controls after seven days of post-heat recovery. For the plasmalemma H^+-ATPase, levels decreased during moderate and severe heat stress, but recovered within seven days. The dramatic difference we observed following one vs. six days of heat stress on the concentration of nutrient-uptake proteins (and nutrient-metabolism proteins, which showed the same pattern) indicates a need to examine the effects of heat stress at more time-points during heat treatments, to elucidate the detailed kinetics and mechanisms of plant responses to heat stress.

While this is the first study, to our knowledge, to examine effects of temperature on the levels of nutrient-uptake proteins in plants, other studies have examined effects of heat on the activity of nutrient-uptake or -assimilation proteins. For example, the uptake rate of P and K in detached corn root pieces during short-term incubations (thus reflecting heat effects on function, rather than concentration, of uptake proteins) increased up to 32 °C and 37 °C, with only small decreases at 37 °C and 42 °C, for K and P uptake, respectively [22]. Chopra [38] found that optimal activity of nitrate reductase in leaves of eight different crop species occurred at >45 °C during short-term incubations. In contrast, Hungria and Kaschuk [21] observed that chronic heat stress (18–38 days at 28 °C vs. 34 °C or 39 °C daytime) decreased the activities of nitrogenase, nitrate reductase, GS, and (especially) GOGAT, in leaves or nodules of the legume, *Phaseolus vulgaris*; heat also decreased the transport of N in xylem. Similarly, chronic heating (10–20 days at 20 °C vs. 35 °C daytime) decreased GS and GOGAT activities in fescue leaves (*Festuca arundinacea*) [39]. These limited results suggest that, unless temperatures reach extremely-high levels, heat stress likely decreases the activities of N metabolism proteins by decreasing their concentration in plant tissues, rather than inhibiting the function of individual N-metabolism proteins, consistent with results from our study.

4. MATERIALS AND METHODS

4.1. Plant Growth Conditions, Temperature Treatments, and Harvesting

Seeds were planted in foam cubes and germinated in a controlled-environment chamber at 400–500 $\mu mol \cdot m^{-2} \cdot s^{-1}$ PAR (photosynthetically-active radiation), 14-h photoperiod, and 24 °C. Plants were watered daily and provided with starter nutrient solution (3 mM N, 1 mM Ca, 1 mM K, 0.5 mM P, 0.5 mM Mg, using $Ca(NO_3)_2$, $K(NO_3)$, KH_2PO_4, $MgSO_4$; and 40 μM Fe, 6 μM Mn, 6 μM Zn, 40 μM B, 4 μM Cu, 0.1 μM Mo, using FeDTPA, $MnCl_2$, $ZnSO_4$, $CuSO_4$, H_3BO_3, Na_2MoO_4).

After producing three to four post-embryonic leaves, plants were transferred to opaque aerated 4 L tubs with lids (one plant per tub) and grown hydroponically in complete nutrient solution [6.2 mM N, 2 mM Ca, 2 mM K, 2 mM Mg, 1 mM P, using NH_4NO_3, $Ca(NO_3)_2$, $K(NO_3)$, KH_2PO_4, $MgSO_4$; and 71 μM Fe, 10 μM Mn, 10 μM Cl, 6 μM Zn, 6 μM Cu, 50 μM B, 0.1 μM Mo, using Fe DTPA, $MnCl_2$, $ZnSO_4$, $CuSO_4$, H_3BO_3, Na_2MoO_4]. Solution pH was monitored daily and maintained at pH 5.6 with addition of 1 N KOH, and solution temperatures were monitored using a thermometer and were typically 3 °C lower than air temperature. Nutrient solution was changed every three to five days to maintain nutrient levels. Plants were moved daily to minimize positions effects inside the chamber. Chamber light, temperature, and CO_2 levels were monitored continuously to ensure that chambers stayed at targeted conditions. We grew plants in hydroponics in order to minimize water stress during heat stress, and thus be able to ascribe treatment responses to heat alone, as well as to avoid confounding plant responses with rhizosphere influences, as would occur in soil.

In the main experiment, plants were grown under the above conditions for five days to allow for post-transfer acclimatization. Then plants ($n = 4$ per treatment combination, per harvest) were randomly assigned one of three temperature treatments: control = 20 °C/25 °C night/day, moderate heating = 30 °C/35 °C night/day, and severe heating = 37 °C/42 °C night/day (light and photoperiod as above). After six days of heat treatment, all plants were then moved to a single chamber and grown under control conditions for seven days to follow post-heating recovery. A random subset ($n = 4$) of plants was harvested after 24 h (1 day) and six days of heat stress, and after seven days of post-heating recovery. Day and night temperatures were both raised the same number of degrees during heat treatments, since climate-change projections indicate that global warming will affect both days and nights (and, in fact, will increase night, more than day, temperatures on average [4]). Our moderate heat treatment was intended to mimic a severe heat-wave under today's climate, while the severe heat treatment was intended to mimic a severe heat-wave in the future, following significant global warming. In a second confirmatory experiment, plants were germinated and grown as above, but at the following temperatures and for 15 days before harvest: 28 °C/23 °C, 32/27, 36 °C/31 °C, and 40 °C/35 °C (day/night). At harvest, plants were separated into leaves, stems, and roots (roots after washing with DI water). Biomass was determined after oven drying at 70 °C for at least 48 h. Sub-samples of fresh root tissue for protein analysis were immediately frozen in liquid N_2after harvest and stored at −80 °C.

4.2. Leaf Gas-Exchange

To monitor the effects of heat treatment on both shoot metabolism and plant C assimilation, we measured steady-state net photosynthesis (P_n; net CO_2 exchange) immediately before harvesting, using an infrared gas analyzer (IRGA) (Model 6400, LiCOR, Lincoln, NE, USA) equipped with a 6-cm^2 leaf-area cuvette which controlled environment conditions (CO_2, light, and temperature); this system also simultaneously measures leaf transpiration, stomatal conductance to water vapor (G_s), and internal CO_2 concentration (C_i). Gas exchange was measured on the most-recently-expanded attached leaf on day 1, and this same now-older leaf was measured on day 6; however, on day 13, the next-newest leaf was measured, since the previously-measured leaf was beginning to senesce, and this newer leaf was not yet fully-expanded. Measurements were made on leaves receiving direct light prior to measurement, at 370 ppm CO_2, under saturating light (1500 $\mu mol \cdot m^{-2} \cdot s^{-1}$PAR), and at the same temperature as plants were experiencing in the growth chamber (25 °C, 35 °C, or 42 °C during heating, 25 °C during recovery). Preliminary light-response curves were generated to determine optimal light levels during measurements and ensure that measurement light levels were not photoinhibitory.

4.3. Nutrient Relations

Total protein was extracted from tissue as in [40], by grinding 400 mg of frozen root tissue in liquid N_2 in a mortar and pestle and then in 2 mL of extraction buffer of the following composition: 0.5 M Tris pH-8, 0.1 M potassium chloride, 0.9 M sucrose, 50 mM ethylene diamine tetra-acetic acid, 2% (*v*:*v*) β-mercaptoethanol, 10 μM leupeptin, and 1 mM phenyl methyl sulfonyl fluoride. The homogenate was transferred to a 15-mL tube, to which was added 2 mL volume of phenol, and then the tubes were incubated for 20 min at room temperature and centrifuged at 5600 rcf (relative centrifugal force) for 15 min at 4 °C to separate aqueous and organic phases. The upper phenol phase was recovered, and after addition of an equal volume of extraction buffer, centrifuged as above. Supernatant obtained was stored overnight at −20 °C in five volumes of 0.1 M ammonium acetate, to precipitate protein. Precipitated protein was then rinsed two times with ammonium acetate and three times with 80% acetone, followed by a final rinse with 100% acetone. Protein samples were dried and then re-solubilized in sample buffer (400 μL) containing 100 mM Tris pH 6.8, 0.5% (*v*/*v*) sodium dodecyl sulphate, and 1% (*v*/*v*) glycerol. Total protein concentration in the sample was determined using a colorimetric assay (DC Protein Assay, BioRad, Hercules, CA, USA), using bovine serum albumin as a standard.

The concentration of each macro- and micro- mineral nutrients was determined separately for powdered dry leaves, stems, and roots by the combustion-MS technique for C and N and by ICP-OES (Inductively Coupled Plasma Optical Emission Spectroscopy; model IRIS Intrepid II; Thermo Corp, Waltham, MA, USA) for remaining nutrients as in [41]. Total nutrient content in the entire plant was calculated from the concentration of each nutrient multiplied by the biomass of each tissue, and then the tissues were summed. Then the root specific uptake rate of each nutrient (x) (total g plant $nutrient_x$ per g dry root per day) was calculated from the total amount of $nutrient_x$ taken up during days 1–6 (heat stress) or from days 7–13 (recovery) (uptake during days 1–6 = total plant $nutrient_x$ at day 6 minus total $nutrient_x$ at day 1, uptake for days 7–13 = total $nutrient_x$ at day 13 minus that at day 6).

The relative amount of nutrient-uptake proteins per unit total root protein was determined by quantitative ELISA (enzyme-linked immunosorbent assay), using protein-specific antibodies to oligopeptides of conserved domains of target proteins; conserved domains determined using bioinformatics techniques, as described in [40]. Antiserum specificity was confirmed with immune (western) blotting (immune vs. pre-immune serum, and antigen-purified vs. crude serum, in both tomato and arabidopsis), and pre-immune serum was used during ELISA to subtract signal attributable to non-specific binding. Then, the relative amount of each uptake protein per g root was calculated from total root protein per g root. Using commercially-available antibodies (Agrisera, Vännäs, Sweden) and ELISA, we also determined the relative level per unit root protein of the following key nutrient assimilatory proteins: NR (Nitrate Reductase), GOGAT (Glutamine OxoGlutarate Amino Transferase), GS (Glutamine Synthetase), and GDH1 (Glutamate DeHydrogenase), as well as the plasmalemma H^+-ATPase, which generates the electrochemical gradient used for uptake of most nutrients.

4.4. Statistical Analysis

Experimental results were analyzed statistically using two-way (temperature x day) analysis-of-variance (ANOVA), with temperature and day as fixed factors (main experiment), or using one-way ANOVA with temperature as a fixed factor (second experiment) (Sigma-plot software, version 12). Following significant ANOVA results, the Holm-Sidak test was used to identify significant differences among the three temperature treatments. Treatment effects were considered significant if $p < 0.05$. Data were transformed (log-transformed for concentration and content data) if they did not meet assumptions of ANOVA (normality, equal variance), though this was rare, and results presented are untransformed means and error bars.

5. CONCLUSIONS

In summary, this study showed that abrupt or short-term heat stress can decrease root (relative to shoot) growth, the concentration of nutrient-uptake and -assimilation proteins, as well as total protein, in roots, and the rate of nutrient uptake by roots. Heat effects on roots and nutrient relations were often long lasting, with incomplete recovery in severely-stressed plants even after seven days of post-heat recovery. The relative effects of moderate vs. severe heat stress on plant nutrient content and concentration were correlated to different degrees with relative effects on root-to-shoot mass, nutrient-uptake rate per g of root, and levels of nutrient-uptake and N-assimilatory proteins. Notably, we did not examine the effects of heat stress on fruit yield or its nutritional quality, and future experiments should include these measures, since in tomato it is the fruit which is harvested for food.

Though heat stress was not accompanied by water stress in this study (as indicated by C_i results), in natural settings, heat stress is often accompanied by water stress caused by drought, high leaf transpiration, or decreases in xylem hydraulic conductivity, and water stress may also contribute to decreases in nutrient uptake during heat stress (e.g., by decreasing water uptake or transport from roots to shoots) [12]. So, increases in abrupt or short-term heat stress with global warming in the future will likely have overall negative effects on plant nutrient relations that will become more severe as temperatures rise, which will contribute to decreases in both crop productivity, as well as nutritional quality. Efforts to develop crop genotypes which maintain nutrient uptake and assimilation during heat stress will likely need to include a focus on increasing the thermotolerance of both root growth and protein synthesis, including synthesis of nutrient-metabolism proteins.

ACKNOWLEDGMENTS

This work was supported by grants from the U.S. Department of Agriculture (SCA 58-3607-4-026) and the Ohio Plant Biotechnology Consortium (N-123169-01) to Scott A Heckathorn. We thank Jonathan Frantz, John Gray, and Jennifer Boldt for helpful comments on the manuscript, and the U.S. Department of Agriculture for assistance with nutrient analysis.

AUTHOR CONTRIBUTIONS

Anju Giri, Scott Heckathorn, and Sasmita Mishra were involved in the conceptual design of the study. Anju Giri performed the experiments, collected and analyzed data, generated figures, and wrote the draft manuscript. Scott Heckathorn was the main faculty advisor of Anju Giri, and was involved in data interpretation and manuscript revision; S.M. was a secondary faculty advisor to Anju Giri. Sasmita Mishra and Charles Krause provided valuable feedback on the manuscript. Charles Krause also provided resources for nutrient analysis.

CONFLICTS OF INTEREST

The authors declare no conflicts of interest.

REFERENCES

1. Boyer, J.S. Plant productivity and environment. Science 1982, 218, 443–448.
2. Williams, A.P.; Allen, C.D.; Macalady, A.K.; Griffin, D.; Woodhouse, C.A.; Meko, D.M.; Swetnam, T.W.; Rauscher, S.A.; Seager, R.; Grissino-Mayer, H.D. Temperature as a potent driver of regional forest drought stress and tree mortality. Nat. Clim. Chang. 2012, 2, 1–6.
3. Zinn, K.E.; Tunc-Ozdemir, M.; Harper, J.F. Temperature stress and plant sexual reproduction: Uncovering the weakest links. J. Exp. Bot. 2010, 61, 1959–1968.
4. Intergovernmental Panel on Climate Change (IPCC). Intergovernmental Panel on Climate Change: 5th Assessment Report. Available online: http://www.ipcc.ch/report/ar5/index.shtml (accessed on 27 September 2013).
5. Meehl, G.A.; Tebaldi, C. More intense, more frequent, and longer lasting heat waves in the 21st century. Science 2004, 305, 994–997.
6. Gutschick, V.P.; BassiriRad, H. Biological extreme events: A research framework. Eos Trans. Am. Geophys. Union 2010, 91, 85–86.
7. Huang, B.; Rachmilevitch, S.; Xu, J. Root carbon and protein metabolism associated with heat tolerance. J. Exp. Bot.2012, 63, 3455–3465.
8. Rennenberg, H.; Loreto, F.; Polle, A.; Brilli, F.; Fares, S.; Beniwal, R.; Gessler, A. Physiological responses of forest trees to heat and drought. Plant Biol. 2006, 8, 556–571.
9. Wahid, A.; Gelani, S.; Ashraf, M.; Foolad, M. Heat tolerance in plants: An overview. Environ. Exp. Bot. 2007, 61, 199–223.
10. Wang, D.; Heckathorn, S.A.; Wang, X.; Philpott, S.M. A meta-analysis of plant physiological and growth responses to temperature and elevated CO2. Oecologia 2012, 169, 1–13.
11. Heckathorn, S.A.; Giri, A.; Mishra, S.; Bista, D. Heat stress and roots. In Climate Change and Plant Abiotic Stress Tolerance; Tuteja, N., Gill, S., Eds.; Wiley Blackwell: Weinheim, Germany, 2014; pp. 109–136.
12. Larcher, W. Physiological Plant Ecology; Springer: Berlin, Germany, 1995.
13. Zheng, D.; Hunt, E.R., Jr.; Running, S.W. A daily soil temperature model based on air temperature and precipitation for continental applications. Clim. Res. 1993, 2, 183–191.
14. Zibilske, L.M.; Makus, D.J. Black oat cover crop management effects on soil temperature and biological properties on a Mollisol in Texas, USA. Geoderma 2009, 149, 379–385.
15. Wang, H.; Lemke, R.; Goddard, T.; Sprout, C. Tillage and root heat stress in wheat in central Alberta. Can. J. Soil Sci.2007, 87, 3–10.

16. Dalmago, G.; Bergamaschi, H.; Comiran, F.; Bianchi, C.; Bergonci, J.; Heckler, B. Soil temperature in maize crops as function of soil tillage systems. In Proceedings of the International Soil Conservation Organisation Conference, Brisbane, Australia, 4–8 July 2004; p. 777.
17. Jordan, P.W.; Nobel, P.S. Thermal and water relations of roots of desert succulents. Ann. Bot. 1984, 54, 705–717.
18. Hao, H.P.; Jiang, C.D.; Zhang, S.R.; Tang, Y.D.; Shi, L. Enhanced thermal-tolerance of photosystem II by elevating root zone temperature in Prunus mira Koehne seedlings. Plant Soil 2012, 353, 367–378.
19. Bassirirad, H. Kinetics of nutrient uptake by roots: Responses to global change. New Phytol. 2000, 147, 155–169.
20. Klimenko, S.; Peshkova, A.; Dorofeev, N. Nitrate reductase activity during heat shock in winter wheat. J. Stress Physiol. Biochem. 2006, 2, 50–55.
21. Hungria, M.; Kaschuk, G. Regulation of N2 fixation and NO3−/NH4+ assimilation in nodulated and N-fertilized Phaseolus vulgaris L. exposed to high temperature stress. Environ. Exp. Bot. 2014, 98, 32–39.
22. Bravo-F, P.; Uribe, E.G. Temperature dependence of the concentration kinetics of absorption of phosphate and potassium in corn roots. Plant Physiol. 1981, 67, 815–819.
23. Nacry, P.; Bouguyon, E.; Gojon, A. Nitrogen acquisition by roots: Physiological and developmental mechanisms ensuring plant adaptation to a fluctuating resource. Plant Soil 2013, 370, 1–29.
24. Nussaume, L.; Kanno, S.; Javot, H.; Marin, E.; Pochon, N.; Ayadi, A.; Nakanishi, T.M.; Thibaud, M.C. Phosphate import in plants: Focus on the PHT1 transporters. Front. Plant Sci. 2011, 2, 1–12.
25. Gierth, M.; Mäser, P. Potassium transporters in plants–involvement in K+ acquisition, redistribution and homeostasis. FEBS Lett. 2007, 581, 2348–2356.
26. Kobayashi, T.; Nishizawa, N.K. Iron uptake, translocation, and regulation in higher plants. Annu. Rev. Plant Biol.2012, 63, 131–152.
27. Miwa, K.; Fujiwara, T. Boron transport in plants: Co-ordinated regulation of transporters. Ann. Bot. 2010, 105, 1103–1108.
28. Lam, H.M.; Coschigano, K.; Oliveira, I.; Melo-Oliveira, R.; Coruzzi, G. The molecular-genetics of nitrogen assimilation into amino acids in higher plants. Annu. Rev. Plant Biol. 1996, 47, 569–593.
29. Preczewski, P.; Heckathorn, S.; Downs, C.; Coleman, J. Photosynthetic thermotolerance is quantitatively and positively correlated with production of specific heat-shock proteins among nine genotypes of Lycopersicon (tomato). Photosynthetica 2000, 38, 127–134.
30. Wang, D.; Heckathorn, S.A.; Barua, D.; Joshi, P.; Hamilton, E.W.; LaCroix, J.J. Effects of elevated CO2 on the tolerance of photosynthesis to acute heat stress in C3, C4, and CAM species. Am. J. Bot. 2008, 95, 165–176.
31. Moyle, L.C. Ecological and evolutionary genomics in the wild tomatoes (Solanum sect. Lycopersicon). Evolution 2008, 62, 2995–3013.
32. Díaz-Pérez, J.C.; Batal, K.D. Colored plastic film mulches affect tomato growth and yield via changes in root-zone temperature. J. Am. Soc. Hort. Sci. 2002, 127, 127–135.
33. Teasdale, J.R.; Abdul-Baki, A.A. Soil temperature and tomato growth associated with black polyethylene and hairy vetch mulches. J. Am. Soc. Hort. Sci. 1995, 120, 848–853.
34. Tindall, J.A.; Mills, H.A.; Radcliffe, D.E. The effect of root zone temperature on nutrient uptake of tomato. J. Plant Nutr. 1990, 13, 939–956.
35. Mainali, K.P.; Heckathorn, S.A.; Wang, D.; Weintraub, M.N.; Frantz, J.M.; Hamilton, E.W., III. Impact of a short-term heat event on C and N relations in shoots vs. roots of the stress-tolerant C4 grass, Andropogon gerardii. J. Plant Physiol.2014, 171, 977–985.

36. Giri, A. Effect of Acute Heat Stress on Nutrient Uptake by Tomato Plants. Master's Thesis, University of Toledo, Toledo, OH, USA, December 2013.
37. Dordas, C.; Chrispeels, M.J.; Brown, P.H. Permeability and channel-mediated transport of boric acid across membrane vesicles isolated from squash roots. Plant Physiol. 2000, 124, 1349–1362.
38. Chopra, R.K. Effects of temperature on the in vivo assay of nitrate reductase in some C3 and C4 Species. Ann. Bot.1983, 51, 617–620.
39. Cui, L.; Cao, R.; Li, J.; Zhang, L.; Wang, J. High temperature effects on ammonia assimilation in leaves of two Festuca arundinacea cultivars with different heat susceptibility. Plant Growth Regul. 2006, 49, 127–136.
40. Mishra, S.; Heckathorn, S.A.; Frantz, J.M. Elevated CO2 affects plant responses to variation in boron availability. Plant Soil 2012, 350, 117–130.
41. Mishra, S.; Heckathorn, S.; Frantz, J.; Yu, F.; Gray, J. Effects of boron deficiency on geranium grown under different nonphotoinhibitory light levels. J. Am. Soc. Hort. Sci. 2009, 134, 183–193.

CHAPTER 9

Enhancing Crop Resilience To Combined Abiotic and Biotic Stress Through The Dissection of Physiological and Molecular Crosstalk

Christos Kissoudis, Clemens van de Wiel, Richard G. F. Visser and Gerard van der Linden*

Laboratory of Plant Breeding, Wageningen University, Wageningen, Netherlands

ABSTRACT

Plants growing in their natural habitats are often challenged simultaneously by multiple stress factors, both abiotic and biotic. Research has so far been limited to responses to individual stresses, and understanding of adaptation to combinatorial stress is limited, but indicative of non-additive interactions. Omics data analysis and functional characterization of individual genes has revealed a convergence of signaling pathways for abiotic and biotic stress adaptation. Taking into account that most data originate from imposition of individual stress factors, this review summarizes these findings in a physiological context, following the pathogenesis timeline and highlighting potential differential interactions occurring between abiotic and biotic stress signaling across the different cellular compartments and at the whole plant level. Potential effects of abiotic stress on resistance components such as extracellular receptor proteins, R-genes and systemic acquired resistance will be elaborated, as well as crosstalk at the levels of hormone, reactive oxygen species, and redox signaling. Breeding targets and strategies are proposed focusing on either manipulation and deployment of individual common regulators such as transcription factors or pyramiding of non- (negatively) interacting components such as R-genes with abiotic stress resistance genes. We propose that dissection of broad spectrum stress tolerance conferred by priming chemicals may provide an insight on stress cross regulation and additional candidate genes for improving crop performance under combined stress. Validation of the proposed strategies in lab and field experiments is a first step toward the goal of achieving tolerance to combinatorial stress in crops.

Keywords: salinity, drought, disease resistance, R-genes, crosstalk, hormones, transcription factors, post-translational modifications

INTRODUCTION

Plants are sessile and cannot escape stressful conditions originating from the physical environment (abiotic stress) and from interactions with insects and microorganisms such as fungi and bacteria (biotic stress). The on-going change in climate conditions due to mostly anthropogenic causes such as the increase in CO_2 emissions (Peters et al., 2011) exaggerates agricultural land deterioration due to temperature rise. This results in increased evapotranspiration, intensifying drought episodes (Zhao and Running, 2010) and increasing soil salinization, augmenting the 7% of the total and 30% of the irrigated agricultural land already affected by salinity (Munns and Tester, 2008). Available data and projections on the effect of climate change on pathogen spread are not conclusive, although the evidence points to increased reproductive potential and geographic expansion that will lead to interactions with both more hosts and different pathogen strains, increasing the chances for the rise of more virulent strains (Garrett et al., 2006). Therefore, the chances of plants encountering abiotic and/or biotic stress in the future are likely to be higher, with more frequent stress interactions.

Plants have developed a multitude of defense responses that allow them to adapt, survive and reproduce under stress conditions (Pieterse et al., 2009). With the advancement of *~omics* technologies and on-going functional characterizations of individual genes, it has become apparent that environmental adaptation is under tight regulation, which is critical for plant survival (López et al., 2008). Many components of this regulatory network are involved in responses to different stresses but may function antagonistically or some responses are prioritized over others, compromising plant resistance to multiple stresses simultaneously (Glazebrook, 2005; Yasuda et al., 2008).

Major components of the regulatory networks underlying environmental stress adaptation, pathogen recognition, and defense include reactive oxygen species (ROS) signaling (Miller et al., 2008), plant hormones (Bari and Jones, 2009; Peleg and Blumwald, 2011), changes in redox status (Munne-Bosch et al., 2013), and inorganic ion fluxes, such as Ca^{2+} (Martí et al., 2013). Based on *~omics* data analyses these components appear to be at least partly shared between both abiotic and biotic stress signaling, indicating crosstalk and convergence of mechanisms in these pathways and the existence of a general stress response (Walley et al., 2007).

The nature of pathogen perception dictates that physical barriers such as the cuticle, stomata, and cell walls are also critical for timely pathogen recognition and interception (Asselbergh et al., 2007). As data generated by *~omics* analyses derive from a mixture of different cell types and tissues, these spatially important interactions may be missed and these datasets may lead to erroneous conclusions about components shared and their significance in abiotic and biotic stress crosstalk. Moreover, as combinatorial stress potentially results in novel interactions between signaling components, extrapolation of results from studies with single stress conditions should be done with care.

Here we will elaborate on the mechanisms involved in adaptation and tolerance to combinatorial abiotic and biotic stress, with a focus on dehydration/salt stress and fungal and bacterial pathogens interaction. This review will particularly emphasize interactions that potentially arise during the pathogenesis timeline and were as yet given little attention. We will discuss molecular components with potentially critical roles in abiotic and biotic stress tolerance crosstalk, and propose breeding approaches toward effective crop improvement against combinatorial stress.

EVIDENCE OF CROSSTALK

Evidence at the Phenotypic and Physiological Level

Studies on the commonly occurring combination of drought and heat stress have revealed that physiological and molecular responses of plants exposed to both stresses are markedly different from their response to the individual stresses (Rizhsky et al., 2004). Similarly, there are numerous reports about abiotic stress (mostly drought and salinity) affecting pathogen resistance, which is indicative of interaction between abiotic and biotic stress. There are reports of disease resistance attenuation by high humidity and high temperature (Wang et al., 2005, 2009). In most cases, abiotic stress predisposes plants to subsequent pathogen infection (Sanogo, 2004; Triky-Dotan et al., 2005; You et al., 2011), although positive effects on resistance to foliar pathogens have also been reported (Wiese et al., 2004; Achuo et al., 2006).

There is evidence that different levels of abiotic stress have a significantly different impact on disease susceptibility (Soliman and Kostandi, 1998; Desprez-Loustau et al., 2006). Salinity stress, in particular, exerts its damaging effect through both osmotic effects and ion toxicity resulting from ion accumulation (mainly Na^+ and Cl^-). As NaCl is an antifungal agent (Blomberg and Adler, 1993) it could potentially exert a direct toxic effect on fungal growth after accumulation inside the plants (Figure 1). In line with this argument are the many examples of reduction of fungal pathogenicity by metal accumulation (Poschenrieder et al., 2006; Fones et al., 2010), and a similar trend is observed for NaCl accumulation (Soliman and Kostandi, 1998). Therefore salt stress–pathogen interactions may be highly influenced by stress intensity, which affects the degree of accumulation of salt in the plant. The different tolerance strategies of the host against ion toxicity (ion exclusion at the roots and/or ion compartmentalization in the above ground organs inside the vacuoles) can impact on the outcome of plant–pathogen interactions under salt stress. Therefore, it appears that the outcome of the interaction in most occasions is plant, genotype, pathogen, and stress intensity dependent. Moreover abiotic stress, except for potentially dampening or strengthening signaling responses for pathogen defense deployment, could create more or less favorable conditions for pathogen growth by additionally influencing the physiological status of the plant such as water and ion content.

Vice versa, plant responses to abiotic stress can be affected by prior interactions with pathogenic fungi. Pathogen infection has been shown to reduce photosynthesis and water use efficiency (WUE) and induce abnormal stomata opening patterns, and all of these are critical for plant tolerance to abiotic stress (Bilgin et al., 2010; Grimmer et al., 2012). Salicylic acid (SA) signaling, induced after infection with biotrophic fungi, can attenuate abscisic acid (ABA) signaling that is orchestrating plant adaptive responses to abiotic stress (Kim et al., 2011b). Infection by a root pathogen was shown to increase shoot Na^+ and Cl^- content under saline conditions in *Phaseolus vulgaris* (You et al., 2011; Figure 1). Finally genetically heightened resistance to pathogens is often accompanied by aFITNESS cost that may generally affect the plant performance under both abiotic stress and stress-free conditions (Huang et al., 2010; Todesco et al., 2010).

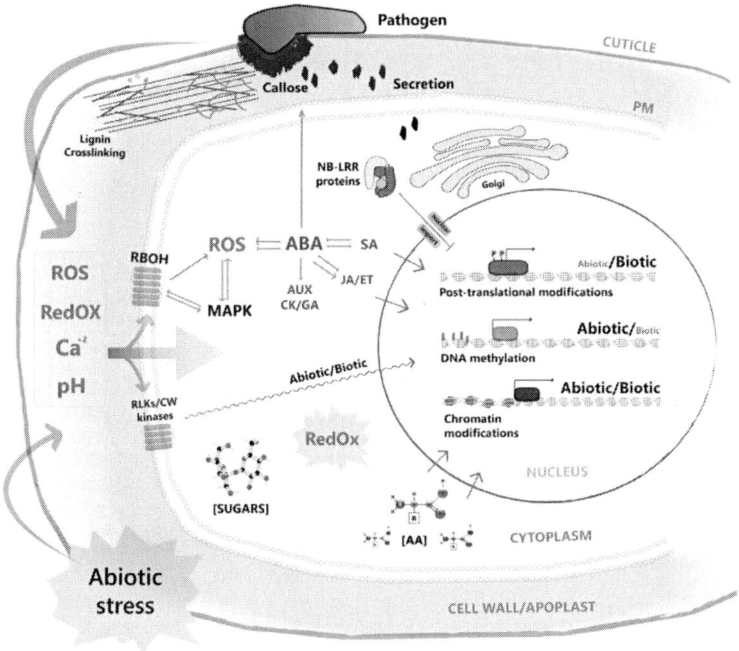

FIGURE 1. A SCHEME FOR THE INTERACTION INTERFACE AND OVERLAPPING SIGNALING PATHWAYS OF ABIOTIC AND BIOTIC STRESS AT THE CELLULAR LEVEL. Both stress factors affect the homeostasis of chemical signals at the apoplastic space such as Ca^{2+}, ROS, and pH levels. Abiotic stress potentially affects the structure and properties of preformed and inducible physical barriers that function against pathogen penetration. Signaling nodes such as RBOHs and RLKs and other cell wall (CW) kinases localized at the plasma membrane, and MAPKs are shared by both stressors, with downstream signal specificity under stress combination remaining elusive. ABA signaling, central for adaptation to abiotic stress, negatively impinges on defense hormone signaling, while, pathogen dependent, positive interactions are observed for JA signaling. ABA–SA interaction is two sided, as activation of SA signaling by pathogen challenge attenuates ABA responses. ABA positively contributes to pre-invasion defense, enhancing callose deposition. Rewiring of secretory machinery under abiotic stress potentially affects its function in the exocytosis of antimicrobial compounds at the site of infection. Nuclear translocation of R-genes is negatively affected under abiotic stress. Redox state, as well as metabolite concentration such as sugars and amino acids (AA), function as drivers for post-translational modifications, modulating the activity of target proteins/transcription factors. Previously/simultaneously encountered stress effect on chromatin and DNA methylation status, potentially impacts on expression patterns of the recipient genes under stress combination. Transcription factor activation and binding to stress responsive gene promoters is a convergence point regulating the signal output under combinatorial stress with diverse outcomes.

A direct interaction of pathogen virulence factors with stress tolerance components of the plant host was demonstrated for the *Pseudomonas syringae* type III effector HopAM1 that targets HSP70 (Jelenska et al., 2010) involved in heat tolerance and stomata closure under stress (Clement et al., 2011). Overexpression of HopAM1 in *Arabidopsis thaliana* results in increased sensitivity to ABA and salt stress, providing proof of direct manipulation of abiotic stress signaling components (Goel et al., 2008).

Interaction of plants with microorganisms can also be beneficial to abiotic tress tolerance. For instance, infection of plants with RNA viruses improved tolerance to drought (Xu et al., 2008). Infection with the vascular pathogen *Verticillium* spp. increased *Arabidopsis thaliana* drought tolerance due to *de novo* xylem formation, which enhances water flow (Reusche et al., 2012). Symbiosis with fungal endophytes (Marquez et al., 2007) as well as association of plant roots with non-pathogenic rhizobacteria and mycorrhizal fungi increases plant vigor under stress conditions through, among others, interactions with hormonal pathways and the sustainment of water and source-sink relations (Dodd and Perez-Alfocea, 2012). Remarkably, rhizobacteria colonization is also shown to enhance plant resistance to fungal pathogens and insects, via systemic signaling that triggers immunity (induced systemic resistance, ISR; Berendsen et al., 2012).

Further evidence for abiotic and biotic stress resistance crosstalk comes from studies of the effects of exogenous application of chemicals that sensitize plant defense responses, a phenomenon called priming (Goellner and Conrath, 2008). For example, application in *Arabidopsis thaliana* of β-aminobutyric acid (β-ABA), a non-protein amino acid, results in enhanced resistance to a wide range of stresses including heat, drought, and salinity stress, as well as enhanced resistance to biotrophic as well as necrotrophic fungi (Ton et al., 2005). Exogenous application of SA renders many crop plants more tolerant to an extensive array of abiotic stresses (Horváth et al., 2007), and similar observations have also been reported after treatment with jasmonates (Walia et al., 2007).

Evidence for Crosstalk From Whole Genome Expression Analyses

Evidence for regulatory crosstalk between abiotic and biotic stress response at the molecular level comes mostly from observations of expression patterns of genes under independent imposition of the single stress conditions. In *Arabidopsis thaliana*, a significant number of genes up-regulated by salinity stress are also induced in response to biotic stresses (Ma et al., 2006). Whole genome expression meta-analysis experiments under different abiotic and biotic stress treatments revealed a significant number of genes that are commonly regulated under abiotic and biotic stress conditions (Ma and Bohnert, 2007; Shaik and Ramakrishna, 2013, 2014). Functional categories enriched in the 197 commonly regulated genes identified by (Ma and Bohnert, 2007) include response to ABA, SA, jasmonic acid (JA), and ethylene (ET), major stress hormones controlling adaptation to abiotic and biotic stress. Several members of signaling pathways involving mitogen-activated protein kinase (MAPK), Ca^{2+}, ROS, phospholipids, mitochondrial functions, vesicle trafficking, and apoptosis were induced under biotic as well as abiotic stresses (Ma and Bohnert, 2007). Transcription factors (TFs) appear to be major orchestrators of stress crosstalk with members of WRKY, MYB, ERF, NAC, and HSF displaying similar induction patterns across stress treatments (Ma and Bohnert, 2007; Shaik and Ramakrishna, 2013). On the other hand, another study using co-expression data to identify *cis*-regulatory elements (CREs) of stress responses identified distinct CREs for the response to abiotic and biotic stressors (Zou et al., 2011). In addition, a number of CREs identified for both types of stress appear to oppositely regulate the expression of downstream genes in response to abiotic or biotic stress.

A different approach, yeast two-hybrid assays targeting major regulators of rice abiotic and biotic stress response, identified proteins that are present in multiple interactomes (Seo et al., 2011; Sharma et al., 2013). These include OsMPK5, the wall-associated kinase 25 (WAK25), sucrose non-fermenting-1-related protein kinase-1 (SnRK1), and WRKY family TFs.

Recently, examination of the transcriptional response of different *Arabidopsis thaliana* accessions to combinations of abiotic and biotic stressors revealed that across the treatments on average 60% of expression changes under combinatorial stress could not be predicted by the changes in response to the individual stresses (Rasmussen et al., 2013). The functional categories enriched in the affected genes were similar to those discovered after transcriptome meta-analyses of individual stressors, i.e., stress hormone responses, ROS, and MAPK signaling and regulation of hypersensitivity response. The response of many of these transcripts was canceled or prioritized under stress combination in comparison with the individual stress pointing to potential antagonistic interactions with detrimental effects on plant adaptation under combinatorial stress. In a similar study, the increased susceptibility to a virus after simultaneous application of drought and heat stress was accompanied by down regulation of pathogenesis-related (PR) genes and R-genes, which were otherwise induced under single viral stress (Prasch and Sonnewald, 2013). This indicates a direct negative effect of abiotic stress on major defense executors, that adds up to the antagonistic regulation observed in other signaling pathways. These studies clearly emphasize that even though regulatory pathways overlap between stresses, combinatorial stress needs to be treated and studied as a unique condition. Further functional characterization of individual gene members playing key roles in these pathways is required to extract meaningful conclusions.

ABIOTIC–BIOTIC STRESS INTERACTION INTERFACE

As mentioned above, abiotic and biotic stress interactions can occur at multiple levels, depending on the type of the stress (osmotic, ionic), the lifestyle, and infection strategy of the pathogen (biotroph/necrotroph, infection by direct penetration/through stomata, etc.) as well as the pathogenesis stage. We will summarize molecular components that according to evidence mentioned above participate in stress crosstalk. We will follow the pathogenesis timeline highlighting first extracellular interactions taking place at the epidermis and the apoplast during the initial stages of pathogenesis and moving on to the interactions in the intracellular environment during pathogen colonization (Figure 2). As information under combined stress is limited, and a detailed coverage of all potential interactions is not possible, our intention is to provide leads for future research that will aid to further dissect plant adaptive responses and tolerance under combined abiotic and biotic stress.

EXTRACELLULAR INTERFACE

Cuticular layer

The cuticle and cell wall constitute the first layers of defense against microbial attackers. They not only serve as physical barriers against pathogen penetration, but also as sensitive sensors for the timely activation of the intracellular and systemic defense responses. *Arabidopsis thaliana* mutants in long-chain acyl-CoA synthetase 2 (LACS2), a gene that is involved in cuticle biosynthesis, exhibit increased permeability of the cuticular layer which leads to increased resistance to *Botrytis cinerea* (Bessire et al., 2007). Interestingly, ABA deficiency causes similar cuticular defects and heightened resistance to *B. cinerea* through faster induction of defense responses and H_2O_2 production both in *Arabidopsis* and tomato, indicating a link between abiotic stress signaling, cuticle structure and defense responses (Curvers et al., 2010). In the study

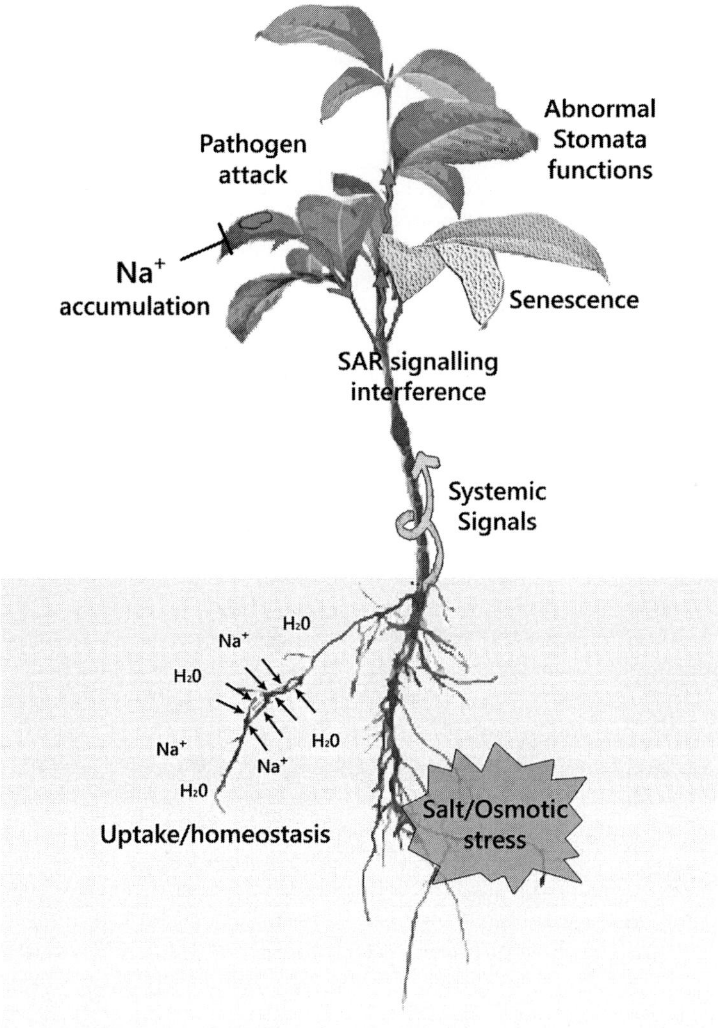

FIGURE 2. A SCHEME FOR THE EFFECTS OF ABIOTIC AND BIOTIC STRESS AT THE PLANT LEVEL. **A** combination of abiotic stress with pathogen infection potentially derails hormone and systemic ROS homeostasis. Pathogen infection has been shown to impair stomata closure under non-stress conditions, with the dynamics of this interaction under abiotic stress being unknown. Senescence is a common component of both abiotic and biotic stress that can potentially be amplified under combinatorial stress. Systemic ROS signals generated after pathogen encounter may alter water relation and salt uptake through their effects in root hydraulic conductance and ion transport. Abiotic stress through ABA signaling negatively affects signals that trigger systemic acquired resistance, enhancing pathogen spread from the initial site of infection. Ion accumulation (Na^+, Cl^-) under salt stress can have a direct toxic effect on pathogen growth.

by Xiao et al. (2004), however, *lacs2 Arabidopsis* mutants show no alteration in the resistance against the necrotroph *Alternaria brassicicola* and biotrophs, and even increased susceptibility against *P. syringae*. The latter observation points to a positive contribution of a thicker cuticle to resistance against *P. syringae*, indicating that the effects may be pathogen-specific (Tang et al., 2007). The well-documented increase in cuticular thickness under conditions of water deficiency (Kosma et al., 2009) may thus result in alteration in the deployment of the pathogen defense response. The cuticle does appear to be a sensor of the osmotic status and to be essential for the up-regulation of ABA biosynthesis genes under osmotic stress (Wang et al., 2011) through a yet not clearly defined mechanism; cuticle disruption by pathogens may therefore affect osmotic stress acclimation.

Cell wall-apoplastic space

Cell walls similarly appear to be an integrated signaling component for the defense against pathogens. Changes in pectin properties and composition in the *Arabidopsis* powdery mildew-resistant (*pmr*) mutants *pmr5* and *pmr6* resulted in a SA, JA, and ET independent increase in resistance to powdery mildew species (Vogel et al., 2004). Cellulose deficiency caused either by non-functional cellulose synthase genes or by chemical treatment enhances the synthesis of the defense hormones SA, JA, and ET and signaling and results in increased resistance to pathogens (Hématy et al., 2009). Intriguingly, these responses were attenuated when plants were grown under high osmotic pressure which reduced the turgor pressure (Hamann et al., 2009), suggesting that the defense response may be initiated by sensing the increased turgor pressure as a result of cell wall weakening. Osmotic stress, which is a common component of many abiotic stresses, may therefore interfere with the ability of plants to sense damage to the cell wall, due to already reduced turgor, resulting in inadequate activation of defense mechanisms.

The above-mentioned alterations in plant pathogen interactions in cell wall component biosynthesis mutants may be the consequence of the erroneous activation of integral receptor proteins such as RLKs and RLPs (receptor-like kinases and receptor-like proteins, respectively) which survey the cell wall integrity and bind to MAMPs and DAMPs (microbial- and damage-associated molecular patterns, respectively). Upon activation these transmembrane proteins (e.g., the RLK family WAK), send signals for the elicitation of downstream defense responses. Changes of cell wall structure and adherence to the plasma membrane upon exposure to abiotic stresses may affect their functional integrity. This is emphasized by the observation that NDR1, an essential component of disease resistance mediated by CC-

NB-LRR genes (McHale et al., 2006), is functioning in cell wall-plasma membrane adhesion. Down-regulation of NDR1 resulted in alterations in the cell wall-plasma membrane interaction and compromised resistance to virulent *P. syringae* (Knepper et al., 2011). Abiotic stress may also affect the abundance of cell wall receptors by influencing their transcript levels. THE1 is a member of the CrRLK1L RLK family that is involved in cell wall damage sensing and subsequent control of the downstream accumulation of ROS, and its expression is down-regulated under abiotic stress but up-regulated after pathogen challenge (Lindner et al., 2012), while similar expression patterns are observed for the WAK gene family (Shaik and Ramakrishna, 2013).

Pathogen recognition activates a battery of defense responses that target the apoplastic space. These include local cell wall enforcement, secretion of antifungal compounds at the site of intended penetration and up-regulation of enzymes with fungal cell wall degrading activities (Van Loon et al., 2006). These events are characterized and regulated by signature changes in pH, ROS homeostasis, and the redox state. Simultaneous exposure to abiotic stress can potentially impinge on the generation and decoding of these signatures, affecting subsequent responses. For example, apoplastic pH is transiently decreased following fungal infection (Felle et al., 2004), while an increase in pH is observed under salt stress (Geilfus and Muhling, 2011). Moreover the downregulation of cell wall peroxidases under abiotic stress (Shaik and Ramakrishna, 2014) can potentially dampen the production of ROS signatures that trigger defense responses (Daudi et al., 2012). Physical barriers enforcement after pathogen encounter through crosslinking of lignin monomers by ROS, which are produced by apoplastic peroxidases, NADPH oxidases and germin-like proteins, prevent pathogen penetration. Lignin content was found to be reduced under mild drought conditions to facilitate the maintenance of growth under conditions of decreased turgor pressure (Vincent et al., 2005), but severe stress resulted in increased lignin content (Lee et al., 2007). These findings may provide insight on the mechanisms leading to differential responses under combined stress across different abiotic stress intensities.

Vesicular trafficking and callose deposition

Another form of inducible defense response at the site of penetration is the formation of papillae that contain callose, antimicrobial secondary metabolites such as phenolic compounds, and ROS. Antimicrobial compounds are accumulating through vesicles originating from cellular compartments, such as the Golgi apparatus, which become polarized toward the site of infection (Underwood and Somerville, 2008). The significance of vesicle-mediated secretion in plant immunity has been demonstrated by the discovery of mutants defective in exocytosis of vesicles (with mutations in SNARE complex proteins *Hv*ROR2 and *At*PEN1), which display diminished penetration resistance to powdery mildew pathogens (Collins et al., 2003). Vesicular trafficking appears to be rewired in an opposite way under salt stress, as vesicles containing Na^+ are fused with the central vacuole to maximize compartmentalization of Na^+(Hamaji et al., 2009). Interestingly, knockout of different SNARE proteins resulted in increased salt tolerance (Hamaji et al., 2009), indicating possible antagonistic interactions of salt stress and pathogen infection at the level of vesicle trafficking, although further comprehensive experiments are required to substantiate this hypothesis.

Callose is a β-1,3-glucan polymer that is deposited at the sites of attempted fungal penetration in the form of papillae. It is an important inducible defense mechanism, with enhanced deposition being observed after exogenous application of priming chemicals like β-ABA. A mutant screen for plants defective in β-ABA-induced priming identified among others mutants in the ABA biosynthesis gene *zeaxanthin epoxidase* (ABA1; Ton et al., 2005). These mutants failed to exhibit both β-ABA-induced callose deposition against *H. parasitica* and increased tolerance to salt stress, thereby providing a link between the induction of abiotic and biotic stress responses by β-ABA. In accordance with these observations the callose-mediated increased resistance of the *ocp3 Arabidopsis* mutant to necrotrophic pathogens requires ABA (Garcia-Andrade et al., 2011). Moreover *ocp3* mutants accumulate higher levels of ABA, and are more drought tolerant (Ramírez et al., 2009). Therefore ocp3, a homeodomain TF, appears to be a convergence point for ABA and callose regulation that can be manipulated to enhance resistance under combinatorial stress.

Callose accumulation appears to be a point of convergence of abiotic and biotic signaling as variability in environmental conditions, which affect the redox state of the plant, such as light intensity, have a significant impact on the magnitude of callose deposition after pathogen elicitation (Luna et al., 2011). As callose deposition is a major component of the pre-invasion defense of plants (Ellinger et al., 2013), detailed characterization of the regulation of callose accumulation under simultaneous abiotic stress may be invaluable in building combined stress tolerance in crops.

INTRACELLULAR SIGNALING INTERACTIONS

Interconnections between Ca^{2+} and ROS signaling

Changes in calcium fluxes and production of ROS are among the earliest plant responses to abiotic stress and pathogen challenge. The decoding of both signals relies on "signature" spatiotemporal patterns and oscillations specific to the stress encountered (Dodd et al., 2010; Mittler et al., 2011). Moreover, both components are highly interconnected: Ca^{2+} signaling components such as calmodulins (CaMs) and calcium-dependent protein kinases (CDPKs) regulate ROS production by NADPH-oxidases (Takahashi et al., 2011). ROS vice versa affect Ca^{2+} signaling through regulation of Ca^{2+} permeable channels (Demidchik et al., 2007). It is plausible that there are either unique signatures for combinations of stresses, or that there is interference between the abovementioned signals that potentially dampens or strengthens the downstream responses.

Whole genome expression analyses coupled with promoter motif identification provided further evidence that Ca^{2+} orchestrates the early responses to both biotic and abiotic stress as the overrepresented motif "CGCGTT" identified in the promoters of the commonly regulated genes, contains the core "CGCG" Ca^{2+}responsive *cis*-element (Walley et al., 2007). The investigation of mutants defective in the induction of a hypersensitive response after pathogen infection has led to the identification of genes encoding for cyclic nucleotide gated channels (CNGCs) which are non-selective cation transporters (Clough et al., 2000). members of which are also involved in salt and heat stress tolerance through regulation of Ca^{2+} fluxes (Guo et al., 2010; Finka et al., 2012). Furthermore, Ca^{2+} downstream signaling components have been shown to mediate responses to both abiotic and biotic stress stimuli. The CAMTA3 TF is important for cold acclimation of *Arabidopsis* by stimulating the expression of CBF1, CBF2, and ZAT12 that are also involved in adaptation to dehydration and oxidative stress (Doherty et al., 2009). Moreover, *C*AMTA3 negatively regulates SA accumulation and plant defenses through CaM binding (Du et al., 2009). Other proteins interacting with CaM include TF families like NAM, WRKY, and MYB (Popescu et al., 2007) many members of which are involved in abiotic and biotic stress crosstalk.

CDPKs have a unique feature to both bind calcium and functionally decode the message by target protein phosphorylation. They appear to represent a central node in the regulation of abiotic and biotic stress responses (Schulz et al., 2013). For example, *Arabidopsis* CPK4 and CPK11 positively regulate ABA responses and their down-regulation renders plants salt-sensitive (Zhu et al., 2007), and are important for the oxidative burst and defense responses (Boudsocq et al., 2010). In addition, CDPKs regulate ROS production through phosphorylation-mediated regulation of RBOH activity (Dubiella et al., 2013). *St*CDPK4- and *St*CDPK5-mediated phosphorylation increases the activity of StRBOHs and the increased ROS production results in a stronger hypersensitivity response after pathogen challenge, favoring resistance against biotrophic pathogens but compromising resistance against necrotrophic fungi (Kobayashi et al., 2012). Recently, the CDPK OsCPK12 was shown to increase salt stress tolerance and decrease blast disease resistance in rice through reduced ROS production as a result of down-regulation of RBOH expression, enhanced expression of antioxidant genes such as APX (ascorbate peroxidase), and increased sensitivity to ABA (Asano et al., 2012).

Dissecting the spatiotemporal and molecular specificity of Ca^{2+} and ROS signaling components is crucial for determining their precise functions in stress responses (Baxter et al., 2014), as is elegantly demonstrated by the identification of different Ca^{2+} binding affinities regulating the activation of two soybean CaMs (Gifford et al., 2013).

Signal relay by MAPKs

Mitogen-activated protein kinases are centrally positioned in Ca^{2+}–ROS crosstalk and regulation as well as in the signal output after stress exposure. MAPK signaling cascades are relayed through MAPK kinase kinases (MAP3Ks) and MAPK kinases (MAP2Ks). Hydrogen peroxide (H_2O_2) has been shown to mediate activation of the three major and well-studied *Arabidopsis* MAPKs, MAPK3, 4, and 6, through MAP3Ks and other kinases (Rentel et al., 2004; Teige et al., 2004). These MAPKs appear to have an overlapping function in signal transduction upon abiotic stress and pathogen challenge. Activation of *Arabidopsis* MAPK3 and MAPK6 as well as their homologs in tobacco WIPK and SIPK (Segonzac et al., 2011) after PAMP recognition is essential for fungal and bacterial resistance (Asai et al., 2002). The importance of MAPK3 and MAPK6 in plant immune responses is highlighted by the discovery that the *P. syringae* effector HopAI1 directly interacts and inactivates both, promoting virulence (Zhang et al., 2007). Additionally MAPK6 is directly involved in regulating ET biosynthesis in *Arabidopsis* by activation through phosphorylation of ACS2 and ACS6, which results in an increase in ET biosynthesis (Liu and Zhang, 2004). MAPK4 acts as a negative regulator of defense responses and SA accumulation by phosphorylating MEKK2, a MAP3K protein (Kong et al., 2012).

On the other hand, down-regulation of MAPK3 resulted in altered stomata opening patterns in response to ABA and H_2O_2 in *Arabidopsis* (Gudesblat et al., 2007). Moreover, the ABA-induced expression of AtCAT1, which is involved in H_2O_2 homeostasis, is controlled by an AtMKK1–AtMAPK6 signaling cascade (Xing et al., 2008). Constitutive activation of AtMAPK4 and AtMAPK6 rendered plants more tolerant to cold and salt stress (Teige et al., 2004) and CAT2 and tAPX, which are involved in H_2O_2 regulation, appear to be regulated by AtMAPK4 (Pitzschke et al., 2009). In rice, OsMAPK5 appears to be a convergence point of abiotic and biotic stress responses, as its silencing results in sensitized defense responses and resistance to fungal and bacterial pathogens at the expense of salt and drought tolerance (Xiong and Yang, 2003).

These examples emphasize the complexity of MAPK-mediated defense signaling with diverse and sometimes overlapping functions of different members of the signaling pathway. Downstream targets of MAPK6 overlapped 60% with MAPK3 targets, while a 50% overlap was observed between MAPK3 and MAPK4 targets (Popescu et al., 2009). Probably, the one-dimensional overlap can be resolved by multidimensional regulation, such as different spatiotemporal transcription and protein subcellular localization, activation thresholds, feedback loops with phosphatases and scaffolding (Tena et al., 2011; Samajova et al., 2013). Many of the above-mentioned components appear to be an integral part of broad stress tolerance priming by exogenous application of chemicals (Beckers et al., 2009; Xia et al., 2009), and the detailed study of MAPK activation, localization, and substrate affinity under these conditions can increase our understanding of plant responses under stress combinations.

Hormone signaling

Plant hormones are central to the integration of environmental stimuli in the coordination of growth under optimal and stress conditions, including the regulation of defense responses after pathogen attack. Plant hormones do not act independently, and extensive synergistic or antagonistic interaction between hormonal pathways is observed in development and stress responses after exogenous application, or through mutant analysis (Wolters and Jurgens, 2009). Transcriptomic studies have aided in unveiling these interactions (Nemhauser et al., 2006), and it was recently shown that hormonal pathways can be directly connected with each other by protein–protein interactions between their signaling components (Hou et al., 2010; Zhu et al., 2011).

Abscisic acid is the major hormone that positively contributes to adaptation to osmotic stress, a major component of several abiotic stresses. Its involvement in the regulation of defense responses has been a topic of recent comprehensive reviews (Asselbergh et al., 2008; Ton et al., 2009). The consensus is that ABA negatively regulates defense responses against both biotrophic and necrotrophic pathogens through negative interactions with SA and JA/ET biosynthesis and signaling; ABA biosynthesis mutations show sensitization of these signaling pathways after pathogen challenge (Achuo et al., 2006; De Torres Zabala et al., 2009; Sanchez-Vallet et al., 2012). Comprehensive analyses of ABA-deficient mutants revealed further pleiotropic alterations that may be part of ABA-defense crosstalk such as reduced cuticle thickness and sensitized H_2O_2 production in response to *B. cinerea* in tomato (Asselbergh et al., 2007) and altered cell wall composition in *Arabidopsis* (Sanchez-Vallet et al., 2012). Moreover ABA compromised a chemically induced systemic acquired resistance (SAR) through suppression of SA biosynthesis in *Arabidopsis*, while genetically enhanced ABA catabolism reversed this effect (Yasuda et al., 2008).

Nevertheless, ABA signaling can positively contribute to pre-invasive defense responses and to early defense signaling against certain necrotrophic pathogens (Adie et al., 2007). ABA positively contributes to resistance against pathogens that infect through stomata, such as *P. syringae* (Melotto et al., 2006), as well as to other pre-invasion defense mechanisms such as callose deposition (Ton and Mauch-Mani, 2004; Adie et al., 2007; Garcia-Andrade et al., 2011).

Identification of downstream regulatory nodes that channel interactions between hormonal pathways is of great importance in fine-tuning resistance to both abiotic and biotic stress. Besides TFs, which will be discussed in a following section, other regulators of the transcriptional machinery have been uncovered to function in stress crosstalk. RNA chaperones such as RNA helicases are shown to regulate transcription in response to various stressors (Li et al., 2008; Mazzucotelli et al., 2008). MED25, a subunit of the mediator complex which is a component of the transcriptional machinery, is involved in the antagonistic crosstalk between ABA and JA (Chen et al., 2012). In a recent report, the *Arabidopsis* pathogenesis-related protein 2 (PR2), which encodes β-1,3-glucanase involved in callose degradation, was shown to be down regulated in response to ABA, partly elucidating ABA-mediated capacitation of callose deposition. The *ahg2-1* mutant in *Arabidopsis* accumulates both ABA and SA and has increased expression of defense related genes, which is an indication that ABA and SA do not always act antagonistically. Transcriptome analysis of the *ahg2-1* mutant revealed complex interactions between ABA and SA signaling involving altered mitochondrial and RNAMETABOLISM (Nishimura et al., 2009), highlighting multilevel connections between the two signaling pathways that add to the complexity and hinder straightforward conclusions.

Recent research has highlighted the direct involvement of the growth hormones gibberellin, cytokinin, auxin, and brassinosteroid in responses to adverse growth conditions and pathogen attack (Robert-Seilaniantz et al., 2011). For example, GA signaling directly regulates JA signaling, mediated through direct binding of the GA repressor protein DELLA to JAZ proteins and relieving JA signaling repression (Hou et al., 2010). DELLA proteins appear to be central nodes in abiotic and biotic stress cross-talk. ABA and ET signaling promote DELLA stabilization which positively affects ROS detoxification (beneficial for acclimation to abiotic stress) through higher expression of ROS detoxification genes (Achard et al., 2008). DELLAs also sensitize JA signaling (through binding of DELLAs to JAZ) at the expense of SA signaling, enhancing resistance to necrotrophic pathogens (Navarro et al., 2008). This may provide an explanation for the often-observed positive correlation between resistance to abiotic stress and resistance to necrotrophs (Navarro et al., 2008; Abuqamar et al., 2009; Ramírez et al., 2009).

Cytokinins were shown to positively regulate defense responses to biotrophic pathogens (Argueso et al., 2012) via SA accumulation, and increased defense gene expression through interaction of the cytokinin response regulator ARR2 with TGA3, a TF central for defense gene activation (Choi et al., 2010). This suggests that the increased cytokinin catabolism observed under abiotic stress-induced senescence may potentially contribute to further down-regulation of SA responses and increased susceptibility to biotrophic pathogens.

The roles of auxin and brassinosteroids in stress responses and their potential participation in stress crosstalk remains elusive. Auxin signaling shows antagonistic crosstalk with SA (Wang et al., 2007), although auxin contributes to reduced senescence (Kim et al., 2011a) which may be of great importance under exposure to a stress combination. Brassinosteroid (BR) signaling positively affects abiotic stress tolerance, as is evident by both BR exogenous application (Divi et al., 2010) and genetic de-repression of the BR signaling pathway (Koh et al., 2007). BR signaling probably interacts synergistically with ABA signaling and stimulates ROS detoxification (Divi et al., 2010). BR's involvement in defense signaling is rather complicated. In tobacco and rice exogenous application of BRs appeared to clearly enhance resistance to a wide range of pathogens (Nakashita et al., 2003). Similar results were obtained in cucumber, which showed heightened resistance to *Fusarium oxysporum* as a result of activated production of H_2O_2 by NADPH oxidase and expression of defense related genes (Li et al., 2013). On the contrary BRs appear to be negatively regulating resistance to the root-infecting oomycete *Pythium graminicola* by antagonizing SA and GA related defense responses (De Vleesschauwer et al., 2012). BR signaling shares LRR–RLK and BAK1 proteins with PAMP immune signaling (Chinchilla et al., 2009). Contradictory effects of BR signaling on immune responses have been recently reported in *Arabidopsis* (Albrecht et al., 2012; Belkhadir et al., 2012; Lin et al., 2013), which require further study.

It is clear that hormonal crosstalk is extensive and occurs in multiple combinations. Further understanding of plant responses under combined stress exposure is required to dissect the multilevel responses under these conditions. As an example of the underlying complexity, both drought stress and exogenous ABA application result in an increased endogenous ABA content in tomato, but they differentially affect resistance to powdery mildew and *Botrytis*, with drought enhancing and ABA application compromising resistance (Achuo et al., 2006). Notably the ABA-deficient tomato mutant *sitiens* exhibited increased resistance similar to the effect of drought (Achuo et al., 2006). The complexity of interactions under abiotic stress is further emphasized by transcriptome analyses under abiotic stress in which up-regulation of a significant number of JA/ET-responsive genes and accumulation of their transcripts was observed (Walia et al., 2007; Huang et al., 2008). Besides the effects of direct hormonal interactions on abiotic and biotic stress tolerance mechanisms additional indirect interactions should be considered, such as the alteration of developmental programs and the regulation of senescence which may be critical for evolutionary speciesFITNESS and yield performance in crop plants (Wu et al., 2012a).

Cellular redox state

The cellular redox state is the sum of reducing and oxidizing redox-active molecules (Potters et al., 2010) and it acts both as a sensor of environmental perturbations (as most of them impose oxidative stress) and as a buffer against these perturbations to maintain cellular homeostasis. It acts as a central integrator of ROS, energy andMETABOLIC regulation under stress as well as optimal conditions. Its major constituents are ascorbate, glutathione (GSH), NADP(H), small proteins acting as antioxidants like thioredoxin and glutaredoxins as well as many diverse metabolites such phenolics, amino acids, carotenoids, and tocopherols. The cellular redox state is dependent on both their accumulation and their reduction-oxidation state (Potters et al., 2010). Genetic manipulation of redox homeostasis results in altered hormone homeostasis and responses to pathogens and abiotic stresses (Mhamdi et al., 2010), exemplifying its significance. As abiotic and biotic stress commonly impinge on the redox status (albeit not in a similar manner; Foyer and Noctor, 2005), redox homeostasis is potentially a central orchestrator of the phenotypic response to stress combinations. Redox perturbations after imposition of a stress factor may affect responses to subsequent challenges by additional stressors, thereby shaping the response to combined stresses. For example, a transient increase in GSH content drives the antagonistic crosstalk between SA and JA signaling (Koornneef et al., 2008) and GSH oxidation appears to drive the induction of both SA and JA pathways (Mhamdi et al., 2013).

Plant hormone signaling can directly perturb the redox status by modifying the expression and activities of antioxidant enzymes. ABA induces the expression of catalase, activating also at the same time the production of the ROS hydrogen peroxide through AtMAPK6 signaling (Xing et al., 2008). SA inhibits the function of catalase and cytosolic ascorbate peroxidase (Corina Vlot et al., 2009) and several glutathione transferases (Tian et al., 2012).

Programed cell death (PCD) is a plant response to developmental and environmental stimuli (e.g., in senescence) and pathogen defense (in the form of HR) that is initiated and regulated by redox changes, like an increased oxidation ratio of GSH and ascorbate (De Pinto et al., 2012). APX appears to be central in the redox regulation leading to PCD. Decreased activity of APX isoforms was observed in heat-induced PCD (Locato et al., 2009), and overexpression or down-regulation in *Arabidopsis* of a thAPX increased or decreased, respectively, sensitivity to NO-induced cell death (Tarantino et al., 2005). APX isoforms are also commonly up-regulated under abiotic stress (Miller et al., 2008). Considering the important role of APX in the drought–heat stress interaction (Koussevitzky et al., 2008) it is of great interest to explore APX enzyme regulation under combinatorial stress.

Redox status changes can directly impact protein function through post-translational modifications. One pronounced example of post-translational modifications controlling protein activity and localization is the interplay of *S*-nitrosylation and thioredoxin-mediated reduction in the control of the oligomeric and monomeric state of NPR1 (Tada et al., 2008), a master regulator of SA-mediated defense responses and recently proposed as a SA receptor (Wu et al., 2012b). The function of many more proteins appears to be regulated by *S*-nitrosylation, among them AtRBOHD as mentioned above (Yun et al., 2011), SA binding protein 3 (SABP3), methionine adenosyltransferase 1, theMETABOLIC enzymes glyceraldehyde 3-phosphate dehydrogenase (GAPDH) and glycine decarboxylase (GDC), as well as metacaspase 9 (Astier et al., 2011). Identification of the dynamics of post-translational modifications on these and newly identified proteins under various stress combinations will shed light on their significance for plant adaptation responses to these conditions.

NO was recently found to exhibit biphasic control over cell death triggered by pathogens and pro-oxidants in *Arabidopsis*. In initial stages *S*-nitrosothiol (SNO) accumulation results in enhanced and accelerated cell death (Yun et al., 2011). However, constitutively high SNO levels decreased cell death through *S*-nitrosylation-mediated reduction in AtRBOHD activity

(Chen et al., 2009; Yun et al., 2011). This differential regulation might have implications in conditions of combined abiotic and biotic stress as both result in increased NO levels. At a certain plateau concentration of NO, signaling components may be desensitized or inversely regulated, as exemplified by AtRBOHD, with detrimental effects on stress acclimation.

Redox changes and post-translational modification appear to be integral in priming for stress tolerance after exogenous application of chemicals (Tanou et al., 2009). This provides a potential explanation of the mechanism of action of diverse chemicals in plant defense sensitization. H_2O_2 and NO priming for salt tolerance in citrus moderately increased the abundance of oxidized and *S*-nitrosylated proteins, which then remained relatively similar after the application of stress. Non-treated plants were more stress sensitive and exhibited increased protein carbonylation and oxidation (Tanou et al., 2012). As both compounds provide increased tolerance to both abiotic and biotic stress, further characterization including the timing and magnitude of these post-translational modifications under different stress treatments and under stress combination may help to better understand the redox changes leading to stress cross-tolerance.

Metabolite homeostasis and signaling

Metabolites are the end products of gene expression and protein activities and therefore are the penultimate regulatory component for the phenotypic expression under stress conditions. As metabolites can have multiple functions such as being energy carriers, structural molecules and redox regulators or exerting direct antimicrobial activity against pathogens, uncovering their regulation and homeostasis under combined stress is of great significance.

Adaptation to both abiotic and biotic stress impinges significantly on primaryMETABOLISM homeostasis. Synthesis of antimicrobial metabolites and defense proteins is energy demanding (Bolton, 2009), while abiotic stress potentially leads to energy deprivation as photosynthesis is reduced under abiotic stress (De Block et al., 2005). As a result, it is fair to assume that under stress combinations these strong antagonistic effects will result in disturbed energy balance. However, recent results challenge the carbohydrate deprivation notion under mild dehydration stress (Hummel et al., 2010) and further experimental data under combined stress are required for firm conclusions. More evidence that sugar homeostasis and signaling drives defense responses are demonstrated by the down regulation of cell wall invertases. This results in dampening of defense responses and increased susceptibility to pathogens as a result of decreased availability of carbohydrates to fuel the defense responses at the site of infection (Essmann et al., 2008). Cell wall invertases appear to be down regulated under abiotic stress (Wingler and Roitsch, 2008) and as the regulation of their activity is a convergence point of hormonal and sugar signals for stress tolerance and senescence progression (Wingler and Roitsch, 2008), fine tuning of their expression might be a focal point in enhancing combined stress tolerance. The metabolic status of the host is also crucial for pathogen growth as it appears that pathogens manipulate different aspects of plant metabolism to achieve optimal conditions for their growth (Chen et al., 2010a).

The significance of amino acid homeostasis for the induction and regulation of defense responses was recently highlighted (Zeier, 2013). Amino acids may function as precursors in hormone biosynthesis and affect the redox state through their chemical properties or as precursors of redox regulators such as GSH. Amino acid abundance can impact hormone signaling through conjugation-mediated regulation of hormone activity (Woldemariam et al., 2012). Amino acid concentration appears to be significantly perturbed by abiotic stress as is revealed by metabolomics studies (Obata and Fernie, 2012). On the other hand a direct link between amino acid abundance and activation of SA-induced defense responses was recently demonstrated with heat-shock factor HsfB1, the translation of which is initiated under conditions of phenylalanine starvation (Pajerowska-Mukhtar et al., 2012). Phenylalanine appears to be accumulated under abiotic stress conditions (Urano et al., 2009; Widodo et al., 2009) and its potential as a molecular switch between abiotic and biotic stress responses should be explored.

METABOLIC alterations under abiotic stress include the accumulation of compounds such as the raffinose family oligosaccharides raffinose and galactinol and the amino acid proline. These exhibit osmoprotective and antioxidant functions and have been positively correlated with abiotic stress tolerance (Korn et al., 2010). Galactinol overproduction was recently associated with increased resistance to necrotrophic pathogens (Mi et al., 2008). Moreover, proline metabolic regulation at the site of pathogen infection is important for both HR deployment and containment, probably through modulation of ROS levels as shown by expression and functional studies of proline dehydrogenase (Senthil-Kumar and Mysore, 2012). Myo-inositol metabolic regulation appears to be a convergence point for abiotic and biotic stress responses. Myo-inositol is accumulating under most abiotic stress conditions and is positively contributing to tolerance as a compatible solute (Tan et al., 2013). A negative relationship between myo-inositol accumulation and pathogen resistance and PCD initiation was found in *Arabidopsis*, with a positive correlation between myo-inositol depletion and increased SA production and cell death (Chaouch and Noctor, 2010).

Analysis of mutants that exhibit qualitative and quantitative alterations in the accumulation of fatty acid metabolites demonstrated that fatty acids are not only structural components of the cellular membranes, but they also exert a multitude of signaling functions. Fatty acid release from the membranes after pathogen encounter triggers the defense response (Savchenko et al., 2010). Linolenic acid (18:3) is a precursor for the production of the major cellular signaling components JA and oxylipins (Reinbothe et al., 2009). A reduction of the levels of oleic acid (18:1) triggers constitutive defense responses that are independent of SA signaling (Kachroo et al., 2001), but dependent on NO production (Mandal et al., 2012). Fatty acid homeostasis is disturbed under abiotic stress, as membrane composition changes are vital for the maintenance of membrane rigidity and functionality. Dehydration stress is shown to result in a reduction in 18:3 and increase in 18:1 lipid levels (Upchurch, 2008), and increased 18:3 levels by FAD3 or FAD8 overexpression enhanced drought tolerance in tobacco (Zhang et al., 2005). Manipulation of fatty acid composition can provide further insight into their function under stress combination.

Transcription factors

Regulatory modules like MAPKs-based pathways and core hormone signaling modules control the expression of a vast number of genes and therefore their manipulation in most cases have severe pleiotropic effects. Identification of downstream regulators involved in abiotic and biotic stress crosstalk such as TFs is important for more targeted manipulation and adaptation of plants to multiple stresses. The appropriate fine-tuning of their expression is an important aspect toward translation of scientific knowledge in crop plant improvement (Kasuga et al., 2004).

Bioinformatics and functional analyses have demonstrated that TFs involved in stress crosstalk comprise a diverse collection of members of the largest TF families in plants, such as NAC, MYB, AP2/ERF, WRKY, and others, reflecting the complexity of the genetic regulatory networks underlying stress crosstalk (Atkinson and Urwin, 2012; Shaik and Ramakrishna, 2014). Many members of these families are involved in regulation of leaf senescence, an integral component of both abiotic and biotic stress (Breeze et al., 2011). Moreover, in most cases the TFs identified are stress hormone-regulated, and therefore potentially act as molecular switches for the fine-tuning of hormonal responses.

Characterization of the mechanism of action of the candidate TFs involved in stress crosstalk is of great importance. For example, a TF with positive contribution to both abiotic and biotic stress tolerance can be directly useful for breeding combined stress tolerance. Functional characterization of several TFs has revealed various members that confer both abiotic and biotic stress tolerance. Overexpression of the rice OsNAC6 conferred tolerance to salt and dehydration stress as well as resistance to blast disease (Nakashima et al., 2007). Similarly in wheat, overexpression of the R2R3MYB gene TaPIMP1 results in drought stress tolerance and resistance to *Bipolaris sorokiniana* through increased expression of abiotic stress (many of them ABA inducible) and defense-related genes (Zhang et al., 2012). Members of the AP2/ERF TF family have been shown to be positive regulators of both abiotic and biotic stress (Jung et al., 2007; Zhang et al., 2009). DREB TFs are also members of the AP2/ERF family and important contributors to abiotic stress tolerance (Liu et al., 2013a) that may have additional signaling functions for biotic stress tolerance. AtDREB2A was upregulated in plants overexpressing the CC-NB-LRR gene ADR1 which conferred pathogen resistance and drought tolerance (Chini et al., 2004). Overexpression of OsDREB1B in tobacco resulted in increased resistance to abiotic stress and also virus infection (Gutha and Reddy, 2008).

Overexpression of AtHSFA1b provided stress hormone independent, but H_2O_2 signaling dependent increased tolerance to drought and resistance to bacterial and oomycete pathogens (Bechtold et al., 2013). It appears that the HSF TF gene family has broad biological functions in ROS signaling and defense responses and SAR regulation (Miller et al., 2008; Pick et al., 2012), which can be further exploited for building broad stress tolerance into crops. Whole genome expression meta analyses can provide evidence of potential antagonistic regulation in different stress responses for a given TF, by analyzing expression patterns under different stress conditions (Shaik and Ramakrishna, 2014). Detailed characterization of spatiotemporal expression and *cis*-element binding patterns is, however, required for the understanding of the underlying mode of regulation. This was recently elegantly demonstrated in the characterization of OsWRKY13 which exhibits tissue specific expression and condition specific binding to *cis*-elements of downstream genes and thereby inversely regulated resistance to drought and bacterial infection of rice (Xiao et al., 2013).

Functional conservation of TF functions across species can be exploited to take advantage of the wealth of experimental data generated in the model plant *Arabidopsis thaliana*. For example, the *Arabidopsis* AtBOS1, an R2R3MYB TF, as well as its homolog in tomato SlAIM1 appear to regulate tolerance to abiotic and biotic stress in the same way, as mutant plants exhibit reduced tolerance to salt stress as well as to *Botrytis* infection (Mengiste et al., 2003; Abuqamar et al., 2009). Further similar efforts should be undertaken to accelerate the translation of experimental observations obtained in model plants species to crops.

The results obtained by the functional characterization of TFs are encouraging as many of them appear to regulate cross-resistance in a unidirectional manner, in contrast to the observations at the level of hormonal regulation that point to antagonistic relationships. Therefore, their manipulation offers many opportunities to bypass the antagonistic effects on abiotic and biotic stress tolerance observed in the more upstream regulatory nodes.

Epigenetic modifications

Epigenetic modifications such as DNA cytosine methylation and histone residues methylation and acetylation contribute to the transcriptional control of amongst others adaptive responses to environmental stimuli (Mirouze and Paszkowski, 2011). A significant portion of these modifications appears to be persistent across generations and significantly contributes to phenotypic variation (Johannes et al., 2009). While cytosine methylation generally has repressive effects on gene transcription, leading to gene silencing, histone modifications can lead to transcriptional activation through local chromatin de-condensation which facilitates the accessibility of TFs (Liu et al., 2010). Recently, epigenetic modifications and specifically chromatin-regulated gene activation have been proposed to govern priming responses (Conrath, 2011). Genome wide approaches studying DNA methylation under abiotic and biotic stress have demonstrated widespread methylation alterations (Bilichak et al., 2012; Dowen et al., 2012). It would be of particular interest to further examine the occurrence of differential alterations and their impact under combinatorial stress.

Functional studies of chromatin remodeling enzymes have revealed a functional involvement of these enzymes in the regulation of both abiotic and biotic stress responses. Histone deacetylase 19 (HDA19) mutants exhibit enhanced basal expression of many SA-responsive genes (Kim et al., 2008) but decreased expression of ABA and JA/ET-responsive genes, and the mutants are hypersensitive to salt stress (Chen et al., 2010b). The histone lysine methyltransferase ATX1 is likely to be involved in dehydration stress signaling, as atx1 mutants were sensitive to drought and ATX1 methyltransferase activity positively regulated the expression of the ABA biosynthesis enzyme NCED3 (Ding et al., 2011). Interestingly, down-regulation of the TF WRKY70 during dehydration stress coincided with decreased presence of ATX1 at the WRKY70 gene locus (Ndamukong et al., 2010).

Chromatin structure can also be altered by the active deposition of variants of the canonical histones. Deposition of one of the these variants, H2A.Z, is linked to transcriptional activation in response to environmental stimuli (Coleman-Derr and Zilberman, 2012), and disruption of this mechanism leads to misregulated responses to both pathogens and elevated temperature (March-Diaz et al., 2008; Kumar and Wigge, 2010).

It would be highly interesting to investigate how a previously imposed stress predisposes plants at the methylation and chromatin level for the encounter of a subsequent stress, (de)sensitizing subsequent responses. This type of acclimation/predisposition may even be a useful tool for preparing seeds and propagated material for stressful environments.

R-gene resistance and systemic acquired resistance

The plant immune system consists of successive layers counteracting suppression of defense responses by pathogens through secretion of effector proteins (Hemetsberger et al., 2012). Recognition of the effectors by corresponding R-genes belonging to NB-LRR protein family or the effect of effectors on intracellular host proteins (guarded proteins) results in effector-triggered immunity (ETI). This is usually but not always manifested by localized cell death, termed the hypersensitivity response (Coll et al., 2011). The complexity in the regulation of ETI is outlined by network analyses of individual and combined hormone mutants, which revealed compensatory interactions in contrast to synergistic interaction observed in PTI (PAMP-triggered immunity; Tsuda et al., 2009), and which may explain the robustness of ETI to genetic perturbations. This robustness may be ideal in building tolerance to combinatorial stress through pyramiding R-genes with genes conferring abiotic stress tolerance.

However, it is becoming clear that there are multiple aspects of regulation at the NB-LRR protein level that are indispensable for the deployment of R-gene resistance (Heidrich et al., 2012). These include spatial regulation of NB-LRR accumulation in cellular compartments (e.g., the nucleus). Reduction of nuclear NB-LRR accumulation was shown to be responsible for the heat stress attenuation of disease resistance conferred by the proteins SNC1 and RPS4 in *Arabidopsis* (Zhu et al., 2010; Mang et al., 2012). Interestingly, mutants with reduced sensitivity to heat-induced defense inhibition were found to be based on changes in among others ABA biosynthesis enzymes, indicating that abiotic stress factors may affect R-gene compartmentation through ABA biosynthesis and signaling, although no further evidence is available. In addition, chaperone-mediated transport and folding of NB-LRR protein is important for their activity (Hubert et al., 2009). The heat shock protein HSP90 is a component of this chaperone machinery. HSP90 is also required for the maintenance of folding of other proteins under stress conditions (Wang et al., 2004), and could potentially become limiting for proper R-gene signaling or stress protection under combined stress conditions. The recent discovery that NB-LRR protein accumulation is controlled by microRNAs (Zhai et al., 2011) adds a novel layer of regulation that would be interesting to investigate under different stress conditions (Kulcheski et al., 2011).

Initial pathogen perception and interception through PTI or ETI triggers systemic signals that prime plant defense responses to effectively counter subsequent infection attempts and limit spreading of the disease. This is referred to as SAR. Many compounds and genes have been identified that function in mobile signal generation and transport. Conversion of MeSA produced at the infection site to SA at the systemic tissues appears to be a prerequisite for SAR manifestation (Park et al., 2007). Additional metabolites such as pipecolic acid, dehydroabietinal, azelaic acid, and glyceraldehyde 3-phosphate probably function in the amplification of the signal, with no clear conclusions yet on their precise placement in the SAR circuit pathway (Dempsey and Klessig, 2012). SAR has been shown to be affected by environmental conditions such as exposure to light (Griebel and Zeier, 2008) and abiotic stresses such as salinity, through ABA suppression of SA biosynthesis (Yasuda et al., 2008). The further investigation of the patterns of accumulation and transport of these metabolites under conditions of combined abiotic and biotic stress may reveal potential connections between their regulation and plant phenotypic responses to combined stress.

APPROACHES FOR GENE IDENTIFICATION AND BREEDING FOR TOLERANCE TO STRESS COMBINATION

In accordance with individual abiotic and biotic stressors, each abiotic stress/pathogen/host combination should be treated independently as, despite the potential universal applicability of some interactions that were characterized in *Arabidopsis*, many unique interactions may be crucial for the phenotypic response. As a result improving crops to these complex stress conditions first requires an extensive phenotypic characterization at different levels of cellular regulation, i.e., transcription, translation, post-translation, and metabolites, as well as at different stages of plant development. As evidence from research on individual abiotic and biotic stress responses points to a strong dependency on developmental (Skirycz et al., 2010) as well as environmental factors (Luna et al., 2011), the environmental conditions and developmental stages of the plants should be appropriately defined before any interpretation of the phenotypic and molecular response can be done. Finally the different layers of defense can be differentially affected by abiotic stress imposition (Figure 2); therefore, the outcome of the interaction will vary with the defense mechanisms employed and on the pathogens involved.

Breeding for resistance to combinatorial stress is challenging. However, various novel approaches can aid in dissecting interactions between various types of stressors and identifying genetic components that can be breeding targets. The combination of different ~omics technologies has enabled the molecular dissection of plant phenotypes (Baerenfaller et al., 2012; Nagano et al., 2012). They provide information about the biological function of the whole gene set of an organism, and overlapping expression patterns might imply participation in common pathways (Quackenbush, 2003), enabling more efficient reverse genetic approaches. Utilization of'~omics in combination with forward genetic approaches like association mapping (Chan et al., 2011) may narrow down the candidate genes responsible for the observed phenotypes and provide targets for functional characterization, further manipulation and improvement of crops through breeding. As mentioned previously, currently there are limited studies on the ~omics characterization of combined abiotic and biotic stress tolerance, however, functional characterization of differentially regulated genes is starting to provide interesting candidates for combined stress tolerance and their mode of action (Atkinson et al., 2013).

Manipulations that induce resistance to abiotic and biotic stress such as application of priming chemicals, followed by comprehensive phenotypic characterization can be used for candidate gene identification and molecular processes underlying stress cross-tolerance. Utilization of pre-existing chemical libraries for compounds that can prime abiotic and/or biotic stress tolerance and identification of their mode of action through chemical genetics approaches can both provide biotechnological targets for crop stress improvement and an opportunity to directly use the identified chemical in agricultural practice if no unintended side effects are observed (Hicks and Raikhel, 2009; McCourt and Desveaux, 2010; Okamoto et al., 2013). Moreover as the effects of chemical priming are shown to, in part, be exerted through induction of phosphorylation and other post-translational modifications (Beckers et al., 2009), probing these modifications and genetically manipulating the underlying codons to constitutively mimic them (Riano-Pachon et al., 2010) can result in altered responses under combinatorial stress.

Breeding for resistance to exposure to combined abiotic and biotic stress by incorporation of genetic components regulating the response to both stresses faces various challenges. For example, TFs can have thousands of binding sites across the genome (Lu et al., 2013), increasing the chance of unwanted pleiotropic effects and therefore more sophisticated deployment should be employed. Both expression regulation and binding specificity can be altered through promoter and binding domain engineering (Desai et al.,

2009; Cox et al., 2013) which can be aided by comparative genomic approaches (Korkuc et al., 2014) and applied through novel site-specific mutagenesis techniques (Liu et al., 2013b). As selective and stimulus specific TF binding drives stress responses regulation (Xiao et al., 2013), implementation of the above methods will aid to fine-tune downstream targets toward the desired phenotypic response. A potential drawback of TF utilization is that resistance typically achieved by this approach is partial, and potentially prone to numerous antagonistic effects between stresses that cannot be predicted and can hinder efficient deployment for crop improvement to combined stresses.

Pyramiding genes that provide increased tolerance to either stress and do not (negatively) interact with each other offers an alternative route. Strong resistance mediated by R-genes, that appear to be robust to perturbations, can be pyramided with well-characterized genes conferring abiotic stress tolerance (Hu and Xiong, 2013; Kissoudis et al., unpublished data). R-gene robustness can be assessed by testing resistance responses under different abiotic stressors prior to pyramiding. The drawback of this approach is the quick breakdown of resistance due to evolving pathogens, and the fact that necrotrophic fungi resistance cannot be acquired with these genes. R-gene stacking aided by novel biotechnological approaches can reduce the risk of breakdown of R-gene-mediated resistance.

Pre-invasion defense mechanisms can be exploited, especially the one that is conferred by preformed or inducible physical barriers such as callose and antimicrobial compound deposition at the site of attempted penetration. As discussed earlier, callose deposition appears to be positively regulated by ABA signaling, therefore positive or no interaction should be expected under abiotic stress. Genes such as the OCP3 TF can be utilized, and for instance pyramiding abiotic stress tolerance with resistance conferred by *mlo* loss of function which sensitizes callose deposition at the site of infection for resistance against powdery mildew (Buschges et al., 1997) may be a viable route (Kissoudis et al. unpublished data). However, pleiotropic effects reported in *mlo* mutants such as compromised resistance against necrotrophic pathogens (Kumar et al., 2001) and accelerated senescence (Piffanelli et al., 2002) can have adverse consequences under stress combination.

The mechanisms through which abiotic stress tolerance is conferred can have a differential effect on disease resistance. As mentioned earlier, drought tolerance through ABA upregulation at the whole-plant level is expected to have antagonistic effects with SA signaling and therefore compromises resistance to biotrophs. Localized ABA sensitization in stomata (Bauer et al., 2013) can overcome these drawbacks and offer an advantage for resistance against pathogen that infect through stomata.

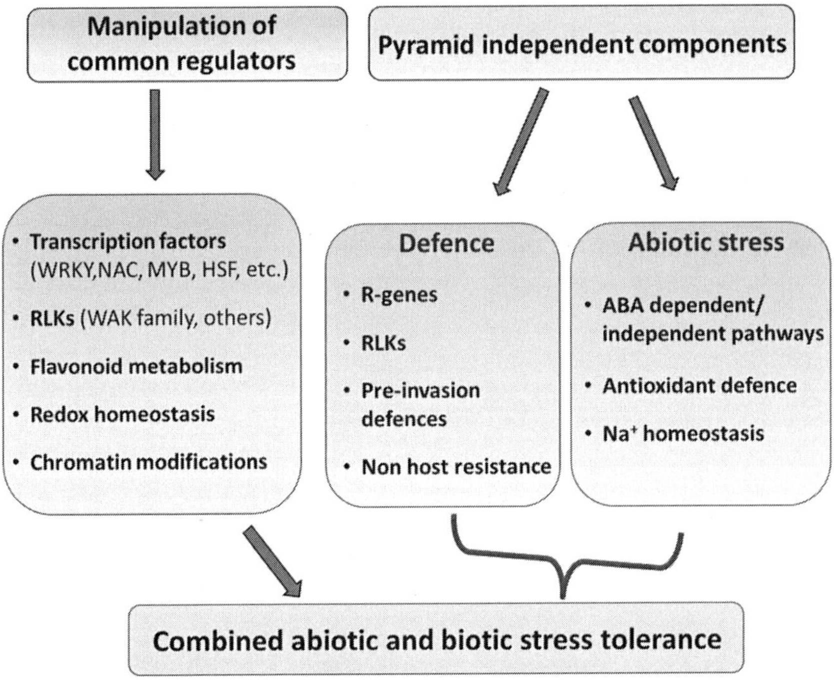

FIGURE 3. APPROACHES FOR BUILDING COMBINED ABIOTIC AND BIOTIC STRESS TOLERANCE IN PLANTS. Two strategies are proposed through either the manipulation of genetic components which potentially regulate resistance to both stresses in a preferentially unidirectional manner, or the pyramiding of genes that independently confer abiotic or biotic stress resistance and do not (negatively) interact. The selection of individual components might differ depending on the pathogen and the abiotic stress scenario.

Manipulation of developmental traits such as root system architecture can be beneficial for drought tolerance (Uga et al., 2013) with potentially no adverse effects on disease resistance, as they employ cell type specific signaling. Deployment of genes that have a protective function on proteins and cellular components under abiotic stress, such as dehydrins, LEA proteins or RNA chaperones (Kang et al., 2013) that apparently are downstream components of abiotic stress adaptation and mostly function through their structural properties, can minimize interaction with biotic stress signaling. Moreover, under salt stress, increased tolerance through Na^+ compartmentalization in the vacuoles may offer an advantage in comparison with Na^+ exclusion, as Na^+ at high concentrations can have adverse effects on pathogen feeding and development.

Approaches that result in greater antioxidant capacity such as the accumulation of flavonoids appear to confer resistance to abiotic and oxidative stress (Nakabayashi et al., 2014) while overproduction of their derivatives, anthocyanins, increase resistance to the necrotrophic pathogen *B. cinerea* in tomato by minimizing ROS burst (Zhang et al., 2013). Therefore engineering for increased flavonoid accumulation can be promising in conferring resistance to multiple stressors, however, it is unknown how it can affect the deployment of hypersensitivity response due to disturbed ROS homeostasis and thus resistance against biotrophic pathogens.

Exploitation and deployment of different strategies (Figure 3) under different abiotic stress/pathogen combinations will demonstrate their feasibility and applicability, further leading toward the goal of breeding for crops that maintain their robustness and yield performance under diverse environmental conditions.

CONFLICT OF INTEREST STATEMENT

The authors declare that the research was conducted in the absence of any commercial or financial relationships that could be construed as a potential conflict of interest.

ACKNOWLEDGMENT

Christos Kissoudis is supported by "Alexander S. Onassis" Public Benefit Foundation scholarship.

REFERENCES

1. Abuqamar, S., Luo, H. L., Laluk, K., Mickelbart, M. V., and Mengiste, T. (2009). Crosstalk between biotic and abiotic stress responses in tomato is mediated by the AIM1 transcription factor. *Plant J.* 58, 347–360.
2. Achard, P., Renou, J. P., Berthome, R., Harberd, N. P., and Genschik, P. (2008). Plant DELLAs restrain growth and promote survival of adversity by reducing the levels of reactive oxygen species. *Curr. Biol.* 18, 656–660.
3. Achuo, E. A., Prinsen, E., and Hofte, M. (2006). Influence of drought, salt stress and abscisic acid on the resistance of tomato to *Botrytis cinerea* and *Oidium neolycopersici*. *Plant Pathol.* 55, 178–186.
4. Adie, B. A. T., Perez-Perez, J., Perez-Perez, M. M., Godoy, M., Sanchez-Serrano, J. J., Schmelz, E. A., et al. (2007). ABA is an essential signal for plant resistance to

pathogens affecting JA biosynthesis and the activation of defenses in *Arabidopsis*. *Plant Cell* 19, 1665–1681.

5. Albrecht, C., Boutrot, F., Segonzac, C., Schwessinger, B., Gimenez-Ibanez, S., Chinchilla, D., et al. (2012). Brassinosteroids inhibit pathogen-associated molecular pattern-triggered immune signaling independent of the receptor kinase BAK1. *Proc. Natl. Acad. Sci. U.S.A.* 109, 303–308.
6. Argueso, C. T., Ferreira, F. J., Epple, P., To, J. P. C., Hutchison, C. E., Schaller, G. E., et al. (2012). Two-component elements mediate interactions between cytokinin and salicylic acid in plant immunity. *PLoS Genet.* 8:e1002448.
7. Asai, T., Tena, G., Plotnikova, J., Willmann, M. R., Chiu, W. L., Gomez-Gomez, L., et al. (2002). MAP kinase signalling cascade in *Arabidopsis* innate immunity. *Nature* 415, 977–983.
8. Asano, T., Hayashi, N., Kobayashi, M., Aoki, N., Miyao, A., Mitsuhara, I., et al. (2012). A rice calcium-dependent protein kinase OsCPK12 oppositely modulates salt-stress tolerance and blast disease resistance. *Plant J.* 69, 26–36.
9. Asselbergh, B., Curvers, K., Franca, S. C., Audenaert, K., Vuylsteke, M., Van Breusegem, F., et al. (2007). Resistance to *Botrytis cinerea* in sitiens, an abscisic acid-deficient tomato mutant, involves timely production of hydrogen peroxide and cell wall modifications in the epidermis. *Plant Physiol.* 144, 1863–1877.
10. Asselbergh, B., De Vleesschauwer, D., and Hofte, M. (2008). Global switches and fine-tuning-ABA modulates plant pathogen defense. *Mol. Plant Microbe Interact.* 21, 709–719.
11. Astier, J., Rasul, S., Koen, E., Manzoor, H., Besson-Bard, A., Lamotte, O., et al. (2011). *S*-nitrosylation: an emerging post-translational protein modification in plants. *Plant Sci.* 181, 527–533.
12. Atkinson, N. J., Lilley, C. J., and Urwin, P. E. (2013). Identification of genes involved in the response of *Arabidopsis* to simultaneous biotic and abiotic stresses. *Plant Physiol.* 162, 2028–2041.
13. Atkinson, N. J., and Urwin, P. E. (2012). The interaction of plant biotic and abiotic stresses: from genes to the field. *J. Exp. Bot.* 63, 3523–3543.
14. Baerenfaller, K., Massonnet, C., Walsh, S., Baginsky, S., Buhlmann, P., Hennig, L., et al. (2012). Systems-based analysis of *Arabidopsis* leaf growth reveals adaptation to water deficit. *Mol. Syst. Biol.* 8, 606.
15. Bari, R., and Jones, J. D. (2009). Role of plant hormones in plant defence responses. *Plant Mol. Biol.* 69, 473–488.
16. Bauer, H., Ache, P., Lautner, S., Fromm, J., Hartung, W., Al-Rasheid, K. A., et al. (2013). The stomatal response to reduced relative humidity requires guard cell-autonomous ABA synthesis. *Curr. Biol.* 23, 53–57.
17. Baxter, A., Mittler, R., and Suzuki, N. (2014). ROS as key players in plant stress signalling. *J. Exp. Bot.* 65, 1229–1240.
18. Bechtold, U., Albihlal, W. S., Lawson, T., Fryer, M. J., Sparrow, P. A., Richard, F., et al. (2013). *Arabidopsis* HEAT SHOCK TRANSCRIPTION FACTORA1b overexpression enhances water productivity, resistance to drought, and infection. *J. Exp. Bot.* 64, 3467–3481.
19. Beckers, G. J., Jaskiewicz, M., Liu, Y., Underwood, W. R., He, S. Y., Zhang, S., et al. (2009). Mitogen-activated protein kinases 3 and 6 are required for full priming of stress responses in *Arabidopsis thaliana*. *Plant Cell* 21, 944–953.
20. Belkhadir, Y., Jaillais, Y., Epple, P., Balsemao-Pires, E., Dangl, J. L., and Chory, J. (2012). Brassinosteroids modulate the efficiency of plant immune responses to microbe-associated molecular patterns. *Proc. Natl. Acad. Sci. U.S.A.* 109, 297–302.
21. Berendsen, R. L., Pieterse, C. M., and Bakker, P. A. (2012). The rhizosphere microbiome and plantHEALTH. *Trends Plant Sci.* 17, 478–486.
22. Bessire, M., Chassot, C., Jacquat, A. C., Humphry, M., Borel, S., Petetot, J. M., et al. (2007). A permeable cuticle in *Arabidopsis* leads to a strong resistance to *Botrytis cinerea*. *EMBO J.* 26, 2158–2168.

23. Bilgin, D. D., Zavala, J. A., Zhu, J., Clough, S. J., Ort, D. R., and Delucia, E. H. (2010). Biotic stress globally downregulates photosynthesis genes. *Plant Cell Environ.* 33, 1597–1613.
24. Bilichak, A., Ilnystkyy, Y., Hollunder, J., and Kovalchuk, I. (2012). The progeny of *Arabidopsis thaliana* plants exposed to salt exhibit changes in DNA methylation, histone modifications and gene expression. *PLoS ONE* 7:e30515.
25. Blomberg, A., and Adler, L. (1993). "Tolerance of fungi to NaCl," in *Stress Tolerance of Fungi*, ed. D. H. Jenings (New York, NY: Marcel Dekker Inc.), 209–232.
26. Bolton, M. D. (2009). Current review: primaryMETABOLISM and plant defense-fuel for the fire. *Mol. Plant Microbe Interact.* 22, 487–497.
27. Boudsocq, M., Willmann, M. R., Mccormack, M., Lee, H., Shan, L., He, P., et al. (2010). Differential innate immune signalling via Ca(2+) sensor protein kinases. *Nature* 464, 418–422.
28. Breeze, E., Harrison, E., Mchattie, S., Hughes, L., Hickman, R., Hill, C., et al. (2011). High-resolution temporal profiling of transcripts during *Arabidopsis* leaf senescence reveals a distinct chronology of processes and regulation. *Plant Cell* 23, 873–894.
29. Buschges, R., Hollricher, K., Panstruga, R., Simons, G., Wolter, M., Frijters, A., et al. (1997). The barley Mlo gene: a novel control element of plant pathogen resistance. *Cell* 88, 695–705.
30. Chan, E. K., Rowe, H. C., Corwin, J. A., Joseph, B., and Kliebenstein, D. J. (2011). Combining genome-wide association mapping and transcriptional networks to identify novel genes controlling glucosinolates in *Arabidopsis thaliana*. *PLoS Biol.* 9:e1001125.
31. Chaouch, S., and Noctor, G. (2010). Myo-inositol abolishes salicylic acid-dependent cell death and pathogen defence responses triggered by peroxisomal hydrogen peroxide. *New Phytol.* 188, 711–718.
32. Chen, L. Q., Hou, B. H., Lalonde, S., Takanaga, H., Hartung, M. L., Qu, X. Q., et al. (2010a). Sugar transporters for intercellular exchange and nutrition of pathogens. *Nature* 468, 527–532.
33. Chen, L. T., Luo, M., Wang, Y. Y., and Wu, K. (2010b). Involvement of *Arabidopsis* histone deacetylase HDA6 in ABA and salt stress response. *J. Exp. Bot.* 61, 3345–3353.
34. Chen, R., Jiang, H. L., Li, L., Zhai, Q. Z., Qi, L. L., Zhou, W. K., et al. (2012). The *Arabidopsis* mediator subunit MED25 differentially regulates jasmonate and abscisic acid signaling through interacting with the MYC2 and ABI5 transcription factors. *Plant Cell* 24, 2898–2916.
35. Chen, R. Q., Sun, S. L., Wang, C., Li, Y. S., Liang, Y., An, F. Y., et al. (2009). The *Arabidopsis* PARAQUAT RESISTANT2 gene encodes an *S*-nitrosoglutathione reductase that is a key regulator of cell death. *Cell Res.* 19, 1377–1387.
36. Chinchilla, D., Shan, L., He, P., De Vries, S., and Kemmerling, B. (2009). One for all: the receptor-associated kinase BAK1. *Trends Plant Sci.* 14, 535–541.
37. Chini, A., Grant, J. J., Seki, M., Shinozaki, K., and Loake, G. J. (2004). Drought tolerance established by enhanced expression of the CC-NBS-LRR gene, ADR1, requires salicylic acid, EDS1 and ABI1. *Plant J.* 38, 810–822.
38. Choi, J., Huh, S. U., Kojima, M., Sakakibara, H., Paek, K. H., and Hwang, I. (2010). The cytokinin-activated transcription factor ARR2 promotes plant immunity via TGA3/NPR1-dependent salicylic acid signaling in *Arabidopsis*. *Dev. Cell* 19, 284–295.
39. Clement, M., Leonhardt, N., Droillard, M. J., Reiter, I., Montillet, J. L., Genty, B., et al. (2011). The cytosolic/nuclear HSC70 and HSP90 molecular chaperones are important for stomatal closure and modulate abscisic acid-dependent physiological responses in *Arabidopsis*. *Plant Physiol.* 156, 1481–1492
40. Clough, S. J., Fengler, K. A., Yu, I. C., Lippok, B., Smith, R. K., and Bent, A. F. (2000). The *Arabidopsis* dnd1 "defense, no death" gene encodes a mutated cyclic nucleotide-gated ion channel. *Proc. Natl. Acad. Sci. U.S.A.* 97, 9323–9328.
41. Coleman-Derr, D., and Zilberman, D. (2012). Deposition of histone variant H2A.Z within gene bodies regulates responsive genes. *PLoS Genet.* 8:e1002988.

42. Coll, N. S., Epple, P., and Dangl, J. L. (2011). Programmed cell death in the plant immune system. *Cell Death Differ.* 18, 1247–1256.
43. Collins, N. C., Thordal-Christensen, H., Lipka, V., Bau, S., Kombrink, E., Qiu, J. L., et al. (2003). SNARE-protein-mediated disease resistance at the plant cell wall. *Nature* 425, 973–977.
44. Conrath, U. (2011). Molecular aspects of defence priming. *Trends Plant Sci.* 16, 524–531.
45. Cox, R. S. III, Nishikata, K., Shimoyama, S., Yoshida, Y., Matsui, M., Makita, Y., et al. (2013). PromoterCAD: data-driven design of plant regulatory DNA. *Nucleic Acids Res.* 41, W569–W574.
46. Curvers, K., Seifi, H., Mouille, G., De Rycke, R., Asselbergh, B., Van Hecke, A., et al. (2010). Abscisic acid deficiency causes changes in cuticle permeability and pectin composition that influence tomato resistance to *Botrytis cinerea*. *Plant Physiol.* 154, 847–860.
47. Daudi, A., Cheng, Z. Y., O'Brien, J. A., Mammarella, N., Khan, S., Ausubel, F. M., et al. (2012). The apoplastic oxidative burst peroxidase in *Arabidopsis* is a major component of pattern-triggered immunity. *Plant Cell* 24, 275–287.
48. De Block, M., Verduyn, C., De Brouwer, D., and Cornelissen, M. (2005). Poly(ADP-ribose) polymerase in plants affects energy homeostasis, cell death and stress tolerance. *Plant J.* 41, 95–106.
49. Demidchik, V., Shabala, S. N., and Davies, J. M. (2007). Spatial variation in H_2O_2 response of *Arabidopsis thaliana* root epidermal Ca^{2+} flux and plasma membrane Ca^{2+} channels. *Plant J.* 49, 377–386.
50. Dempsey, D. A., and Klessig, D. F. (2012). SOS – too many signals for systemic acquired resistance? *Trends Plant Sci.* 17, 538–545.
51. De Pinto, M. C., Locato, V., and De Gara, L. (2012). Redox regulation in plant programmed cell death. *Plant Cell Environ.* 35, 234–244.
52. Desai, T. A., Rodionov, D. A., Gelfand, M. S., Alm, E. J., and Rao, C. V. (2009). Engineering transcription factors with novel DNA-binding specificity using comparative genomics. *Nucleic Acids Res.* 37, 2493–2503.
53. Desprez-Loustau, M. L., Marcais, B., Nageleisen, L. M., Piou, D., and Vannini, A. (2006). Interactive effects of drought and pathogens in forest trees. *Ann. For. Sci.* 63, 597–612.
54. De Torres Zabala, M., Bennett, M. H., Truman, W. H., and Grant, M. R. (2009). Antagonism between salicylic and abscisic acid reflects early host–pathogen conflict and moulds plant defence responses. *Plant J.* 59, 375–386.
55. De Vleesschauwer, D., Van Buyten, E., Satoh, K., Balidion, J., Mauleon, R., Choi, I. R., et al. (2012). Brassinosteroids antagonize gibberellin- and salicylate-mediated root immunity in rice. *Plant Physiol.* 158, 1833–1846.
56. Ding, Y., Avramova, Z., and Fromm, M. (2011). The *Arabidopsis* trithorax-like factor ATX1 functions in dehydration stress responses via ABA-dependent and ABA-independent pathways. *Plant J.* 66, 735–744.
57. Divi, U. K., Rahman, T., and Krishna, P. (2010). Brassinosteroid-mediated stress tolerance in *Arabidopsis* shows interactions with abscisic acid, ethylene and salicylic acid pathways. *BMC Plant Biol.* 10:151.
58. Dodd, A. N., Kudla, J., and Sanders, D. (2010). The language of calcium signaling. *Annu. Rev. Plant Biol.* 61, 593–620.
59. Dodd, I. C., and Perez-Alfocea, F. (2012). Microbial amelioration of crop salinity stress. *J. Exp. Bot.* 63, 3415–3428.
60. Doherty, C. J., Van Buskirk, H. A., Myers, S. J., and Thomashow, M. F. (2009). Roles for *Arabidopsis* CAMTA transcription factors in cold-regulated gene expression and freezing tolerance. *Plant Cell* 21, 972–984.
61. Dowen, R. H., Pelizzola, M., Schmitz, R. J., Lister, R., Dowen, J. M., Nery, J. R., et al. (2012). Widespread dynamic DNA methylation in response to biotic stress. *Proc. Natl. Acad. Sci. U.S.A.* 109, E2183–E2191.

62. Du, L. Q., Ali, G. S., Simons, K. A., Hou, J. G., Yang, T. B., Reddy, A. S. N., et al. (2009). Ca^{2+}/calmodulin regulates salicylic-acid-mediated plant immunity. *Nature* 457, 1154–1116.
63. Dubiella, U., Seybold, H., Durian, G., Komander, E., Lassig, R., Witte, C. P., et al. (2013). Calcium-dependent protein kinase/NADPH oxidase activation circuit is required for rapid defense signal propagation. *Proc. Natl. Acad. Sci. U.S.A.* 110, 8744–8749.
64. Ellinger, D., Naumann, M., Falter, C., Zwikowics, C., Jamrow, T., Manisseri, C., et al. (2013). Elevated early callose deposition results in complete penetration resistance to powdery mildew in *Arabidopsis*. *Plant Physiol.* 161, 1433–1444.
65. Essmann, J., Schmitz-Thom, I., Schon, H., Sonnewald, S., Weis, E., and Scharte, J. (2008). RNA interference-mediated repression of cell wall invertase impairs defense in source leaves of tobacco. *Plant Physiol.* 147, 1288–1299.
66. Felle, H. H., Herrmann, A., Hanstein, S., Huckelhoven, R., and Kogel, K. H. (2004). Apoplastic pH signaling in barley leaves attacked by the powdery mildew fungus *Blumeria graminis* f. sp. *hordei*. *Mol. Plant Microbe Interact.* 17, 118–123.
67. Finka, A., Cuendet, A. F., Maathuis, F. J., Saidi, Y., and Goloubinoff, P. (2012). Plasma membrane cyclic nucleotide gated calcium channels control land plant thermal sensing and acquired thermotolerance. *Plant Cell* 24, 3333–3348.
68. Fones, H., Davis, C. A., Rico, A., Fang, F., Smith, J. A., and Preston, G. M. (2010). Metal hyperaccumulation armors plants against disease. *PLoS Pathog.* 6:e1001093.
69. Foyer, C. H., and Noctor, G. (2005). Redox homeostasis and antioxidant signaling: aMETABOLIC interface between stress perception and physiological responses. *Plant Cell* 17, 1866–1875.
70. Garcia-Andrade, J., Ramirez, V., Flors, V., and Vera, P. (2011). *Arabidopsis* ocp3 mutant reveals a mechanism linking ABA and JA to pathogen-induced callose deposition. *Plant J.* 67, 783–794.
71. Garrett, K. A., Dendy, S. P., Frank, E. E., Rouse, M. N., and Travers, S. E. (2006). Climate change effects on plant disease: genomes to ecosystems. *Annu. Rev. Phytopathol.* 44, 489–509.
72. Geilfus, C. M., and Muhling, K. H. (2011). Real-time imaging of leaf apoplastic pH dynamics in response to NaCl stress. *Front. Plant Sci.* 2:13.
73. Gifford, J. L., Jamshidiha, M., Mo, J., Ishida, H., and Vogel, H. J. (2013). Comparing the calcium binding abilities of two soybean calmodulins: towards understanding the divergent nature of plant calmodulins. *Plant Cell* 25, 4512–4524
74. Glazebrook, J. (2005). Contrasting mechanisms of defense against biotrophic and necrotrophic pathogens. *Annu. Rev. Phytopathol.* 43, 205–227.
75. Goel, A. K., Lundberg, D., Torres, M. A., Matthews, R., Akimoto-Tomiyama, C., Farmer, L., et al. (2008). The *Pseudomonas syringae* type III effector HopAM1 enhances virulence on water-stressed plants. *Mol. Plant Microbe Interact.* 21, 361–370.
76. Goellner, K., and Conrath, U. (2008). Priming: it's all the world to induced disease resistance. *Eur. J. Plant Pathol.* 121, 233–242.
77. Griebel, T., and Zeier, J. (2008). Light regulation and daytime dependency of inducible plant defenses in *Arabidopsis*: phytochrome signaling controls systemic acquired resistance rather than local defense. *Plant Physiol.* 147, 790–801.
78. Grimmer, M. K., John Foulkes, M., and Paveley, N. D. (2012). Foliar pathogenesis and plant water relations: a review. *J. Exp. Bot.* 63, 4321–4331.
79. Gudesblat, G. E., Iusem, N. D., and Morris, P. C. (2007). Guard cell-specific inhibition of *Arabidopsis* MPK3 expression causes abnormal stomatal responses to abscisic acid and hydrogen peroxide. *New Phytol.* 173, 713–721.
80. Guo, K. M., Babourina, O., Christopher, D. A., Borsic, T., and Rengel, Z. (2010). The cyclic nucleotide-gated channel AtCNGC10 transports Ca^{2+} and Mg^{2+} in *Arabidopsis*. *Physiol. Plant.* 139, 303–312.
81. Gutha, L. R., and Reddy, A. R. (2008). Rice DREB1B promoter shows distinct stress-specific responses, and the overexpression of cDNA in tobacco confers improved abiotic and biotic stress tolerance. *Plant Mol. Biol.* 68, 533–555.

82. Hamaji, K., Nagira, M., Yoshida, K., Ohnishi, M., Oda, Y., Uemura, T., et al. (2009). Dynamic aspects of ion accumulation by vesicle traffic under salt stress in *Arabidopsis*. *Plant Cell Physiol.* 50, 2023–2033.
83. Hamann, T., Bennett, M., Mansfield, J., and Somerville, C. (2009). Identification of cell-wall stress as a hexose-dependent and osmosensitive regulator of plant responses. *Plant J.* 57, 1015–1026.
84. Heidrich, K., Blanvillain-Baufume, S., and Parker, J. E. (2012). Molecular and spatial constraints on NB-LRR receptor signaling. *Curr. Opin. Plant Biol.* 15, 385–391.
85. Hématy, K., Cherk, C., and Somerville, S. (2009). Host–pathogen warfare at the plant cell wall. *Curr. Opin. Plant Biol.* 12, 406–413.
86. Hemetsberger, C., Herrberger, C., Zechmann, B., Hillmer, M., and Doehlemann, G. (2012). The *Ustilago maydis* effector Pep1 suppresses plant immunity by inhibition of host peroxidase activity. *PLoS Pathog.* 8:e1002684.
87. Hicks, G. R., and Raikhel, N. V. (2009). Opportunities and challenges in plant chemical biology. *Nat. Chem. Biol.* 5, 268–272.
88. Horváth, E., Szalai, G., and Janda, T. (2007). Induction of abiotic stress tolerance by salicylic acid signaling. *J. Plant Growth Regul.* 26, 290–300.
89. Hou, X., Lee, L. Y., Xia, K., Yan, Y., and Yu, H. (2010). DELLAs modulate jasmonate signaling via competitive binding to JAZs. *Dev. Cell* 19, 884–894.
90. Hu, H., and Xiong, L. (2013). Genetic engineering and breeding of drought-resistant crops. *Annu. Rev. Plant Biol.* 65, 715–774.
91. Huang, D., Wu, W., Abrams, S. R., and Cutler, A. J. (2008). The relationship of drought-related gene expression in *Arabidopsis thaliana* to hormonal and environmental factors. *J. Exp. Bot.* 59, 2991–3007.
92. Huang, X., Li, J., Bao, F., Zhang, X., and Yang, S. (2010). A gain-of-function mutation in the *Arabidopsis* disease resistance gene RPP4 confers sensitivity to low temperature. *Plant Physiol.* 154, 796–809.
93. Hubert, D. A., He, Y., Mcnulty, B. C., Tornero, P., and Dangl, J. L. (2009). Specific *Arabidopsis* HSP90.2 alleles recapitulate RAR1 cochaperone function in plant NB-LRR disease resistance protein regulation. *Proc. Natl. Acad. Sci. U.S.A.* 106, 9556–9563.
94. Hummel, I., Pantin, F., Sulpice, R., Piques, M., Rolland, G., Dauzat, M., et al. (2010). *Arabidopsis* plants acclimate to water deficit at low cost through changes of carbon usage: an integrated perspective using growth, metabolite, enzyme, and gene expression analysis. *Plant Physiol.* 154, 357–372.
95. Jelenska, J., Van Hal, J. A., and Greenberg, J. T. (2010). *Pseudomonas syringae* hijacks plant stress chaperone machinery for virulence. *Proc. Natl. Acad. Sci. U.S.A.* 107, 13177–13182.
96. Johannes, F., Porcher, E., Teixeira, F. K., Saliba-Colombani, V., Simon, M., Agier, N., et al. (2009). Assessing the impact of transgenerational epigenetic variation on complex traits. *PLoS Genet.* 5:e1000530.
97. Jung, J., Won, S. Y., Suh, S. C., Kim, H., Wing, R., Jeong, Y., et al. (2007). The barley ERF-type transcription factor HvRAF confers enhanced pathogen resistance and salt tolerance in *Arabidopsis*. *Planta* 225, 575–588.
98. Kachroo, P., Shanklin, J., Shah, J., Whittle, E. J., and Klessig, D. F. (2001). A fatty acid desaturase modulates the activation of defense signaling pathways in plants. *Proc. Natl. Acad. Sci. U.S.A.* 98, 9448–9453.
99. Kang, H., Park, S. J., and Kwak, K. J. (2013). Plant RNA chaperones in stress response. *Trends Plant Sci.* 18, 100–106.
100. Kasuga, M., Miura, S., Shinozaki, K., and Yamaguchi-Shinozaki, K. (2004). A combination of the *Arabidopsis* DREB1A gene and stress-inducible rd29A promoter improved drought- and low-temperature stress tolerance in tobacco by gene transfer. *Plant Cell Physiol.* 45, 346–350.
101. Kim, J. I., Murphy, A. S., Baek, D., Lee, S. W., Yun, D. J., Bressan, R. A., et al. (2011a). YUCCA6 over-expression demonstrates auxin function in delaying leaf senescence in *Arabidopsis thaliana*. *J. Exp. Bot.* 62, 3981–3992.

102. Kim, T. H., Hauser, F., Ha, T., Xue, S., Bohmer, M., Nishimura, N., et al. (2011b). Chemical genetics reveals negative regulation of abscisic acid signaling by a plant immune response pathway. *Curr. Biol.* 21, 990–997.

103. Kim, K. C., Lai, Z., Fan, B., and Chen, Z. (2008). *Arabidopsis* WRKY38 and WRKY62 transcription factors interact with histone deacetylase 19 in basal defense. *Plant Cell* 20, 2357–2371.

104. Knepper, C., Savory, E. A., and Day, B. (2011). *Arabidopsis* NDR1 is an integrin-like protein with a role in fluid loss and plasma membrane-cell wall adhesion. *Plant Physiol.* 156, 286–300.

105. Kobayashi, M., Yoshioka, M., Asai, S., Nomura, H., Kuchimura, K., Mori, H., et al. (2012). StCDPK5 confers resistance to late blight pathogen but increases susceptibility to early blight pathogen in potato via reactive oxygen species burst. *New Phytol.* 196, 223–237.

106. Koh, S., Lee, S. C., Kim, M. K., Koh, J. H., Lee, S., An, G., et al. (2007). T-DNA tagged knockout mutation of rice OsGSK1, an orthologue of *Arabidopsis* BIN2, with enhanced tolerance to various abiotic stresses. *Plant Mol. Biol.* 65, 453–466.

107. Kong, Q., Qu, N., Gao, M., Zhang, Z., Ding, X., Yang, F., et al. (2012). The MEKK1-MKK1/MKK2-MPK4 kinase cascade negatively regulates immunity mediated by a mitogen-activated protein kinase kinase kinase in *Arabidopsis*. *Plant Cell* 24, 2225–2236.

108. Koornneef, A., Leon-Reyes, A., Ritsema, T., Verhage, A., Den Otter, F. C., Van Loon, L. C., et al. (2008). Kinetics of salicylate-mediated suppression of jasmonate signaling reveal a role for redox modulation. *Plant Physiol.* 147, 1358–1368.

109. Korkuc, P., Schippers, J. H., and Walther, D. (2014). Characterization and identification of cis-regulatory elements in *Arabidopsis* based on single-nucleotide polymorphism information. *Plant Physiol.* 164, 181–200.

110. Korn, M., Gartner, T., Erban, A., Kopka, J., Selbig, J., and Hincha, D. K. (2010). Predicting *Arabidopsis* freezing tolerance and heterosis in freezing tolerance from metabolite composition. *Mol. Plant* 3, 224–235.

111. Kosma, D. K., Bourdenx, B., Bernard, A., Parsons, E. P., Lu, S., Joubes, J., et al. (2009). The impact of water deficiency on leaf cuticle lipids of *Arabidopsis*. *Plant Physiol.* 151, 1918–1929.

112. Koussevitzky, S., Suzuki, N., Huntington, S., Armijo, L., Sha, W., Cortes, D., et al. (2008). Ascorbate peroxidase 1 plays a key role in the response of *Arabidopsis thaliana* to stress combination. *J. Biol. Chem.* 283, 34197–34203.

113. Kulcheski, F. R., De Oliveira, L. F., Molina, L. G., Almerao, M. P., Rodrigues, F. A., Marcolino, J., et al. (2011). Identification of novel soybean microRNAs involved in abiotic and biotic stresses. *BMC Genomics* 12:307.

114. Kumar, J., Huckelhoven, R., Beckhove, U., Nagarajan, S., and Kogel, K. H. (2001). A compromised Mlo pathway affects the response of barley to the necrotrophic fungus *Bipolaris sorokiniana* (Teleomorph: *Cochliobolus sativus*) and its toxins. *Phytopathology* 91, 127–133.

115. Kumar, S. V., and Wigge, P. A. (2010). H2A.Z-containing nucleosomes mediate the thermosensory response in *Arabidopsis*. *Cell* 140, 136–147.

116. Lee, B. R., Kim, K. Y., Jung, W. J., Avice, J. C., Ourry, A., and Kim, T. H. (2007). Peroxidases and lignification in relation to the intensity of water-deficit stress in white clover (*Trifolium repens* L.). *J. Exp. Bot.* 58, 1271–1279.

117. Li, D., Liu, H., Zhang, H., Wang, X., and Song, F. (2008). OsBIRH1, a DEAD-box RNA helicase with functions in modulating defence responses against pathogen infection and oxidative stress. *J. Exp. Bot.* 59, 2133–2146.

118. Li, P. F., Chen, L., Zhou, Y. H., Xia, X. J., Shi, K., Chen, Z. X., et al. (2013). Brassinosteroids-induced systemic stress tolerance was associated with increased transcripts of several defence-related genes in the phloem in *Cucumis sativus*. *PLoS ONE* 8:e66582.

119. Lin, W., Lu, D., Gao, X., Jiang, S., Ma, X., Wang, Z., et al. (2013). Inverse modulation of plant immune and brassinosteroid signaling pathways by the receptor-like cytoplasmic kinase BIK1. *Proc. Natl. Acad. Sci. U.S.A.* 110, 12114–12119.
120. Lindner, H., Müller, L. M., Boisson-Dernier, A., and Grossniklaus, U. (2012). CrRLK1L receptor-like kinases: not just another brick in the wall. *Curr. Opin. Plant Biol.* 15, 659–669.
121. Liu, C., Lu, F., Cui, X., and Cao, X. (2010). Histone methylation in higher plants. *Annu. Rev. Plant Biol.* 61, 395–420.
122. Liu, S., Wang, X., Wang, H., Xin, H., Yang, X., Yan, J., et al. (2013a). Genome-wide analysis of *ZmDREB* genes and their association with natural variation in drought tolerance at seedling stage of *Zea mays* L. *PLoS Genet.* 9:e1003790.
123. Liu, W. S., Yuan, J. S., and Stewart, C. N. (2013b). Advanced genetic tools for plant biotechnology. *Nat. Rev. Genet.* 14, 781–793.
124. Liu, Y., and Zhang, S. (2004). Phosphorylation of 1-aminocyclopropane-1-carboxylic acid synthase by MPK6, a stress-responsive mitogen-activated protein kinase, induces ethylene biosynthesis in *Arabidopsis. Plant Cell* 16, 3386–3399.
125. Locato, V., De Pinto, M. C., and De Gara, L. (2009). Different involvement of the mitochondrial, plastidial and cytosolic ascorbate-glutathione redox enzymes in heat shock responses. *Physiol. Plant.* 135, 296–306.
126. López, M. A., Bannenberg, G., and Castresana, C. (2008). Controlling hormone signaling is a plant and pathogen challenge for growth and survival. *Curr. Opin. Plant Biol.* 11, 420–427.
127. Lu, Z., Yu, H., Xiong, G., Wang, J., Jiao, Y., Liu, G., et al. (2013). Genome-wide binding analysis of the transcription activator ideal plant architecture1 reveals a complex network regulating rice plant architecture. *Plant Cell* 25, 3743–3759.
128. Luna, E., Pastor, V., Robert, J., Flors, V., Mauch-Mani, B., and Ton, J. (2011). Callose deposition: a multifaceted plant defense response. *Mol. Plant Microbe Interact.* 24, 183–193.
129. Ma, S., and Bohnert, H. J. (2007). Integration of *Arabidopsis thaliana* stress-related transcript profiles, promoter structures, and cell-specific expression. *Genome Biol.* 8, R49.
130. Ma, S., Gong, Q., and Bohnert, H. J. (2006). Dissecting salt stress pathways. *J. Exp. Bot.* 57, 1097–1107.
131. Mandal, M. K., Chandra-Shekara, A. C., Jeong, R. D., Yu, K., Zhu, S., Chanda, B., et al. (2012). Oleic acid-dependent modulation of NITRIC OXIDE ASSOCIATED1 protein levels regulates nitric oxide-mediated defense signaling in *Arabidopsis. Plant Cell* 24, 1654–1674.
132. Mang, H. G., Qian, W. Q., Zhu, Y., Qian, J., Kang, H. G., Klessig, D. F., et al. (2012). Abscisic acid deficiency antagonizes high-temperature inhibition of disease resistance through enhancing nuclear accumulation of resistance proteins SNC1 and RPS4 in *Arabidopsis. Plant Cell* 24, 1271–1284.
133. March-Diaz, R., Garcia-Dominguez, M., Lozano-Juste, J., Leon, J., Florencio, F. J., and Reyes, J. C. (2008). Histone H2A.Z and homologues of components of the SWR1 complex are required to control immunity in *Arabidopsis. Plant J.* 53, 475–487.
134. Marquez, L. M., Redman, R. S., Rodriguez, R. J., and Roossinck, M. J. (2007). A virus in a fungus in a plant: three-way symbiosis required for thermal tolerance. *Science* 315, 513–515.
135. Martí, M. C., Stancombe, M. A., and Webb, A. A. R. (2013). Cell- and stimulus type-specific intracellular free Ca^{2+} signals in *Arabidopsis. Plant Physiol.* 163, 625–634.
136. Mazzucotelli, E., Mastrangelo, A. M., Crosatti, C., Guerra, D., Stanca, A. M., and Cattivelli, L. (2008). Abiotic stress response in plants: when post-transcriptional and post-translational regulations control transcription. *Plant Sci.* 174, 420–431.
137. McCourt, P., and Desveaux, D. (2010). Plant chemical genetics. *New Phytol.* 185, 15–26.
138. McHale, L., Tan, X., Koehl, P., and Michelmore, R. W. (2006). Plant NBS-LRR proteins: adaptable guards. *Genome Biol.* 7, 212.

139. Melotto, M., Underwood, W., Koczan, J., Nomura, K., and He, S. Y. (2006). Plant stomata function in innate immunity against bacterial invasion. *Cell* 126, 969–980.

140. Mengiste, T., Chen, X., Salmeron, J., and Dietrich, R. (2003). The BOTRYTIS SUSCEPTIBLE1 gene encodes an R2R3MYB transcription factor protein that is required for biotic and abiotic stress responses in *Arabidopsis*. *Plant Cell* 15, 2551–2565.

141. Mhamdi, A., Hager, J., Chaouch, S., Queval, G., Han, Y., Taconnat, L., et al. (2010). *Arabidopsis* GLUTATHIONE REDUCTASE1 plays a crucial role in leaf responses to intracellular hydrogen peroxide and in ensuring appropriate gene expression through both salicylic acid and jasmonic acid signaling pathways. *Plant Physiol.* 153, 1144–1160.

142. Mhamdi, A., Han, Y., and Noctor, G. (2013). Glutathione-dependent phytohormone responses: teasing apart signaling and antioxidant functions. *Plant Signal. Behav.* 8, e24181.

143. Mi, S. K., Song, M. C., Eun, Y. K., Yang, J. I., Hwangbo, H., Young, C. K., et al. (2008). Galactinol is a signaling component of the induced systemic resistance caused by *Pseudomonas chlororaphis* O6 root colonization. *Mol. Plant Microbe Interact.* 21, 1643–1653.

144. Miller, G., Shulaev, V., and Mittler, R. (2008). Reactive oxygen signaling and abiotic stress. *Physiol. Plant.* 133, 481–489.

145. Mirouze, M., and Paszkowski, J. (2011). Epigenetic contribution to stress adaptation in plants. *Curr. Opin. Plant Biol.* 14, 267–274.

146. Mittler, R., Vanderauwera, S., Suzuki, N., Miller, G., Tognetti, V. B., Vandepoele, K., et al. (2011). ROS signaling: the new wave? *Trends Plant Sci.* 16, 300–309.

147. Munne-Bosch, S., Queval, G., and Foyer, C. H. (2013). The impact of global change factors on redox signaling underpinning stress tolerance. *Plant Physiol.* 161, 5–19.

148. Munns, R., and Tester, M. (2008). Mechanisms of salinity tolerance. *Annu. Rev. Plant Biol.* 59, 651–681.

149. Nagano, A. J., Sato, Y., Mihara, M., Antonio, B. A., Motoyama, R., Itoh, H., et al. (2012). Deciphering and prediction of transcriptome dynamics under fluctuating field conditions. *Cell* 151, 1358–1369.

150. Nakabayashi, R., Yonekura-Sakakibara, K., Urano, K., Suzuki, M., Yamada, Y., Nishizawa, T., et al. (2014). Enhancement of oxidative and drought tolerance in *Arabidopsis* by overaccumulation of antioxidant flavonoids. *Plant J.* 77, 367–379.

151. Nakashima, K., Tran, L. S., Van Nguyen, D., Fujita, M., Maruyama, K., Todaka, D., et al. (2007). Functional analysis of a NAC-type transcription factor OsNAC6 involved in abiotic and biotic stress-responsive gene expression in rice. *Plant J.* 51, 617–630.

152. Nakashita, H., Yasuda, M., Nitta, T., Asami, T., Fujioka, S., Arai, Y., et al. (2003). Brassinosteroid functions in a broad range of disease resistance in tobacco and rice. *Plant J.* 33, 887–898.

153. Navarro, L., Bari, R., Achard, P., Lison, P., Nemri, A., Harberd, N. P., et al. (2008). DELLAs control plant immune responses by modulating the balance of jasmonic acid and salicylic acid signaling. *Curr. Biol.* 18, 650–655.

154. Ndamukong, I., Jones, D. R., Lapko, H., Divecha, N., and Avramova, Z. (2010). Phosphatidylinositol 5-phosphate links dehydration stress to the activity of ARABIDOPSIS TRITHORAX-LIKE factor ATX1. *PLoS ONE* 5:e13396.

155. Nemhauser, J. L., Hong, F. X., and Chory, J. (2006). Different plant hormones regulate similar processes through largely nonoverlapping transcriptional responses. *Cell* 126, 467–475.

156. Nishimura, N., Okamoto, M., Narusaka, M., Yasuda, M., Nakashita, H., Shinozaki, K., et al. (2009). ABA hypersensitive germination2-1 causes the activation of both abscisic acid and salicylic acid responses in *Arabidopsis*. *Plant Cell Physiol.* 50, 2112–2122.

157. Obata, T., and Fernie, A. R. (2012). The use of metabolomics to dissect plant responses to abiotic stresses. *Cell. Mol. Life Sci.* 69, 3225–3243.

158. Okamoto, M., Peterson, F. C., Defries, A., Park, S. Y., Endo, A., Nambara, E., et al. (2013). Activation of dimeric ABA receptors elicits guard cell closure, ABA-regulated

gene expression, and drought tolerance. *Proc. Natl. Acad. Sci. U.S.A.* 110, 12132–12137.
159. Pajerowska-Mukhtar, K. M., Wang, W., Tada, Y., Oka, N., Tucker, C. L., Fonseca, J. P., et al. (2012). The HSF-like transcription factor TBF1 is a major molecular switch for plant growth-to-defense transition. *Curr. Biol.* 22, 103–112.
160. Park, S. W., Kaimoyo, E., Kumar, D., Mosher, S., and Klessig, D. F. (2007). Methyl salicylate is a critical mobile signal for plant systemic acquired resistance. *Science* 318, 113–116.
161. Peleg, Z., and Blumwald, E. (2011). Hormone balance and abiotic stress tolerance in crop plants. *Curr. Opin. Plant Biol.* 14, 290–295.
162. Peters, G. P., Marland, G., Le Quéré, C., Boden, T., Canadell, J. G., and Raupach, M. R. (2011). Rapid growth of CO_2 emission after the 2008–2009 global financial crisis. *Nat. Clim. Change* 2, 1–3.
163. Pick, T., Jaskiewicz, M., Peterhansel, C., and Conrath, U. (2012). Heat shock factor HsfB1 primes gene transcription and systemic acquired resistance in *Arabidopsis. Plant Physiol.* 159, 52–55.
164. Pieterse, C. M., Leon-Reyes, A., Van Der Ent, S., and Van Wees, S. C. (2009). Networking by small-molecule hormones in plant immunity. *Nat. Chem. Biol.* 5, 308–316.
165. Piffanelli, P., Zhou, F. S., Casais, C., Orme, J., Jarosch, B., Schaffrath, U., et al. (2002). The barley MLO modulator of defense and cell death is responsive to biotic and abiotic stress stimuli. *Plant Physiol.* 129, 1076–1085.
166. Pitzschke, A., Djamei, A., Bitton, F., and Hirt, H. (2009). A major role of the MEKK1-MKK1/2-MPK4 pathway in ROS signalling. *Mol. Plant* 2, 120–137.
167. Popescu, S. C., Popescu, G. V., Bachan, S., Zhang, Z. M., Gerstein, M., Snyder, M., et al. (2009). MAPK target networks in *Arabidopsis thaliana* revealed using functional protein microarrays. *Genes Dev.* 23, 80–92.
168. Popescu, S. C., Popescu, G. V., Bachan, S., Zhang, Z. M., Seay, M., Gerstein, M., et al. (2007). Differential binding of calmodulin-related proteins to their targets revealed through high-density *Arabidopsis* protein microarrays. *Proc. Natl. Acad. Sci. U.S.A.* 104, 4730–4735.
169. Poschenrieder, C., Tolrà, R., and Barceló, J. (2006). Can metals defend plants against biotic stress? *Trends Plant Sci.* 11, 288–295.
170. Potters, G., Horemans, N., and Jansen, M. A. (2010). The cellular redox state in plant stress biology – a charging concept. *Plant Physiol. Biochem.* 48, 292–300.
171. Prasch, C. M., and Sonnewald, U. (2013). Simultaneous application of heat, drought, and virus to *Arabidopsis* plants reveals significant shifts in signaling networks. *Plant Physiol.* 162, 1849–1866.
172. Quackenbush, J. (2003). Genomics. Microarrays – guilt by association. *Science* 302, 240–241.
173. Ramírez, V., Coego, A., López, A., Agorio, A., Flors, V., and Vera, P. (2009). Drought tolerance in *Arabidopsis* is controlled by the OCP3 disease resistance regulator. *Plant J.* 58, 578–591.
174. Rasmussen, S., Barah, P., Suarez-Rodriguez, M. C., Bressendorff, S., Friis, P., Costantino, P., et al. (2013). Transcriptome responses to combinations of stresses in *Arabidopsis. Plant Physiol.* 161, 1783–1794.
175. Reinbothe, C., Springer, A., Samol, I., and Reinbothe, S. (2009). Plant oxylipins: role of jasmonic acid during programmed cell death, defence and leaf senescence. *FEBS J.* 276, 4666–4681.
176. Rentel, M. C., Lecourieux, D., Ouaked, F., Usher, S. L., Petersen, L., Okamoto, H., et al. (2004). OXI1 kinase is necessary for oxidative burst-mediated signalling in *Arabidopsis. Nature* 427, 858–861
177. Reusche, M., Thole, K., Janz, D., Truskina, J., Rindfleisch, S., Drubert, C., et al. (2012). Verticillium infection triggers VASCULAR-RELATED NAC DOMAIN7-dependent de novo xylem formation and enhances drought tolerance in *Arabidopsis. Plant Cell* 24, 3823–3837.

178. Riano-Pachon, D. M., Kleessen, S., Neigenfind, J., Durek, P., Weber, E., Engelsberger, W. R., et al. (2010). Proteome-wide survey of phosphorylation patterns affected by nuclear DNA polymorphisms in *Arabidopsis* thaliana. *BMC Genomics* 11:411.
179. Rizhsky, L., Liang, H., Shuman, J., Shulaev, V., Davletova, S., and Mittler, R. (2004). When defense pathways collide. The response of *Arabidopsis* to a combination of drought and heat stress 1[w]. *Plant Physiol.* 134, 1683–1696.
180. Robert-Seilaniantz, A., Grant, M., and Jones, J. D. G. (2011). Hormone crosstalk in plant disease and defense: more than just JASMONATE-SALICYLATE antagonism. *Annu. Rev. Phytopathol.* 49, 317–343.
181. Samajova, O., Komis, G., and Samaj, J. (2013). Emerging topics in the cell biology of mitogen-activated protein kinases. *Trends Plant Sci.* 18, 140–148.
182. Sanchez-Vallet, A., Lopez, G., Ramos, B., Delgado-Cerezo, M., Riviere, M. P., Llorente, F., et al. (2012). Disruption of abscisic acid signaling constitutively activates *Arabidopsis* resistance to the necrotrophic fungus *Plectosphaerella cucumerina*. *Plant Physiol.* 160, 2109–2124.
183. Sanogo, S. (2004). Response of chile pepper to *Phytophthora capsici* in relation to soil salinity. *Plant Dis.* 88, 205–209.
184. Savchenko, T., Walley, J. W., Chehab, E. W., Xiao, Y., Kaspi, R., Pye, M. F., et al. (2010). Arachidonic acid: an evolutionarily conserved signaling molecule modulates plant stress signaling networks. *Plant Cell* 22, 3193–3205.
185. Schulz, P., Herde, M., and Romeis, T. (2013). Calcium-dependent protein kinases: hubs in plant stress signaling and development. *Plant Physiol.* 163, 523–530.
186. Segonzac, C., Feike, D., Gimenez-Ibanez, S., Hann, D. R., Zipfel, C., and Rathjen, J. P. (2011). Hierarchy and roles of pathogen-associated molecular pattern-induced responses in *Nicotiana benthamiana*. *Plant Physiol.* 156, 687–699.
187. Senthil-Kumar, M., and Mysore, K. S. (2012). Ornithine-delta-aminotransferase and proline dehydrogenase genes play a role in non-host disease resistance by regulating pyrroline-5-carboxylate metabolism-induced hypersensitive response. *Plant Cell Environ.* 35, 1329–1343.
188. Seo, Y. S., Chern, M., Bartley, L. E., Han, M., Jung, K. H., Lee, I., et al. (2011). Towards establishment of a rice stress response interactome. *PLoS Genet.* 7:e1002020.
189. Shaik, R., and Ramakrishna, W. (2013). Genes and co-expression modules common to drought and bacterial stress responses in *Arabidopsis* and rice. *PLoS ONE* 8:e77261.
190. Shaik, R., and Ramakrishna, W. (2014). Machine learning approaches distinguish multiple stress conditions using stress-responsive genes and identify candidate genes for broad resistance in rice. *Plant Physiol.* 164, 481–495.
191. Sharma, R., De Vleesschauwer, D., Sharma, M. K., and Ronald, P. C. (2013). Recent advances in dissecting stress-regulatory crosstalk in rice. *Mol. Plant* 6, 250–260.
192. Skirycz, A., De Bodt, S., Obata, T., De Clercq, I., Claeys, H., De Rycke, R., et al. (2010). Developmental stage specificity and the role of mitochondrialMETABOLISM in the response of *Arabidopsis* leaves to prolonged mild osmotic stress. *Plant Physiol.* 152, 226–244.
193. Soliman, M. F., and Kostandi, S. F. (1998). Effect of saline environment on yield and smut disease severity of different corn genotypes (*Zea mays* L.). *J. Phytopathol.* 146, 185–189.
194. Tada, Y., Spoel, S. H., Pajerowska-Mukhtar, K., Mou, Z. L., Song, J. Q., Wang, C., et al. (2008). Plant immunity requires conformational charges of NPR1 via *S*-nitrosylation and thioredoxins. *Science* 321, 952–956.
195. Takahashi, F., Mizoguchi, T., Yoshida, R., Ichimura, K., and Shinozaki, K. (2011). Calmodulin-dependent activation of MAP kinase for ROS homeostasis in *Arabidopsis*. *Mol. Cell.* 41, 649–660.
196. Tan, J. L., Wang, C. Y., Xiang, B., Han, R. H., and Guo, Z. F. (2013). Hydrogen peroxide and nitric oxide mediated cold- and dehydration-induced myo-inositol phosphate synthase that confers multiple resistances to abiotic stresses. *Plant Cell Environ.* 36, 288–299.

197. Tang, D., Simonich, M. T., and Innes, R. W. (2007). Mutations in LACS2, a long-chain acyl-coenzyme A synthetase, enhance susceptibility to avirulent *Pseudomonas syringae* but confer resistance to *Botrytis cinerea* in *Arabidopsis*. *Plant Physiol.* 144, 1093–1103.
198. Tanou, G., Filippou, P., Belghazi, M., Job, D., Diamantidis, G., Fotopoulos, V., et al. (2012). Oxidative and nitrosative-based signaling and associated post-translational modifications orchestrate the acclimation of citrus plants to salinity stress. *Plant J.* 72, 585–599.
199. Tanou, G., Molassiotis, A., and Diamantidis, G. (2009). Hydrogen peroxide- and nitric oxide-induced systemic antioxidant prime-like activity under NaCl-stress and stress-free conditions in citrus plants. *J. Plant Physiol.* 166, 1904–1913.
200. Tarantino, D., Vannini, C., Bracale, M., Campa, M., Soave, C., and Murgia, I. (2005). Antisense reduction of thylakoidal ascorbate peroxidase in *Arabidopsis* enhances paraquat-induced photooxidative stress and nitric oxide-induced cell death. *Planta* 221, 757–765.
201. Teige, M., Scheikl, E., Eulgem, T., Doczi, R., Ichimura, K., Shinozaki, K., et al. (2004). The MKK2 pathway mediates cold and salt stress signaling in *Arabidopsis*. *Mol. Cell* 15, 141–152.
202. Tena, G., Boudsocq, M., and Sheen, J. (2011). Protein kinase signaling networks in plant innate immunity. *Curr. Opin. Plant Biol.* 14, 519–529.
203. Tian, M. Y., Von Dahl, C. C., Liu, P. P., Friso, G., Van Wijk, K. J., and Klessig, D. F. (2012). The combined use of photoaffinity labeling and surface plasmon resonance-based technology identifies multiple salicylic acid-binding proteins. *Plant J.* 72, 1027–1038.
204. Todesco, M., Balasubramanian, S., Hu, T. T., Traw, M. B., Horton, M., Epple, P., et al. (2010). Natural allelic variation underlying a majorFITNESS trade-off in *Arabidopsis thaliana*. *Nature* 465, 632–636.
205. Ton, J., Flors, V., and Mauch-Mani, B. (2009). The multifaceted role of ABA in disease resistance. *Trends Plant Sci.* 14, 310–317.
206. Ton, J., Jakab, G., Toquin, V., Flors, V., Iavicoli, A., Maeder, M. N., et al. (2005). Dissecting the β-aminobutyric acid-induced priming phenomenon in *Arabidopsis*. *Plant Cell* 17, 987–999.
207. Ton, J., and Mauch-Mani, B. (2004). β-amino-butyric acid-induced resistance against necrotrophic pathogens is based on ABA-dependent priming for callose. *Plant J.* 38, 119–130.
208. Triky-Dotan, S., Yermiyahu, U., Katan, J., and Gamliel, A. (2005). Development of crown and root rot disease of tomato under irrigation with saline water. *Phytopathology* 95, 1438–1444.
209. Tsuda, K., Sato, M., Stoddard, T., Glazebrook, J., and Katagiri, F. (2009). Network properties of robust immunity in plants. *PLoS Genet.* 5:e1000772.
210. Uga, Y., Sugimoto, K., Ogawa, S., Rane, J., Ishitani, M., Hara, N., et al. (2013). Control of root system architecture by DEEPER ROOTING 1 increases rice yield under drought conditions. *Nat. Genet.* 45, 1097–1102.
211. Underwood, W., and Somerville, S. C. (2008). Focal accumulation of defences at sites of fungal pathogen attack. *J. Exp. Bot.* 59, 3501–3508.
212. Upchurch, R. G. (2008). Fatty acid unsaturation, mobilization, and regulation in the response of plants to stress. *Biotechnol. Lett.* 30, 967–977.
213. Urano, K., Maruyama, K., Ogata, Y., Morishita, Y., Takeda, M., Sakurai, N., et al. (2009). Characterization of the ABA-regulated global responses to dehydration in *Arabidopsis* by metabolomics. *Plant J.* 57, 1065–1078.
214. Van Loon, L. C., Rep, M., and Pieterse, C. M. J. (2006). Significance of inducible defense-related proteins in infected plants. *Annu. Rev. Phytopathol.* 44, 135–162.
215. Vincent, D., Lapierre, C., Pollet, B., Cornic, G., Negroni, L., and Zivy, M. (2005). Water deficits affect caffeate *O*-methyltransferase, lignification, and related enzymes in maize leaves. A proteomic investigation. *Plant Physiol.* 137, 949–960.

216. Vlot, A. C., Dempsey, D. A., and Klessig, D. F. (2009). Salicylic acid, a multifaceted hormone to combat disease. *Annu. Rev. Phytopathol.* 47, 177–206.

217. Vogel, J. P., Raab, T. K., Somerville, C. R., and Somerville, S. C. (2004). Mutations in PMR5 result in powdery mildew resistance and altered cell wall composition. *Plant J.* 40, 968–978.

218. Walia, H., Wilson, C., Condamine, P., Liu, X., Ismail, A. M., and Close, T. J. (2007). Large-scale expression profiling and physiological characterization of jasmonic acid-mediated adaptation of barley to salinity stress. *Plant Cell Environ.* 30, 410–421.

219. Walley, J. W., Coughlan, S., Hudson, M. E., Covington, M. F., Kaspi, R., Banu, G., et al. (2007). Mechanical stress induces biotic and abiotic stress responses via a novel *cis*-element. *PLoS Genet.* 3:e172.

220. Wang, C., Cai, X., and Zheng, Z. (2005). High humidity represses Cf-4/Avr4- and Cf-9/Avr9-dependent hypersensitive cell death and defense gene expression. *Planta* 222, 947–956.

221. Wang, D., Pajerowska-Mukhtar, K., Culler, A. H., and Dong, X. N. (2007). Salicylic acid inhibits pathogen growth in plants through repression of the auxin signaling pathway. *Curr. Biol.* 17, 1784–1790.

222. Wang, W., Vinocur, B., Shoseyov, O., and Altman, A. (2004). Role of plant heat-shock proteins and molecular chaperones in the abiotic stress response. *Trends Plant Sci.* 9, 244–252.

223. Wang, Y., Bao, Z., Zhu, Y., and Hua, J. (2009). Analysis of temperature modulation of plant defense against biotrophic microbes. *Mol. Plant Microbe Interact.* 22, 498–506.

224. Wang, Z. Y., Xiong, L. M., Li, W. B., Zhu, J. K., and Zhu, J. H. (2011). The plant cuticle is required for osmotic stress regulation of abscisic acid biosynthesis and osmotic stress tolerance in *Arabidopsis. Plant Cell* 23, 1971–1984.

225. Widodo, Patterson, J. H., Newbigin, E., Tester, M., Bacic, A., and Roessner, U. (2009).METABOLIC responses to salt stress of barley (*Hordeum vulgare* L.) cultivars, Sahara and Clipper, which differ in salinity tolerance. *J. Exp. Bot.* 60, 4089–4103.

226. Wiese, J., Kranz, T., and Schubert, S. (2004). Induction of pathogen resistance in barley by abiotic stress. *Plant Biol. (Stuttg.)* 6, 529–536.

227. Wingler, A., and Roitsch, T. (2008). Metabolic regulation of leaf senescence: interactions of sugar signalling with biotic and abiotic stress responses. *Plant Biol. (Stuttg.)* 10(Suppl. 1), 50–62.

228. Woldemariam, M. G., Onkokesung, N., Baldwin, I. T., and Galis, I. (2012). Jasmonoyl-L-isoleucine hydrolase 1 (JIH1) regulates jasmonoyl-L-isoleucine levels and attenuates plant defenses against herbivores. *Plant J.* 72, 758–767.

229. Wolters, H., and Jurgens, G. (2009). Survival of the flexible: hormonal growth control and adaptation in plant development. *Nat. Rev. Genet.* 10, 305–317.

230. Wu, X. Y., Kuai, B. K., Jia, J. Z., and Jing, H. C. (2012a). Regulation of leaf senescence and crop genetic improvement. *J. Integr. Plant Biol.* 54, 936–952.

231. Wu, Y., Zhang, D., Chu, J. Y., Boyle, P., Wang, Y., Brindle, I. D., et al. (2012b). The *Arabidopsis* NPR1 protein is a receptor for the plant defense hormone salicylic acid. *Cell Rep.* 1, 639–647.

232. Xia, X. J., Wang, Y. J., Zhou, Y. H., Tao, Y., Mao, W. H., Shi, K., et al. (2009). Reactive oxygen species are involved in brassinosteroid-induced stress tolerance in cucumber. *Plant Physiol.* 150, 801–814.

233. Xiao, F., Goodwin, S. M., Xiao, Y., Sun, Z., Baker, D., Tang, X., et al. (2004). Arabidopsis CYP86A2 represses Pseudomonas syringae type III genes and is required for cuticle development. *EMBO J.* 23, 2903–2913.

234. Xiao, J., Cheng, H., Li, X., Xiao, J., Xu, C., and Wang, S. (2013). Rice WRKY13 regulates cross talk between abiotic and biotic stress signaling pathways by selective binding to different *cis*-elements. *Plant Physiol.* 163, 1868–1882.

235. Xing, Y., Jia, W., and Zhang, J. (2008). AtMKK1 mediates ABA-induced CAT1 expression and H2O2 production via AtMPK6-coupled signaling in *Arabidopsis. Plant J.* 54, 440–451.

236. Xiong, L., and Yang, Y. (2003). Disease resistance and abiotic stress tolerance in rice are inversely modulated by an abscisic acid-inducible mitogen-activated protein kinase. *Plant Cell* 15, 745–759.
237. Xu, P., Chen, F., Mannas, J. P., Feldman, T., Sumner, L. W., and Roossinck, M. J. (2008). Virus infection improves drought tolerance. *New Phytol.* 180, 911–921.
238. Yasuda, M., Ishikawa, A., Jikumaru, Y., Seki, M., Umezawa, T., Asami, T., et al. (2008). Antagonistic interaction between systemic acquired resistance and the abscisic acid-mediated abiotic stress response in *Arabidopsis*. *Plant Cell* 20, 1678–1692.
239. You, M. P., Colmer, T. D., and Barbetti, M. J. (2011). Salinity drives host reaction in *Phaseolus vulgaris* (common bean) to *Macrophomina phaseolina*. *Funct. Plant Biol.* 38, 984–992.
240. Yun, B. W., Feechan, A., Yin, M., Saidi, N. B., Le Bihan, T., Yu, M., et al. (2011). *S*-nitrosylation of NADPH oxidase regulates cell death in plant immunity. *Nature* 478, 264–268.
241. Zeier, J. (2013). New insights into the regulation of plant immunity by amino acid metabolic pathways. *Plant Cell Environ.* 36, 2085–2103.
242. Zhai, J. X., Jeong, D. H., De Paoli, E., Park, S., Rosen, B. D., Li, Y. P., et al. (2011). MicroRNAs as master regulators of the plant NB-LRR defense gene family via the production of phased, trans-acting siRNAs. *Genes Dev.* 25, 2540–2553.
243. Zhang, G., Chen, M., Li, L., Xu, Z., Chen, X., Guo, J., et al. (2009). Overexpression of the soybean GmERF3 gene, an AP2/ERF type transcription factor for increased tolerances to salt, drought, and diseases in transgenic tobacco. *J. Exp. Bot.* 60, 3781–3796. d
244. Zhang, J., Shao, F., Li, Y., Cui, H., Chen, L., Li, H., et al. (2007). A *Pseudomonas syringae* effector inactivates MAPKs to suppress PAMP-induced immunity in plants. *Cell Host Microbe* 1, 175–185.
245. Zhang, M., Barg, R., Yin, M., Gueta-Dahan, Y., Leikin-Frenkel, A., Salts, Y., et al. (2005). Modulated fatty acid desaturation via overexpression of two distinct ω-3 desaturases differentially alters tolerance to various abiotic stresses in transgenic tobacco cells and plants. *Plant J.* 44, 361–371.
246. Zhang, Y., Butelli, E., De Stefano, R., Schoonbeek, H. J., Magusin, A., Pagliarani, C., et al. (2013). Anthocyanins double the shelf life of tomatoes by delaying overripening and reducing susceptibility to gray mold. *Curr. Biol.* 23, 1094–1100.
247. Zhang, Z. Y., Liu, X., Wang, X. D., Zhou, M. P., Zhou, X. Y., Ye, X. G., et al. (2012). An R2R3 MYB transcription factor in wheat, TaPIMP1, mediates host resistance to *Bipolaris sorokiniana* and drought stresses through regulation of defense- and stress-related genes. *New Phytol.* 196, 1155–1170.
248. Zhao, M. S., and Running, S. W. (2010). Drought-induced reduction in global terrestrial net primary production from 2000 through 2009. *Science* 329, 940–943.
249. Zhu, S. Y., Yu, X. C., Wang, X. J., Zhao, R., Li, Y., Fan, R. C., et al. (2007). Two calcium-dependent protein kinases, CPK4 and CPK11, regulate abscisic acid signal transduction in *Arabidopsis*. *Plant Cell* 19, 3019–3036.
250. Zhu, Y., Qian, W. Q., and Hua, J. (2010). Temperature modulates plant defense responses through NB-LRR proteins. *PLoS Pathog.* 6:e1000844.
251. Zhu, Z., An, F., Feng, Y., Li, P., Xue, L., A, M., et al. (2011). Derepression of ethylene-stabilized transcription factors (EIN3/EIL1) mediates jasmonate and ethylene signaling synergy in *Arabidopsis*. *Proc. Natl. Acad. Sci. U.S.A.* 108, 12539–12544.
252. Zou, C., Sun, K., Mackaluso, J. D., Seddon, A. E., Jin, R., Thomashow, M. F., et al. (2011). *Cis*-regulatory code of stress-responsive transcription in *Arabidopsis thaliana*. *Proc. Natl. Acad. Sci. U.S.A.* 108, 14992–14997.

CHAPTER 10

Stress Responsive Non-protein Coding RNAs

Agata Tyczewska[1], Kamilla Bąkowska-Żywicka[1], Joanna Gracz[1] and Tomasz Twardowski[1]

[1] *Institute of Bioorganic Chemistry, Polish Academy of Sciences, Poznań, Poland*

ABSTRACT

Environmental or growth conditions are constantly influencing the development and maturation of organisms. To ensure precise control of gene expression under these normal as well as stress conditions, organisms developed very accurate molecular regulatory mechanisms at transcriptional and posttranscriptional levels. Key roles of small noncoding RNAs (miRNAs and siRNAs) in regulation of gene expression are no longer astonishing us nor are constantly emerging new classes of non-protein coding RNAs (tRNA-derived small RNAs, snoRNA-derived small RNAs, mRNA-derived small RNAs, and long non-coding RNAs). Although there is evidence of involvement of those various classes of noncoding RNAs in response to stress conditions, in most cases, we are still lacking basic information on their targets and interplay between different regulatory pathways. In this review, we focus on "canonical" and "noncanonical" noncoding RNAs and their involvement in abiotic stress responses.

Keywords: Noncoding RNAs, abiotic stress, crops, yeast

1. INTRODUCTION

Biological organisms exposed to any external circumstances that exert a detrimental influence on them are capable of establishing mechanisms of protection and adaptation. To minimize stress influence, several different strategies can be applied, these are tolerance, resistance, and avoidance or ultimately escape. Because of their sedentary lifestyle, plants are restricted to tolerance, resistance, and avoidance mechanisms only, and thus require short-term strategies to quickly and efficiently readapt their metabolism [1]. Therefore, to ensure their survival under unfavorable conditions plants have established refined adjustments to stresses at all levels (anatomical, morphological, cellular, biochemical, and molecular) [2]. On the other hand, microorganisms, such as industrially important baking yeast, have specific and delicately balanced internal conditions for optimal growth and function. Industrial applications expose yeast to multiple unfavorable environmental conditions, therefore accurate stress response is key for yeast cells. When environmental conditions change abruptly, yeast cells rapidly adjust its internal balance to that required for growth in the new conditions.

The physiological and biochemical changes in plants under stress conditions are a result of alteration in abundance of many transcripts and therefore proteins [3] pinpointing the role of TGS (transcriptional gene expression) and PTGS (post-transcriptional gene expression) in the adaptation to environmental changes. To achieve precise and timely regulation of stress-responsive gene networks in addition to regulatory DNA sequences and different types of proteins, plants, as well as many other organisms, including yeast, have employed the gene expression regulation by RNA molecules.

For many decades RNA was considered to be a passive intermediate in the flow of information from DNA to protein, apart from other well-known and described functions in translation (tRNAs, rRNAs) or splicing (UsnRNA). In 1998, its new role has been discovered and regulatory capacity of RNA molecules immensely broadened our understanding of gene expression processes. Moreover, new sequencing technologies and high-resolution microarray analysis have revealed genome-wide pervasive transcription in many eukaryotes, generating a large number of RNAs of no coding capacity. Now, nearly 20 years after the first discovery of the RNA interference, we know that small non-coding RNAs not only regulate gene expression on post-transcriptional and transcriptional levels, but can also affect the organization and modification of chromatin. Moreover, they regulate growth and development of organisms, organ development, hormone signalling, and the defense against pathogens [4, 5]. They also function during environmental stress adaptation, providing precise regulation of gene expression.

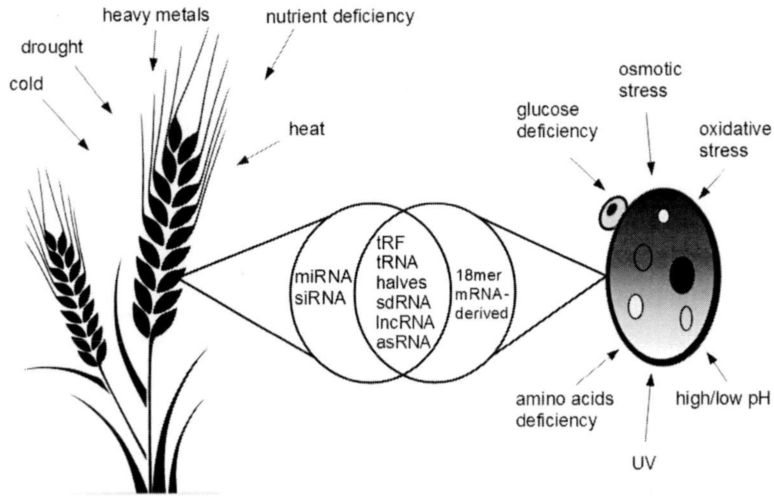

Differentially expressed RNAs are involved in regulation of plant and yeast metabolism pathways during development as well as stress conditions. Baking yeast are depleted of RNAi dependent molecules (miRNA and siRNA) that are present in plants.

FIGURE 1. Overview of different classes of ncRNAs associated with response to abiotic stress conditions in plants and yeast.

Herein, we wish to describe the roles of non-protein coding RNA molecules during abiotic stresses in industrially important organisms, such as crops and yeasts. We wish to focus on "canonical" small RNAs, such as miRNAs and siRNAs, but also new classes such as mRNA-derived RNAs, tRNA-derived small RNAs, snoRNA-derived small RNAs (sdRNAs), mRNA-derived small RNAs, and long non-coding RNAs (lncRNAs) (Figure 1).

2. CANONICAL SMALL RNAS

In plants, small RNAs exhibit unexpected complexity, but two main categories can be distinguished based on their biogenesis and function and these are miRNAs (microRNAs) and siRNAs (short interfering RNAs).

Baking yeast lack these two categories of small RNAs since their biogenesis and functions involve RNAi machinery, which is absent in *S. cerevisiae.*

2.1. MIRNAS

miRNAs are predominantly 21 nucleotides in length [6] and they are derived from 70–500 nucleotide long, imperfect, hairpin-like structured primary transcripts called pri-miRNAs by the activity of RNase III dicer-like (DCL) enzyme [7]. DCL1 processes primary miRNA transcript into an miRNA-miRNA* duplex with 2nt overhangs at the 3' end [6]. With the help from hyponastic leaves (HYL1) and SERRATE proteins miRNA duplex is released and then Hua Enhancer 1 (HEN1) methylates 3' ends of the duplex and therefore stabilizes it by preventing uridylation and subsequent degradation. HASTY, the plant homolog of exportin 5, transports the miRNA duplex into the cytoplasm [8, 9]. The mature miRNA is loaded into RISC (RNA-induced silencing complex) and guides the translational inhibition or cleavage of target mRNAs through the action of AGO proteins in a sequence-specific manner [8–10].

To date, many stress-responsive miRNAs have been identified and the vast majority of their targets are transcriptional factors or proteins playing important roles not only in developmental and growth processes but also in biotic and abiotic stress responses [11]. There appears to exist common pathways of stress responses based on the induction/reduction of similar small noncoding RNA entities. It has been shown that different stress conditions trigger similar responses in various plant species; but on the other hand, the plant's response to similar stress condition could trigger the expression of different pools of miRNAs even in different varieties of the same plant species.

Drought and salinity are major external factors influencing crop productivity and seed quality worldwide. Understanding plant's responses to these stresses is therefore of high importance and one of the main research topics nowadays. Several studies showed differential expression profiles of many genes during water deficit conditions [12, 13]. Recent efforts have also led to the identification of many miRNAs altered in response to drought in many plant species such as cowpea [14], soybean [15], or *Triticum dicoccoides* [16]. Strikingly, the plant response varies in different species, depending on the habitat or plant family, resulting in majority of miRNAs having differential accumulation patterns; nevertheless, some miRNAs share common accumulation pattern as shown below. The differences observed are not only attributable to plant species or growth conditions but could also result from different genotypes in the same species. Soybean seedlings representing two different genotypes (drought-sensitive and tolerant) showed an increase in the level of five miRNAs (miR166-5p, miR169f-3p, miR1513c, miR397ab, and miR-seq13) [15]. In wild emmer wheat, 13 miRNAs were found to be differentially regulated upon drought (miR1867, miR896, miR398, miR528, miR474, miR396, miR894, miR156, miR1432, miR166, and miR171) [16]. Interestingly, some miRNAs share common accumulation pattern; for example, miR474 has also been shown to be upregulated in maize under water deficit [17]. Genome-wide approach allowed to identify 30 miRNA families significantly down- (16: miR156, miR159, miR168, miR170, miR171, miR172, miR319, miR396, miR397, miR408, miR529, miR896, miR1030, miR1035, miR1050, miR1088, and miR1126) or upregulated (14: miR159, miR169, miR171, miR319, miR395, miR474, miR845, miR851, miR854, miR896, miR901, miR903, miR1026, and miR1125) under the drought stress in *Oryza sativa*; strikingly 9 miRNAs (miR156, miR168, miR170, miR171, miR172, miR319, miR396, miR397, and miR408) showed opposite expression patterns to previously described in drought-stressed Arabidopsis [18]. Some miRNA gene families were identified in both down- and upregulated groups (miR171, miR319, miR896). One miRNA that is consistently being upregulated during drought is miR393. The increase in the level of this microRNA has been observed for rice, Arabidopsis, *Medicago truncatula*, and *Phaseolus vulgaris* [19–23].

It is established that somewhere between 6% to 20% of worldwide cultivated arable land is affected by excessive concentrations of salt [24–26]. Some cellular and metabolic processes observed in plants during salt stress conditions (as a result of the limitation of the plant's ability to uptake water) are similar to those observed during drought [24]. Nevertheless, there are many genes and pathways that discriminate these two stress conditions [12, 28]. In salt-tolerant and salt-sensitive *Zea mays* lines, microarray hybridization experiments led to the identification of 98 salt-responsive miRNAs representing 27 families. Several microRNAs showed differential accumulation levels upon salinity in roots: members of miR156, miR164, miR167, and miR396 families were downregulated while miR162, miR168, miR395 and miR474 were upregulated [28]. In rice, other microRNAs appear to be involved in response to salt stress, these are miR169g, miR169n, miR169o, as well as miR393 [29, 30]. These microRNAs target the transcript of NF-YA gene encoding for a subunit A of a conserved transcription factor previously shown to be downregulated in drought-affected wheat [31]. Three microRNAs (miR414, miR164e, and miR408) have been established to be downregulated under salinity in rice [32]. Although, their target genes OsABP (ATP-binding protein), OsDBH (DEAD-box helicase), and OsDSHCT (DOB1/SK12/helY-like DEAD-box helicase), were for the first time validated and upregulated in response to salt stress their functions still need to be experimentally confirmed.

Cold stress is one of the most severe abiotic stresses. The disorders triggered by low temperatures depend on the duration of stress condition and the rate of temperature decrease. Cold stress includes chilling (above 0°C) that inhibits water uptake and freezing (below 0°C) that induces cellular dehydration and thus osmotic stress and hyper accumulation of reactive oxygen species [33]. There are several factors (genes and transcription factors) playing important roles in cold stress acclimatization [10, 34–37]. Post-transcriptional mechanisms of cold acclimation include pre-mRNA processing, mRNA stability, and export from the nucleus. During the last few years, an emerging role of microRNAs in cold stress acclimatization and responses began to unveil. Cold, as well as ABA, dehydration, and salt stress upregulate the expression of miR393, miR397b, and miR402. In accordance with the upregulation of miR393 under cold stress, its target, putative E3 ubiquitin ligase SCF complex F-box protein, appears to be downregulated by cold stress in Arabidopsis [38]. Thus, during cold acclimation, the cleavage of E3 ubiquitin ligase mRNA would lead to diminished proteolysis of its targets (probably positive regulators of cold stress). Moreover, one of the targets of miR393 encodes an F-box protein (At4g03190) [39, 40], which is similar to glucose repression resistance 1 (GRR1), a yeast protein involved in glucose repression. Plants can use sugar status as a signal to modulate growth and development in response to abiotic stresses since it has been observed that sugar metabolism is affected by various abiotic stresses. It has been speculated therefore that miR393 could integrate sugar signaling with cold stress responses [41]. Moreover, in sugarcane, it has been observed that miR319 is differentially expressed during cold treatment for different periods of time, in both shoots and roots [42]. These data are in accordance with previous reports showing that miR319 was upregulated in response to cold stress in Arabidopsis [39, 40]. Genome-wide experiments in *Prunus persica* led to the identification of miR156, miR159, miR164, miR167, miR172, miR393, miR396, miR414, miR2275, and miR5021 as cold responsive [43, 44]. They are regulating genes involved in flower and leaf developmental processes. In rice, 18 miRNAs have been shown to respond rapidly to cold stress conditions (miR156, miR166, miR166m, miR167a,b,c, miR168b, miR169e, miR169f, miR169h, miR171a, miR535, miR319a,b, miR1884b, miR444a.1, miR1850, miR1868, miR1320, miR1435, and miR1876). Most of them are downregulated, which suggests that their targets are being upregulated in adaptation to cold stress. Interestingly, there were four miRNAs (miR1435, miR1876, miR1320, and miR1884) present in rice, but not Arabidopsis, in response to cold stress, which suggests a species-specific mechanisms [10]. miR171 is a large family consisting of 12 known members. In Arabidopsis, miR171a was induced 6 h after cold treatment [40]. On the other hand, miR171a in rice (rice and Arabidopsis share the same mature sequence) was significantly downregulated. Moreover, the expression levels of the other three miRNAs (e, f, i) were also decreased in rice in response to cold. In contrast, miR171 c, d, and h were induced after 6 h [10]. The difference in expression pattern as in the case of miR171 could be reflective of divergence in response of particular plant species to cold stress.

miRNAs also appear to play crucial roles in regulation of nutrient homeostasis. There are 14 mineral elements essential for plant growth and development. Phosphate (Pi) is one of the most prominent ones; it is involved in phosphorylation reactions, synthesis of nucleic acids, membrane lipids, but also in energy delivery and often is a limiting nutrient for plant growth. The phosphate acquisition from soil particles requires the secretion of plant enzymes that release phosphate, then phosphate transport through Pht1 family of transporters (phosphate transporters protein family 1). Phosphate homeostasis in cells is regulated through the action of PHO1 and PHO2 proteins, transcription factors PHR1, WRK75, ZAT6, bHLH32, mYB62, in Arabidopsis and PTF1 in rice and microRNAs. MiR399, upregulated in low-phosphate stressed plants [45–47] targets two genes; a phosphate transporter PHO2 and a putative ubiquitin conjugating enzyme UBC24 transcript [6, 48] have been isolated from the phloem and their level increases in the phloem upon phosphate starvation, suggesting their role as phosphate starvation signals being translocated from shoots to roots where they promote phosphate uptake by downregulating PHO2 [49, 50]. Recently, miR-827-3p was shown to be upregulated in nitrogen and phosphorus deficiency conditions [51, 52]. In rice, it has been shown to target two genes encoding proteins containing SPX and MFS (major facilitator superfamily) domains. SPX-domain proteins are involved in Pi sensing and/or Pi transport [53]; whereas MFS-domain proteins are membrane proteins functioning in small molecule transport [54]. It has been suggested that SPX-MFS proteins might be involved in intracellular or intercellular Pi transport through regulating the expression of Pi transporter (PT) genes [55] and/or Pi storage or metabolism [56]. Similarly, in Arabidopsis, it targets the nitrogen/phosphate balance regulator Nitrogen Limitation Adaptation (NLA, AT1g02860) transcripts [57]. Overexpression of miR827 under Pi-starvation conditions causes the downregulation of NLA (together with downregulation of PHO2 through the action of miR399) leading to increased levels of phosphate transporter 1 (PHT1) and elevated activity of Pi uptake at the plasma membrane.

Sulfur (S) is another fundamental macronutrient necessary for proper growth and development. For plants, it is available from the soil in sulfate form, it is taken up by the roots, and distributed within the whole plant by sulfate transporters. It is found in amino acids, peptides and proteins, iron-sulfur clusters, and several co-factors, but also in an oxidized form in polysaccharides, lipids, and modified proteins [58, 59]. Sulfate deprivation induces the accumulation of miR395 that targets ATP sulfurylases (APSs) and sulpfate transporter AST68 [6, 60], thus reducing their levels. In Brassica plants, deprived of sulfur, the level of miR395 increases in the root, stem, and leaf tissues, it is however strongest in the phloem [61, 62].

The regulation of copper (Cu) homeostasis is also managed through the action of microRNAs. This micronutrient is indispensable for photosynthesis and plant protection against reactive oxygen species [63, 64]. Under copper-deficient conditions, upregulation of miR398 leads to the downregulation of cytosol and chloroplastic Cu/Zn superoxidase dismutase transcripts level (CSD1 and CSD2) and copper chaperone for superoxide dismutase CCS1 [65, 67]. Cu/Zn SOD can be, therefore, replaced with FeSOD in chloroplasts [68]. The level of miR398 is elevated in Brassica under copper deprivation in leaf, stem, and root tissue but most importantly in the phloem [62], such as in the case of sulfur deprivation, which is in accordance with other research suggesting that the phloem has a distinct set of microRNAs from leaves and roots and responds specifically to stress conditions [63]. Several other microRNAs have also been recently linked with copper-deficient conditions (miR397, miR408, and miR857). They are predicted to target Cu-containing proteins (laccases and plantacyanin) [66].

Under high, heavy metal concentrations, plants are subjected to cation imbalances that lead to alterations in plant physiology and biochemistry. It is not surprising, therefore, that in response to these kind of stresses a set of microRNAs is differentially regulated. It has been observed in rice [68], *M. truncatula* [69], *Brassica napus* [70], and Arabidopsis [71]. In *Brassica*, several microRNAs have been linked with cadmium exposure (miR156, miR160, miR164b, miR171, miR393, miR394a,b,c and miR396a,) [70, 72]. In rice, assorted microRNAs have been upregulated in roots (miR601, miR602, and miR603) while some were downregulated in leaves (miR602 and miR606) or in roots (miR604) [68]. The levels of miR171, miR319, miR393, and miR529 were all heightened in response to mercury (Hg), cadmium (Cd), and aluminium (Al) exposure in Medicago [69].

2.2. SIRNAS

siRNAs are derived from double-stranded RNAs that result from transcription of inverted-repeat sequences, natural cis-antisense transcript pairs, or by the action of RNA-dependent RNA polymerases (RDRs) that convert single-stranded RNA transcripts (e.g., virus replication) into double-stranded RNAs [8]. They also derive from heterochromatic regions and DNA repeats and mediate the silencing maintenance of the regions from which they originate [73]. These are then cleaved by dicer-like (DCL) enzymes to produce 21-24-nucleotide long species, the size depends on the specific catalytic activity of the respective DCL protein. Similarly to miRNAs, siRNAs are then loaded into Argonaute (AGO) protein-containing RISC that guides target regulation at the transcriptional level via the so called RNA-directed DNA methylation (RdDM) or at post-transcriptional level [8, 26, 74]. Several classes of siRNA can be distinguished, such as trans-acting siRNAs (tasiRNA), heterochromatic RNAs (hc-siRNA), and natural antisense siRNAs (nat-siRNA).

Tasi-RNAs are derived from specific genetic loci called TAS genes through a microRNA-dependent pathway [75, 76]. Cleaved TAS transcripts are converted into dsRNAs by RNA-dependent RNA polymerase 6 (RDR6), which are in turn cleaved into 21-nt siRNAs. Four families of TAS genes have been identified in Arabidopsis. TAS2 siRNAs target PPR mRNAs (pentatricopeptide repeat); TAS3 siRNAs target ARF2, 3 and 4 transcription factors; TAS4 siRNAs decrease the level of MYB transcription factors [78]. TAS1 and TAS2 are recognized by miR173, TAS3 by miR390 and TAS4 by miR828 [64]. TasiRNAs derived from TAS4 have been implicated in Pi deficiency conditions. They are directly involved in the biosynthesis of anthocyanins in response to low Pi [78]. A putative siRNA pta22 from *Pinus taeda* has been implicated in the response to pathogens, targeting transcripts of two genes encoding for disease resistance proteins [79]. The levels of TAS1, TAS2, and TAS3 tasiRNAs have been elevated in hypoxia-treated samples in Arabidopsis suggesting their role in stress responses [80]. These changes in tasiRNA levels have been proved to be correlated with the levels of TAS-targeting miRNAs (miR173 and miR390). The decrease in PPR transcripts level can be associated with the protection of mitochondria during hypoxia stress.

The usage of high-throughput sequencing methods allowed to identify thousands of genes in convergent overlapping pairs that can generate complementary transcripts in rice and *Arabidopisis* [41, 81, 82]. A natural cis-antisense transcript pair SRO5-P5CDH nat-siRNA plays an important role in osmoprotection and oxidative stress management that results from high salinity [83]. P5CDH (pyrroline-5-carboxylate dehydrogenase) and SRO5 (similar to RCD One) are both involved in regulation of proline metabolism. The downregulation of P5CDH transcripts causes accumulation of proline, which is an important mechanism involved in the increased tolerance to salt stress conditions. But the decrease in the level of P5CDH causes an increase in a toxic metabolite P5C and reactive oxygen species that are detoxified by SRO5 proteins in the mitochondria [83].

There has also been a report showing changes in 4 siRNA levels in wheat seedlings under cold, heat, salinity, and dehydration stresses. Two of them (named 002061_0636_3054.1 and 005047_0654_1904.1) were downregulated in heat, salinity, and dehydration. The latter one was also responding to the cold stress by being upregulated. The level of another one (080621_1340_0098.1) was decreased under heat but upregulated under cold stress. The fourth one (997927_0100_2975.1) was not differentiating only under heat stress, but downregulated when plants were treated with salinity, cold, and dehydration conditions [84].

Heterochromatic 24-nt siRNAs are essential components of RdDM pathway [85]. The pathway has been established to protect plants from invasive nucleic acids, such as transposable elements and repetitive sequences, through DNA methylation and chromatin modifications [86–88]. It has been shown that 21-nt and 24-nt hcsiRNAs are the most abundant class (80%) of all small RNAs in plants [60, 89, 90]. The role of hcsiRNAs in abiotic stress responses is not well understood; although in the 1980s, McClintock has already suggested that all kinds of stresses could potentially reshape a plant genome via transpose activation [91]. Global analysis in plants such as Arabidopsis and rice suggests that the vast majority of transposons are inactive, methylated, and targeted by siRNAs [92]. Recently, few researches suggested that hcsiRNAs could play an important role in the adaptation to stress conditions. In Arabidopsis seedlings subjected to heat stress, ONSEN (copia-type retrotransposon) levels have been shown to increase in mutants affected in RdDM pathway [93]. After the recovery period, the level of ONSEN transcripts decreased gradually suggesting a role for hcsiRNAs. What is interesting is that the insertion of new ONSEN copies has been observed during flower development but before gametogenesis. Surprisingly, ONSEN insertions were also able to confer heat stress responsiveness to nearby genes. Together, all these data suggest that hcsiRNAs can contribute to control of transgenerational retrotransposition triggered by abiotic stresses and can participate in new regulatory networks established to protect plants against stress conditions. The detection of mPing transposition into a rice homologue of flowering time gene-CONSTANS in stressed cultivars supported this intriguing hypothesis. It turns out that stress-activated transposons could positively contribute to genome adaptation to growth in colder climates (reviewed in [1]). Moreover, stress-mediated induction was shown for Tos17 (rice), Tto1 (tobacco), Tnt1 (tobacco), and BARE-1 (barley) retrotransposons. Low temperature treatment was shown to decrease methylation and increase the excision rate of Tam3 transposon by binding its transposase to GCHCG sites immediately after DNA replication, thus preventing de novo sequence methylation (reviewed in [1]).

3. NON-CANONICAL SMALL RNAS

In plants, RNAi-dependent as well long ncRNAs-dependent mechanisms constitute the vast majority of stress response mechanisms. However, other non-protein coding RNAs are continuingly being discovered. At the same time, *S. cerevisiae* lacks the RNA interference pathway, and thus the ncRNA-based gene regulation is performed solely by the RNAi-independent mechanisms. Newly identified small non-coding RNAs other than miRNAs and siRNAs, their functions in gene regulation in yeast and plants, as well as possible mechanisms of action are discussed below.

3.1. TRNA-DERIVED SMALL RNAS

In the expanding repertoire of small noncoding RNAs, tRNA-derived RNA fragments have been identified in all domains of life [reviewed in 94]. According to the nomenclature based on tRNA fragment size and the part of the tRNA molecule from which fragments are derived [95], tRNA fragments can be separated in two major classes: tRNA halves and smaller tRNA fragments (tRFs). In most of the cases reported till now, full-length tRNA levels do not decline significantly and tRNA fragment levels are consistently lower than those of full-length tRNAs, suggesting that only a small proportion of tRNAs is subjected to cleavage. This is in contrast to the complete depletion of tRNAs targeted by colicins in bacteria [96]. Recent evidences suggest that tRNA-derived fragments are not just a random degradation products but rather stable entities that may have major biological functions. The direct influence on protein biosynthesis has been revealed since: (i) tRF binds to 30S and inhibit translation in *Archaea* [97] and (ii) tRNA halves in human cell lines inhibit protein synthesis by displacing eIF4G/eIF4A from uncapped>capped RNAs [98].

So far, only few reports on plant tRNA-derived fragments are published. Such small RNAs have been observed under non-stress conditions in plants [99]; however, the basal levels of tRNA halves are low and often increase during stress conditions. In 2009, it has been reported that tRNA fragments are present abundantly under phosphate (Pi) deficiency in *Arabidopis* [78]. In this study, a significantly higher amount (almost six-fold) of small RNAs derived from tRNAs was present in the roots than in the shoots. The accumulation of tRNA fragments in the −Pi roots was 1.4 times higher than that in the +Pi roots. When the origin of these RNAs was further analyzed, no correlation was observed between the abundance of small RNAs from specific tRNA species and their codon usage. What gained a special interest is the observation that a 19-nucleotide sequence processed from the 5′ end of $tRNA^{Gly}{}_{TCC}$represented over 80% of all tRNA-derived small RNAs in the roots and accounted for up to 18.44% and 27.70% of total sequence reads in the +Pi and −Pi root libraries, respectively, compared with only 1.00% to 1.79% in the shoot libraries. These results revealed a spatial and temporal expression pattern of small RNAs derived from the specific cleavage on tRNA molecules. The authors speculated that such differential accumulation of tRNA fragments between roots and shoots may represent the consequence of long-distance movement. Similar movement phenomenon of tRNA halves was observed in the phloem sap of *Curbita maxima* [100]. In this study, the presence of all tRNA anticodon families was inspected by northern blot hybridization assays. In 12 out of 20 probed tRNA cases, stable processing products were detected in the phloem sap but not in leaf tissue extracts. The size of tRNA fragments ranged from 31 to ~60 nt and they derived from both 3' or 5' ends of particular tRNA molecule. The authors proposed that these phloem-delivered tRNA fragments may be a long-distance signal to coordinate the metabolic status between source and sink tissues.

A comprehensive expression profile of non-coding small RNAs was also performed in barley (*Hordeum vulgare* L.) under phosphorous-deficient and -sufficient conditions [101]. The deficiency in phosphorus decreases the plant growth and grain yields. Hackenberg et al., using hight-throughput sequencing techniques, found that 56 out of the total 61 tRNAs were generating stable fragments in both P-deficient and P-sufficient shoots. Six tRNA-derived fragments were significantly upregulated, whereas four were significantly downregulated in P-deficient shoots. Notably, the read count of tRNA fragments was two-fold greater in P-deficient shoots than in P-sufficient shoots. The most abundant of these small RNAs was tRF derived from $tRNA^{Gly}_{TCC}$ (58.6% of the total tRFs in P-deficient shoots and 58.2% in P-sufficient shoots). $tRNA^{Ala}_{AGC}$-derived sRNAs (previously reported as the most abundant in rice meristems) were the second most abundant species in P-sufficient shoots, but the third in P-deficient shoots. Such differential accumulation of tRFs between the two P treatments indicate that P has a great impact on the tRNA processing to small RNAs.

Deep sequencing technologies also provided evidence that novel subset of small RNAs are derived from the chloroplast genome (csRNAs) of Chinese cabbage [102]. The chloroplast small RNAs (csRNAs) included, among others, those that derive from tRNAs. Wang et al. found that the csRNAs derived from most of the chloroplast tRNA sequences constituted 5' parts of the molecules. Moreover, the first nucleotide of these csRNAs were predominantly located at the first nucleotide of the mature chloroplast tRNA sequences, revealing that the biogenesis of csRNA in tRNA molecules was specific to the 5' end. The tRNA-derived csRNAs only slightly declined in the heat-treated seedlings. However, the length of these csRNAs was related to heat stress response. In the heat-treated seedlings, the abundance of longer csRNAs (29–32 nt) decreased, but the of the shorter ones (16–25 nt) increased. csR-trnA-1 and csR-trnA-2 were the two most predominant tRNA-derived csRNA families and originated from chloroplast $tRNA^{Ala}$. Such reduced abundance of tRNA-derived fragments 1 h after the heat treatment (the time period before the appearance of leaf etiolation) may suggest that the heat-responsive csRNAs play roles in the maintenance of subcellular structures and photosynthetic capacity of chloroplasts. The csRNAs derived from tRNAs may, therefore, play a role similar to the reported tRNA halves under various stress conditions.

In 2013, Loss-Morais et al. summarized all existing sequencing data aiming at the characterization of plant tRNA-derived fragments, their accumulation patterns under abiotic and biotic stresses, the identification of their putative targets, as well as possible association with Argonaute (AGO) proteins [103]. They have inspected 34 Arabidopsis deep sequencing libraries, including 25 AGO-IP libraries and found tRFs in the AGO1, 2, 4, and 7 IP libraries. Both, 5' and 3' tRFs were associated with AGO, mirroring previous results in mammalian systems [104–105]. Interestingly, tRFs from the central part of tRNAs were also detected, although 5' tRFs formed the most abundant class and showed the highest sequence diversity. The AGO-associated 5' tRFs were predominantly 19-mers. Then, to investigate if the 5' tRFs associated with AGOs act in the RNAi pathway in plants, as has been suggested in animals [104], the authors searched for tRF targets in Arabidopsis and identified four possible target genes. They have found that drought conditions enhanced the expression of the four tRFs, including the 5'-tRF^{Gly}_{TCC}, which was already known to be up-regulated in response to phosphate deprivation [78].

The existence of transgenerationally transmitted, heat-responsive tRNA-derived fragments in plants was revealed for the first time by Bilichak et al. in *Brassica rapa* [106]. In the Bilichak study, small RNA sequencing was performed to compare alterations in RNAs in somatic and reproductive tissues of *B. rapa* plants and in their progeny in response to heat treatment. The authors have demonstrated high tissue-specific alterations in the small RNA accumulation profiles in tissues that were not directly exposed to stress, namely, in the endosperm and pollen. Importantly, they have revealed that the progeny of stressed plants exhibited the highest fluctuations in the small RNA accumulation levels (tRNA fragments among them). It has been speculated that perturbations in the expression of small RNAs in somatic tissues caused by environmental fluctuations would eventually be imprinted in the transcriptome patterns in gametes and progeny.

The presence of tRNA-derived fragments has also been reported in industrially important budding yeast *S. cerevisiae* [107, 108]. Thompson et al. were the first to demonstrate (already in 2008) that *S. cerevisiae* contain a small RNA population consisting primarily of tRNA halves and rRNA fragments [106]. tRNA fragmentation to stable shorter pieces occur in yeast cells in the absence and in the presence of stress conditions. However, the accumulation of tRNA fragments in yeast was most pronounced during oxidative stress conditions, especially during entry into stationary growth phase. Both 5′ and 3′ fragments of tRNAs were detectable suggesting the occurrence of endonucleolytic cleavage. The nuclease involved in this cleavage was characterized one year later [109]. It has been shown in yeast that tRNAs are cleaved by RNase T2 family member Rny1p, which is released from the vacuole into the cytosol during oxidative stress. Rny1p modulates yeast cell survival during oxidative stress independent of its catalytic ability. This suggests that upon release to the cytosol, Rny1p promotes cell death by direct interactions with downstream components.

The possible function of tRNA fragments in yeast has been described in 2012 [108]. Although RNA interference (RNAi) is conserved in diverse eukaryotic species, including budding yeast *Saccharomyces castellii* and *Candida albicans* [110], it has been lost in *S. cerevisiae*. Thus, this organism provides an ideal system for studying the RNAi-independent mechanisms of tRF-based gene expression regulation. The potential novel function of tRNA fragments is supposed to appear as a response to specific environmental conditions and includes tRNA processing and association with the ribosomes. As a result of high throughput sequencing of ribosome-associated small RNAs, it appeared that one of the most abundant classes of processed RNAs were tRNAs. In addition to previously reported cleavage in the anticodon loop in yeast tRNAs [107], other breakage points were also detected (e.g., in the D- and T-loop regions), reminiscent to those observed previously in higher eukaryotes [111]. Moreover, an obvious differential stability of tRNA halves was recognized. Northern blot analysis confirmed the presence of two stable processing products derived from $tRNA^{His}$ and revealed that cleavage is stress-dependent. Similar to previous findings tRNA processing was mainly detected during amino acid and sugar starvation conditions. On the contrary, experimental results obtained for $tRNA^{Ser}$ suggested that only the 3′ part of this tRNA is stable. The association of tRNA fragments with yeast ribosomes suggest its potential regulatory function in protein biosynthesis, as already reported for *Archea* [97] and humans [98].

3.2. SNORNA-DERIVED SMALL RNAS

Small nucleolar RNAs are a highly evolutionarily conserved class of RNAs, which are present throughout the eukaryotes and whose origin lies in the Archaea. There are two classes of snoRNAs (C/D and H/ACA box) that function as ribonucleoprotein (RNP) complexes to guide enzymatic modification of target RNAs at sites determined by RNA:RNA antisense interactions. Generally, C/D box snoRNAs are ~70–120 nucleotides (nt) long and guide the methylation of target RNAs, while H/ACA box snoRNAs are ~100–200 nt long and guide the pseudouridulation of their targets. These RNAs were initially discovered in the nucleolus and thought to exclusively target ribosomal RNAs inside this sub-nuclear compartment.

However, numerous snoRNAs do not possess target RNAs—therefore, they are called "orphan snoRNAs".

Small nucleolar RNAs are localized in the nucleoli and Cajal bodies in eukaryotic cells. Components of the RNA silencing pathway associate with these structures, and two recent reports have revealed that a human and a protozoan snoRNA can be processed into microRNA-like RNAs [112–113]. By systematic analyses of deep-sequencing libraries from diverse eukaryotic organisms, Taft et al. revealed that small RNAs with evolutionary conservation of size and position are derived from the vast majority of snoRNA loci in animals (human, mouse, chicken, fruit fly), Arabidopsis, and fission yeast [114]. These small RNAs derived from snoRNAs (sdRNAs) in Arabidopsis were strongly associated with AGO7 proteins. Arabidopsis Argonaute proteins preferentially load small RNAs with specific 5′ nucleotides and AGO7 is selective for 5′ uracil (U) and 5′ adenine (A). Intriguingly, H/ACA sdRNAs are dominantly 5′A, while C/D box are dominantly 5′U. Therefore, it is likely that sdRNAs play roles in the regulation of gene expression and transcriptional silencing. In particular, the fact that both miRNAs and sdRNAs are 5′U biased strengthens the link between them and suggests that some snoRNAs, including "orphan" snoRNAs whose targets are unknown, may function solely as intermediates in the sdRNA pathway.

However, snoRNA-mediated change in gene expression should be treated not as an example but as an addition to siRNA-based mechanisms. The reason for this statement is that canonical snoRNAs interact with their target RNAs through their 10–20 nt long antisense-box elements. In contrast, snoRNA-derived small RNAs can interact with other sequence elements with their targets, as they form a different ribonucleoprotein complexes.

The presence of snoRNA processing products was experimentally verified by high throughput sequencing, as well as northern blot hybridization in *S. cerevisiae* [108]. The results confirmed the presence of a shortened version of both types of snoRNAs. The processing events were most prominent under most tested yeast growth conditions, which included UV radiation, anaerobic growth, high or low pH treatment, amino acid starvation and sugar starvation. Like in the case of tRNA-derived fragments, the association of sdRNAs with the ribosomes implicates their possible regulatory role in protein biosynthesis in *S. cerevisiae*. In this organism, RNAi silencing pathway has been lost during evolution, therefore the mechanism of sdRNAs action in *S. cerevisiae* is likely to be different from microRNA-like sdRNA action in plants.

Despite the differences in possible mechanism of action of sdRNAs, for both plants and yeast, the presence of small regulatory RNAs in the nucleolus allows us to speculate that the nucleolus is involved in the regulation of expression, possibly in response to cellular conditions.

3.3. MRNA-DERIVED RNAS

Small mRNA fragments have been reported in diverse eukaryotes and a portion of them is stress-related [reviewed in 115]. A part of them is annotated as antisense RNAs (asRNAs), overlapping protein coding transcripts on the opposite strand. Some asRNAs were found to repress sense RNAs and lead to inverse expression between asRNAs and their corresponding sense RNAs in respect to growth phases, stress conditions, or environmental changes. Therefore, the corresponding protein-coding transcripts (mRNAs) represent a mixture of cell-cycle factors, chromatin remodellers, and metabolism related proteins. It has been shown recently that about one-fifth of the ORF genes in *S. cerevisiae* are coupling with asRNAs; however, the ratio of genes coupling with asRNAs shows a negative association with gene regulatory complexity [116]. This observation supports previous hypothesis that some asRNAs belong simply to a transcriptional noise. Nevertheless, asRNAs evolve more slowly when their sense genes are under more complex regulation. Older genes coupling with asRNAs are more likely to demonstrate inverse expression, reflecting the role of these asRNAs as repressors, especially under heat stress stimuli.

Recent tiling assays in plants (mostly in Arabidopsis) also revealed the majority of stress-responsive novel antisense transcripts [117]. Despite a linear correlation between the expression ratios of selected sense and antisense transcripts, biological functions of most antisense RNAs remains unclear. In plants, it has been postulated that antisense transcripts might have the potential to produce endogenous siRNAs (nat-siRNAs) as mentioned in Paragraph 2.2. Recent genome-wide analysis reported an accumulation of sRNAs in their overlapping region, suggesting the occurrence of an RNA interference event [82]. However, the biological processes of generating nat-siRNAs is not completely understood at this time.

The power of mRNA-derived ncRNAs for rapid global translation attenuator in stress was recently demonstrated in *S. cerevisiae* [118]. mRNA exon-derived 18-residue-long ncRNA (picked up in previous genomic screen for ribosome-bound small RNAs in *S. cerevisiae*) has been functionally characterized [108]. This ribosome-bound ncRNA, originating as a short sense fragment of TRM10 open reading frame, is needed for the rapid shutdown of global translation and is capable of adjusting translation rates by interacting with polysomes under hyperosmotic growth conditions. Therefore, the existence of a largely unexplored mechanism of translation control has been clearly demonstrated.

4. LONG NON-CODING RNAS

Long non-coding RNAs (lncRNAs) are a wide group of molecules identified in yeast, plants, and mammals. In general, lncRNAs can be defined as polyadenylated or nonpolyadenylated, more than 200 nt long transcripts with low protein coding potential (coding for less than 100 amino acids). Acceleration in transcriptome research, achieved thanks to the development of high-throughput technologies such as microarrays or next-generation sequencing methods, allow us now to fully appreciate complicated interactions that lead to precise gene expression regulation. It turned out that apart from genes encoding proteins, transposons, genes for housekeeping RNAs (like ribosomal RNAs), intergenic regions, which lie between protein-coding sequences, are also being expressed. Intriguingly, similarly to mRNA, those transcripts, namely lncRNAs, are capped at the 5' end and many of them are targets of the splicing process [119]. However, in contrast to mRNA, their expression level is very low and they do not have long open reading frames, which are evolutionary conserved [120]. LncRNA also differ from protein-coding transcripts in ribosome occupancy [121]. Large-scale genomics projects, such as Encyclopedia of DNA Elements (ENCODE), proved that lncRNAs are not only transcriptome noise, but indeed transcripts with biological functions. As a result of the ENCODE project, it has been shown that 75% of human genome is transcribed and about 80% of those RNA molecules have some biochemical function [122]. Most of the intergenic regions of Arabidopsis, rice or corn are shown to be transcribed as well as human and constitute a source of lncRNAs that are polyadenylated. Such polyadenylated stable lncRNAs are transcribed by RNA polymerase II and can be divided into four groups based on their genomic origin and relationship with adjacent protein-coding genes:

1. intergenic lncRNAs (lincRNAs) that are transcribed from sequences between two genes
2. intronic ncRNAs (incRNAs) that overlap with intronic sequences within another transcript
3. natural antisense transcripts (NATs) derived from complementary DNA strand of their associated genes
4. sense lncRNAs overlapping with one or more exon sequences of the transcript on the same strand.

Recently in Arabidopsis and rice [123, 124], another category of lncRNAs has been described — these are nonpolyadenylated transcripts, 50-300 nt in length with low coding potential, but without any sequence similarity to known ncRNAs. This novel group is referred to as intermediate-sized ncRNAs (im-ncRNAs).

Despite the effort of 20 years of investigation [125], the elusive role of lncRNAs is still not fully described nor understood. Thus far functions of only few such molecules are characterized. We know that lncRNAs are engaged mainly in transcriptional gene expression regulation by acting as scaffolds for transcriptional factors and genetic modifiers, molecular signals, decoys or guides. Moreover lncRNAs can also encode for miRNA and target specific mRNAs for decay or function as miRNA sponges. Most studies has been performed on animal systems, but although research on plants are limited, the emerging picture is that the regulatory functions of plant lncRNAs are similar to animal ones [126]. Till today, almost 40,000 putative lncRNAs have been identified in *A. thaliana* [127] and thousands of them in *Oryza sativa* [128], *Zea mays* [129], *Medicago truncatula* [130], *Populus trichocarpa* [131], and other plant species [132, 133]. With rapid development of bioinformatics tools and transcriptome analysis methodologies, genome-wide identifications of plant lncRNAs have been conducted. In maize implementation of SVM tools (support vector machines), together with Python pipeline on cDNA dataset resulted in the identification of 2,492 potential ncRNAs, which represent 13.3% of initial sequences. In total, 237 ncRNAs were classified as shRNA precursors, and 1,225 as siRNA precursors, which constituted 59.4% of predicted ncRNAs particles. The remaining 1,011 was considered to be potential long non-coding transcripts [134]. Recently, new gold standard to study the complexity of eukaryotic transcriptomes emerged — the RNA-sequencing technology (RNA-seq). It allows an accurate quantification of expression levels of transcripts and also reveals transcripts that are missing or incomplete from the reference genome. Computational prediction based on RNA-seq data from rice anthers, pistils, seeds, and shoots, together with 40 available rice RNA-seq libraries led to the identification of 2,224 reliably expressed lncRNAs, including 1,624 lincRNAs and 600 long non-coding natural antisense transcripts (NATs). Further verification of rice insertional mutants allowed to set a pool of lncRNAs that are preferentially expressed at the reproductive stage. Several lncRNAs were identified as competing endogenous RNAs (ceRNAs), which sequester miR160 or miR164 in a type of target mimicry [135].

Another feature that complicates the retrieval of true lncRNAs is their weak sequence conservation. It is estimated that only from 2% to 5.5% of lncRNAS are conserved in their primary sequence and only some of them may be associated with short conserved elements. Most likely, it is a result of rapid evolution—lncRNAs are frequent targets of positive selection [136]. Some lncRNAs and their target genes can be distinguished by their conserved synteny across species — those lncRNAs play roles in cis-functions [136]. Other lncRNAs may be recognized by conserved secondary structures, which allow them to interact with RNA-binding proteins [124]. Genome-wide analyses carried out so far determined that expression of different groups of lncRNAs is highly tissue-specific and many of them are responsive to biotic and abiotic stress conditions.

In conjunction with the climatic changes, drought is the condition that has been recently extensively studied, and thus many drought-responsive lncRNAs were identified. In *Populus trihocarpa,* a model tree species, RNA-seq experiments conducted on control- and drought-treated plants revealed 504 drought-responsive lincRNAs and allowed for basic annotation set of 2,542 of them. Mutual interaction of miRNA and lncRNAs was also reported; a total of 30 miRNAs were predicted to target the sense strand of lincRNAs, 21 were found to target the antisense strand, and 20 target mimicry events was predicted of known *Populus* miRNA [130]. A potential new model organism of the family Poaceae, *Foxtail millet,* was also subjected to water deficient conditions. Deep transcriptome sequencing revealed 585 lncRNAs responding to PEG-induced drought stress. Those stress conditions induced the expression of 17 lincRNAs and 2 NATs at different expression levels. Qi et al. [138] identified one lncRNA, whose sequence was shared with its counterpart in sorghum. In maize, one of most important crop species, genome-wide identification of differentially expressed lncRNAs during drought conditions led to the identification of 567 upregulated and 97 downregulated lncRNAs, among them 538 particles were considered to be novel. Moreover, 8 lncRNAs molecules were homologous to the miRNA precursors, 62 were classified as both shRNA and siRNA precursors, and 279 were classified as siRNA precursors [139].

In the best known model plant, *Arabidopsis thaliana,* genome-wide characterization of lncRNAs was performed as well. A correlated expression of lncRNAs with its epigenetic and structural features in response to four stresses (heat, cold, drought, and salt) has been described [140]. The authors identified 245 polyadenylated and 58 nonpolyadenylated lncRNAs that are differentially expressed under stress stimuli, and most of the selected candidates were further validated by qRT-PCR. From experiments on Arabidopsis came best studied cases of plant lncRNAs functions such as: *COLDAIR, COOLAIR, At4/IPS1, npc48,* and *npc536* [141–145].

One of best described mechanism of lncRNAs action is lncRNA transcript IPS1 (Induced by Phosphate Starvation 1). IPS1 can interact with miRNA as a competitor and function as miRNA target mimics, which resembles the miRNA sponges from animal systems. Maintaining the phosphate balance is a complicated mechanism in plants, regulated, among others, by miR399 as described in Paragraph 2.1. Low activity of PHO2, ruled by mRNA cleavage mediated by miR399, causes the elevation of phosphate uptake by increasing the expression of two root phosphate transporters. Phosphate starvation also increases the level of IPS1 transcript that has a 23-nt conserved domain, partially complementary to miR399 with 3-nt mismatch overlapping with the miR399-mediated cleavage site. As a non-cleavable product, IPS1 competes with PHO2 and can therefore weaken the miR399-mediated repression of PHO2 [142]. The miRNA sponge strategy is used in the therapy of human diseases and similar processes in plants (target mimic) and can be a very useful tool in plant research as well as in agricultural applications. As mentioned before, in Arabidopsis to date about 20 putative target mimicry events were predicted, which suggest the potential role of this mechanism in other pathways than the maintenance of phosphate homeostasis [146].

Another model organism, *Saccharomyces cerevisiae*, allows researchers to define and clarify a large number of new and unexpected roles of lncRNAs, such as promoting the timing of gene expression [147], cell cycle regulation during stress conditions [148], or local reduction of histone density and chromatin remodeling in response to glucose starvation [149]. Upon osmostress in yeast, hundreds of stress-responsive genes are induced by the stress-activated protein kinase (SAPK) p38/Hog1. Whole-genome tiling arrays were used to identify a set of Hog1-induced lncRNAs. One of the genes expressing a Hog1-dependent lncRNA in antisense orientation is CDC28, the cyclin-dependent kinase 1 (CDK1) that controls the cell cycle in yeast. Cdc28 lncRNA mediates the establishment of gene looping and the relocalization of Hog1 and RSC from the 3' UTR to the +1 nucleosome to induce CDC28 expression. The increase in expression level of Cdc28 makes cells able to re-enter the cell cycle more efficiently after stress conditions occur. This may represent a more general mechanism to prime the expression of genes needed after stresses [148].

5. CONCLUSIONS

The rapidly growing human population is constantly stressing agro-ecosystems. In 2007, a FAO (Food and Agriculture Organization of the United Nations) report stated that only about 3.5% of the global agricultural area is not affected by any of the environmental stresses. Since recent advances in biotechnology and molecular biology have dramatically changed our understanding of gene expression regulation and responses of plants to abiotic stresses, we are now armed in new tools that could help us fight various stress conditions and improve the growth and yield of crops and other industrially important species. Our journey in the small RNA world has begun in the past century, in the 90s, with the discovery of RNA interference pathway, followed by siRNAs and their role in DNA methylation and chromatin modifications. Consequently, new and relatively surprising small RNA groups (tRNA-derived small RNAs, snoRNA-derived small RNAs, mRNA-derived small RNAs) have been discovered and slowly but surely their role in cells and in the adaptation to various environmental conditions is being established. Concurrently, thousands of lncRNAs have already been identified and thousands of them are still waiting to be discovered in different organisms, tissues, or in response to different stimuli. With the advent of new high-throughput techniques, it became easier and much faster to collect a vast amount of different types of molecular data. The analysis of such a high amount of data and understanding of mutual correlations and relationships between genes, their roles, and small RNAs is now a challenge that researchers and scientists must face. Nevertheless, we believe that by combining new methodologies with the help of bioinformatic approaches and wet lab experiments, we can shed light on a genomic "dark matter" and start to appreciate important physiological roles of various types of non-protein coding RNAs, as well as their potential applications.

6. ACKNOWLEDGEMENTS

This work was supported by grants from National Science Center Poland: 7690/B/P01/2011/40,DEC-2011/01/D/NZ9/03631, DEC-2011/01/N/ NZ9/ 02900, DEC-2012/06/A/NZ9/00125 and 2014/13/D/ NZ1/00061. This publication was also supported by the Polish Ministry of Science and Higher Education, under the KNOW program.

REFERENCES

1. Boyko A, Kovalchuk I. Epigenetic control of plant stress response. Environ Mol Mutagen. 2008;49:61-72.
2. Grativol C, Hemerly AS, Ferreira PC. Genetic and epigenetic regulation of stress responses in natural plant populations. Biochim Biophys Acta. 2012;1819:176-85.
3. Saibo NJ, Lourenço T, Oliveira MM. Transcription factors and regulation of photosynthetic and related metabolism under environmental stresses. Ann Bot. 2009;103:609-23.
4. Ni FT, Chu LY, Shao HB, Liu ZH. Gene expression and regulation of higher plants under soil water stress. Curr Genomics. 2009;10:269-80.
5. Moazed D. Small RNAs in transcriptional gene silencing and genome defence. Nature. 2009;22;457:413-20.
6. Jones-Rhoades MW, Bartel DP. Computational identification of plant microRNAs and their targets, including a stress-induced miRNA. Mol Cell. 2004;14:787-99.
7. Reinhart BJ, Weinstein EG, Rhoades MW, Bartel B, Bartel DP. MicroRNAs in plants. Genes Dev. 2002;16(13):1616-26.
8. Voinnet O.Origin, biogenesis, and activity of plant microRNAs. Cell. 2009; 136:669-87.
9. Sunkar R, Jagadeeswaran G. In silico identification of conserved microRNAs in large number of diverse plant species. BMC Plant Biol. 2008;8:37.
10. Lv DK, Bai X, Li Y, Ding XD, Ge Y, Cai H, Ji W, Wu N, Zhu YM. Profiling of cold-stress-responsive miRNAs in rice by microarrays. Gene. 2010;459:39-47.

11. Guleria P, Mahajan M, Bhardwaj J, Yadav SK. Plant small RNAs: Biogenesis, mode of action and their roles in abiotic stresses. Genomics Proteomics Bioinformatics. 2011;9:183-99.
12. Bartels D, Sunkar R. Drought and salt tolerance in plants. Crit Rev Plant Sci. 2005;24:23-58.
13. Shinozaki K, Yamaguchi-Shinozaki K. Gene networks involved in drought stress response and tolerance. J Exp Bot. 2007;58:221-7.
14. Barrera-Figueroa BE, Gao L, Diop NN, Wu Z, Ehlers JD, Roberts PA, Close TJ, Zhu JK, Liu R. Identification and comparative analysis of drought-associated microRNAs in two cowpea genotypes. BMC Plant Biol. 2011;11:127.
15. Kulcheski FR, de Oliveira LF, Molina LG, Almerão MP, Rodrigues FA, Marcolino J, Barbosa JF, Stolf-Moreira R, Nepomuceno AL, Marcelino-Guimarães FC, Abdelnoor RV, Nascimento LC, Carazzolle MF, Pereira GA, Margis R. Identification of novel soybean microRNAs involved in abiotic and biotic stresses. BMC Genomics. 2011;12:307.
16. Kantar M, Lucas SJ, Budak H. miRNA expression patterns of Triticum dicoccoides in response to shock drought stress. Planta. 2011;233:471-84.
17. Wei, L., Zhang, D., Xiang, F. and Zhang, Z. Differentially expressed miRNAs potentially involved in the regulation of defense mechanism to drought stress in maize seedlings. Int. J. Plant Sci. 2009;170:979-989.
18. Zhou L, Liu Y, Liu Z, Kong D, Duan M, Luo L. Genome-wide identification and analysis of drought-responsive microRNAs in Oryza sativa. J Exp Bot. 2010;61:4157-68.
19. Zhao B, Liang R, Ge L, Li W, Xiao H, Lin H, Ruan K, Jin Y. Identification of drought-induced microRNAs in rice. Biochem Biophys Res Commun. 2007;354:585-90.
20. Jian X, Zhang L, Li G, Zhang L, Wang X, Cao X, Fang X, Chen F. Identification of novel stress-regulated microRNAs from Oryza sativa L. Genomics. 2010;95:47-55.
21. Liu HH, Tian X, Li YJ, Wu CA, Zheng CC. Microarray-based analysis of stress-regulated microRNAs in Arabidopsis thaliana. RNA. 2008;14:836-43.
22. Arenas-Huertero C, Pérez B, Rabanal F, Blanco-Melo D, De la Rosa C, Estrada-Navarrete G, Sanchez F, Covarrubias AA, Reyes JL. Conserved and novel miRNAs in the legume Phaseolus vulgaris in response to stress. Plant Mol Biol. 2009;70:385-401.
23. Trindade I, Capitão C, Dalmay T, Fevereiro MP, Santos DM. miR398 and miR408 are up-regulated in response to water deficit in Medicago truncatula. Planta. 2010;231:705-16.
24. Munns R, Tester M. Mechanisms of salinity tolerance. Annu Rev Plant Biol. 2008;59:651-81.
25. Kruszka K, Pieczynski M, Windels D, Bielewicz D, Jarmolowski A, Szweykowska-Kulinska Z, Vazquez F. Role of microRNAs and other sRNAs of plants in their changing environments. J Plant Physiol. 2012;169:1664-72.
26. Khraiwesh B, Zhu JK, Zhu J. Role of miRNAs and siRNAs in biotic and abiotic stress responses of plants. Biochim Biophys Acta. 2012;1819:137-48.
27. Golldack D, Lüking I, Yang O. Plant tolerance to drought and salinity: Stress regulating transcription factors and their functional significance in the cellular transcriptional network. Plant Cell Rep. 2011;30:1383-91.
28. Ding D, Zhang L, Wang H, Liu Z, Zhang Z, Zheng Y. Differential expression of miRNAs in response to salt stress in maize roots. Ann Bot. 2009;103:29-38.
29. Zhao B, Ge L, Liang R, Li W, Ruan K, Lin H, Jin Y. Members of miR-169 family are induced by high salinity and transiently inhibit the NF-YA transcription factor. BMC Mol Biol. 2009;10:29.
30. Gao P, Bai X, Yang L, Lv D, Pan X, Li Y, Cai H, Ji W, Chen Q, Zhu Y. osa-MIR393: A salinity- and alkaline stress-related microRNA gene. Mol Biol Rep. 2011;38:237-42.
31. Stephenson TJ, McIntyre CL, Collet C, Xue GP. Genome-wide identification and expression analysis of the NF-Y family of transcription factors in Triticum aestivum. Plant Mol Biol. 2007;65:77-92.

32. Macovei A, Tuteja N. microRNAs targeting DEAD-box helicases are involved in salinity stress response in rice (Oryza sativa L.). BMC Plant Biol. 2012;12:183.
33. Chinnusamy V, Zhu J, Zhu JK. Cold stress regulation of gene expression in plants. Trends Plant Sci. 2007;12:444-51.
34. Stockinger EJ, Gilmour SJ, Thomashow MF. Arabidopsis thaliana CBF1 encodes an AP2 domain-containing transcriptional activator that binds to the C-repeat/DRE, a cis-acting DNA regulatory element that stimulates transcription in response to low temperature and water deficit. Proc Natl Acad Sci U S A. 1997;94:1035-40.
35. Liu Q, Kasuga M, Sakuma Y, Abe H, Miura S, Yamaguchi-Shinozaki K, Shinozaki K. Two transcription factors, DREB1 and DREB2, with an EREBP/AP2 DNA binding domain separate two cellular signal transduction pathways in drought- and low-temperature-responsive gene expression, respectively, in Arabidopsis. Plant Cell. 1998;10:1391-406.
36. Chinnusamy V, Ohta M, Kanrar S, Lee BH, Hong X, Agarwal M, Zhu JK. ICE1: A regulator of cold-induced transcriptome and freezing tolerance in Arabidopsis. Genes Dev. 2003;17:1043-54.
37. Fowler S, Thomashow MF. Arabidopsis transcriptome profiling indicates that multiple regulatory pathways are activated during cold acclimation in addition to the CBF cold response pathway. Plant Cell. 2002;14:1675-90.
38. Zimmermann P, Hirsch-Hoffmann M, Hennig L, Gruissem W. GENEVESTIGATOR. Arabidopsis microarray database and analysis toolbox. Plant Physiol. 2004;136:2621-32.
39. Sunkar R, Girke T, Jain PK, Zhu JK. Cloning and characterization of microRNAs from rice. Plant Cell. 2005;17:1397-411.
40. Liu HH, Tian X, Li YJ, Wu CA, Zheng CC. Microarray-based analysis of stress-regulated microRNAs in Arabidopsis thaliana. RNA. 2008;14:836-43.
41. Sunkar R, Chinnusamy V, Zhu J, Zhu JK. Small RNAs as big players in plant abiotic stress responses and nutrient deprivation. Trends Plant Sci. 2007;12:301-9.
42. Thiebaut F, Rojas CA, Almeida KL, Grativol C, Domiciano GC, Lamb CR, Engler Jde A, Hemerly AS, Ferreira PC. Regulation of miR319 during cold stress in sugarcane. Plant Cell Environ. 2012;35:502-12.
43. Lee H, Yoo SJ, Lee JH, Kim W, Yoo SK, Fitzgerald H, Carrington JC, Ahn JH. Genetic framework for flowering-time regulation by ambient temperature-responsive miRNAs in Arabidopsis. Nucleic Acids Res. 2010;38:3081-93.
44. Barakat A, Sriram A, Park J, Zhebentyayeva T, Main D, Abbott A. Genome wide identification of chilling responsive microRNAs in Prunus persica. BMC Genomics. 2012;13:481.
45. Fujii H, Chiou TJ, Lin SI, Aung K, Zhu JK. A miRNA involved in phosphate-starvation response in Arabidopsis. Curr Biol. 2005;15:2038-43.
46. Bari R, Datt Pant B, Stitt M, Scheible WR. PHO2, microRNA399, and PHR1 define a phosphate-signaling pathway in plants. Plant Physiol. 2006;141:988-99.
47. Chiou TJ, Aung K, Lin SI, Wu CC, Chiang SF, Su CL. Regulation of phosphate homeostasis by MicroRNA in Arabidopsis. Plant Cell. 2006;18:412-21.
48. Sunkar R, Zhu JK. Novel and stress-regulated microRNAs and other small RNAs from Arabidopsis. Plant Cell. 2004;16:2001-19.
49. Aung K, Lin SI, Wu CC, Huang YT, Su CL, Chiou TJ. PHO2, a phosphate overaccumulator, is caused by a nonsense mutation in a microRNA399 target gene. Plant Physiol. 2006;141:1000-11.
50. Bari R, Datt Pant B, Stitt M, Scheible WR. PHO2, microRNA399, and PHR1 define a phosphate-signaling pathway in plants. Plant Physiol. 2006;141:988-99.
51. Lin WY, Huang TK, Chiou TJ. Nitrogen limitation adaptation, a target of microRNA827, mediates degradation of plasma membrane-localized phosphate transporters to maintain phosphate homeostasis in Arabidopsis. Plant Cell. 2013;25:4061-74.

52. Hackenberg M, Shi BJ, Gustafson P, Langridge P. Characterization of phosphorus-regulated miR399 and miR827 and their isomirs in barley under phosphorus-sufficient and phosphorus-deficient conditions. BMC Plant Biol. 2013;13:214.
53. Lenburg ME, O'Shea EK. Signaling phosphate starvation. Trends Biochem Sci. 1996;21:383-7.
54. Marger MD, Saier MH Jr. A major superfamily of transmembrane facilitators that catalyse uniport, symport and antiport. Trends Biochem Sci. 1993;18:13-20.
55. Yao ZF, Liang CY, Zhang Q, Chen ZJ, Xiao BX, Tian J, Liao H. SPX1 is an important component in the phosphorus signalling network of common bean regulating root growth and phosphorus homeostasis. J Exp Bot. 2014;65:3299-310.
56. Lin SI, Santi C, Jobet E, Lacut E, El Kholti N, Karlowski WM, Verdeil JL, Breitler JC, Périn C, Ko SS, Guiderdoni E, Chiou TJ, Echeverria M. Complex regulation of two target genes encoding SPX-MFS proteins by rice miR827 in response to phosphate starvation. Plant Cell Physiol. 2010;51:2119-31.
57. Thatcher SR, Burd S, Wright C, Lers A, Green PJ. Differential expression of miRNAs and their target genes in senescing leaves and siliques: insights from deep sequencing of small RNAs and cleaved target RNAs. Plant Cell Environ. 2015;38:188-200.
58. Rausch T, Wachter A. Sulfur metabolism: A versatile platform for launching defence operations. Trends Plant Sci. 2005;10:503-9.
59. Kopriva S. Regulation of sulfate assimilation in Arabidopsis and beyond. Ann Bot. 2006;97(4):479-95.
60. Jones-Rhoades MW, Bartel DP, Bartel B. MicroRNAS and their regulatory roles in plants. Annu Rev Plant Biol. 2006;57:19-53.
61. Buhtz A, Pieritz J, Springer F, Kehr J. Phloem small RNAs, nutrient stress responses, and systemic mobility. BMC Plant Biol. 2010;10:64.
62. Buhtz A, Springer F, Chappell L, Baulcombe DC, Kehr J. Identification and characterization of small RNAs from the phloem of Brassica napus. Plant J. 2008;53:739-49.
63. Burkhead JL, Reynolds KA, Abdel-Ghany SE, Cohu CM, Pilon M. Copper homeostasis. New Phytol. 2009;182:799-816.
64. Allen E, Xie Z, Gustafson AM, Carrington JC. micro-RNA-directed phasing during trans-acting siRNA biogenesis in plants. Cell. 2005;121:207-21.
65. Yamasaki H, Abdel-Ghany SE, Cohu CM, Kobayashi Y, Shikanai T, Pilon M. Regulation of copper homeostasis by micro-RNA in Arabidopsis. J Biol Chem. 2007;282:16369-78.
66. Abdel-Ghany SE, Pilon M. MicroRNA-mediated systemic down-regulation of copper protein expression in response to low copper availability in Arabidopsis. J Biol Chem. 2008;283:15932-45.
67. Quinn JM, Merchant S. Two copper-responsive elements associated with the Chlamydomonas Cyc6 gene function as targets for transcriptional activators. Plant Cell. 1995;7:623-8.
68. Huang SQ, Peng J, Qiu CX, Yang ZM. Heavy metal-regulated new microRNAs from rice. J Inorg Biochem. 2009;103:282-7.
69. Zhou ZS, Huang SQ, Yang ZM. Bioinformatic identification and expression analysis of new microRNAs from Medicago truncatula. Biochem Biophys Res Commun. 2008;374:538-42.
70. Huang SQ, Xiang AL, Che LL, Chen S, Li H, Song JB, Yang ZM. A set of miRNAs from Brassica napus in response to sulphate deficiency and cadmium stress. Plant Biotechnol J. 2010;8:887-99.
71. Ding YF, Zhu C. The role of microRNAs in copper and cadmium homeostasis. Biochem Biophys Res Commun. 2009;386:6-10.
72. Xie FL, Huang SQ, Guo K, Xiang AL, Zhu YY, Nie L, Yang ZM. Computational identification of novel microRNAs and targets in Brassica napus. FEBS Lett. 2007;581:1464-74.

73. Chen X. Small RNAs and their roles in plant development. Annu Rev Cell Dev Biol. 2009;25:21-44.
74. Trindade I, Santos D, Dalmay T, Fevereiro P. Facing the Environment: Small RNAs and the Regulation of Gene Expression Under Abiotic Stress in Plants. In: Shanker A, Venkateswarlu B, editors. Abiotic Stress Response in Plants - Physiological, Biochemical and Genetic Perspectives. Publisher: InTech, 2011. p. 113-136.
75. Vazquez F, Vaucheret H, Rajagopalan R, Lepers C, Gasciolli V, Mallory AC, Hilbert JL, Bartel DP, Crété P. Endogenous trans-acting siRNAs regulate the accumulation of Arabidopsis mRNAs. Mol Cell. 2004;16:69-79.
76. Vazquez F, Legrand S, Windels D. The biosynthetic pathways and biological scopes of plant small RNAs. Trends Plant Sci. 2010;15(6):337-45.
77. Mallory AC, Bouché N. MicroRNA-directed regulation: To cleave or not to cleave. Trends Plant Sci. 2008;13:359-67.
78. Hsieh LC, Lin SI, Shih AC, Chen JW, Lin WY, Tseng CY, Li WH, Chiou TJ. Uncovering small RNA-mediated responses to phosphate deficiency in Arabidopsis by deep sequencing. Plant Physiol. 2009;151:2120-32.
79. Lu S, Sun YH, Amerson H, Chiang VL. MicroRNAs in loblolly pine (Pinus taeda L.) and their association with fusiform rust gall development. Plant J. 2007;51:1077-98. Epub 2007 Jul 17.
80. Moldovan D, Spriggs A, Yang J, Pogson BJ, Dennis ES, Wilson IW. Hypoxia-responsive microRNAs and trans-acting small interfering RNAs in Arabidopsis. J Exp Bot. 2010;61:165-77.
81. Wang XJ, Gaasterland T, Chua NH. Genome-wide prediction and identification of cis-natural antisense transcripts in Arabidopsis thaliana. Genome Biol. 2005;6:R30.
82. Lu C, Jeong DH, Kulkarni K, Pillay M, Nobuta K, German R, Thatcher SR, Maher C, Zhang L, Ware D, Liu B, Cao X, Meyers BC, Green PJ. Genome-wide analysis for discovery of rice microRNAs reveals natural antisense microRNAs (nat-miRNAs). Proc Natl Acad Sci U S A. 2008;105:4951-6.
83. Borsani O, Zhu J, Verslues PE, Sunkar R, Zhu JK. Endogenous siRNAs derived from a pair of natural cis-antisense transcripts regulate salt tolerance in Arabidopsis. Cell. 2005;123:1279-91.
84. Yao Y, Ni Z, Peng H, Sun F, Xin M, Sunkar R, Zhu JK, Sun Q. Non-coding small RNAs responsive to abiotic stress in wheat (Triticum aestivum L.). Funct Integr Genomics. 2010;10:187-90.
85. Zilberman D, Cao X, Jacobsen SE. ARGONAUTE4 control of locus-specific siRNA accumulation and DNA and histone methylation. Science. 2003;299:716-9.
86. Matzke MA, Birchler JA. RNAi-mediated pathways in the nucleus. Nat Rev Genet. 2005;6:24-35.
87. Matzke M, Kanno T, Daxinger L, Huettel B, Matzke AJ. RNA-mediated chromatin-based silencing in plants. Curr Opin Cell Biol. 2009;21:367-76.
88. Law JA, Jacobsen SE. Establishing, maintaining and modifying DNA methylation patterns in plants and animals. Nat Rev Genet. 2010;11:204-20.
89. Chen X. Small RNAs and their roles in plant development. Annu Rev Cell Dev Biol. 2009;25:21-44.
90. Ruiz-Ferrer V, Voinnet O. Roles of plant small RNAs in biotic stress responses. Annu Rev Plant Biol. 2009;60:485-510.
91. McClintock B. The significance of responses of the genome to challenge. Science. 1984; 16;226:792-801.
92. Lisch D. Epigenetic regulation of transposable elements in plants. Annu Rev Plant Biol. 2009;60:43-66.
93. Ito H, Gaubert H, Bucher E, Mirouze M, Vaillant I, Paszkowski J. An siRNA pathway prevents transgenerational retrotransposition in plants subjected to stress. Nature. 2011;472:115-9.
94. Gebetsberger J, Polacek N. Slicing tRNAs to boost functional ncRNA diversity. RNA Biol. 2013;10:1798-806.

95. Sobala A, Hutvagner G.Transfer RNA-derived fragments: Origins, processing, and functions. Wiley Interdiscip Rev RNA. 2011;2:853-62.
96. Masaki H, Ogawa T. The modes of action of colicins E5 and D, and related cytotoxic tRNases. Biochimie. 2002;84:433-8.
97. Gebetsberger J, Zywicki M, Künzi A, Polacek N. tRNA-derived fragments target the ribosome and function as regulatory non-coding RNA in *Haloferax volcanii*. Archaea. 2012;2012:260909.
98. Ivanov P, Emara MM, Villen J, Gygi SP, Anderson P. Angiogenin-induced tRNA fragments inhibit translation initiation. Mol Cell. 2011;43:613-23.
99. Nowacka M, Strozycki PM, Jackowiak P, Hojka-Osinska A, Szymanski M, Figlerowicz M. Identification of stable, high copy number, medium-sized RNA degradation intermediates that accumulate in plants under non-stress conditions. Plant Mol Biol. 2013;83:191-204.
100. Zhang S, Sun L, Kragler F. The phloem-delivered RNA pool contains small noncoding RNAs and interferes with translation. Plant Physiol. 2009;150:378-87.
101. Hackenberg M, Huang PJ, Huang CY, Shi BJ, Gustafson P, Langridge P. A comprehensive expression profile of microRNAs and other classes of non-coding small RNAs in barley under phosphorous-deficient and -sufficient conditions. DNA Res. 2013;20:109-25.
102. Wang L, Yu X, Wang H, Lu YZ, de Ruiter M, Prins M, He YK. A novel class of heat-responsive small RNAs derived from the chloroplast genome of Chinese cabbage (*Brassica rapa*). BMC Genomics. 2011;12:289.
103. Loss-Morais G, Waterhouse PM, Margis R. Description of plant tRNA-derived RNA fragments (tRFs) associated with argonaute and identification of their putative targets. Biol Direct. 2013;8:6.
104. Haussecker D, Huang Y, Lau A, Parameswaran P, Fire AZ, Kay MA. Human tRNA-derived small RNAs in the global regulation of RNA silencing. RNA. 2010;16:673-95.
105. Cole C, Sobala A, Lu C, Thatcher SR, Bowman A, Brown JW, Green PJ, Barton GJ, Hutvagner G. Filtering of deep sequencing data reveals the existence of abundant Dicer-dependent small RNAs derived from tRNAs. RNA. 2009;15:2147-60.
106. Bilichak A, Ilnytskyy Y, Wóycicki R, Kepeshchuk N, Fogen D, Kovalchuk I. The elucidation of stress memory inheritance in *Brassica rapa* plants. Front Plant Sci. 2015;6:5.
107. Thompson DM, Lu C, Green PJ, Parker R. tRNA cleavage is a conserved response to oxidative stress in eukaryotes. RNA. 2008;14:2095-103.
108. Zywicki M, Bakowska-Zywicka K, Polacek N. Revealing stable processing products from ribosome-associated small RNAs by deep-sequencing data analysis. Nucleic Acids Res. 2012;40:4013-24.
109. Thompson DM, Parker R. The RNase Rny1p cleaves tRNAs and promotes cell death during oxidative stress in *Saccharomyces cerevisiae*. J Cell Biol. 2009;185:43-50.
110. Drinnenberg IA, Weinberg DE, Xie KT, Mower JP, Wolfe KH, Fink GR, Bartel DP. RNAi in budding yeast. Science. 2009;326:544-50.
111. Lee YS, Shibata Y, Malhotra A, Dutta A. A novel class of small RNAs: tRNA-derived RNA fragments (tRFs). Genes Dev. 2009;23:2639-49.
112. Ender C, Krek A, Friedländer MR, Beitzinger M, Weinmann L, Chen W, Pfeffer S, Rajewsky N, Meister G. A human snoRNA with microRNA-like functions. Mol Cell. 2008;32:519-28.
113. Saraiya AA, Wang CC. snoRNA, a novel precursor of microRNA in *Giardia lamblia*. PLoS Pathog. 2008;4:e1000224.
114. Taft RJ, Glazov EA, Lassmann T, Hayashizaki Y, Carninci P, Mattick JS. Small RNAs derived from snoRNAs. RNA. 2009;15:1233-40.
115. Tuck AC, Tollervey D. RNA in pieces. Trends Genet. 2011;27:422-32.
116. Lin CH, Tsai ZT, Wang D. Role of antisense RNAs in evolution of yeast regulatory complexity. Genomics. 2013;102:484-90.
117. Matsui A, Ishida J, Morosawa T, Okamoto M, Kim JM, Kurihara Y, Kawashima M, Tanaka M, To TK, Nakaminami K, Kaminuma E, Endo TA, Mochizuki Y, Kawaguchi S,

Kobayashi N, Shinozaki K, Toyoda T, Seki M. Arabidopsis tiling array analysis to identify the stress-responsive genes. Methods Mol Biol. 2010;639:141-55.

118. Pircher A, Bakowska-Zywicka K, Schneider L, Zywicki M, Polacek N. An mRNA-derived noncoding RNA targets and regulates the ribosome. Mol Cell. 2014;54:147-55.

119. Guttman M, Amit I, Garber M, French C, Lin MF, Feldser D, Huarte M, Zuk O, Carey BW, Cassady JP, Cabili MN, Jaenisch R, Mikkelsen TS, Jacks T, Hacohen N, Bernstein BE, Kellis M, Regev A, Rinn JL, Lander ES. Chromatin signature reveals over a thousand highly conserved large non-coding RNAs in mammals. Nature. 2009 12;458:223-7.

120. Zhou P, Silverstein KA, Gao L, Walton JD, Nallu S, Guhlin J, Young ND. Detecting small plant peptides using SPADA (Small Peptide Alignment Discovery Application). BMC Bioinformatics. 2013 20;14:335.

121. Chew GL, Pauli A, Rinn JL, Regev A, Schier AF, Valen E. Ribosome profiling reveals resemblance between long non-coding RNAs and 5' leaders of coding RNAs. Development. 2013;140:2828-34.

122. ENCODE Project Consortium. An integrated encyclopedia of DNA elements in the human genome. Nature. 2012;489:57-74.

123. Liu TT, Zhu D, Chen W, Deng W, He H, He G, Bai B, Qi Y, Chen R, Deng XW. A global identification and analysis of small nucleolar RNAs and possible intermediate-sized non-coding RNAs in Oryza sativa. Mol Plant. 2013;6:830-46.

124. Wang Y, Wang X, Deng W, Fan X, Liu TT, He G, Chen R, Terzaghi W, Zhu D, Deng XW. Genomic features and regulatory roles of intermediate-sized non-coding RNAs in Arabidopsis. Mol Plant. 2014;7:514-27.

125. Brown CJ, Hendrich BD, Rupert JL, Lafrenière RG, Xing Y, Lawrence J, Willard HF. The human XIST gene: Analysis of a 17 kb inactive X-specific RNA that contains conserved repeats and is highly localized within the nucleus. Cell. 1992;71:527-42.

126. Liu J, Wang H, Chua NH. Long noncoding RNA transcriptome of plants. Plant Biotechnol J. 2015;13:319-28.

127. Wang H, Chung PJ, Liu J, Jang IC, Kean MJ, Xu J, Chua NH. Genome-wide identification of long noncoding natural antisense transcripts and their responses to light in Arabidopsis. Genome Res. 2014;24:444-53.

128. Lu T, Zhu C, Lu G, Guo Y, Zhou Y, Zhang Z, Zhao Y, Li W, Lu Y, Tang W, Feng Q, Han B. Strand-specific RNA-seq reveals widespread occurrence of novel cis-natural antisense transcripts in rice. BMC Genomics. 2012 22;13:721.

129. Li L, Eichten SR, Shimizu R, Petsch K, Yeh CT, Wu W, Chettoor AM, Givan SA, Cole RA, Fowler JE, Evans MM, Scanlon MJ, Yu J, Schnable PS, Timmermans MC, Springer NM, Muehlbauer GJ. Genome-wide discovery and characterization of maize long non-coding RNAs. Genome Biol. 2014;15:R40.

130. Wen J, Parker BJ, Weiller GF. In Silico identification and characterization of mRNA-like noncoding transcripts in Medicago truncatula. In Silico Biol. 2007;7:485-505.

131. Shuai P, Liang D, Tang S, Zhang Z, Ye CY, Su Y, Xia X, Yin W. Genome-wide identification and functional prediction of novel and drought-responsive lincRNAs in Populus trichocarpa. J Exp Bot. 2014;65:4975-83.

132. Xin M, Wang Y, Yao Y, Song N, Hu Z, Qin D, Xie C, Peng H, Ni Z, Sun Q. Identification and characterization of wheat long non-protein coding RNAs responsive to powdery mildew infection and heat stress by using microarray analysis and SBS sequencing. BMC Plant Biol. 2011;11:61.

133. Chen J, Quan M, Zhang D. Genome-wide identification of novel long non-coding RNAs in Populus tomentosa tension wood, opposite wood and normal wood xylem by RNA-seq. Planta. 2015;241:125-43.

134. Boerner S, McGinnis KM. Computational identification and functional predictions of long noncoding RNA in Zea mays. PLoS One. 2012;7:e43047.

135. Zhang YC, Liao JY, Li ZY, Yu Y, Zhang JP, Li QF, Qu LH, Shu WS, Chen YQ. Genome-wide screening and functional analysis identify a large number of long noncoding RNAs involved in the sexual reproduction of rice. Genome Biol. 2014;15:512.

136. Pang KC, Frith MC, Mattick JS. Rapid evolution of noncoding RNAs: Lack of conservation does not mean lack of function. Trends Genet. 2006;22:1-5.
137. Ulitsky I, Shkumatava A, Jan CH, Sive H, Bartel DP. Conserved function of lincRNAs in vertebrate embryonic development despite rapid sequence evolution. Cell. 2011;147:1537-50.
138. Qi X, Xie S, Liu Y, Yi F, Yu J. Genome-wide annotation of genes and noncoding RNAs of foxtail millet in response to simulated drought stress by deep sequencing. Plant Mol Biol. 2013;83:459-73.
139. Zhang W, Han Z, Guo Q, Liu Y, Zheng Y, Wu F, Jin W. Identification of maize long non-coding RNAs responsive to drought stress. PLoS One. 2014;9:e98958.
140. Di C, Yuan J, Wu Y, Li J, Lin H, Hu L, Zhang T, Qi Y, Gerstein MB, Guo Y, Lu ZJ. Characterization of stress-responsive lncRNAs in Arabidopsis thaliana by integrating expression, epigenetic and structural features. Plant J. 2010;1:208-11.
141. Shin H, Shin HS, Chen R, Harrison MJ. Loss of At4 function impacts phosphate distribution between the roots and the shoots during phosphate starvation. Plant J. 2006;45:712-26.
142. Franco-Zorrilla JM, Valli A, Todesco M, Mateos I, Puga MI, Rubio-Somoza I, Leyva A, Weigel D, García JA, Paz-Ares J. Target mimicry provides a new mechanism for regulation of microRNA activity. Nat Genet. 2007;39:1033-7.
143. Ben Amor B, Wirth S, Merchan F, Laporte P, d'Aubenton-Carafa Y, Hirsch J, Maizel A, Mallory A, Lucas A, Deragon JM, Vaucheret H, Thermes C, Crespi M. Novel long non-protein coding RNAs involved in Arabidopsis differentiation and stress responses. Genome Res. 2009;19:57-69.
144. Heo JB, Sung S. Vernalization-mediated epigenetic silencing by a long intronic noncoding RNA. Science. 2011;331:76-9.
145. Ding J, Lu Q, Ouyang Y, Mao H, Zhang P, Yao J, Xu C, Li X, Xiao J, Zhang Q. A long noncoding RNA regulates photoperiod-sensitive male sterility, an essential component of hybrid rice. Proc Natl Acad Sci U S A. 2012;109:2654-9.
146. Wu HJ, Wang ZM, Wang M, Wang XJ. Widespread long noncoding RNAs as endogenous target mimics for microRNAs in plants. Plant Physiol. 2013;161:1875-84.
147. Wang S, Tran EJ. Unexpected functions of lncRNAs in gene regulation. Commun Integr Biol. 2013;6:e27610.
148. Solé C, Nadal-Ribelles M, de Nadal E, Posas F. A novel role for lncRNAs in cell cycle control during stress adaptation. Curr Genet. 2014 Sep 28.
149. Oda A, Takemata N, Hirata Y, Miyoshi T, Suzuki Y, Sugano S, Ohta K. Dynamic transition of transcription and chromatin landscape during fission yeast adaptation to glucose starvation. Genes Cells. 2015;20:392-407.

CHAPTER 11

Variation among Soybean Cultivars in Mesophyll Conductance and Leaf Water Use Efficiency

James Bunce

USDA-ARS, Crop Systems and Global Change Laboratory, Beltsville, MD 20705, USA

ABSTRACT

Improving water use efficiency (WUE) may prove a useful way to adapt crop species to drought. Since the recognition of the importance of mesophyll conductance to CO_2 movement from inside stomatal pores to the sites of photosynthetic carboxylation, there has been interest in how much intraspecific variation in mesophyll conductance (g_m) exists, and how such variation may impact leaf WUE within C_3 species. In this study, the g_m and leaf WUE of fifteen cultivars of soybeans grown under controlled conditions were measured under standardized environmental conditions. Leaf WUE varied by a factor of 2.6 among the cultivars, and g_m varied by a factor of 8.6. However, there was no significant correlation ($r = -0.047$) between g_m and leaf WUE. Leaf WUE was linearly related to the sub-stomatal CO_2 concentration. The value of g_m affected the ratio of maximum Rubisco carboxylation capacity calculated from the sub-stomatal CO_2 concentration to that calculated from the CO_2 concentration at the site of carboxylation. That is, variation in g_m affected the efficiency of Rubisco carboxylation, but not leaf WUE. Nevertheless, there is considerable scope for genetically improving soybean leaf water use efficiency.

Keywords: soybean; water use efficiency; mesophyll conductance; stomatal conductance; internal CO_2 concentration; photosynthesis

1. INTRODUCTION

With increasing limitations on the amount of water available to support agriculture, increasing the water use efficiency (WUE) of crops—the ratio of crop dry mass gained to water consumed—is a reasonable objective. Selection for high leaf WUE in wheat improved yield in dry conditions [1]. Leaf WUE is often defined as the ratio of photosynthesis to transpiration. However, the rate of transpiration is directly related to the difference in water vapor pressure between the air inside and outside the leaf (the leaf to air vapor pressure difference, LAVPD), so the LAVPD during the measurement has a large impact on WUE. Because of this, comparisons of leaf WUE often use "intrinsic" water use efficiency [2], the ratio of photosynthesis to stomatal conductance.

Limitations to C_3 photosynthesis at high light and at the current ambient CO_2 concentration have long been quantified by the maximum capacity of rubisco carboxylation (V_{Cmax}) [3]. However, it has been recognized within the last several years that mesophyll conductance to CO_2 movement from inside the stomata to the site of fixation within chloroplasts (g_m) can be a significant limitation to photosynthesis (A) [4,5]. Because of the potential of variation in g_m to affect A independently from stomatal conductance (g_s), there is interest in determining the extent of variation in g_m within species, and its effect on leaf WUE.

Intraspecific variation in g_m has been reported in barley [6], grape [7], tomato [8], wheat [9], and rice [10], and its relationship to leaf WUE examined. In grape and barley, g_m was positively correlated with leaf WUE. In tomato and rice, variation was found in the ratio of g_m to g_s, and that ratio was correlated with leaf WUE. In wheat variation, g_m was correlated with A, but there was no clear relationship between g_m and leaf WUE.

Among these studies, g_m was quantified either from leaf fluorescence changes with CO_2 concentration, or from carbon isotope discrimination, both of which methods are based on assumptions which are sometimes dubious [11,12,13]. A more rapid method for estimating g_m, with fewer assumptions, has been developed based on the response of photosynthesis to oxygen concentration [14], and was used here to screen fifteen cultivars of soybean for g_m and to test for relationships among g_m, g_s, A, and leaf WUE. Genetic improvement of WUE depends upon sufficient genetic variation in WUE and the identification of physiological processes which affect it, which, as indicated by the literature cited, may vary with crop species. The goals of this study were to determine whether variation in intrinsic leaf WUE or g_m occurred in soybean measured under standardized conditions, and to determine whether any variation in g_m was related to variation in intrinsic leaf WUE.

2. RESULTS

The cultivars differed significantly in g_s, g_m, leaf intrinsic WUE, sub-stomatal CO_2 concentration (C_i), and the CO_2concentration at rubisco (C_c), but not in $V_{Cm}C_c$ (V_{Cmax} based on C_c), $V_{Cm}C_i$ (V_{Cmax} based on C_c), or A (Table 1). For cultivars with high and low values of g_m, the values estimated using the initial slope of A vs. C_i agreed with those using the O_2 effect. Intrinsic leaf WUE varied by a factor of 2.6 among the cultivars, with Fiskeby having the highest, and Ford having the lowest WUE (Table 1). Mesophyll conductance varied by a factor of 8.6 among cultivars, with Biloxi having the highest and Clark the lowest mean values. Variation in cultivar means of intrinsic WUE was due mostly to variation in g_s, which ranged from 0.38 to 1.12 mmol H_2O $m^{-2}{\cdot}s^{-1}$, rather than photosynthesis, which ranged from 23.7 to 31.1 μ mol $CO_2 m^{-2}{\cdot}s^{-1}$. Fiskeby, with the highest WUE, had the second lowest g_s, and Ford, with the lowest WUE, had the highest g_s. For all individual leaf measurements, intrinsic WUE was strongly negatively related to g_s, with the reciprocal of WUE linearly related to g_s (Figure 1). C_i ranged from about 225 to 345 μ mol·mol^{-1} and there was a negative linear relationship between C_i and intrinsic WUE (Figure 2).

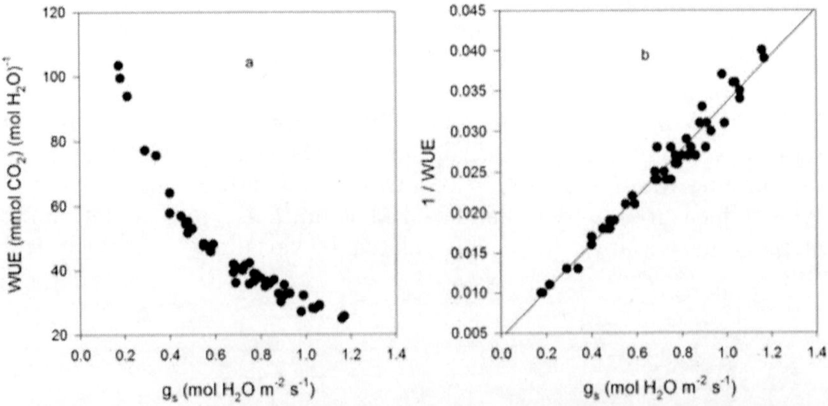

Figure 1. Leaf intrinsic water use efficiency (WUE) (**a**); and the reciprocal of leaf intrinsic WUE (**b**) as a function of stomatal conductance (g_s) for all measurements made on fifteen cultivars of soybean. The linear regression of (1/WUE) = 0.0044 + 0.029 × g_s had an r^2 value of 0.965.

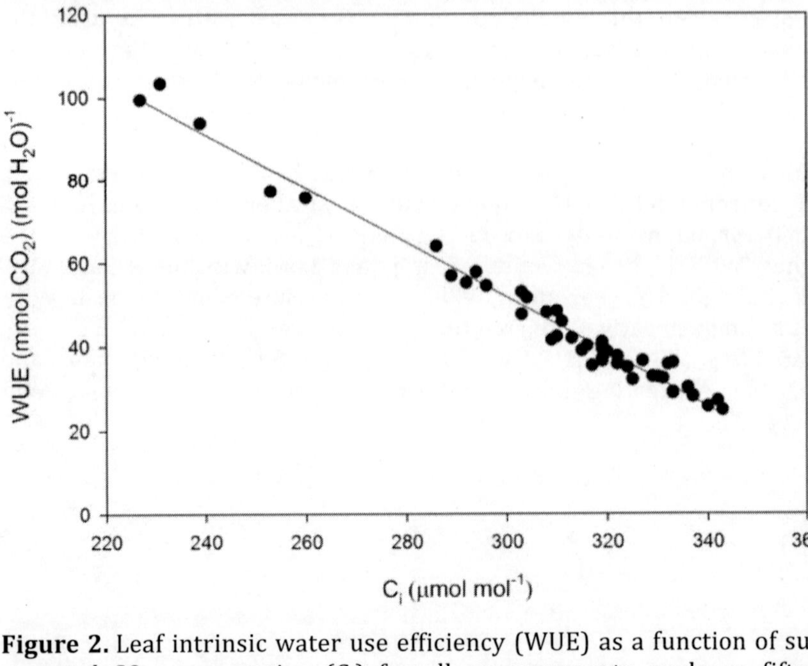

Figure 2. Leaf intrinsic water use efficiency (WUE) as a function of sub-stomatal CO_2concentration (C_i) for all measurements made on fifteen cultivars of soybean. The linear regression: WUE = 248 − 0.654 × C_i had an r^2 of 0.98.

Table 1. Mean leaf gas exchange parameter values of the soybean cultivars tested. Parameters were determined for three or four leaves per cultivar. g_s is stomatal conductance in mmol (H_2O) $m^{-2} \cdot s^{-1}$, g_m is mesophyll conductance in mmol (CO_2) $m^{-2} \cdot s^{-1}$, A is photosynthetic rate in μ mol (CO_2) $m^{-2} \cdot s^{-1}$ measured at 400 μ mol·mol^{-1}CO_2, WUE is intrinsic water use efficiency in mmol (CO_2) per mol (H_2O), $V_{Cm}C_i$ is the maximum rate of rubisco carboxylation based on C_i, and $V_{Cm}C_c$ is the maximum rate of rubisco carboxylation based on C_c. C_i is the sub-stomatal CO_2 concentration, and $C_c$is the CO_2 concentration at rubisco, in μ mol·mol^{-1}. Probability of equal means from ANOVA is given, and the Tukey–Kramer honestly significant difference (HSD) is provided when the probability is <0.05.

Cultivar	g_s	g_m	A	WUE	$V_{Cm}C_i$	$V_{Cm}C_c$	C_i	C_c
A5959	703	2.63	27.7	43.1	113	118	313	299
Biloxi	803	3.60	28.9	36.4	116	123	323	307
Chief	878	2.28	31.1	35.9	127	139	321	297
Clark	677	0.42	26.6	40.3	108	138	324	258
Essex	587	1.50	26.4	56.9	114	121	291	272
Fiskeby V	418	2.56	24.4	67.6	114	122	276	259
Ford	1123	0.60	29.4	26.2	116	138	340	288
Holt	502	2.05	25.0	63.5	105	112	287	270
Kent	820	0.78	28.0	34.7	115	132	328	291
Lincoln	725	0.77	28.4	39.1	118	137	317	277
Perry	883	2.10	30.8	35.7	121	127	322	306
PI-41	430	1.09	25.8	62.0	117	130	283	257
Ripley	380	0.76	23.7	64.0	101	118	279	242
Spencer	728	2.68	29.0	41.6	118	124	314	297
Wabash	910	1.11	28.1	31.8	114	128	333	300
Probability	**0.001**	**0.016**	**0.081**	**0.029**	**0.097**	**0.279**	**0.019**	**0.036**
HSD	592	2.79		31.1			56	65

Mesophyll conductance had no significant correlation with intrinsic WUE (r = −0.047), g_s (r = +0.065), A (r = +0.234), or $V_{Cmax}C_i$ (r = +0.227) (Figure 3). Using means for the cultivars for all of these parameters also produced no significant correlations (r = 0.079, 0.217, 0.040, and 0.301, for g_m versus intrinsic WUE, g_s, A, and $V_{Cmax}C_i$, respectively). The ratio of g_m to g_s in soybean had a small but significant correlation with intrinsic leaf WUE (Figure 4), but there was no significant correlation between C_i and g_m (Figure 4). The ratio of the maximum capacity of rubisco carboxylation modelled based on C_i to that that based on C_c decreased with mesophyll resistance, which is the reciprocal of g_m, from about 0.70 to 0.98 (Figure 5). Photosynthesis at C_a = 400 μ mol·mol^{-1} was positively related to C_i (Figure 6), similar to a saturating A vs. $C_i$curve for an individual leaf. While there was a small but significant correlation between C_i and C_c (Figure 7), there was a wide range of C_i values for a given C_c. For example cultivar means of C_i ranged from about 275 to 325 μ mol·mol^{-1} for mean C_c values of about 256 to 259 μ mol·mol^{-1} (Figure 7).

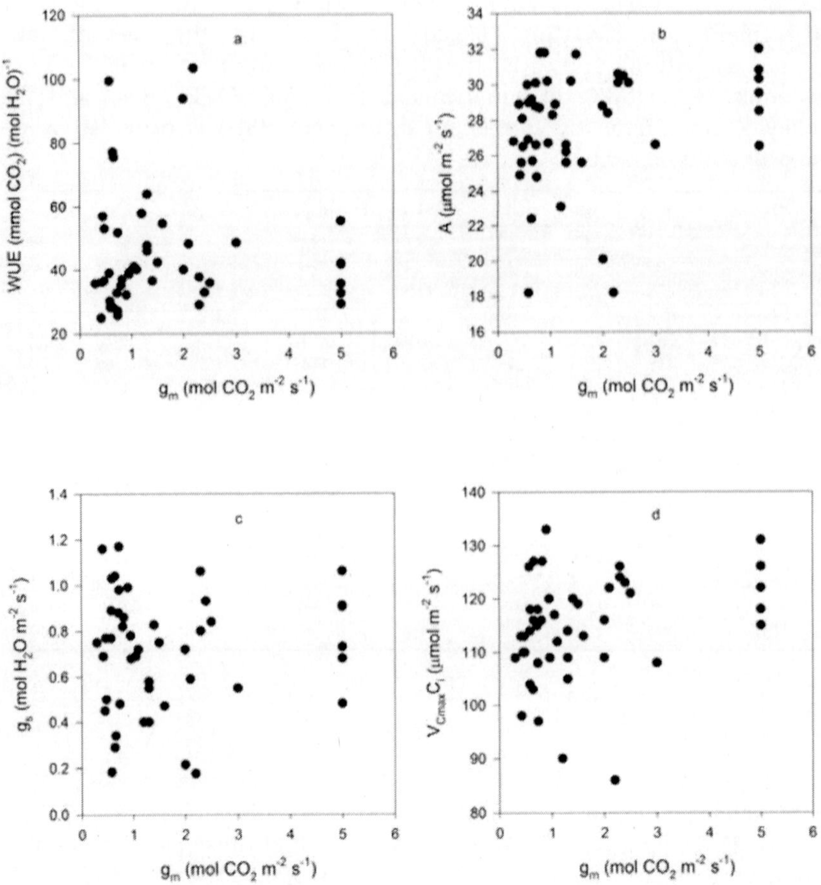

Figure 3. Correlations between g_m and (**a**) leaf intrinsic water use efficiency (WUE); (**b**) CO_2 assimilation rate (A); (**c**) stomatal conductance (g_s); (**d**) and the maximum rate of rubisco carboxylation based on C_i ($V_{Cmax}C_i$) for all measurements made on fifteen cultivars of soybean. Correlation coefficients were not significant at p = 0.05, and are given in the text.

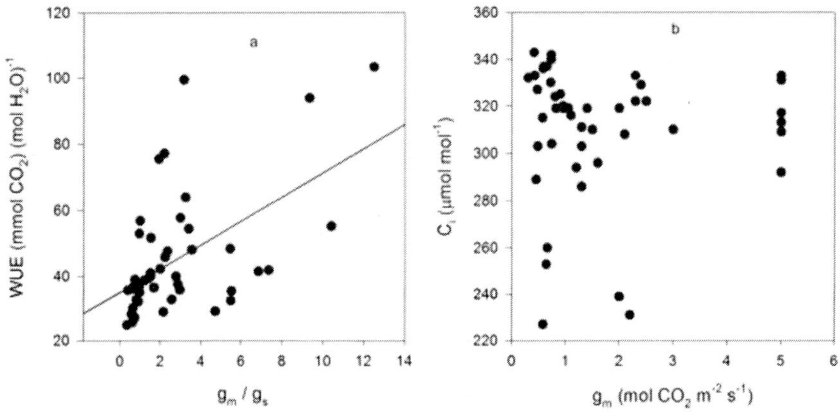

Figure 4. Relationships between the ratio of g_m to g_s and WUE (**a**); and between $g_m$and C_i (**b**) for all measurements made on fifteen cultivars of soybean. The correlation between the g_m to g_s ratio and WUE had an r value of +0.53, and that between g_m and C_i was +0.116, which was not significant at $p = 0.05$.

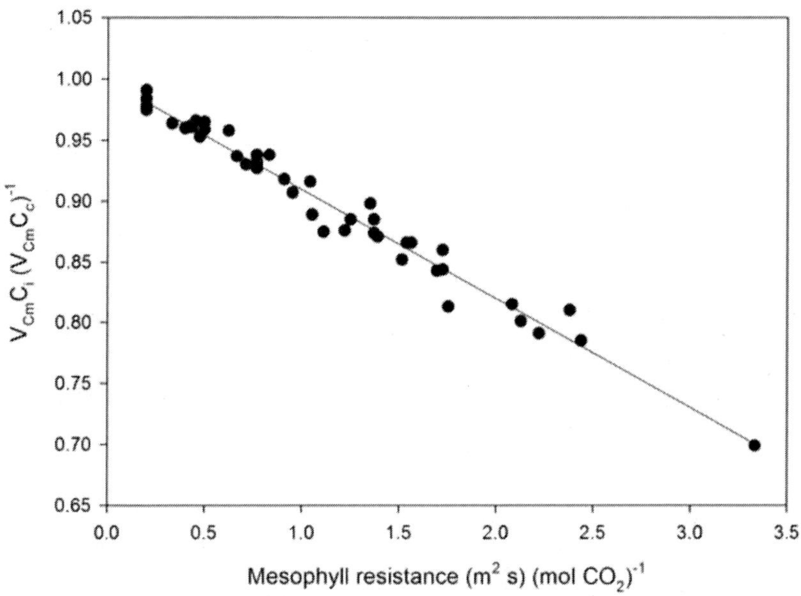

Figure 5. The ratio of V_{Cmax} based on C_i to that based on C_c as a function of mesophyll resistance, the reciprocal of g_m, for all measurements made on fifteen cultivars of soybean. The linear regression: $V_{Cm}C_i/V_{Cm}C_c = 1.00 - 0.090 \times (1/g_m)$ had an r^2 of 0.976.

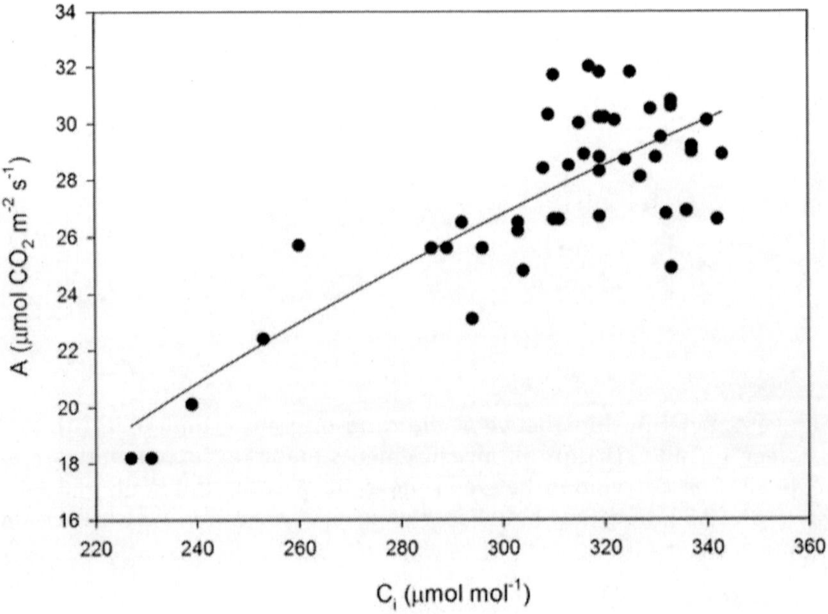

Figure 6. CO_2 assimilation rate (A) as a function of sub-stomatal CO_2 concentration (C_i) for all measurements made on leaves of fifteen cultivars of soybean. The equation $A = -125.6 + 26.7 \ln(C_i)$ had an r^2 value of 0.645.

3. DISCUSSION

The linear relationship between intrinsic leaf WUE and C_i was as expected, since C_a was constant across the measurements. Carbon isotope discrimination has long been used as a surrogate for C_i in screening for leaf WUE in C_3species. It is important to know how much variation in g_m would disrupt the correlation between C_i (and WUE) and carbon isotope discrimination, which should reflect C_c rather than C_i [6]. The ranking of these soybean cultivars based on C_c or carbon isotope discrimination would not provide a reliable ranking of their C_i or intrinsic leaf WUE values (Figure 7).

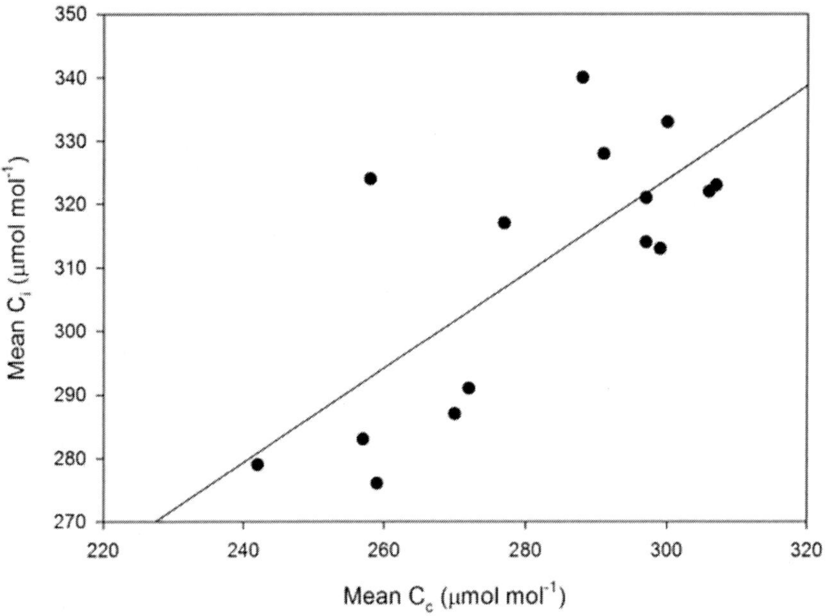

Figure 7. Cultivar mean values of C_i versus mean values of C_c. The linear regression had an r^2 value of 0.532.

While variation among cultivars in C_i was more strongly influenced by variation in g_s than A, there was still the usual [15] penalty in A associated with low g_s, low C_i and high intrinsic WUE (Figure 5). One reason for interest in g_m with regard to WUE is the possibility that high g_m might offset the penalty in A associated with low C_i and high WUE [16]. Certainly, a high g_m results in a higher value of A at a given C_i, and results in $V_{Cm}C_i$ being closer to $V_{Cm}C_c$ (Figure 4). However, g_m was poorly correlated with any other leaf gas exchange parameter (g_s, A, C_i, $V_{Cm}C_i$, WUE) in this data. It is possible that the correlation between the ratio of g_m to g_s and WUE reported in rice and tomato [8,10] has little to do with variation in g_m, but simply reflects a strong relationship between g_s and WUE, as found here in soybean. Furthermore, it remains unknown how much a genetic increase in g_m might cost in terms of leaf nitrogen, which would tend to offset the resulting increased efficiency of rubisco [16]. It is thought that there is a metabolic component to g_m, and that therefore there would be some nitrogen cost to increasing g_m [16]. Reasons for cultivar differences in the operational C_i under these identical and non-stressful environmental conditions remain unknown, but g_s and the operational C_i were clearly the primary determinates of leaf intrinsic WUE among these soybean leaves.

The observed 2.95 fold range in mean g_s among soybean cultivars would have a substantial impact on canopy transpiration, even though relative differences in transpiration decrease as the scale increases from leaf to canopy [17]. Using the mean leaf boundary layer conductance of 1200 mmol (H_2O) $m^{-2}\cdot s^{-1}$ of soybean leaves measured near midday on 9 days in Beltsville, Maryland [18], a 2.95 fold range in leaf g_s would translate into approximately a 1.45 fold range in canopy transpiration. Such a large range in canopy transpiration could have a substantial impact on the rate of development of plant water stress in the field. Thus, there is considerable scope for reduction in canopy transpiration rate and improvement in soybean WUE through breeding. However, the results of this study make it unlikely that screening soybeans for g_m values would be a profitable method of attempting to increase WUE in soybean.

It should be noted that the intrinsic leaf WUE and g_s values measured here were measured under a standardized LAVPD. "Intrinsic" leaf WUE is, of course, not really intrinsic to the leaf, but would be expected to vary with the measurement conditions, as A and g_s vary with environment. The response of g_s to LAVPD in soybean is known to vary among cultivars [19]. Fletcher et al. [20] have identified lines of soybeans differing in the response of whole plant transpiration rate to LAVPD, but did not provide information about either g_s or photosynthetic responses to LAVPD, or information on leaf WUE. Field measurements of g_s and A in response to LAVPD in several soybean cultivars [21] did not present data on operational C_i.

Our results indicated that g_s and C_i were correlated with leaf intrinsic WUE, and may be useful in selecting lines with high WUE. However, the lack of correlation between g_m and leaf intrinsic WUE indicates that g_m may not be useful in ranking lines for WUE, although both factors varied among cultivars. It will be important to assess the operational C_i of soybean lines under a range of field conditions to determine how robust the differences in leaf intrinsic WUE found here are in the field. The results presented here also indicate that carbon isotope discrimination may not reliably identify soybean lines with differences in leaf intrinsic WUE.

4. MATERIALS AND METHODS

Three or four plants of each of fifteen cultivars of soybean (*Glycine max* L. Merr.) obtained from the U.S. Department of Agriculture soybean germplasm collection (Table 1) were grown together in one large controlled environment chamber at the controlled environment facility of the Beltsville Agricultural Research Center, Beltsville, Maryland. All of the cultivars are adapted to the eastern United States. Some (Biloxi, Clark, Kent, PI-471938, A5959) were chosen because they had diverse responses of stomatal conductance or transpiration rate to humidity in prior studies [19,20]. The others were chosen based on my prior measurements of leaf gas exchange rates in various experiments. The chamber had 14 h of light per day at 1000 μ mol $m^{-2}·s^{-1}$ photosynthetic photon flux density (PPFD) from a combination of metal halide and high pressure sodium lamps. This gave a daily total photon flux similar to average mid-summer days at Beltsville, Maryland. The day/night air temperatures were 26/20 °C, with a dew point temperature of about 18 °C. These approximate mean conditions in Beltsville for the soybean growing season. The CO_2 concentration was controlled at 400 μ $mol·mol^{-1}$ during the day and 430 μ $mol·mol^{-1}$ at night by the addition of pure CO_2 or air scrubbed of CO_2, under the control of a WMA-5 (PP Systems, Amesbury, MA, USA) infrared analyzer which sampled chamber air continuously. Plants were grown one per pot in 15 cm diameter plastic pots filled with vermiculite and were fertilized daily with a complete nutrient solution containing 14.5 mN nitrogen.

Leaf gas exchange measurements were conducted on fourth main stem trifoliolate leaves a few days after those leaves reached full area expansion. Leaf gas exchange measurements were made with a recently calibrated Li-6400 portable photosynthesis system (Li-Cor, Inc., Lincoln, NE, USA). Leaf temperature was controlled at 25 °C, the PPFD was 1500 μ mol $m^{-2}·s^{-1}$, and the LAVPD was maintained at 1.0 ± 0.1 kPa. Each leaf was measured under three conditions of CO_2 and O_2. First, the leaf was exposed to 21% O_2, and 400 μ $mol·mol^{-1}$ external CO_2. The external CO_2 was then lowered to 300 μ mol mol^{-1} at the same O_2 concentration. Finally, the oxygen concentration was lowered to 2%, while the CO_2 concentration remained at 300 μ $mol·mol^{-1}$. Care was taken to ensure that gas exchange rates were steady at each condition. These measurements were made on three or four replicate leaves from different plants for each cultivar. A, g_s, C_i, and LAVPD were calculated by the instrument software.

Intrinsic leaf WUE was calculated separately for each leaf as A/g_s measured at 400 ± 5 μ mol mol^{-1} CO_2, and 21% O_2, at 25 °C, and at a LAVPD of 1 kPa. Mesophyll conductance was also calculated for each leaf from the gas exchange rates at 300 μ $mol·mol^{-1}$ CO_2 at 21% and 2% O_2 using the method described in Bunce [14], with the calculation utility developed by Singh [22]. The method is based on the idea that the sensitivity of photosynthesis to oxygen depends on the concentration of CO_2 at the site of Rubisco, rather than the sub-stomatal CO_2 concentration. This method of estimating g_m uses the change in photosynthetic rate between two O_2 concentrations at limiting

CO_2 concentrations to indicate the CO_2concentration at the site of rubisco (C_c), using a standard biochemical C_3 photosynthesis model [14]. This method of estimating g_m also provides estimates of V_{Cmax} based on C_i and based on C_c. For some leaves with the highest and lowest values of g_m, additional points on A vs. C_i curves at 21% O_2 were used to calculate g_m using the Sharkey et al. calculation utility [23] based on the curvature of the initial slope of the A vs. C_i curve.

Analysis of variance was used to test for differences among cultivars in mean leaf gas exchange parameters. For both $g_s$and g_m, homogeneity of variance was violated, so values were log transformed prior to ANOVA. Correlations among gas exchange parameters were calculated using all data on individual leaves, because of the increased statistical power compared with using mean values, given the substantial variability among leaves with cultivars. Where correlations were significant at p = 0.05, linear or simple non-linear regressions are presented.

CONFLICTS OF INTEREST

The author declares no conflict of interest.

REFERENCES

1. Condon, A.G.; Richards, R.A.; Rebetzke, G.L.; Farquhar, G.D. Breeding for high water-use efficiency. J. Exp. Bot. 2004, 55, 2447–2460.
2. Seibt, U.; Rajabi, A.; Griffiths, H.; Berry, J.A. Carbon isotopes and water use efficiency: Sense and sensitivity. Oecologia2008, 155, 441–454.
3. Farquhar, G.D.; von Caemmerer, S.; Berry, J.A. A biochemical model of photosynthetic CO2 assimilation in leaves of C3 species. Planta 1980, 149, 78–90.
4. Warren, C.R. Stand aside stomata, another actor deserves centre stage: The forgotten role of the internal conductance to CO2 transfer. J. Exp. Bot. 2008, 59, 1475–1487.
5. Evans, J.R.; Kaldenhoff, R.; Genty, B.; Terashima, I. Resistances along the CO2 diffusion pathway inside leaves. J. Exp. Bot. 2009, 21, 475–495.
6. Barbour, M.M.; Warren, C.R.; Farquhar, G.D.; Forrester, G.; Brown, H. Variability in mesophyll conductance between barley genotypes, and effects on transpiration efficiency and carbon isotype discrimination. Plant Cell Environ. 2010, 33, 1176–1185.
7. Tomas, M.; Medrano, H.; Brugnoli, E.; Escalona, J.M.; Martorell, S.; Pou, A.; Ribas-Carbo, M.; Flexas, J. Variability of mesophyll conductance in grapevine cultivars under water stress conditions in relation to leaf anatomy and water use efficiency. Aust. J. Grape Wine Res. 2014, 20, 272–280.
8. Galmes, J.; Conesa, M.A.; Ochogavia, J.M.; Perdomo, J.A.; Francis, D.M.; Ribas-Carbo, M.; Save, R.; Flexas, J.; Medrano, H.; Cifre, J. Physiological and morphological adaptations in relation to water use efficiency in Mediterranean accessions of Solanum lycopersicum. Plant Cell Environ. 2011, 34, 245–260.

9. Jahan, E.; Amthor, J.S.; Farquhar, G.D.; Trethowan, R.; Barbour, M.M. Variation in mesophyll conductance among Australian wheat genotypes. Funct. Plant Biol. 2014, 41, 568–580.
10. Gioliani, R.; Koteyeva, N.; Voznesenskaya, E.; Evans, M.A.; Cousins, A.B.; Edwards, G.E. Coordination of leaf photosynthesis, transpiration, and structural traits in rice and wild relatives (genus Oryza). Plant Physiol. 2013, 162, 1632–1651.
11. Pons, T.L.; Flexas, J.; von Caemmeer, S.; Evans, J.R.; Genty, B.; Ribas-Carbo, M.; Brugnoli, E. Estimating mesophyll conducgtance to CO2: Methodology, potential errors, and recommendations. J. Exp. Bot. 2009, 60, 2217–2234.
12. Tholen, D.; Ethier, G.; Genty, B.; Pepin, S.; Zhu, X.-G. Variable mesophyll conductance revisited: Theoretical background and experimental implications. Plant Cell Environ. 2012, 35, 2087–2103.
13. Gu, L.; Sun, Y. Artefactual responses of mesophyll conductance to CO2 and irradiance estimated with the variable J and online isotope discrimination methods. Plant Cell Environ. 2014, 37, 1231–1249.
14. Bunce, J.A. Use of the response of photosynthesis to oxygen to estimate mesophyll conductance to carbon dioxide in water-stressed soybean leaves. Plant Cell Environ. 2009, 32, 875–881.
15. Blum, A. Effective use of water (EUW) and not water-use efficiency (WUE) is the target of crop yield improvement under drought stress. Field Crops Res. 2009, 112, 119–123.
16. Buckley, T.N.; Warren, C.R. The role of mesophyll conductance in the economics of nitrogen and water use in photosynthesis. Photosynth. Res. 2014, 119, 77–88.
17. Wilson, K.B.; Carlson, T.N.; Bunce, J.A. Feedback significantly influences the simulated effect of CO2 on seasonal evapotranspiration from two agricultural species. Glob. Chang. Biol. 1999, 5, 903–917.
18. Bunce, J.A. Use of a minimally invasive method of measuring leaf stomatal conductance to examine stomatal responses to water vapor pressure under field conditions. Agric. Meteorol. 2006, 139, 335–343.
19. Bunce, J.A. Identifying soybean lines differing in gas exchange sensitivity to humidity. Ann. Appl. Biol. 1984, 105, 313–318.
20. Fletcher, A.L.; Sinclair, T.R.; Allen, L.H., Jr. Transpiration responses to vapor pressure deficit in well water "slow-wilting" and commercial soybean. Environ. Exp. Bot. 2007, 61, 145–151.
21. Gilbert, M.E.; Holbrook, N.M.; Zwieniecki, M.A.; Sadok, W.; Sinclair, T.R. Field confirmation of genetic variation in soybean transpiration response to vapor pressure deficit and photosynthetic compensation. Field Crops Res. 2011, 124, 85–92.
22. Singh, S.K.; Reddy, V.R. Methods of mesophyll conductance estimation: Its impact on key biochemical parameters and photosynthetic limitations in phosphorus-stressed soybean across CO2. Physiol. Plant. 2016, 157, 234–254.
23. Sharkey, T.D.; Bernacchi, C.J.; Farquhar, G.D.; Singsaas, E.L. Fitting photosynthetic carbon dioxide response curves for C3 leaves. Plant Cell Environ. 2007, 30, 1035–1040.

CHAPTER 12

Interaction Effect between Elevated CO_2 and Fertilization on Biomass, Gas Exchange and C/N Ratio of European Beech (*Fagus sylvatica* L.)

Neda Lotfiomran [1,*], Michael Köhl [1] and Jörg Fromm[2]

[1]*Institute for World Forestry, University of Hamburg, Leuschnerstrasse 91, 21031 Hamburg, Germany*

[2]*Institute for Wood Biology, University of Hamburg, Leuschnerstrasse 91d, 21031 Hamburg, Germany*

ABSTRACT

The effects of elevated CO_2 and interaction effects between elevated CO_2 and nutrient supplies on growth and the C/N ratio of European beech (*Fagus sylvatica* L.) saplings were studied. One-year-old beech saplings were grown in a greenhouse at ambient (385 ppm) and elevated CO_2 (770 ppm/950 ppm), with or without fertilization for two growing seasons. In this study, emphasis is placed on the combined fertilization including phosphorus, potassium and nitrogen with two level of elevated CO_2. The fertilized plants grown under elevated CO_2 had the highest net leaf photosynthesis rate (Ac). The saplings grown under elevated CO_2 had a significantly lower stomatal conductance (gs) than saplings grown under ambient air. No interaction effect was found between elevated

CO_2 and fertilization on Ac. A interaction effect between CO_2 and fertilization, as well as between date and fertilization and between date and $CO_2$was detected on gs. Leaf chlorophyll content index (CCI) and leaf nitrogen content were strongly positively correlated to each other and both of them decreased under elevated CO_2. At the end of both growing seasons, stem dry weight was greater under elevated CO_2 and root dry weight was not affected by different treatments. No interaction effect was detected between elevated CO_2 and nutrient supplies on the dry weight of different plant tissues (stems and roots). However, elevated CO_2 caused a significant decrease in the nitrogen content of plant tissues. Nitrogen reduction in the leaves under elevated CO_2 was about 10% and distinctly higher than in the stem and root. The interaction effect of elevated CO_2 and fertilization on C/N ratio in plants tissues was significant. The results led to the conclusion that photosynthesis and the C/N ratio increased while stomatal conductance and leaf nitrogen content decreased under elevated CO_2 and nutrient-limited conditions. In general, under nutrient-limited conditions, the plant responses to elevated CO_2 were decreased.

Keywords: climate change; elevated CO_2; fertilization; interaction effect; European beech; biomass; gas exchange; C/N ratio

1. INTRODUCTION

Carbon dioxide is a natural component of the atmosphere and an essential factor for the growth of trees since nearly half of their constituents are composed of carbon. The CO_2 content in the ambient atmosphere has not been constant through time. Since around 1850, it has continuously risen from about 270 ppm to around 400 ppm at present and will continue to do so in future [1]. Carbon dioxide emissions are the most important cause of global warming. As a sink for CO_2, forests play an important role in the discussion about global warming [2]. The recent rise of the CO_2 content in the ambient air and its implications for the biosphere, including global warming, have been a focus of politics for more than two decades, and numerous studies have analyzed the effects of elevated CO_2 on plants [3,4,5,6,7,8,9]. A large body of information has also been produced in response to the effect of elevated CO_2 on trees [6,8,9,10,11,12,13,14]. Despite this vast number of CO_2related papers published to date, however, more research on trees and shrubs under elevated CO_2 is still needed. Nevertheless, a considerable but heterogeneous variety of experimental results and a wide diversity of interpretations have become accessible.

Photosynthesis is directly affected by a varying CO_2 content in the ambient air and is therefore one of the major physiological processes being monitored during all kind of experiments. Numerous short-term studies have showed an increase in the photosynthesis rate caused by an elevated CO_2 concentration [3,5,6,10,11,15,16,17]. However, the photosynthetic behaviour of trees under long-term exposure to elevated CO_2 is less clear [12]. Although it is generally agreed that short-term growth under elevated CO_2 causes a stimulation of the photosynthesis rate in trees, it has frequently been reported that this enhancement may decline or even disappear with time [18,19]. Furthermore, in several studies a "down regulation" of the photosynthesis rate was shown in plants grown under elevated CO_2 over longer periods (weeks or months) [9,11,13,20].

The production of biomass is one of many paths along which carbon is metabolized [9]. Several measurements on seedlings and young trees indicate that total growth increases with an increased carbon uptake [6,8,11,21,22,23]. There is also strong evidence that plant biomass production under elevated CO_2 is mostly larger than in ambient air, although a number of studies showed no significant increase, or even a decrease, of biomass production under elevated CO_2 [7,24,25]. In those studies that do show an increase, the magnitude of biomass enhancement varies widely between observations. The enhancing effect of elevated CO_2 on biomass production can decline or totally disappear with time [11,18,26,27,28]. Growth stimulation, as well as other tree responses to elevated CO_2 depends on various factors such as the developmental stage of the plants [29,30], genetic factors, environmental conditions [6,31] and water and nutrient availability [6,8,32,33,34]. Therefore several investigators have focused on the interaction of elevated CO_2 with, for example, drought [35], temperature [36], and other greenhouse gases such as O_3, SO_2 and NO_3 [37,38,39] as well as nutrient availability [40,41]. The plant responses to elevated CO_2, decreased under nutrient limited conditions but it is reported that this tendency was weak and mostly readily apparent in plants which were grown with severe nutrient limitation [42]. However, it is difficult to generalize the plant responses under elevated CO_2 to nutrient limitations.

Nitrogen is part of all living cells and of all proteins, enzymes and metabolic processes involved in the synthesis and transfer of energy. Trees require a considerable amount of N for their growth, especially when growing faster under elevated CO_2 [33]. One of the most common observations is a lower N concentration in plant components grown under elevated CO_2 than in ambient air [20,27,28,43,44,45].

One of main species in deciduous forests in Europe is the European beech (*Fagus sylvatica* L.). The distribution of the European beech is wide, although concentrated in west and central Europe. With a 10.3 m^3/ha/year growth rate, after Spruce (*Picea abies*), beech is the second fastest growing tree species in European forests [46]. Due to this importance, it is necessary to understand how European beech responses to global warming, caused mainly by enrichment of CO_2concentration in the atmosphere. Elevated CO_2 caused changes in the wood structure and anatomical characteristics of European beech [36,47]. Increases in both the photosynthesis rate and biomass of European beech under elevated CO_2 have been reported in previous studies [18,48,49]. On the other hand, it is now widely accepted that the initial stimulation of photosynthesis under elevated CO_2 is often followed by a decline in photosynthesis, usually accompanied by a decrease in nutrient content, especially nitrogen in plant tissues.

The response of European beech to high CO_2 can be influenced by an interaction effect between elevated CO_2 and other factors such as temperature, drought and other greenhouse gases [36,37,38,50]. Although numerous studies dealt with responses of beech tree growth under elevated CO_2, combined experiments with CO_2 and nutrient application are rare and the fertilizer treatments are mostly based on nitrogen [50]. In the present study, one-year-old potted beech saplings with and without fertilization were grown under different levels of elevated CO_2 (770/950 ppm) in a greenhouse for two growing seasons. The effects of elevated CO_2 and fertilization on the photosynthesis rate (Ac), stomatal conductance (gs) and leaf chlorophyll content were investigated. The variation of biomass allocation and C/N ratio between different treatments were also studied. In this study, emphasis is placed on the combined fertilization including phosphorus, potassium and nitrogen with elevated CO_2. The main objective of the present study was to investigate how different levels of nutrient supplies interact with the effects of elevated CO_2 on European beech.

2. RESULTS

2.1. Variation of Height Growth and Dry Weight of Leaves, Stem, Root and Total Biomass between the Treatments

Regarding height growth, no significant differences were observed between different treatments in both years (Figure 1), as supported by statistical analysis (Table 1). The variation of stem dry weight between treatments showed a similar pattern in both growing seasons (Figure 1). Plants grown under elevated CO_2 conditions produced significantly heavier stems (Figure 1), but the differences between fertilized and unfertilized plants grown under different CO_2 levels (elevated-ambient) were not significant (Table 1). No interaction effect between elevated CO_2 and fertilization on stem dry weight was found. In 2010, no significant changes in the dry weight of leaves were observed between different treatments. In 2011, however, the dry weight of the leaves varied significantly between treatments (Figure 1, Table 1). Unfertilized plants under ambient air (control) had the greatest leaf weight; in contrast, unfertilized saplings under elevated CO_2 had the lowest value. The interaction between CO_2 and fertilization was significant. Hence, fertilization alone had no effect on dry weight but under elevated CO_2, the difference between the dry weight of the leaves of unfertilized and fertilized plants was significant (Table 1). In both years, on average, 50% of the total biomass consisted of root dry mass and differences in root dry mass as well as total biomass of treatments were unaffected by either CO_2 concentration of fertilization (Table 1).

Table 1. Statistical strength (F and *p*-values) of the effects of fertilization and CO_2concentrations on height, leaf dry mass, stem dry mass and root dry mass between treatments; * $p \leq 0.05$; ** $p \leq 0.01$; *** $p \leq 0.001$.

		Height		Leaf		Stem		Root		Total biomass	
		F	*p*-value	F	*p*-value	F	*p*-value	F	*p*-value	F	*p*-value
CO_2 concentration	2010	2.88	0.10	3.27	0.08	5.10 *	0.03	0.02	0.88	2.18	0.15
	2011	0.00	0.98	6.55 **	0.01	4.36 *	0.04	2.32	0.14	1.76	0.19
Fertilization	2010	1.16	0.29	1.35	0.25	0.15	0.70	1.53	0.22	1.05	0.31
	2011	0.69	0.41	1.70	2.00	0.16	0.90	0.42	0.52	0.02	0.88
CO_2 × Fertilization	2010	1.65	0.21	0.87	0.36	1.30	0.26	0.14	0.70	0.25	0.62
	2011	1.13	0.30	15.2 ***	0.00	0.00	0.95	0.58	0.45	2.18	0.15

2.2. Seasonal Variations in Photosynthesis Rate and Stomatal Conductance

The seasonal tendencies in photosynthesis rate (Ac) and stomatal conductance (g_s) for the different treatments in 2011 are summarized in Figure 2. The maximum mean value of Ac occurred with fertilized saplings grown under elevated CO_2 (950 ppm). The CO_2 uptake rate of fertilized plants grown under different CO_2 concentrations was higher than in unfertilized plants during the whole growing season (Figure 2). Repeated measures ANOVA showed a significant effect of elevated $CO_2$and fertilization as well as measurement date on Ac. Both elevated CO_2 and fertilization caused an increase in Ac and during the growing season (May–September), the rate of net photosynthesis decreased significantly (Table 2). Significant interactions also occurred between fertilization and elevated CO_2 but no interaction effects were detected between date and CO_2 and between date and fertilization (Table 2).

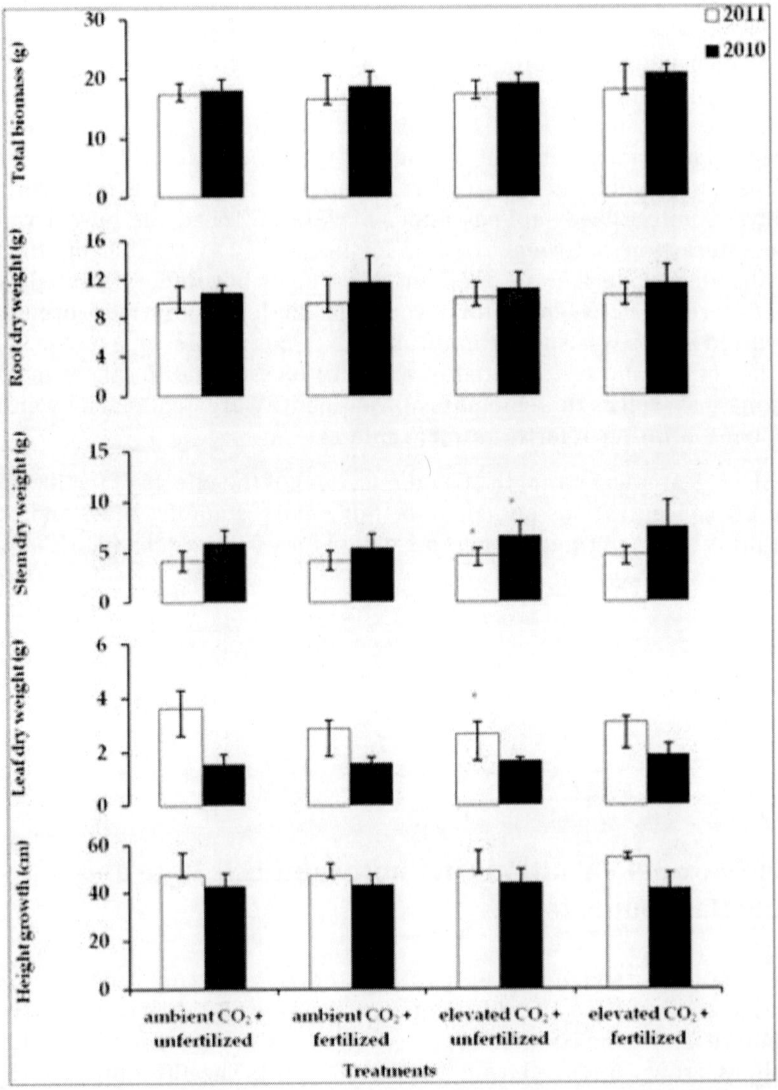

Figure 1. Effect of elevated CO_2 (770 ppm in 2010 (□), 950 ppm in 2011 (■)) and fertilization on mean values (± st.dev.) of height growth, leaf dry mass, stem dry mass, root dry mass and the total biomass at the end of both growing seasons, (n = 10). An asterisk indicates significant differences between treatments ($p \leq 0.05$).

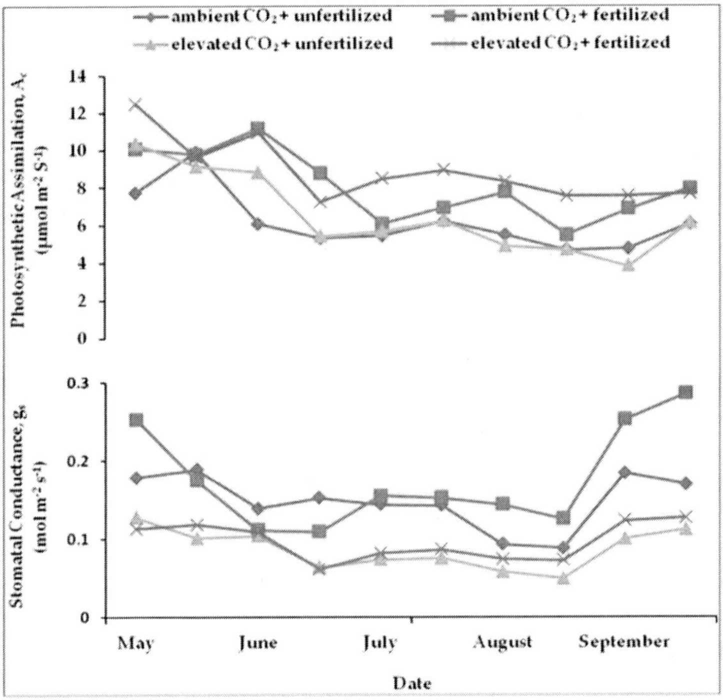

Figure 2. The ratio of net photosynthetic rate (Ac) (above) and stomatal conductance (gs) (below) under ambient CO_2 (385 ppm) or elevated CO_2 (950 ppm) with different levels of nutrient availability during the second growing season (May-September 2011). Values are means of 10 replicates for each treatment. (◊; ambient CO_2 + unfertilized), (■: ambient CO_2 + fertilized), (Δ; elevated CO_2 + unfertilized) and (x; elevated CO_2 + fertilized).

Table 2. Results (MS; Mean Square, F statistics and *p*-values) from repeated sample measure ANOVA for net photosynthetic rate (Ac) and stomatal conductance (gs) of saplings during the growing season. * $p \leq 0.05$; ** $p \leq 0.01$; *** $p \leq 0.001$.

Source	Ac			gs		
	MS	F	*p*-value	MS	F	*p*-value
CO_2 concentration	28.02	5.54 *	0.021	0.490	397.5 ***	0.000
Fertilization	439.24	86.83 ***	0.000	0.036	29.6 ***	0.000
CO_2 × Fertilization	3.22	0.636	0.428	0.008	6.86 *	0.011
Date	620.55	165.25 ***	0.000	0.097	61.394 ***	0.000
Date × CO_2	6.49	1.73	0.193	0.014	8.592 ***	0.000
Date × Fertilization	5.43	1.44	0.233	0.015	9.228 ***	0.000

Statistical analysis showed gs to be significantly affected by CO_2, fertilization and date of samplings (Table 2). Saplings under ambient CO_2 had significantly higher gs during the whole growing season. At the end of the growing season, the respective highest and lowest value of gs was observed in fertilized saplings under ambient CO_2 and unfertilized plants under elevated CO_2, (Figure 2). During the growing season, the increasing trend in gs was significant (Table 2). Elevated CO_2 significantly decreased gs; in contrast, fertilization caused a significant increase in this parameter (Table 2). There was also an interaction effect between CO_2 and fertilization, as well as between date and fertilization and between date and CO_2on gs.

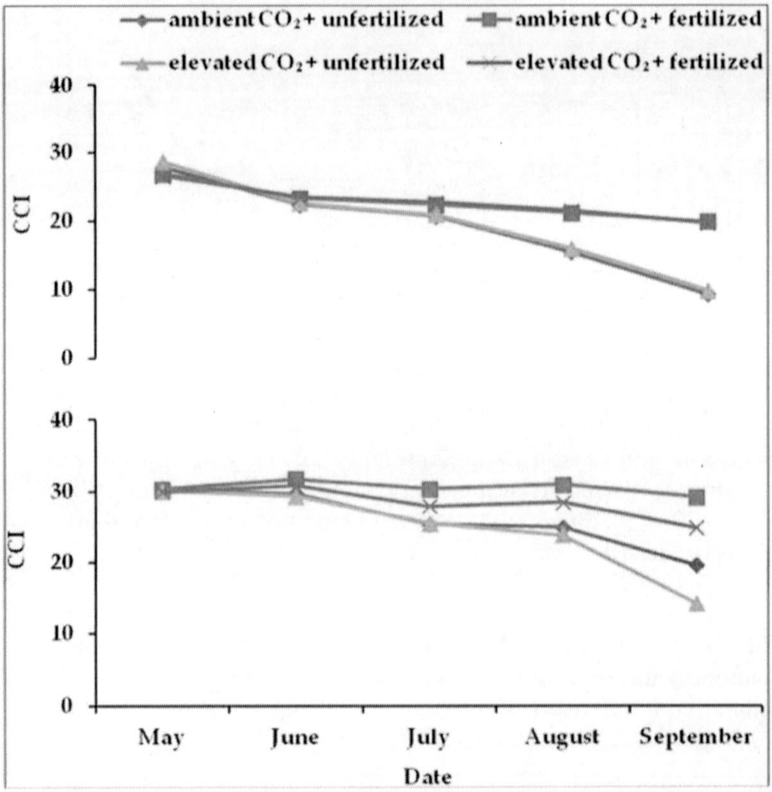

Figure 3. The variation of the mean chlorophyll concentration index (CCI) under ambient CO_2 (385 ppm) or elevated CO_2 (770 ppm during the first growing season (above) and 950 ppm during the second growing season (below)) with different levels of nutrient availability. Values are means of 10 replicates for each treatment. (◊; ambient CO_2 + unfertilized), (■: ambient CO_2 + fertilized), (Δ; elevated CO_2 + unfertilized) and (x; elevated CO_2 + fertilized).

Table 3. Statistical strength (p-values) of the effects of fertilization and CO_2concentrations on the CCI between the treatments and correlations (Pearson Correlation Coefficient) between the CCI and the nitrogen content in leaves; * $p \leq 0.05$; ** $p \leq 0.01$; *** $p \leq 0.001$.

		CO_2 concentration	Fertilization	CO_2 × Fertilization	Leaf nitrogen content
		p-value	p-value	p-value	*values of Pearson correlation*
CCI	2010	0.95	0.000 ***	0.83	0.74 **
	2011	0.000 ***	0.000 ***	0.67	0.85 **

2.3. Chlorophyll Concentration Index (CCI) in Leaves

During both growing seasons, the rate of decline in CCI of unfertilized plants was significantly faster than that in fertilized plants, except in May. In addition, no interaction effect between fertilization and CO_2 concentration could be detected (Table 3, Figure 3). Elevated CO_2 reduced CCI significantly in 2011 (Table 3).

2.4. Correlation between Nitrogen Content and CCI at the end of Growing Seasons

In both years, at the end of the growing season (September), the highest values of leaf nitrogen content were measured in fertilized plants grown under ambient air (Figure 4). Both parameters, CCI and nitrogen, were strongly positively correlated with each other (Table 3).

2.5. C/N Ratio in Plants Tissues

The C/N ratio in the leaves varied significantly from 26.6%–55.3% in 2010 and from 21.0%–44.9% in 2011 (Figure 5). In both growing seasons, the lowest value was measured in fertilized plants growing under ambient air (Figure 5). Fertilization caused a significant decrease in the C/N ratio because of rising N content. Elevated CO_2 increased this ratio in both years (Figure 5), and in both years, the interaction effect between elevated CO_2 and fertilization was significant. Regarding the stems, the C/N ratio varied substantially among treatments in both experimental years (Figure 5). The differences in the mean values between treatments showed the same trend in both growing seasons (Figure 5). Fertilized plants grown under either of the

CO_2 concentrations (ambient air/elevated) had a significantly lower C/N ratio, but the effects of the CO_2 concentration on the C/N ratio were statistically insignificant (Table 4).

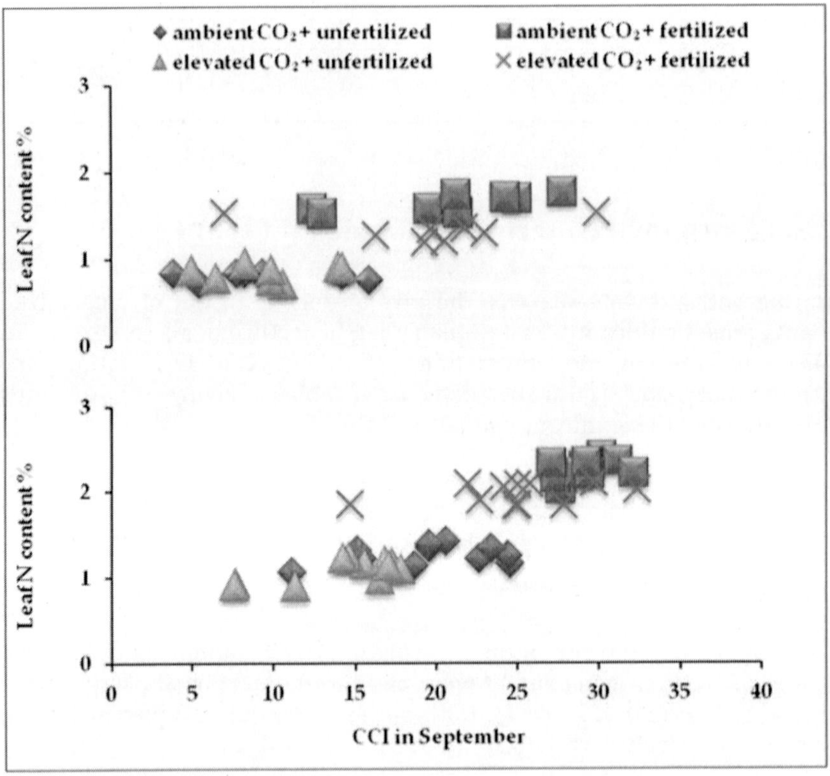

Figure 4. Correlation between CCI and leaf nitrogen content in the leaf at the end of the first (above) and the second (below) growing season.

Table 4. Statistical strength (F and *p*-values) of the effects of fertilization and CO_2concentrations on the C/N ratio of leaf, stem and root between the treatments; * $p \leq 0.05$; ** $p \leq 0.01$; *** $p \leq 0.001$.

		CO_2 concentration	Fertilization	CO_2 × Fertilization	Leaf nitrogen content
		p-value	*p*-value	*p*-value	*values of Pearson correlation*
CCI	2010	0.95	0.000 ***	0.83	0.74 **
	2011	0.000 ***	0.000 ***	0.67	0.85 **

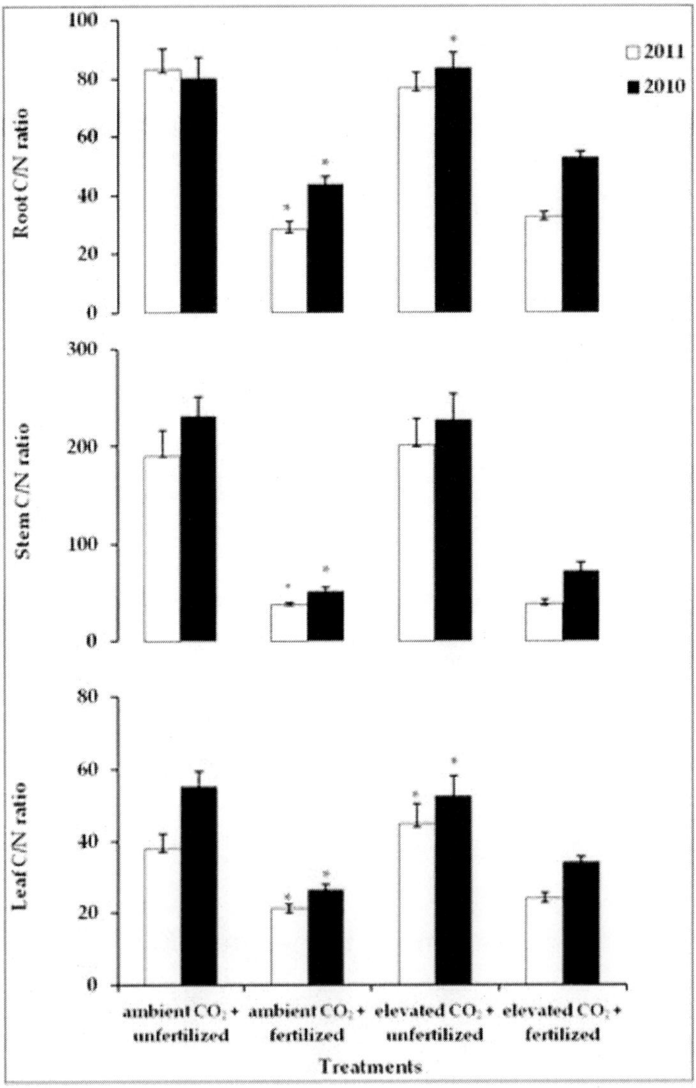

Figure 5. Effect of elevated CO_2 (770 ppm in 2010 (□), 950 ppm in 2011 (■)) and fertilization on mean values of the C/N ratio in leaf, stem and root, (n = 10). An asterisk indicates significant differences between treatments ($p \leq 0.05$).

The mean values of the C/N ratio in the roots ranged from 43.8%–83.7% in 2010 and from 28.3%–83.3% in 2011 (Figure 5). In both growing seasons, the lowest ratio was measured in fertilized plants grown under ambient air. Similar to the leaf and stem, fertilization resulted in a decrease in the C/N ratio in roots (Figure 5). A significant effect of elevated CO_2concentration on the C/N ratio was only detected in 2010 (Table 4).

3. DISCUSSION

3.1. Phenological Responses to Elevated CO_2

The study showed that the height of the saplings varied between the treatments but was not statistically significant. This may be due to the large spread of data acquired per treatment. Although Overdieck et al. [36] mentioned a positive influence of elevated CO_2 on height growth of beech, generalizing from these earlier results for beech appears to be difficult because of the differing experimental conditions and genotypes of the study material. The increase in the stem weight of beech saplings grown under elevated CO_2 was the most obvious and significant response in this study (Figure 1). In both study years, the stems were heavier when grown under elevated CO_2 than under ambient air. In general, it appears that beech produces more biomass under elevated CO_2 [36,49]. In this study, however, the total biomass was not affected by elevated CO_2. This might be due to the fact that half of the biomass consisted of roots, which remained unaffected by the differing growth conditions [50]. No interaction effect was found between elevated CO_2 and fertilization on the dry weight of saplings during either of the growing seasons.

3.2. Physiological Responses to Elevated CO_2

Photosynthesis throughout the growing season was significantly higher in fertilized saplings under both ambient and elevated CO_2 (Figure 2). The positive effects of elevated CO_2 on the rate of photosynthesis of beech have been previously reported [36,51] but a contrary effect or no significant effect at all [49] has also sometimes been observed. At the beginning of the growing season, stimulation of the photosynthesis rate was observed in all saplings (fertilized and unfertilized) under elevated CO_2 but this increase disappeared very soon in unfertilized plants. This reduction was in part due to a decrease in N content and could also be a consequence of the potential sink limitations imposed by pot size. Further, with respect to fertilization, a positive effect under elevated CO_2 has been repeatedly described [52].

Both leaf nitrogen content and the chlorophyll content index in plants under elevated CO_2 were significantly less than in plants grown under ambient air. The decrease in leaf chlorophyll content in beech under elevated CO_2 was reported [38,53], which may be a consequence of leaf nitrogen dilution [8]. Part of the decrease in the photosynthesis rate could be caused by the loss of chlorophyll in plants which were grown under elevated CO_2 during the growing season.

A clear effect of the different treatments on the stomatal conductance of the saplings could be seen in the second year, when elevated CO_2 resulted in a significant reduced stomatal conductance, [51]. In the short term, greater depolarization under elevated CO_2 will result in reduced stomatal aperture [17]. In the long term, changes in stomatal density or stomatal index as well as stomatal aperture can be caused by the reduction of stomatal conductance under elevated CO_2 [17]. On the other hand, no significant reduction of the stomatal conductance of beech under elevated CO_2 was observed by Heath and Kerstiens [50]. Interestingly, the rate of stomatal conductance was significantly higher in fertilized saplings compared to unfertilized ones (Figure 2; Table 2).

3.3. Biochemical Responses to Elevated CO_2

The C/N ratio can be increased, decreased or left unaffected by elevated CO_2, but, on average, increases by about 15% when the CO_2 concentration was doubled [54]. As we know, elevated CO_2 causes an increase in photosynthetic activity [6,11,16]; therefore, an elevated C content was observed in all components of the saplings grown under elevated CO_2. In fertilized saplings, a higher N content was detected in plant organs, but the C content remained unaffected by fertilization. In addition, a reduction in the nitrogen content in plants grown under elevated CO_2 has been observed very often in the past [23,34,43,44,55]. What is the reason for such a reduction? The cause of this reduction is not clear, and the physiological mechanisms responsible for this observation have not been definitely established, although a considerable number of hypotheses have been advanced to account for it [45]. One contributing factor is dilution of N by increased photosynthetic assimilation of C. In addition, studies show strong evidence for a general decrease in the uptake of N per unit mass or length of roots under elevated CO_2. This decreased uptake appears to be the result of both a decreased N demand by shoots and of a decreased ability of the soil-root system to supply the plant with N. The best-supported mechanism for a decreased N supply is a decrease in the transpiration-driven flow of N in the soil due to a decreased stomatal conductance at elevated CO_2. Another hypothesis indicates that altered root-system architecture may also play a role [45]. There is also limited evidence suggesting that under elevated CO_2, plants may exhibit increased rates of N loss through volatilization and/or root exudation, further contributing to lowering the N concentrations [45]. In this study, the N reduction under elevated CO_2 amounted to 10% in the leaves but distinctly less in the stems and roots. Similar observations were reported by Cotrufo et al. [43]. In our saplings, the C/N ratio was higher when grown under elevated CO_2 (Figure 5). Certainly, one contributing factor is the dilution of N by increased CO_2 uptake. Reduction in N uptake can also be due to decreased transpiration and stomatal conductance [56,57,58]. Such a decrease of stomatal conductance under elevated CO_2 was detected in the saplings in the present experiment. Although some evidence suggests that altered root-system architecture may also play a role [41], in this study, no significant differences in root architecture were found between treatments. Due to an increase in N by fertilization, the resulting C/N ratio was lower in fertilized saplings grown under different CO_2concentration.

4. MATERIALS AND METHODS

The seedlings of European beech (*Fagus sylvatica* L.), were collected from a tree nursery near Hamburg and grown in the institute's greenhouse in Hamburg, located at 53°30′ N and 10°12′ E at an elevation of 25 m a.s.l., during two growth seasons (May to September 2010–2011). In spring, the one-year-old saplings were planted in 1 litre tops filled with 50% sand and 50% of a standard commercial substrate (TKS1), including 140 mg/L N, 80 mg/L P_2O and 190 mg/L K_2O; pH 5.6.

4.1. Growth Conditions and Experimental Design

The saplings were randomly assigned to two chambers with different CO_2 concentrations in the greenhouse. In one of them, the CO_2 concentration was the same as in ambient air (on average 385 ppm). In the other one, the CO_2 concentration was raised to either 770 ppm in the first growing season (2010) or 950 ppm in in the second growing season (2011). In the second year, the CO_2 concentration increased by 200% to investigate if this increase reinforced the responses of the saplings to elevated CO_2. In each chamber, the saplings were divided into two groups with different nutrient supply rates (fertilized and unfertilized). For the fertilized group, 0.2% NPK liquid fertilizer (Wuxal Top N) was applied once per week during the growing season.

From each of four treatment combinations of CO_2 concentration and nutrient supply (ambient CO_2 + unfertilized (control), ambient CO_2 +

fertilized, elevated CO_2 + unfertilized, elevated CO_2 + fertilized), 10 plants were analyzed. All plants were irrigated weekly with 80 mL of tap water.

During the growing season, the temperature in the greenhouse was kept constant at about 20 °C. The length of the photoperiod was the same as in the Hamburg area and the relative air humidity (RH) was about 70%. During the experiments, the CO_2 level, temperature, photoperiod and air humidity were monitored by the Computer Climate model CC 600 (RAM Co. Measurement and Control, Germany) every 12 min. Leaf gas exchange parameters were measured many times during the night (from 21:00) and it was found that the photosynthesis activities were about zero. The CO_2 levels (ambient air, 770 and 950 ppm) were maintained between 7:00 and 21:00 each day, which was matched to the photoperiod. Moreover, all plants were inspected once a week during the growing season.

4.2. Seasonal Gas Exchange Measurements

During the second growing season (2011), regular measurements of the net photosynthesis rate (A_C) and stomatal conductance (gs) were made using a portable Infra-Red Analyzer (IRGA; LI-6400, Li-Cor., Lincoln, NE. USA) on one leaf of each sapling. It was considered that the chosen leaves for gas exchange measurements were of similar age. The rate of air flow through the system was set to 600 µmol s^{-1} and the light intensity, provided by a red-blue light source was set to 800 µmol m^{-2} s^{-1}. The CO_2 concentration in the chambers was controlled regularly and measurements were performed at the CO_2 concentrations under which plants were grown.

The air humidity was regulated by adjusting the air flow through a desiccant tube. The chlorophyll content index (CCI) of the leaves of all saplings was non-destructively determined by a chlorophyll meter (SPAD-502 plus Konica Minolta) once a week on 15–20 leaves per sapling during both growing seasons. The values measured correspond to the percentage of chlorophyll in the leaves. They were calculated from the amount of light, transmitted through a leaf at two wavelengths for which the absorbance of chlorophyll is different.

4.3. Growth Measurements

At the end of the each growing season, the tree height was measured and then all trees were harvested and separated into leaves, stems and roots. After the samples were dried at 70 °C for one week, the dry weight of these three fractions was determined and the sum was calculated as a total biomass.

4.4. C/N Ratio

After having determined the dry weight of the various organs, the leaves, stems without bark and roots were ground to a powder using a mill. Then, 5–10 mg of the powder was filled into tin capsules and the total carbon and nitrogen contents of each sample were measured using an element analyzer instrument (Vario EL cube; Hanau, Germany). During analysis, the temperature in the oxidation oven was 1050 °C, while in the thermal conductivity detector and chromatographic column it was 115 °C. The carrier gas pressure was 80 kPa and the flow rate 125 mL/min. The oxygen addition was 20 mL and the oxygen pressure was 50 kPa.

4.5. Statistical Analysis

The main effects of elevated CO_2, fertilization and their interaction were evaluated with a general linear model. To evaluate the gas exchange data during the growing season, repeated measures multivariate ANOVA was applied. The Pearson's correlation test was used to check the correlation between the chlorophyll concentration index and leaf N content at the end of the growing season. The normality of distribution and equality was checked using the Kolmogorov-Smirnov test. The data for each experimental year were analyzed separately and independently from each other. The statistical significance was set at $p \le 0.05$ for all tests. Standard deviation (SD) was used to show the distribution of the data around the mean. All statistical analyses were performed using SPSS 18 (SPSS Inc., Chicago, IL, USA) and a Complete Randomized Design (CRD) was considered for the greenhouse experiments.

5. CONCLUSIONS

To find out how trees respond to elevated CO_2 in combination with nutrient availability, we investigated the interaction between elevated CO_2 and different levels of nutrient supply on European beech saplings. The interaction between elevated CO_2 and fertilization on photosynthetic activities (Ac) of beech was not statistically significant. Generally enhanced tree growth by elevated CO_2 was greater than by fertilization. Negative interaction effects between CO_2 and fertilization were found in the C/N ratio of leaves and roots. The reduction of N content under elevated CO_2 in plant organs was observed, especially in leaves [43]. Such a decrease in N content obviously affects the nutrient composition of plant material used by herbivores. Therefore, elevated CO_2 has complex effects on the whole ecosystem level. Future studies will also focus on changes in other macronutrients, such as potassium, phosphorus and calcium, under increasing CO_2 levels.

ACKNOWLEDGMENTS

The present work was supported by the graduate school C1-chemistry for resource and energy management of Hamburg University.

AUTHOR CONTRIBUTIONS

Experiments and analysis were performed by Neda Lotfiomran and were part of her PhD thesis under the supervision of Jörg Fromm. Statistical analysis was performed by Michael Köhl. The manuscript was written by Neda Lotfiomran and Jörg Fromm.

CONFLICTS OF INTEREST

The authors declare no conflict of interest.

REFERENCES

1. Burschel, P.; Weber, M. Der Treibhauseffekt. Bedrohung und Aufgabe für die Forstwirtschaft. Allg. Forstz. 1988, 43, 1010–1016.
2. Matyssek, R.; Fromm, J.; Rennenberg, H.; Roloff, A. Biologie der Bäume; Ulmer-Verlag: Stuttgart, Germany, 2010.
3. Cure, J.D.; Acock, B. Crop responses to CO2 doubling: A literature survey. Agric. For. Meteorol. 1986, 38, 127–145.
4. Sionit, N.; Kramer, P.J. Woody plants reactions to CO2 enrichment. In CO2 Enrichment and Greenhouse Crops; Enoch, H.Z., Timball, B.A., Eds.; CRC Press: Boca Raton, FL, USA, 1986; Volume II, pp. 69–85.
5. Strain, B.R. Direct effects of increasing atmospheric CO2 on plants and ecosystems. Trends Ecol. 1987, 2, 18–21.
6. Bazzaz, F.A. The response of natural ecosystems to the rising global CO2 levels. Annu. Rev. Ecol. Evol. Syst. 1990, 21, 167–196.
7. Bazzaz, F.A.; Miao, S.L. Successional status, seed size, and responses of tree seedlings to CO2, light, and nutrients. Ecology 1993, 74, 104–112.
8. Ceulemans, R.; Mousseau, M. Tansley Review No. 71. Effects of elevated atmospheric CO2 on woody plants. New Phythol. 1994, 127, 425–446.

9. Körner, C. Plant CO2 responses: An issue of definition, time and resource supply. New Phythol. 2006, 172, 393–411.
10. Curtis, P.S. A meta-analysis of leaf gas exchange and nitrogen in trees grown under elevated carbon dioxide. Plant Cell Environ. 1996, 19, 127–137.
11. Saxe, H.; Ellsworth, D.S.; Heath, D.S. Tansley Review no. 98. Tree and forest functioning in an enriched CO2atmosphere. New Phythol. 1998, 139, 395–436.
12. Medlyn, B.E.; Badeck, F.W.; DePury, D.G.G.; Barton, C.V.M.; Broadmeadow, M.; Ceulemans, R.; De Angelis, P.; Forstreuter, M.; Jach, M.E.; Kellomaki, S.; et al. Effects of elevated CO2 on photosynthesis in European forest species: A meta-analysis of model parameters. Plant Cell Environ. 1999, 22, 1475–1495.
13. Nowak, R.S.; Ellsworth, D.S.; Smith, S.D. Functional responses of plants to elevated atmospheric CO2—Do photosynthetic and productivity data from FACE experiments support early predictions? New Phytol. 2004, 162, 253–280.
14. Taub, D. Effects of rising atmospheric concentrations of carbon dioxide on plants. Nat. Educ. Knowl. 2010, 1, Article 21.
15. Long, S.P.; Drake, B.G. Photosynthetic CO2 assimilation and rising atmospheric CO2 concentrations. In Crop Photosynthesis: Spatial and Temporal Determinants; Baker, N.R., Thomas, H., Eds.; Elsevier: New York, NY, USA, 1992; pp. 69–107.
16. Ceulemans, R.; Janssens, I.A.; Jach, M.E. Effects of CO2 enrichment on trees and forests: Lessons to be learned in view of future ecosystem studies. Ann. Bot. 1999, 84, 577–590.
17. Ainsworth, E.A.; Rogers, A. The response of photosynthesis and stomatal conductance to rising [CO2]: Mechanisms and environmental interactions. Plant Cell Environ. 2007, 30, 258–270.
18. EL Kohen, A.; Venet, L.; Mousseau, M. Growth and photosynthesis of two deciduous forest species at elevated carbon dioxide. Funct. Ecol. 1993, 7, 480–488.
19. Woodward, F.I. Potential impacts of global elevated CO2 concentrations on plants. Curr. Opin. Plant Biol. 2002, 5, 207–211.
20. Koike, T.; Watanabe, M.; Watanabe, Y.; Agathokleous, E.; Eguchi, N.; Takagi, K.; Satoh, F.; Kitaoka, S.; Funada, R. Ecophysiology of deciduous trees native to Northeast Asia grown under FACE (Free Air CO2 Enrichment). J. Agric. Meteorol. 2015, 71, 174–184.
21. Pritchard, S.; Rogers, H.; Prior, S.; Peterson, C. Elevated CO2 and Plant structure: A review. Glob. Chang. Biol. 1999, 5, 807–837.
22. Poorter, H.; van Berkel, Y.; Baxter, R.; den Hertog, J.; Dijkstra, P.; Gifford, R.M.; Griffin, K.L.; Roy, R.J.; Wong, S.C. The effect of elevated CO2 on the chemical composition and construction costs of leaves of 27 C3 species. Plant Cell Environ. 1997, 20, 472–482.
23. Ainsworth, E.A.; Long, S.P. What have we learned from 15 years of free-air CO2 enrichment (FACE)? A meta-analytic review of the responses of photosynthesis, canopy properties and plant production to rising CO2. New Phythol. 2005, 165, 351–372.
24. Reekie, E.G.; Bazzaz, F.A. Competition patterns of resource use among seedlings of five tropical trees grown at ambient and elevated CO2. Oecologia 1989, 79, 212–222.
25. Bazzaz, F.A.; Garbutt, K. The response of annuals in competitive neighborhoods: Effects of elevated CO2. Ecology1998, 69, 937–946.
26. Sionit, N.; Hellmers, H.; Strain, B.R. Interaction of atmospheric CO2 enrichment and irradiance on plant growth. Agron. J. 1982, 74, 721–725.
27. Garbutt, K.; Williams, W.E.; Bazzaz, F.A. Analysis of the differential response of five annuals to elevated CO2 during growth. Ecology 1990, 71, 1185–1194.
28. Bloom, A.J. As carbon dioxide rises, food quality will decline without careful nitrogen management. Calif. Agric.2009, 63, 67–72.
29. Norby, R.J.; Gunderson, C.A.; Wullschleger, S.D.; O'Neill, E.G.; McCracken, M.K. Productivity and compensatory response of yellow poplar trees in elevated CO2. Nature 1992, 357, 322–324.

30. Hättenschwiler, S.; Miglietta, F.; Raschi, A.; Körner, C. Thirty years of in situ growth under elevated CO2: A model for future forest responses? Glob. Chang. Biol. 1997, 3, 463–471.
31. Poorter, H.; Navas, M.L. Plant growth and competition at elevated CO2: On winners, losers and functional groups. New Phythol. 2003, 157, 175–198.
32. Roberntz, P.; Stockfors, J. Effects of elevated CO2 and nitrogen on net photosynthesis, stomatal conductance and needle respiration of field-grown Norway spruce trees. Tree Physiol. 1998, 18, 233–241.
33. Leakey, A.D.B.; Ainsworth, E.A.; Bernacchi, C.J.; Rogers, A.; Long, S.P.; Ort, D.R. Elevated CO2 effects on plant carbon, nitrogen, and water relations: Six important lessons from FACE. J. Exp. Bot. 2009, 60, 2859–2879.
34. Reddy, A.R.; Rasineni, G.K.; Raghavendra, A.S. Concentration on photosynthesis and plant productivity. Curr. Sci.2010, 99, 46–57.
35. Beerling, D.J.; Heath, J.; Woodward, F.I.; Mansfield, T.A. Drought—CO2 interactions in trees observations and mechanisms. New Phythol. 1996, 134, 235–242.
36. Overdieck, D.; Ziche, D.; Böttcher-Jungclaus, K. Temperature responses of growth and wood anatomy in European beech saplings grown in different carbon dioxide concentrations. Tree Physiol. 2007, 27, 261–268.
37. Grams, T.E.E.; Anegg, S.; Häberle, K.H.; Langebartels, C.; Matyssek, R. Interactions of chronic exposure to elevated CO2 and O3 levels in the photosynthetic light and dark reactions of European beech (Fagus sylvatica L.). New Phythol.1999, 144, 95–107.
38. Lütz, C.; Anegg, S.; Gerant, D.; Slaoui-Sosse, B.; Gerard, J.; Dizengremel, P. Beech trees exposed to high CO2 and to simulated summer ozone levels: Effects on photosynthesis, chloroplast components and leaf enzyme activity. Physiol. Plant. 2000, 109, 252–259.
39. Liu, X.; Kozovits, A.R.; Grams, T.E.E.; Blaschke, H.; Rennenberg, H.; Matyssek, R. Competition modifies effects of enhanced ozone/carbon dioxide concentrations on carbohydrate and biomass accumulation in juvenile Norway spruce and European beech. Tree Physiol. 2004, 24, 1045–1055.
40. Norby, R.J.; Warren, J.M.; Iversen, C.M.; Medlyn, B.E.; McMurtrie, R.E. CO2 enhancement of forest productivity constrained by limited nitrogen availability. Proc. Natl. Acad. Sci. USA 2010, 107, 19368–19373.
41. De Oliveira, E.A.D.; Approbato, A.U.; Legracie, J.R.; Martinez, C.A. Soil-nutrient availability modifies the response of young pioneer and late successional trees to elevated carbon dioxide in a Brazilian tropical environment. Environ. Exp. Bot. 2012, 77, 53–62.
42. Kirschbaum, M.U.; Lambie, S.M. Re-analysis of plant CO2 responses during the exponential growth phase: Interactions with light, temperature, nutrients and water availability. Funct. Plant Biol. 2015, 42, 989–1000.
43. Cotrufo, M.F.; Ineson, P.; Scott, A. Elevated CO2 reduces the nitrogen concentration of plant tissues. Glob. Chang. Biol.1998, 4, 43–54.
44. Yin, X. Responses of leaf nitrogen concentration and specific leaf area to atmospheric CO2 enrichment: A retrospective synthesis across 62 species. Glob. Chang. Biol. 2002, 8, 631–642.
45. Taub, D.R.; Wang, X. Why are nitrogen concentrations in plant tissues lower under elevated CO2? A critical examination of the hypotheses. J. Integr. Plant. Biol. 2008, 50, 1365–1374.
46. Spiecker, H.; Mieläikinen, K.; Köhl, M.; Skovsgaard, J.P. Discussion. In Growth Trends in European Forests; Springer: Berlin/Heidelberg, Germany, 1996; pp. 355–367.
47. Lotfiomran, N.; Fromm, J.; Luinstra, G.A. Effects of elevated CO2 and different nutrient supplies on wood structure of European beech (Fagus sylvatica) and gray poplar (Populus× canescens). IAWA J. 2015, 36, 84–97.
48. Overdieck, D. Effects of atmospheric CO2 enrichment on CO2 exchange rates of beech stands in small model ecosystems. Water Air Soil Pollut. 1993, 70, 259–277.

49. Epron, D.; Liozon, R.; Mousseau, M. Effects of elevated CO2 concentration on leaf characteristics and photosynthetic capacity of beech (Fagus sylvatica) during the growing season. Tree Physiol. 1996, 16, 425–432.
50. Heath, J.; Kerstiens, G. Effects of elevated CO2 on leaf gas exchange in beech and oak at two levels of nutrient supply: Consequences for sensitivity to drought in beech. Plant Cell Environ. 1997, 20, 57–67.
51. Leverenz, J.W.; Bruhn, D.; Saxe, H. Responses of two provenances of Fagus sylvatica seedlings to a combination of four temperature and two CO2 treatments during their first growing season: Gas exchange of leaves and roots. New Phythol. 1999, 144, 437–454.
52. Liozon, R.; Badeck, F.W.; Genty, B.; Meyer, S.; Saugier, B. Leaf photosynthetic characteristics of beech (Fagus sylvatica L.) saplings during three years of exposure to elevated CO2 concentration. Tree Physiol. 2000, 20, 239–247.
53. Mousseau, M.; Dufrene, E.; El Kohen, A.; Epron, D.; Godard, D.; Liozon, R.; Pontailler, J.Y.; Saugier, B. Growth Strategy and Tree Response to Elevated CO2: A Comparison of Beech (Fagus sylvatica) and Sweet Chestnut. In Carbon Dioxide and Terrestrial Ecosystems; Academic Press: San Diego, CA, USA, 1996; p. 71.
54. Gifford, R.M.; Barrett, D.J.; Lutze, J.L. The effects of elevated [CO2] on the C:N and C:P mass ratios of plant tissues. Plant Soil 2000, 224, 1–14.
55. De Graaff, M.A.; van Groenigen, K.J.; Six, J.; Hungate, B.; van Kessel, C. Interactions between plant growth and soil nutrient cycling under elevated CO2: A meta-analysis. Glob. Chang. Biol. 2006, 12, 2077–2091.
56. Van Vuuren, M.M.I.; Robinson, D.; Fitter, A.H.; Chasalow, S.D.; Williamson, L.; Raven, J.A. Effects of elevated atmospheric CO2 and soil water availability on root biomass, root length and N, P and K uptake by wheat. New Phytol. 1997, 135, 455–465.
57. McDonald, E.P.; Erickson, J.E.; Kruger, E.L. Can decreased transpiration limit plant nitrogen acquisition in elevated CO2? Funct. Plant Biol. 2002, 29, 1115–1120.
58. Del Pozo, A.; Pérez, P.; Gutiérrez, D.; Alonso, A.; Morcuende, R.; Martínez-Carrasco, R. Gas exchange acclimation to elevated CO2 in upper-sunlit and lower-shaded canopy leaves in relation to nitrogen acquisition and partitioning in wheat grown in field chambers. Environ. Exp. Bot. 2007, 59, 371–380.

CHAPTER 13

Calmodulin Gene Expression in Response to Mechanical Wounding and *Botrytis cinerea* Infection in Tomato Fruit

Hui Peng [1,2], Tianbao Yang [1,*] and Wayne M. Jurick II [1]

[1]*Food Quality Laboratory, Beltsville Agricultural Research Center, Agricultural Research Service, United States Department of Agriculture, Beltsville, MD 20705, USA*

[2]*College of Life Sciences, Guangxi Normal University, Guilin 541004, China*

ABSTRACT

Calmodulin, a ubiquitous calcium sensor, plays an important role in decoding stress-triggered intracellular calcium changes and regulates the functions of numerous target proteins involved in various plant physiological responses. To determine the functions of calmodulin in fleshy fruit, expression studies were performed on a family of six calmodulin genes (*SlCaMs*) in mature-green stage tomato fruit in response to mechanical injury and *Botrytis cinerea* infection. Both wounding and pathogen inoculation triggered expression of all those genes, with *SlCaM2* being the most responsive one to both treatments. Furthermore, all calmodulin genes were upregulated by salicylic acid and methyl jasmonate, two signaling molecules involved in plant immunity.

In addition to *SlCaM2*, *SlCaM1* was highly responsive to salicylic acid and methyl jasmonate. However, *SlCaM2* exhibited a more rapid and stronger response than *SlCaM1*. Overexpression of *SlCaM2* in tomato fruit enhanced resistance to *Botrytis*-induced decay, whereas reducing its expression resulted in increased lesion development. These results indicate that calmodulin is a positive regulator of plant defense in fruit by activating defense pathways including salicylate- and jasmonate-signaling pathways, and *SlCaM2* is the major calmodulin gene responsible for this event.

Keywords: calcium signaling; plant defense; salicylic acid; jasmonic acid; postharvest decay

1. INTRODUCTION

Calcium is a universal second messenger involved in growth, development and mediating responses to a variety of abiotic and biotic stresses in plants [1,2,3]. Cellular changes in calcium are captured by calcium sensors containing the EF-hand motif. Calmodulin (CaM) is a ubiquitous calcium sensor in plants, and plays an important role in almost all aspects of cell activity [4,5,6]. In contrast to animals which have one or a few *CaM* genes encoding identical isoforms, plants have multiple genes encoding more diversified isoforms. In all plants examined, *CaM* genes, even genes encoding the same isoform, are differentially expressed in response to numerous external stimuli such as touch, heat shock, cold, light, pathogens, and to phytohormones. In *Arabidopsis*, seven *CaM* genes encode four highly conserved isoforms. A loss-of-function mutant in *Arabidopsis AtCaM2* affects pollen germination [7]. The *atcam3* knockout mutant exhibits reduced thermo-tolerance after heat treatment, whereas overexpression of *AtCaM3* significantly increases thermo-tolerance [8]. In soybean, specific CaM isoforms *SCaM-4* and *SCaM-5*, are highly induced either by a fungal elicitor or pathogen attack, whereas three other *SCaMs* show no response to these stimuli [9]. Transgenic tobacco plants overexpressing *SCaM-4* and *SCaM-5* display spontaneous lesions and constitutive expression of systemic acquired resistance-associated genes. Thus, the level of individual CaM proteins is differentially regulated in plants upon exposure to various stimuli.

Tomato (*Solanum lycopersicum* L.) is the second most important vegetable crop, and the total worldwide production in 2012 was 161.8 million tons with a farm gate value of $80 billion. However, over 25% of fresh produce including tomato fruit are lost after harvest due to the mechanical damage during handling, transportation, and decay [10]. Previous studies on the function of tomato *CaMs* have focused on their role in vegetative tissues. Bergey and Ryan (1999) reported an accumulation of *CaM* mRNA and CaM protein in tomato leaves after wounding or systemin treatment, suggesting that it plays a role in plant defense [11]. Zhao *et al.* (2013) reported that there are six *CaM* genes in tomato encoding for four isoforms [12]. SlCaMs in leaves were highly responsive to a variety of biotic and abiotic stimuli. Silencing SlCaM2 and SlCaM6 altered expression of defense-related genes and reduced resistance to pathogens. However, there is a gap in the literature concerning the importance of CaM in the ripening of fleshy fruit, as well as in response to stresses that can be encountered during postharvest handling and storage. In this study, we report an expression analysis of *CaM* gene family in response to postharvest stresses, and characterization of the functional significance of a specific *CaM* gene responsible for disease resistance in tomato fruit.

2. RESULTS

2.1. SlCaMs Are Responsive to Mechanical Wounding

To study the effect of wounding on *SlCaMs* expression, mature green stage fruit were selected because this is the specific stage that is routinely harvested by the tomato industry, Without wounding, the most abundantly expressed genes were *SlCaM1* and *SlCaM5* (Figure 1). After wounding, the expression of all *SlCaM*s were stimulated within one hour after treatment and peaked following 2–4 h. Among them, *SlCaM2* showed the most profound stimulation. Its expression increased more than 10-fold in one hour, and peaked at 32-fold within two hours. Wounding also increased the expression of *SlCaM3*, *SlCaM4* and *SlCaM6* 2–4 fold, albeit at lower levels. However, wounding had little effect on the expression of *SlCaM1* and *SlCaM5*. *SlPR2b* and *SlLAP-A1*, two genes known to be induced by wounding [13,14] were used as a positive control, and their expression was triggered by wounding as expected. These results collectively suggest that *SlCaMs* in tomato fruit are wound-responsive.

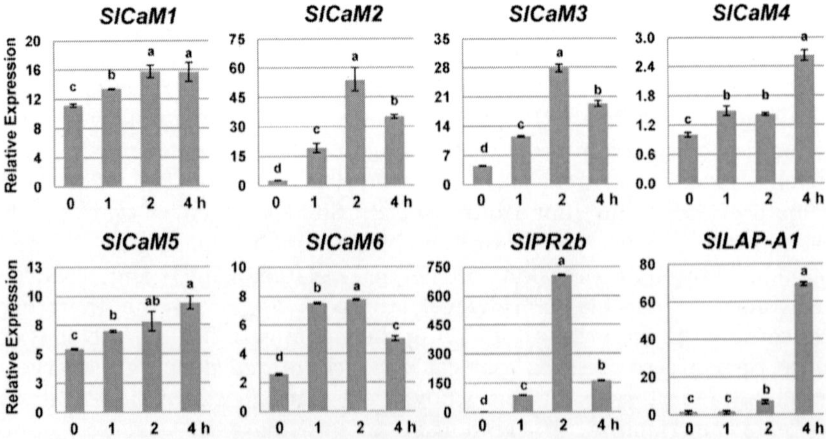

Figure 1. Expression of *SlCaMs* in wounded tomato fruit. Transcription levels of *SlCaMs* were measured by RT-qPCR. Relative gene expression levels are shown following normalization with actin transcript values. Error bars represent standard error of the mean. For each gene, different letters indicate statistically significant differences among mean values ($p < 0.05$). The results are based on at least three repeats in three independent experiments.

2.2. Pathogen Infection Triggers Calmodulin Gene Expression

Wounding is a prerequisite for most fungal postharvest pathogens that cause decay during storage. To investigate the effects of pathogen infection on *SlCaMs* expression, wounded tomato fruit were inoculated with the necrotrophic fungal pathogen, *Botrytis cinerea.* The growth of the fungus was observed two days postinoculation. The inoculated wounds displayed extensive water soaking, mycelial growth, and necrosis. In contrast, the control treatments (mock inoculation), wounds containing water only, were asymptomatic. Additionally, the expression levels of *SlCaMs* in response to pathogen infection were investigated. Since wounding triggered *SlCaMs* gene expression, much of the increased *SlCaMs* in early time points after both mock and pathogen-inoculation may have resulted from wounding (Figure 2). However, the pathogen-treatment had a more profound and prolonged stimulation of all *SlCaMs* than the mock inoculation. After 24 h of treatment, expression differences between pathogen- and mock inoculation were more distinguishable. For instance, *SlCaM2* exhibited the most dramatic

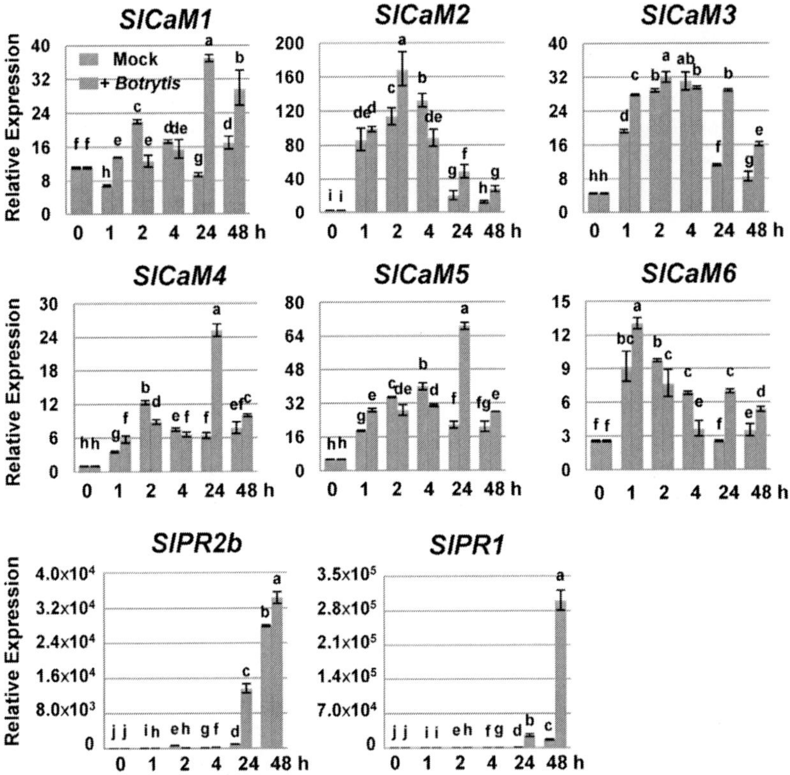

Figure 2. Expression of *SlCaMs* in response to fungal infection. Tomato fruit at the mature-green stage were mechanically wounded and immediately treated with water (mock inoculation) or inoculated with *Botrytis cinerea* conidia. The wounded and wounded-plus-inoculated areas were excised after 0 to 48 h of incubation at 20 °C. Total RNA samples for RT-PCR were isolated from pericarp tissue and transcript levels of *SlCaMs* genes were determined by RT-qPCR. Relative gene expression levels are shown following normalization with actin transcript values. For each gene, different letters indicate statistically significant differences among mean values ($p < 0.05$). The results are based on at least three repeats in three independent experiments.

stimulation in response to pathogen infection and mock inoculation immediately after treatment, suggesting that this stimulatory effect resulted mainly from physical wounding. However, following 24 h inoculation, the wound effect on *SlCaM2* expression declined, and the pathogen effect was evident. Similarly, pathogen treatment enhanced the expression of *SlCaM1*, *SlCaM3*, *SlCaM4* and *SlCaM5* 24 h post inoculation. *SlPR2b* was stimulated by 87–106 fold after 1 h of treatments mainly due to the effect of wounding. However, it was increased over 13,600 fold at 24 h post inoculation. Interestingly, the expression of *SlPR1* was remarkably high after 24 h post inoculation. These results suggest that *SlCaMs* could be involved in signaling events during fungal infection by *B. cinerea*.

2.3. Calmodulin Genes Are Salicylic Acid-Responsive

Salicylic acid (SA) is a key signaling molecule for the activation of genes involved in systemic acquired resistance to both biotic and abiotic stresses [13,15]. Treating harvested fruits with SA can help reduce decay incidence by activating defense genes such as *PR-1* and *PR-2* [16,17,18,19,20]. To determine whether SA affects the expression of *SlCaMs*, fruits were treated with salicylic acid for different time periods ranging from 0 to 48 h. Previously, it was observed that *SlPR1*, a marker of SA-regulated gene expression, showed the highest induction at 4 mM [21], and thus this concentration of SA was utilized in the following experiments. All *SlCaMs* were upregulated by SA (Figure 3). There were two peaks for the expression of *SlCaM3*, *SlCaM4* and *SlCaM5*. The first peak (minor) occurred one hour after treatment, and second peak (major) appeared at or after 24 h. This pattern was similar to that of the SA-responsive gene *SlPR1*. In comparison, only one peak was observed for *SlCaM1* and *SlCaM6* after 8 or 24 h of treatment. However, the highest induction was observed after 24 h of treatment for all CaMs.

2.4. Methyl Jasmonate Stimulates Calmodulin Gene Expression

Jasmonic acid (JA) is another signal molecule that regulates plant responses to wounding/insect stress and necrotrophic pathogen attack [13,22]. To investigate the expression patterns of *SlCaMs* in response to JA, fruits were treated with 20 μM methyl jasmonate (MeJA), the methyl ester of JA. The expression of all *SlCaMs* increased after applying MeJA (Figure 4). The stimulatory response for *SlCaM2*, *SlCaM3* and *SlCaM6* was detected one hour after treatment. However, significant changes were evident 4 h post treatment for *SlCaM1*, and at 8 h post treatment for *SlCaM3*, *SlCaM4* and *SlCaM5*. The most JA-responsive genes were *SlCaM1* and *SlCaM2*. It is interesting to note that the expression of *SlCaM2* exhibited a wave-like pattern in response to MeJA. The first peak appeared at one hour, the second at 8 h, and the third at 48 h, which was similar to the expression pattern of the JA-responsive gene *SlPR2b*.

These results suggest that *SlCaM2* is both a JA early responsive and late responsive gene. Taken together, *SlCaMs* positively respond to JA in tomato fruit, which is opposite to JA's effect on those genes in leaf tissue. Interestingly, the expression of all six *SlCaMs* in tomato leaves was inhibited by MeJA treatment, suggesting that there is tissue differential expression for *SlCaMs*' response to JA [12].

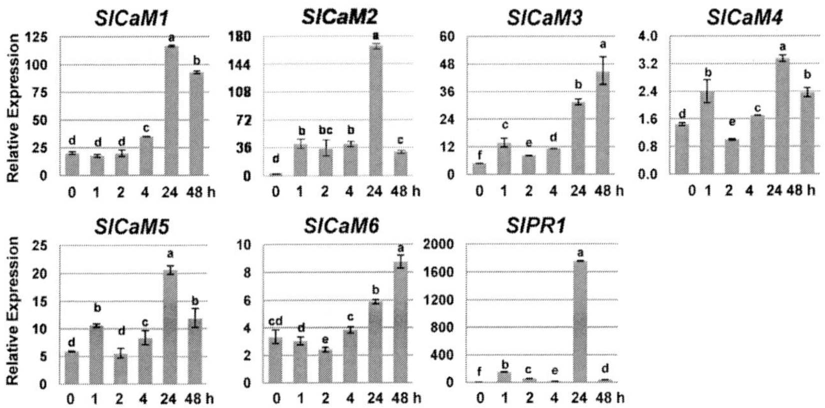

Figure 3. *SlCaMs* expression levels in response to salicylic acid treatment. Transcription levels of *SlCaMs* genes were investigated by quantitative real time RT-qPCR. Mature green stage fruit were treated with 4 mM salicylic acid for different time periods as indicated. Relative gene expression levels are shown following normalization with actin transcript values. Error bars represent the standard error of the mean. For each gene, different letters indicate statistically significant differences among the means ($p < 0.05$). The results are based on at least three repeats in three independent experiments.

2.5. Overexpression of the SlCaM2 Gene Reduces Symptoms Incited by Botrytis Cinerea

Since *SlCaM2* was the only gene highly responsive to wounding and pathogen infection, as well as SA and McJA, it was selected for further functional analysis *in planta*. To quickly assess transgene expression in fruit, we selected the agroinjection method to transiently express *SlCaM2* in the early mature-green stage tomato fruit. We introduced *SlCaM2*-sense and antisense constructs into tomato fruit via *Agrobacterium tumefaciens*. Forty-eight hours after injection, the expression level of *SlCaM2* in control (vector alone) was ~2–3 fold higher than non-transgenic fruit (WT) (Figure 5A), suggesting that *Agrobacterium* itself could stimulate the expression of the endogenous *SlCaM2* expression. However, as compared to empty vector alone, the expression level of *SlCaM2* in the sense fruit was over five fold higher 48 h after injection. In contrast, the expression in the antisense fruit was reduced 4–5 times as compared to the vector only control. These results demonstrate that the transient *Agrobacterium*-mediated transformation methodology was a viable choice to study candidate gene function in tomato fruits.

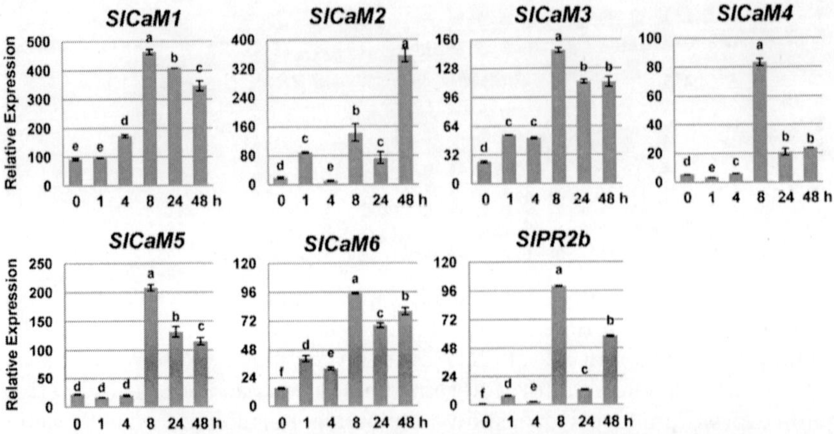

Figure 4. *SlCaMs* gene expression in response to methyl jasmonate treatment. Transcription levels of *SlCaMs* genes were investigated by quantitative real time RT-qPCR. Mature green stage fruit were treated with 20 μM methyl jasmonate for different periods of time. Relative gene expression levels are shown following normalization with actin transcript values. Error bars represent standard error of the mean. For each gene, different letters indicate statistically significant differences among mean values ($p < 0.05$). The results are based on at least three repeats in three independent experiments.

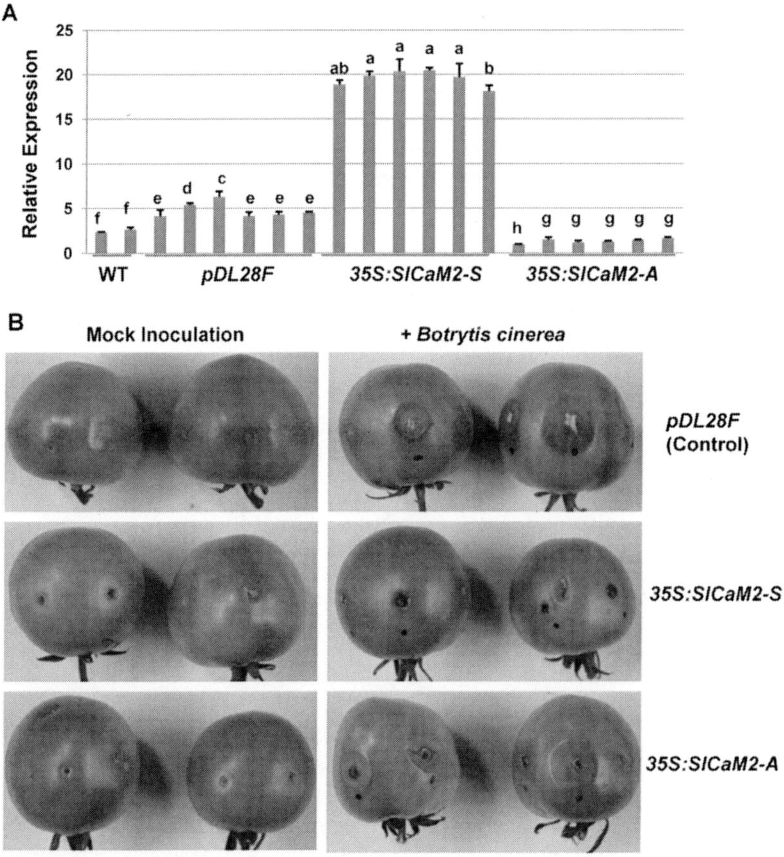

Figure 5. Expression level of *SlCaM2* in tomato fruit in relation to *B. cinerea* infection. Mature green stage fruits were agroinjected with *A. tumefaciens* carrying different constructs. *pDF28F*, empty vector (control); *35S:SlCaM2-S*, pDL28F carrying *SlCaM2* in the sense orientation; *35S:SlCaM2-A*, pDL28F carrying *SlCaM2* in the antisense orientation. (**A**) Examination of expression levels of *SlCaM2* in different transgenic fruits. 48 h after agroinjection, a piece of pericarp tissue from each fruit was used for RT-qPCR. The results are based on at least three repeats in three independent experiments. For each gene, different letters indicate statistically significant differences among mean values ($p < 0.05$). (**B**) Pathogen resistance and susceptibility test shows that overexpression of *SlCaM2* enhanced resistance to *B. cinerea*. The fruits agroinjected with different constructs were inoculated into a wound with water (mock) or *B. cinerea* conidia, and put in a moisture saturated container to observe decay development. The photos were taken 72 h post inoculation.

At 48 h post agroinjection, *B. cinerea* conidial suspensions or water were pipetted into wounds (via nail punctures) on transgenic fruit, and monitored for decay development. In the first 24 h, no obvious symptoms were observed. At 2 days postinoculation, control and antisense fruit started to display water soaking and necrosis, while the pathogen inoculated wounds in sense fruit showed some slight necrosis. Three days postinoculation, the control fruit exhibited severe decay, as lesion development was evident at the inoculation site which included mycelial proliferation, accompanied by soft watery circular lesions that were delineated by healthy tissue. The antisense fruit had similar symptoms of decay compared to the control (Figure 5B). However, sense fruit showed limited necrosis and limited water-soaking. All the mock inoculated fruit were asymptomatic at the inoculation sites. These results indicate that increased expression of *SlCaM2* enhanced resistance to *B. cinerea*, and ultimately limited fungal growth, lesion development, and colonization in tomato fruit.

3. DISCUSSION

Fleshy fruits are an important part of the human diet, providing fiber, minerals, nutrients and various other substances beneficial to human health. Nearly a quarter of all fresh fruits and vegetables in the U.S. market are lost after harvest due to damage caused by abiotic and biotic stresses such as wounding and pathogen infection [10]. The importance of calcium in fruit ripening and postharvest handling has been recognized for many years. Calcium is most frequently associated with delayed ripening maintaining fruit quality (particularly firmness), and preventing decay through the strengthening of cell walls by cross-linking de-esterified pectic acid residues [23,24,25,26,27,28,29,30]. However, it is well-recognized that calcium mediated signaling is important for plant responses to a wide variety of stresses including mechanical touch and pathogen attack [1,2,3,31]. Intracellular calcium changes in response to those stimuli are perceived and decoded by calcium sensors including CaM [4,5,6]. Further, *CaM* genes themselves respond to the biotic and abiotic signals. In tomato, wounding or systemin increased the accumulation of *CaM* in leaves [11]. Zhao *et al.* (2013) reported that SlCaMs in tomato leaves were highly responsive to a variety of biotic and abiotic stimuli [12]. Silencing of *SlCaM2* and *SlCaM6* led to the reduced resistance to tobacco rattle virus and *Pythium aphanidermatum*. However, it did not affect the resistance to *Pseudomonas syringae* and *Xanthomonas oryzae*. In this study, we found that six *SlCaMs*, especially *SlCaM2*, in tomato fruits were upregulated by mechanical wounding, necrotrophic fungal infection, SA and JA. It is important to note that *SlCaMs* genes respond to JA in leaves and fruits very differently. Thus, there may be tissue specificities for *SlCaMs* expression in response to stresses that accompany functional changes in different tissues.

SA and JA are two major signal molecules that mediate plant defense responses. In general, the SA-dependent pathway is activated by biotrophic pathogens, while the JA pathway is triggered by necrotrophs and herbivore attack/wounding [13,22,32,33]. The interactions between those two pathways can be antagonistic or synergistic [34,35]. It has been reported that the necrotrophic pathogens such as *Botrytis cinerea* can produce an exopolysaccharide, which acts as an elicitor of SA pathway in tomato [36,37]. We observed that SA-responsive gene *SlPR1*was remarkably stimulated by *Botrytis* infection which supported their findings (Figure 2). It has been well documented that the synthesis of SA and SA signaling are under extensive regulation by calcium signaling [38,39,40,41,42]. Cross-talk between the JA signaling pathway and calcium signaling also occurs [43,44]. A calcium/CaM-regulated transcription factor AtSR1/CAMTA3 has been shown to be involved in both SA- and JA-dependent pathways [45,46]. Analysis of the AtSR1/CAMTA3 gene knockout mutant showed enhanced resistance to both biotrophic and necrotrophic pathogens [41,47]. Previously, the expression patterns of *SlSRs*, the orthologs of *AtSRs/CaMTAs* in tomato, were analyzed. All of them displayed a positive response to both SA and MeJA in tomato fruit [21]. Therefore *SlSRs* could be candidate targets for *SlCaMs* in controlling disease resistance. In addition, there are quite a few other CaM-target proteins involved in SA-signaling that respond both positively and negatively [38,39,40,41,42]. Considering six *SlCaM* genes encoding four isoforms, the regulation of wounding and disease resistance in tomato fruit by calcium/CaM is hypothesized to be complex. Nevertheless, *SlCaM2* will be a first choice for further expression and functional analysis studies in tomato fruits. Introducing *SlCaM2* into tomato, by classical breeding or through molecular means, might enhance the tolerance/resistance to mechanical injury, pathogens, and other stresses leading to the availability of high quality fresh fruit for consumption.

4. EXPERIMENTAL SECTION

4.1. Plant Materials

Tomato plants (*S. lycopersicum* cv. Moneymaker) were grown in a greenhouse at 28 °C with a 16 h/8 h (light/dark) cycle. Fruit were harvested at the mature green stage (MG), as defined by USDA-ARS criteria [48], when the fruits were physiologically mature but not yet ripe. In the industry, tomatoes are often harvested at this stage for packing and shipment, and subsequently treated with ethylene to promote ripening prior to sale. The greenhouse-grown MG fruit were held under ambient conditions overnight to reduce harvest shock prior to treatment.

4.2. Mechanical Injury, Methyl Jasmonate and Salicylic Acid Treatments

Wounding was executed by manually cutting fruit pericarp into one inch pieces with a sharp knife at room temperature. The fruit pieces were put in a plastic box with the wet towel to maintain high humidity for the indicated time period. For SA treatment, fruits were immersed in 4 mM SA solution for 0 to 48 h. For MeJA treatment, fruits were sealed in a jar with 20 μM MeJA. After each treatment, pericarp samples were immediately frozen in liquid nitrogen and stored at −80 °C.

4.3. Pathogen Infection and Decay Assay

Fruits were mechanically injured and infected by inoculation with a conidial suspension of the fungal pathogen *B. cinerea* strain 22B which was isolated from naturally infected apple fruit [49]. The fungus was propagated via single spore isolate and maintained on Potato Dextrose Agar. The inoculation was done essentially as described previously [21]. Briefly, fruit were punctured (3 mm depth, 2 mm diameter) at six sites around the equator of each fruit; 3 sites with 10 μL of conidial suspension (1×10^4 conidia/mL), and the other three with 10 μL of sterile Tween20-treated water for mock inoculation. After inoculation, the fruit were stored in plastic sealed containers with moist towels to maintain high humidity and kept at 20 °C. Pericarp tissue samples were obtained from inoculated fruits by using a cork borer to isolate the tissue immediately surrounding the inoculated area at different intervals after treatment. The pericarp tissue was collected from fruit at the different time points, frozen in liquid nitrogen and stored at −80 °C.

4.4. RNA Extraction and RT-qPCR

Total RNA was isolated from frozen tissue using RNeasy Plant Mini Kit following the manufacturer's instruction (Qiagen, Valencia, CA, USA). Reverse transcription and qPCR were performed as described [21]. Briefly, one μg of total RNA was used to synthesize cDNA with iScript™ kit (Bio-RAD, Hercules, CA, USA), RT-qPCR analysis of cDNA was performed on a CFX96 Real-time System (Bio-RAD). Gene specific primers listed in Table 1 were designed with the Primer3 software [50]. The efficiency coefficient E was calculated for all primer pairs individually by plotting the relationship between Cq value (threshold cycle) and log[cDNA]. All reactions were performed in triplicate from three independent samples. Cq was used for relative quantification of the input target number. Relative fold difference (*N*) was the number of the treated target gene copies relative to the untreated control gene copies and is calculated as follows: $N = 2^{\Delta Cq}$. ΔCq was the difference in threshold cycles for *SlCaMs* targets and the *actin* internal reference. Relative gene expression (fold changes) was calculated based on *N* with the lowest value as 1. Student's *t* test ($p < 0.05$) was used to determine the significant difference of relative expression of individual genes among different treatments and controls (Microsoft Excel 2007).

Table 1. Primers used for qPCR and cloning.

Primer Name	Oligonucleotides	Gene ID
SlCaM1-a	CCAGAGTTCCTTAACCTGATGG	Solyc01g008950
SlCaM1-b	CTTTTCGCCTAGGTTTGTCATC	
SlCaM2-a	TCTGAGGAGGAGTTGAAAGAGG	Solyc10g081170
SlCaM2-b	TCAACATCAGCTTCCCTAATCA	
SlCaM3-a	GATGGTAATGGAACCATCGACT	Solyc10g077010
SlCaM3-b	CATCAGTGAGCTTCTCACCAAG	
SlCaM4-a	TCAGATCTCGGAGTTCAAAGAAG	Solyc11g072240
SlCaM4-b	CAGGTTAAGGAACTCAGGGAAGT	
SlCaM5-a	TTAACTTGATGGCTCGGAAGAT	Solyc12g099990
SlCaM5-b	ACGAATCATCTCGTCAACCTCT	
SlCaM6-a	ATCACTTGGTCAGAATCCCACT	Solyc03g098050
SlCaM6-b	AGCTGCAGAAATAAAGCCATTC	
SlPR1-a	CTGTGAAGATGTGGGTTGATGAG	NM-001247429
SlPR1-b	TCTCCAGTTACCTGGTGGATCAT	
SlPR2b-a	TCTTGCCCCATTTCAAGTTC	M80608
SlPR2b-b	TGCACGTGTATCCCTCAAAA	
SlLAPA1-a	TGTCGCAGCATGTGAAAATATG	Solyc12g010020
SlLAPA1-b	AGCACCAGTTAATGTTGCCAGA	
SlActin-a	GAAATAGCATAAGATGGCAGACG	X55749
SlActin-b	ATACCCACCATCACACCAGTAT	
SlCaM2-S1 *	<u>ggtggtacc</u>ATGGCGGATCAGCTGACGG	
SlCaM2-S2 *	<u>ggaggatcc</u>CTTGGCCATCATGACCTTAAC	
SlCaM2-A1 *	<u>ggtggtacc</u>CTTGGCCATCATGACCTTAAC	
SlCaM2-A2 *	<u>ggaggatcc</u>ATGGCGGATCAGCTGACGG	

* Primers used for cloning full length *SlCaM2*. The underlined portion indicates the restriction enzyme site.

4.5. Construction of Ti Plasmids Carrying Sense- and Antisense-SlCaM2 Gene

The full length tomato *SlCaM2* was amplified from a mixture of fruit tissues by Pfx DNA polymerase, and subcloned to TA cloning Kit (Life Technology, Grand Island, NY, USA) using gene specific primers (Table 1). The nucleotide sequences of the positive clones were confirmed by sequencing. The full length of *SlCaM2* was subcloned into pDL28F, a derivative of pCambia1300 [41] in the sense- and antisense- orientations in the sites of *Kpn* I and *Bam*H I downstream of 35S promoter, and introduced into *Agrobacterium tumefaciens* strain GV3101. The positive clones were verified by PCR using gene-specific primers.

4.6. Fruit Agroinjection

Agroinjection of tomato fruit was carried as described [51]. Briefly, agrobacterium cultures were grown overnight from individual colonies at 28 °C in YEB medium plus selective antibiotics, transferred to induction medium (0.5% beef extract, 0.1% yeast extract, 0.5% peptone, 0.5% sucrose, 2 mM $MgSO_4$, 20 μM acetosyringone, 10 mM MES, pH 5.6) plus antibiotics, and grown again overnight. The next day, cultures were resuspended with infiltration medium (10 mM $MgCl_2$, 10 mM MES, 200 μM acetosyringone, pH 5.6 with OD_{600} of 1.0), and incubated at room temperature with gentle agitation (20 rpm) for a minimum of 2 h. Cultures were injected into the early mature green fruit using a syringe with a 0.5- × 16-mm needle. The needle was introduced into the fruit tissue through the stylar apex and blossom end. Because of differences in fruit size, the injection was terminated when the solution started running off the injection site.

5. CONCLUSIONS

Calmodulin, a ubiquitous calcium receptor in the eukaryotic cell, plays an important role in almost all aspects of cell activity in plants [4,5,6]. Previous studies have suggested that calmodulins in tomato vegetative tissues respond to a variety of biotic and abiotic stimuli, such as wounding or systemin treatment [11] and pathogen attack [12]. We have observed that the expression levels of six *SlCaMs* in mature green tomato fruit are stimulated by mechanical injury and *B. cinerea* infection. Among the six genes, *SlCaM2* was the most responsive to both pathogen infection and wounding. Further, expressions of all *SlCaMs* were upregulated by SA and MeJA which occurs 24 and 8 h after treatment, respectively. However, *SlCaM2* also had a detectable peak as early as one hour after treatment by SA and MeJA. In general, *SlCaMs* in tomato fruit are early wound-responsive genes, but late responsive genes to pathogen attack, JA and SA. *SlCaMs* may regulate the wound response and disease resistance via SA- and/or JA-dependent signaling in fruit. Based on its gene expression level, *SlCaM2* is the major *CaM* gene in tomato fruit that responds to mechanical injury and pathogen attack. Transient expression of *SlCaM2* into mature green tomato fruit significantly increases resistance to *B. cinerea.* Conversely, reducing its expression facilitates pathogen growth in the host. Since *SlCaM2* is an SA-responsive gene, it will be interesting to test whether increasing its expression also confers resistance to biotrophic pathogens (*i.e., Cladosporium fulvum*). Moreover, it will be of great utility to identify target(s) of *SlCaM2* to improve overall fruit quality by strengthening tolerance to wounding and concomitantly increasing disease resistance to fungal postharvest plant pathogens.

ACKNOWLEDGMENTS

We thank Ernest Paroczay and Ziying Zou for their excellent technical support, Verneta Gaskins for assisting with the pathogen infection assays and Liqun Du for providing pDL28F plasmid. Use of a company or product name by the U.S. Department of Agriculture does not imply approval or recommendation of the product to the exclusion of others that may also be suitable. This research was funded by USDA-ARS NP306 project no. 1245-43000-012-00D.

AUTHOR CONTRIBUTIONS

H. Peng performed gene expression experiments and statistical analysis of the expression data, and helped to evaluate the data. W. Jurick designed and performed pathogen inoculation and decay experiments. T. Yang established all the experiments, analyzed data and wrote the manuscript. H. Peng and W. Jurick edited the manuscript.

CONFLICTS OF INTEREST

The authors declare no conflict of interest.

REFERENCES

1. Dodd, A.N.; Kudla, J.; Sanders, D. The language of calcium signaling. Annu. Rev. Plant Biol. 2010, 61, 593–620.
2. Batistic, O.; Kudla, J. Analysis of calcium signaling pathways in plants. BBA Gen Subj. 2012, 1820, 1283–1293.
3. DeFalco, T.A.; Bender, K.W.; Snedden, W.A. Breaking the code: Ca2+ sensors in plant signalling. Biochem. J. 2010, 425, 27–40.
4. Reddy, A.S.N.; Ali, G.S.; Celesnik, H.; Day, I.S. Coping with stresses: Roles of calcium- and calcium/calmodulin-regulated gene expression. Plant Cell 2011, 23, 2010–2032.
5. Yang, T.; Poovaiah, B.W. Calcium/calmodulin-mediated signal network in plants. Trends Plant Sci. 2003, 8, 505–512.
6. Bouche, N.; Yellin, A.; Snedden, W.A.; Fromm, H. Plant-specific calmodulin-binding proteins. Annu. Rev. Plant Biol.2005, 56, 435–466.
7. Landoni, M.; de Francesco, A.; Galbiati, M.; Tonelli, C. A loss-of-function mutation in calmodulin2 gene affects pollen germination in Arabidopsis thaliana. Plant Mol. Biol. 2010, 74, 235–247.
8. Zhang, W.; Zhou, R.G.; Gao, Y.J.; Zheng, S.Z.; Xu, P.; Zhang, S.Q.; Sun, D.Y. Molecular and genetic evidence for the key role of AtCaM3 in heat-shock signal transduction in Arabidopsis. Plant Physiol. 2009, 149, 1773–1784.
9. Heo, W.D.; Lee, S.H.; Kim, M.C.; Kim, J.C.; Chung, W.S.; Chun, H.J.; Lee, K.J.; Park, C.Y.; Park, H.C.; Choi, J.Y.; et al. Involvement of specific calmodulin isoforms in salicylic acid-independent activation of plant disease resistance responses. Proc. Natl. Acad. Sci. USA 1999, 96, 766–771.
10. Hodges, R.J.; Buzby, J.C.; Bennett, B. Postharvest losses and waste in developed and less developed countries: Opportunities to improve resource use. J. Agric. Sci. 2011, 149, 37–45.
11. Bergey, D.R.; Ryan, C.A. Wound- and systemin-inducible calmodulin gene expression in tomato leaves. Plant Mol. Biol. 1999, 40, 815–823.
12. Zhao, Y.; Liu, W.; Xu, Y.P.; Cao, J.Y.; Braam, J.; Cai, X.Z. Genome-wide identification and functional analyses of calmodulin genes in Solanaceous species. BMC Plant Biol. 2013, 13.
13. Robert-Seilaniantz, A.; Grant, M.; Jones, J.D.G. Hormone crosstalk in plant disease and defense: More than just jasmonate-salicylate antagonism. Annu. Rev. Phytopathol. 2011, 49, 317–343.
14. Chao, W.S.; Gu, Y.Q.; Pautot, V.; Bray, E.A.; Walling, L.L. Leucine aminopeptidase RNAs, proteins, and activities increase in response to water deficit, salinity, and the wound signals systemin, methyl jasmonate, and abscisic acid. Plant Physiol. 1999, 120, 979–992.
15. Vlot, A.C.; Dempsey, D.A.; Klessig, D.F. Salicylic acid, a multifaceted hormone to combat disease. Annu. Rev. Phytopathol. 2009, 47, 177–206.
16. Asghari, M.; Aghdam, M.S. Impact of salicylic acid on post-harvest physiology of horticultural crops. Trends Food Sci. Technol. 2010, 21, 502–509.
17. Tornero, P.; Gadea, J.; Conejero, V.; Vera, P. Two PR-1 genes from tomato are differentially regulated and reveal a novel mode of expression for a pathogenesis-related gene during the hypersensitive response and development. Mol. Plant Microbe Interact. 1997, 10, 624–634.
18. Uquillas, C.; Letelier, I.; Blanco, F.; Jordana, X.; Holuigue, L. NPR1-independent activation of immediate early salicylic acid-responsive genes in Arabidopsis. Mol. Plant Microbe Interact. 2004, 17, 34–42.
19. Van Loon, L.C.; Rep, M.; Pieterse, C.M.J. Significance of inducible defense-related proteins in infected plants. Annu. Rev. Phytopathol. 2006, 44, 135–162.

20. Rohwer, C.L.; Erwin, J.E. Jasmonate-induced changes in polyphenol oxidase, peroxidase, and proteinase inhibitors in horticultural species. Hortscience 2006, 41, 1072–1072.
21. Yang, T.B.; Peng, H.; Whitaker, B.D.; Jurick, W.M. Differential expression of calcium/calmodulin-regulated SlSRs in response to abiotic and biotic stresses in tomato fruit. Physiol. Plantarum. 2013, 148, 445–455.
22. Kazan, K.; Manners, J.M. Jasmonate signaling: Toward an integrated view. Plant Physiol. 2008, 146, 1459–1468.
23. Saftner, R.A.; Bai, J.H.; Abbott, J.A.; Lee, Y.S. Sanitary dips with calcium propionate, calcium chloride, or a calcium amino acid chelate maintain quality and shelf stability of fresh-cut honeydew chunks. Postharvest. Biol. Technol. 2003, 29, 257–269.
24. Abbott, J.A.; Conway, W.S.; Sams, C.E. Postharvest calcium-chloride infiltration affects textural attributes of apples. J. Am. Soc. Hortic. Sci. 1989, 114, 932–936.
25. Park, S.; Cheng, N.H.; Pittman, J.K.; Yoo, K.S.; Park, J.; Smith, R.H.; Hirschi, K.D. Increased calcium levels and prolonged shelf life in tomatoes expressing Arabidopsis H+/Ca2+ transporters. Plant Physiol. 2005, 139, 1194–1206.
26. Ritenour, M.A.; Stoffella, P.J.; He, Z.L.; Narciso, J.A.; Salvatore, J.J. Postharvest calcium chloride dips of whole tomato fruit reduce postharvest decay under commercial conditions. Hortscience 2006, 41, 1016–1017.
27. Saftner, R.A.; Conway, W.S.; Sams, C.E. Postharvest calcium infiltration alone and combined with surface coating treatments influence volatile levels, respiration, ethylene production, and internal atmospheres of "Golden Delicious" apples. J. Am. Soc. Hortic. Sci. 1999, 124, 553–558.
28. Poovaiah, B.W.; Shekhar, V.C. Effects of calcium infiltration of golden delicious apples on fruit firmness and senescence. Hortscience 1978, 13, 357–357.
29. Sams, C.E.; Conway, W.S.; Abbott, J.A.; Lewis, R.J.; Benshalom, N. Firmness and decay of apples following postharvest pressure infiltration of calcium and heat-treatment. J. Am. Soc. Hortic. Sci. 1993, 118, 623–627.
30. Conway, W.S.; Sams, C.E. The effects of postharvest infiltration of calcium, magnesium, or strontium on decay, firmness, respiration, and ethylene production in apples. J. Am. Soc. Hortic. Sci. 1987, 112, 300–303.
31. Lecourieux, D.; Raneva, R.; Pugin, A. Calcium in plant defence-signalling pathways. New Phytol. 2006, 171, 249–269.
32. Nimchuk, Z.; Eulgem, T.; Holt, B.F., 3rd; Dangl, J.L. Recognition and response in the plant immune system. Annu. Rev. Genet. 2003, 37, 579–609.
33. Fu, Z.Q.; Dong, X. Systemic acquired resistance: Turning local infection into global defense. Annu. Rev. Plant Biol.2013, 64, 839–863.
34. Mur, L.A.J.; Kenton, P.; Atzorn, R.; Miersch, O.; Wasternack, C. The outcomes of concentration-specific interactions between salicylate and jasmonate signaling include synergy, antagonism, and oxidative stress leading to cell death. Plant Physiol. 2006, 140, 249–262.
35. Koornneef, A.; Leon-Reyes, A.; Ritsema, T.; Verhage, A.; den Otter, F.C.; van Loon, L.C.; Pieterse, C.M.J. Kinetics of salicylate-mediated suppression of jasmonate signaling reveal a role for redox modulation. Plant Physiol. 2008, 147, 1358–1368.
36. Abd el Rahman, T.; el Oirdi, M.; Gonzalez-Lamothe, R.; Bouarab, K. Necrotrophic pathogens use the salicylic acid signaling pathway to promote disease development in tomato. Mol. Plant Microbe Interact. 2012, 25, 1584–1593.
37. El Oirdi, M.; Abd el Rahman, T.; Rigano, L.; el Hadrami, A.; Rodriguez, M.C.; Daayf, F.; Vojnov, A.; Bouarab, K. Botrytis cinerea manipulates the antagonistic effects between immune pathways to promote disease development in tomato. Plant Cell 2011, 23, 2405–2421.
38. Wang, L.; Tsuda, K.; Truman, W.; Sato, M.; Nguyen, L.V.; Katagiri, F.; Glazebrook, J. CBP60g and SARD1 play partially redundant critical roles in salicylic acid signaling. Plant J. 2011, 67, 1029–1041.
39. Zhang, Y.; Xu, S.; Ding, P.; Wang, D.; Cheng, Y.T.; He, J.; Gao, M.; Xu, F.; Li, Y.; Zhu, Z.; et al. Control of salicylic acid synthesis and systemic acquired resistance by two

members of a plant-specific family of transcription factors. Proc. Natl. Acad. Sci. USA 2010, 107, 18220–18225.

40. Wang, L.; Tsuda, K.; Sato, M.; Cohen, J.D.; Katagiri, F.; Glazebrook, J. Arabidopsis CaM binding protein CBP60g contributes to MAMP-induced SA accumulation and is involved in disease resistance against Pseudomonas syringae. PLoS Pathogens 2009, 5, e1000301.
41. Du, L.Q.; Ali, G.S.; Simons, K.A.; Hou, J.G.; Yang, T.B.; Reddy, A.S.N.; Poovaiah, B.W. Ca2+/calmodulin regulates salicylic-acid-mediated plant immunity. Nature 2009, 457, 1154–1158.
42. Poovaiah, B.W.; Du, L.Q.; Wang, H.Z.; Yang, T.B. Recent advances in calcium/calmodulin-mediated signaling with an emphasis on plant-microbe interactions. Plant Physiol. 2013, 163, 531–542.
43. Leon, J.; Rojo, E.; Titarenko, E.; Sanchez-Serrano, J.J. Jasmonic acid-dependent and -independent wound signal transduction pathways are differentially regulated by Ca2+/calmodulin in Arabidopsis thaliana. Mol. Genet. Gen. 1998, 258, 412–419.
44. Dombrowski, J.E.; Bergey, D.R. Calcium ions enhance systemin activity and play an integral role in the wound response. Plant Sci. 2007, 172, 335–344.
45. Qiu, Y.J.; Xi, J.; Du, L.Q.; Suttle, J.C.; Poovaiah, B.W. Coupling calcium/calmodulin-mediated signaling and herbivore-induced plant response through calmodulin-binding transcription factor AtSR1/CAMTA3. Plant Mol. Biol. 2012, 79, 89–99.
46. Laluk, K.; Prasad, K.V.S.K.; Savchenko, T.; Celesnik, H.; Dehesh, K.; Levy, M.; Mitchell-Olds, T.; Reddy, A.S.N. The calmodulin-binding transcription factor signal responsive1 is a novel regulator of glucosinolate metabolism and herbivory tolerance in Arabidopsis. Plant Cell Physiol. 2012, 53, 2008–2015.
47. Galon, Y.; Nave, R.; Boyce, J.M.; Nachmias, D.; Knight, M.R.; Fromm, H. Calmodulin-binding transcription activator (CAMTA) 3 mediates biotic defense responses in Arabidopsis. FEBS Lett. 2008, 582, 943–948.
48. Sargent, S.A.; Moretti, C.L. Tomato. Agricultural Handbook Number 66 (HB66); Draft-Updated August 2014. USDA-ARS: USA. Available online: http://www.ba.ars.usda.gov/hb66/tomato.pdf (accessed on 11 August 2014).
49. Yan, H.J.; Jurick, W.M., II; Lou, Y.G.; Gaskins, V.L.; Kim, Y.-K. First report of pyrimethanil resistance in Botrytis cinerea from stored apples in pennsylvania. Plant Dis. 2014, 98, 999.
50. Untergrasser, A.; Cutcutache, I.; Koressaar, T.; Ye, J.; Faircloth, B.C.; Remm, M.; Rozen, S.G. Primer 3—New capabilities and interfaces. Nucleic Acids Res. 2012, 40, e115.
51. Orzaez, D.; Mirabel, S.; Wieland, W.H.; Granell, A. Agroinjection of tomato fruits. A tool for rapid functional analysis of transgenes directly in fruit. Plant Physiol. 2006, 140, 3–11.

CHAPTER 14

Rootstock Breeding for Abiotic Stress Tolerance in Citrus

Berken Cimen[1] and Turgut Yesiloglu[1]

[1] Çukurova University, Faculty of Agriculture, Department of Horticulture, Turkey

ABSTRACT

Citrus species are the most widely produced fruit crops in the world. Citrus fruits are mainly produced in coastal areas in several countries as well as Mediterranean region, and production in these regions is affected by both biotic and abiotic stresses, including drought, extreme temperature, salinity, citrus canker, citrus tristeza virus, citrus greening, and others. The use of rootstocks in fruit production includes not only stronger resistance against pathogens but also a higher tolerance to abiotic stress conditions such as salinity, heavy metals, nutrient stress, water stress, and alkalinity. There is extensive genetic diversity in citrus which provides several materials to be used as rootstocks against abiotic stress. In this work, we tried to provide an overview of the abiotic stresses in citrus by combining literature with our studies, role of citrus rootstocks commercially used against abiotic stresses and rootstock breeding in citrus.

Keywords: Citrus, rootstock, salinity, alkalinity, drought, breeding

1. INTRODUCTION

Citrus is the most important tree fruit crop in the world, and citrus fruits are regarded as major household items in more than 100 countries around the world as well as the world juice industry which is also led by citrus juices. Citrus industry is regarded as a leading industry in some regions, such as the mountainous regions of China and coastal plains in several countries, such as California and Florida in USA, Valencia in Spain, and Adana in Turkey. According to the data published in 2013, the world's total citrus fruit production is 135.761.181 tons [1], which consists of 71.445.352 tons of oranges (*Citrus sinensis* (L.) Osb.), 28.678.213 tons of mandarins (*Citrus reticulata* Blanco), 15.191.482 tons of lemons (*Citrus limon* Burm. F.) and limes (*Citrus latifolia* Tan. and *Citrus aurantifolia Swingle)*, 8.453.446 tons of grapefruits (*Citrus paradisi* Macf.) and pummelos [*Citrus maxima* (Burm.) Merr.], and 11.992.686 tons of other citrus fruits (Figure 1).

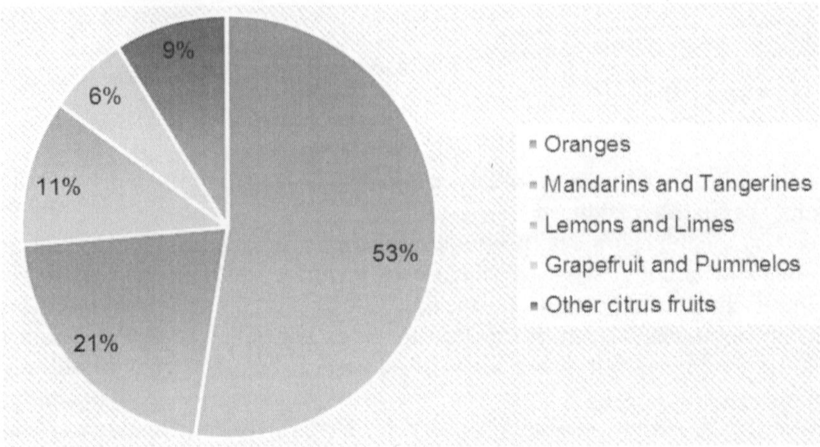

Figure 1. World's citrus production quantity (tons)

The role of citrus fruits in providing nutrients and medicinal value has been recognized since ancient times. Citrus fruits, belonging to the genus Citrus of the family Rutaceae, are well known for their refreshing fragrance, thirst-quenching ability, and providing adequate vitamin C as per recommended dietary allowance. In addition to ascorbic acid, these fruits contain several phytochemicals, which play the role of nutraceuticals, such as carotenoids (lycopene and β-carotene), limonoids, flavanones (naringins and rutinoside), and vitamin B complex and related nutrients (thiamine, riboflavin, nicotinic acid, pantothenic acid, pyridoxine, folic acid, biotin, choline, and inositol). These substances greatly contribute to the supply of anticancer agents and other nutraceutical compounds with antioxidant, inflammatory, cholesterol, and allergic activities, all of them essential to prevent cardiovascular and degenerative diseases, thrombosis, cancer, atherosclerosis, and obesity. In spite of these beneficial traits, there is still a major need to improve fruit quality to meet current consumers' demands [2, 3].

Being vegetatively propagated, a citrus tree is normally composed of the rootstock and scion. Rootstocks play an important role in the rapid development of citrus in the world as well as breeding new cultivars. The necessity of using rootstocks for citrus fruits is to have a profitable production against some limiting factors such as climate, bad soil conditions, and diseases. Besides these factors, the use of citrus rootstocks provides a large number of choice to the growers to increase fruit quality and yield, obtain early fruiting and uniform cropping, avoid juvenility, controlling the tree size, have the opportunity for high-density planting, etc. These factors give many economic important advantages to the growers and, as a result, the citrus fruits are the most produced fresh fruits in the world for several decades [4]. Choosing a rootstock is an important decision, and local climatic and soil conditions are important factors in rootstock selection. Although any citrus variety can be used as a rootstock, some of them are better suited to specific conditions than the others [5, 6]. Some characteristics in a desirable citrus rootstock should be listed such as a good adaptation to all kinds of soils, tolerance to salinity, iron chlorosis, flooding, drought, high affinity with commercial species/cultivars, high yields of good fruit quality, reduced tree size, resistance to citrus tristeza virus (CTV), resistance to citrus blight, resistance to fungal diseases affecting citrus (*Phytophthora* spp., *Armillaria mellea,* etc.), and resistance to nematodes.

Citrus production is affected by both biotic and abiotic stresses, including drought, extreme temperature, salinity, citrus canker, citrus tristeza virus, citrus greening, and others. These stresses can severely influence growth and development of both rootstocks and/or scions of citrus trees, thus reducing both fruit production and fruit quality. The use of rootstocks in fruit production includes not only stronger resistance against pathogens but also a higher tolerance to abiotic stress conditions such as salinity, heavy metals, nutrient stress, water stress, and alkalinity [7]. Soil salinity is a major factor reducing crop production among the world as well as citrus production. Responses to salinity in terms of citrus production are affected by the amount of irrigation, climate, soil type, and fertilization [8]. Citrus rootstocks differ in terms of their tolerance to salinity conditions; many soils and sources of water contain high amounts of salts that can inhibit the growth and yield salt-sensitive citrus [9]. The high level of bicarbonate ions in the soil affects metabolic processes in roots and leaves, decreasing soil and plant Fe availability, leading to the condition known as lime-induced iron chlorosis. The most evident effect of Fe chlorosis is a decrease in photosynthetic pigments, resulting in a relative enrichment of carotenoids over chlorophylls, and production of yellow, chlorotic leaves resulting as a decrease in fruit yield and quality [10, 11, 12]. Drought is considered as the principal factor that limits global agricultural production, among environmental constraints. Species greatly differ in the ability to overcome water deficiency. Drought in citrus trees causes reductions in stomatal conductance (g_S), leaf transpiration rate (E), and net CO_2 assimilation; decreases fruit quality and yield in long-term periods of stress; and increases fruit abscission [13].

Although the sour orange has many excellent horticultural advantages in terms of abiotic stress, it has a very important disadvantage for its susceptibility to citrus tristeza virus. This problem has severely reduced the use of this rootstock in many places especially the Western Mediterranean. Castle and Gmitter [14] reported that sour orange no longer has a secure place in today's rootstock portfolios because of its susceptibility to CTV. Castle [15] indicated that sour orange is an excellent rootstock for areas free of CTV. Ollitrault et al. [16] reported that the arrival of tristeza radically called into question of using sour orange in the Mediterranean area, whereas it had been almost the only rootstock in the region.

The need to produce stress-tolerant crops was evident even in ancient times [17]. Searching rootstock alternative to sour orange keeps on all over the world. In addition, it is a known fact that every rootstock cannot be used in every ecology. Due to the limiting effects of different ecological factors and diseases, a rootstock which is suitable for a country could not succeed in another one. Accordingly, the rootstocks that will be offered as alternative to sour orange is supposed to show some characteristics such as tolerance to lime-induced iron chlorosis in calcareous soils, tolerance or resistance to *Phytophthora citrophthora.* In addition to these, the alternative rootstocks should have CTV tolerance which does not exist in sour orange. On the contrary, all the required traits are present in the citrus germplasm. For instance, *Poncirus trifoliata* has tolerance to mainly biotic stress and cold hardness and some *Citrus* species have adaptation to both abiotic and biotic stresses. However, the complexity of citrus biology and genetics makes it difficult to combine them through traditional breeding.

Consequently, in this chapter, we try to provide an overview of the abiotic stresses in citrus, role of citrus rootstocks commercially used against abiotic stresses, and rootstock breeding for tolerance to abiotic stress.

2. TYPES OF ABIOTIC STRESSES IN CITRICULTURE

Many different citrus genotypes are commercially grown in a wide diversity of soil and climatic conditions; therefore, trees are subjected to important abiotic and biotic stresses that limit the production and, in some instances, the use of certain rootstocks and varieties [18]. Citrus trees are subjected to several abiotic constraints such as acid, alkaline, and salty soils; flooding and drought; and freezing and high temperatures.

Related to the global warming, drought problems have occurred in many countries. In addition, salinity became a major cause for citrus production in the coastal regions of Mediterranean by increasing the use of fertilizers and decreasing precipitations. Salinity and drought in the calcareous soil of Mediterranean region can lead to major problems in citriculture in terms of fruit yield and quality. Thus it is necessary to breed new rootstocks that are genetically tolerant to abiotic stress conditions and alternatives to existing rootstocks.

In a recent work, we have tried to handle three common abiotic stresses (salinity, drought, and alkalinity) occurring in citriculture especially in the coastal Mediterranean region by combining the literature with the studies conducted at Çukurova University.

2.1. SALINITY

Salinity is a major environmental factor affecting the performance of many crop plants and reducing agricultural productivity [19, 20]. It is estimated that more than a third of all of the irrigated soils in the world is affected by salinity. The loss of farmable soils due to salinization is directly in conflict with the needs of the world population which is increasing continuously. Salt stress is a major stress problem in arid and semiarid regions and irrigated areas. Almost 7% of the world land area, 20% of the cultivated land, and nearly half of the irrigated land are affected by high salt concentrations [21].

Salinity affects the crop during both the vegetative and the reproductive stage and therefore causes reduction in plant growth and development with low water potential in the root medium (osmotic effect), too high internal ion concentration (ion excess/toxicity), and nutritional imbalance by depression in uptake and/or shoot transport (ion deficiency). Most of the salt stress in nature is due to sodium salts, particularly NaCl [22, 23]. High concentrations of Na^+ and Cl^- in the root medium saturation depress nutrient–ion activities and produce extreme rations of Na^+/Ca^{2+}, Na^+/K^+, Ca^{2+}/Mg^{2+}, and Cl^-/NO_3^- [24]. As a result, plants become susceptible to osmotic and specific ion injury as well as nutritional disorders that may result in reduced yield and quality. These processes may be occurring at the same time, but whether they ultimately affect crop yield and quality depends on the salinity level, composition of salt, exposed period to salinity, the crop species and cultivars, the growth stage of plants, and a number of environmental factors [25, 26, 27, 28]. When the salt concentration reaches a harmful level for plant growth, a salinity condition is said to have developed. The degree to which growth and normal metabolism can be maintained is described as salt tolerance. Salt tolerance of vegetable crops varies considerably among species and depends upon the cultural conditions under which the crops are grown. Soil, water, plant, and environment can affect the salt tolerance of a plant. Therefore, plant response to a given salt concentration cannot be predicted on an absolute basis but on relative performance [29].

The effects of salinity on plants are evidenced by a severe reduction in plant growth and yield and, if the saline conditions persist, plant death can occur [30]. Salinity causes a deficiency of water in plant tissue, and low water potential reduces growth by inhibiting cell division and cell expansion [31]. The reduction in growth is mainly due to an osmotic effect of the accumulation of salts near the root zone, whereas the buildup of toxic saline ions in plant tissues is responsible for the progressive impairment of several physiological processes [20].

Osmotic effects resulting from salinity may cause disturbances in the water balance of the plant, including a reduction of turgor and an inhibition of growth, as well as stomatal closure and reduction of photosynthesis [32, 33]. The primary effect of high salt concentration in plants is stomatal closure. This causes a low transpiration rate and reduces the CO_2 availability for photosynthesis [34]. Hussain et al. [35] indicated that salinity reduced the photosynthetic availability of some citrus species and genera. As a result, plants become susceptible to osmotic and specific ion injury as well as to nutritional disorders that may result in reduced yield and quality.

Many researchers so far have reported citrus trees as salt-sensitive plants [30, 36, 37]. Salinity reduces citrus tree growth and fruit yield [9]. Growth reduction and some physiological and biochemical disturbances due to excessive concentrations of Cl^- and Na^+ in leaves are the main problems that are caused by salinity stress [37]. Also, salt stress has a dramatic impact on the citrus industry by decreasing the growth of trees and fruit yield and quality. Salinity may also cause nutrient deficiencies or imbalances, due to the competition of Na^+ and Cl^- with nutrients such as K^+, Ca^{2+}, Mg^{2+} and NO_3^-. In addition to osmotic effect, high K^+ concentration in a salinized nutrient solution increased the absorption of Cl^- citrus roots. Salt tolerance in citrus has been linked to the exclusion of toxic ions from the shoot [38]. Thus, citrus rootstocks have a great influence on the amount of

Cl^- and/or Na^+accumulated in the foliage of grafted trees [30]. For instance, in Fino lemon trees, sour orange rootstock is considered a good Cl^- and Na^+ excluder, whereas the *Citrus macrophylla* rootstock is a Cl^- and Na^+ accumulator [39].

Citrus trees, under salinity stress, suffer growth reduction and some physiological and biochemical disturbances due to excessive concentrations of Cl^- and Na^+ in leaves [37]. Depending on the soil type, irrigation method, and frequency, the soil solution salinity might also rise several fold between irrigations. Continual improvement of rootstocks and/or scions will be necessary to sustain irrigated citrus in increasingly salinized environments [30].

Sour orange is one of the most frequently used rootstocks in Mediterranean countries. It is known to be tolerant to salinity and calcareous soil among citrus rootstocks. However, it is highly susceptible to tristeza disease [37], that the disease should take into consideration the citriculture in Turkey and other countries which use sour orange as rootstock owing to be threatened by it. Genetically improved with favorable agronomical characteristics, such as resistance to pests and diseases such as the citrus tristeza virus, and salinity tolerance, rootstocks may be a long-term approach. Hence, screening studies based on physiological responses of genotypes to salinity stress should be used. In addition, continuous improvement of rootstocks is necessary to sustain cultivating citrus trees under salinized environments.

Yesiloglu et al. [40] established a screening study for the physiological evaluation of global tolerance to salinity rootstock collection of Çukurova University in the frame of the CIBEWU project, No: 015453. They screened 29 different genotypes that can be used as citrus rootstock under salinity stress assessed by several growth and physiological parameters such as fresh and dry weights of shoot and root; leaf chlorophyll concentration; and fluorescence, chloride, and Na content. High concentration of Cl and/or Na in the leaves of Citrus has been frequently related to disturbances in nutrition, gas exchange, and water relations. Unpublished data of the project regarding the genotypes used in the screening study and Cl^- and Na^+ concentrations in root and leaves of the genotypes are presented below. 34-12 N citremon, 4475 SRA citrumelo, CRC 4475 Swingle citrumelo, *Severinia buxifolia* have accumulated higher chloride in leaves, while Tuzcu Cleopatra mandarin, Gou Tou, Rangpur lime, CRC 02 Volkameriana, Tuzcu 31-31 sour orange have lower chloride in leaves. Rubidoux trifoliate with trifoliate hybrids and *Severinia buxifolia* were poor Cl^- excluder. It was found that Tuzcu Cleopatra mandarin and Rangpur lime are the best Cl^- excluders. Gou Tou, Antalya Cleopatra mandarin have the lowest amount chloride in the roots, whereas *Citrus ichangensis,* Benecke trifoliata, Volkameriana, 08 A 3015 Rubidoux, and SRA Pomeroy have the highest values. According to the results, Tuzcu Cleopatra mandarin, Rangpur lime, Gou Tou, and Antalya Cleopatra mandarin were found to be the most tolerant to salt stress. *Severinia buxifolia*, CRC-4475 Swingle citrumelo, Local trifoliata, and Benecke trifoliata were the most sensitive group to salt stress. Data were evaluated by using modified "weighted-ranking" method based on the parameters of chlorophyll fluorescence (*Fv*'/*Fm*'), leaf chlorophyll concentration by SPAD readings, leaf Cl and Na concentrations, leaf K and Ca concentrations, root K and Ca concentrations, growth parameters (shoot and root fresh and dry weights, shoot length, and leaf number), and visual ratings of leaf chlorosis. A classification which belongs to the screening study for salinity stress was performed as a result of the work and reported as follows in Table 1 (unpublished data). Genotypes were classified as very sensitive, sensitive, acceptable, tolerant, and very tolerant to salinity.

Table 1. Classification of rootstocks in collection of Çukurova University in respect of salinity tolerance

Genotypes	Tolerance to salt stress
Rubidoux trifoliate 08A 30.15	2
Citremon 34 12 N	1
Citrumelo 4475 SRA	1
AREC Swingle citrumelo	3
Antalya Cleopatra mandarin	3
Benecke trifoliate	2
C-35 Citrange	4
Swingle citrumelo 4475 CRC	1
Volkameriana CRC 01	4
Volkameriana CRC 02	4
Citrumelo CRC 1452	3
Carrizo citrange	3
Citrus ichangensis CRC	3
Citrus sulcata	3
Citrus sunki	3
Macrophylla	2
Rangpur lime	5
Gou Tou sour orange SRA 506	4
Pomeroy trifoliate SRA	1
Sacaton citrumelo	4
Severinia buxifolia SRA	1
Smooth Seville sour orange	4

Troyer citrange	3
Tuzcu 31-31 sour orange	4
Tuzcu 891 sour orange	4
Tuzcu Cleopatra mandarin	5
Volkameriana	4
Local trifoliate	2

*1: very sensitive, 2: sensitive, 3: acceptable, 4: tolerant, 5: very tolerant

Khoshbakht et al. [41] reported that the effects of salinity on photosynthesis range from the restriction of CO_2 diffusion into the chloroplast, via limitations on stomatal opening mediated by shoot- and root-generated hormones, and on the mesophyll transport of CO_2, to alterations in leaf photochemistry and carbon metabolism. The authors conducted a study and investigated the NaCl effects on gas exchange parameters of nine citrus rootstocks and reported that sour orange and Cleopatra mandarin were the rootstocks most tolerant to salinity of all the nine citrus rootstocks studied. Also, Cimen et al. [42] determined that the tolerances of Sarawak bintangor, Shekwasha, Fuzhu and Cleopatra mandarin to salt stress were determined by investigating the photosynthetic parameters and significant salinity effects on the photosynthetic performances of these rootstocks were reported (Figure 2).

2.2. Alkalinity

It is well known that iron is an essential micronutrient for all higher plants including citrus. Most of the iron existing in rhizosphere cannot be taken up by plants because iron is highly insoluble. Two main working hypotheses have been put forward for this chlorosis. In the first working hypothesis, the main cause of Fe deficiency chlorosis is thought to be the inhibition of Fe acquisition by HCO_3^- in the rhizosphere. For the second hypothesis, Fe inactivation in the leaf apoplast by an alkalization process properly noted by the "distant effect" of HCO_3^- is thought to be the main trigger for Fe deficiency chlorosis in leaves [43]. The Mediterranean Basin is characterized by the prevalence of calcareous surface horizon. In these soils, iron (Fe) chlorosis can lead to diminished yields and even plant death, particularly in semiarid areas where irrigation water has high bicarbonate contents, soil pH is high (7.0–9.0), and organic matter content is low [44]. Citrus production is increasing throughout the Mediterranean countries and more and more citrus orchards are being planted on marginal soils.

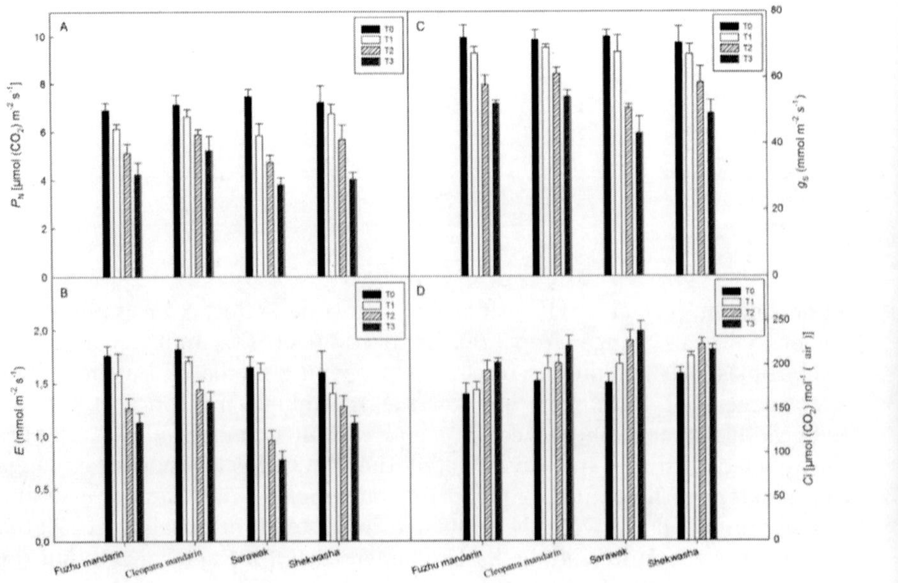

Figure 2. Effects of different salinity levels on photosynthetic rate (A), transpiration rate (B), stomatal conductance (C), and initial CO_2 concentrations (D) of four genotypes. The bars indicate the standard deviation. (T0 = 0 mM NaCl, T1 = 50 mM NaCl, T2 = 75 mM NaCl, T4 = 100 mM NaCl). Data presented from Cimen et al. [**42**].

Mediterranean countries have a suitable climate for citrus production, but it is estimated that 20–50% of fruit trees grown in the Mediterranean basin suffer from iron (Fe) deficiency. The most prevalent cause of Fe deficiency in this region is the presence of high levels of carbonate ions in calcareous soils, characterized by a high pH [45]. These soils often have more than 20% of calcium and magnesium carbonates and are strongly buffered, with a pH between 7.5 and 8.5 [11]. Fe uptake is highly dependent on soil pH, and Fe activity in solution decreases 1000-fold for each pH unit rise to reach a minimum within the range from 7.4 to 8.5 [46]. Leaf Fe chlorosis in plants is an old problem occurring in areas of calcareous and/or alkaline soils. Yield reductions from Fe-induced leaf chlorosis have been found in tomato, raspberry, kiwifruit, pineapple, vines, and citrus [47]. Moreover, the severity of leaf chlorosis and the differential behavior of genotypes can be determined by the chlorophyll concentration in leaves [11, 46, 48]. The high level of bicarbonate ions in the soil affects metabolic processes in roots and leaves, decreasing soil and plant Fe availability, leading to the condition known as lime-induced iron chlorosis. The most evident effect of Fe chlorosis is a decrease in photosynthetic pigments, resulting in a relative enrichment of carotenoids over chlorophylls and producing yellow, chlorotic leaves. The loss of pigmentation is caused by decreased chlorophyll content in chloroplasts. This negatively affects the rate of photosynthesis and, therefore, the development of biomass. Fe deficiency affects the physiology and biochemistry of the whole plant, as Fe is an important cofactor of many enzymes, including those involved in the biosynthetic pathway of chlorophylls [10, 11].

The use of rootstocks in fruit production includes not only a stronger resistance against pathogens but also a higher tolerance to abiotic stress conditions such as salinity, heavy metal, nutrient stress, water stress, and alkalinity [7]. Recent studies showed that citrus rootstocks had different tolerance levels to iron deficiency [48, 49]. Studies emphasized that high pH conditions reduced iron uptake in citrus rootstocks [48, 49, 50, 51, 49]. Also, rootstocks affect tree growth, fruit quality, and yield [15, 48, 52]. Moreover, scion behavior depends in part on the rootstock-induced effects on leaf gas exchange [53]. González-Mas et al. [53] indicated that in calcareous soils, citrus production depends on the availability of suitable rootstocks that are tolerant to low Fe soil conditions. Studies have found that "Volkameriana,"' and "sour orange" plants were tolerant; "Carrizo and Troyer" citranges were intermediate, whereas the *"Poncirus trifoliata"* rootstock was more sensitive to iron chlorosis [46, 48, 49, 51, 54]. In addition, Cimen et al. [55] indicated that Young "Navelina" orange trees budded on Tuzcu 31-31 sour orange, and Gou Tou sour oranges performed best under Fe deprived conditions in plant growth chamber. Navelina on Volkameriana and Cleopatra mandarin was moderate; C-35 citrange and local trifoliate were poorly adapted to lime-induced Fe deficiency (Figure 3).

Figure 3. Response of Navelina Nave orange budded on eight different rootstocks in response to iron deficiency. Data presented from Cimen et al. [55].

Among physiological processes, photosynthesis is the basic determinant of plant growth and productivity, and the ability to maintain the rate of carbon assimilation under environmental stress is of fundamental importance to plant production [56]. Since Fe catalyzes chlorophyll biosynthesis [57], it would be expected to promote the photosynthetic rate (*P*n) while Fe deficiency to reduce it [58, 59]. Most of the knowledge concerning the effect of Fe deficiency on the photosynthetic parameters has been obtained with annual plants. However, relatively few studies have focused on the consequences of induced Fe deficiency on photosynthesis in evergreen fruit trees and especially in citrus. On the contrary, Fe is a component of several metalloenzymes, including peroxidase and catalase. Although both enzymes could be used as biochemical indicators of Fe availability in citrus [36], there are contradictory reports concerning the effect of Fe deficiency on catalase and peroxidase activity [60]. Cimen et al. [55] reported that at sufficient Fe supply, plants had higher activity of catalase (CAT) than the plants with Fe deprived conditions. Slight decreases were recorded on the Navelina orange leaves of Tuzcu 31-31 and Gou Tou sour oranges, while decreases were remarkable in the leaves of C-35 citrange and TGK0633 (obtained by a selection of trifoliate orange in Turkey) under short supply of Fe. In addition, Navelina leaves of C-35 citrange and TGK0633 displayed maximum decreases in APX (ascorbate peroxidase) activity, similarly CAT activity. There were no significant APX activity decreases in the leaves of Tuzcu 31 31, Gou Tou sour oranges, Volkameriana, and Antalya Cleopatra mandarin (Figure 4).

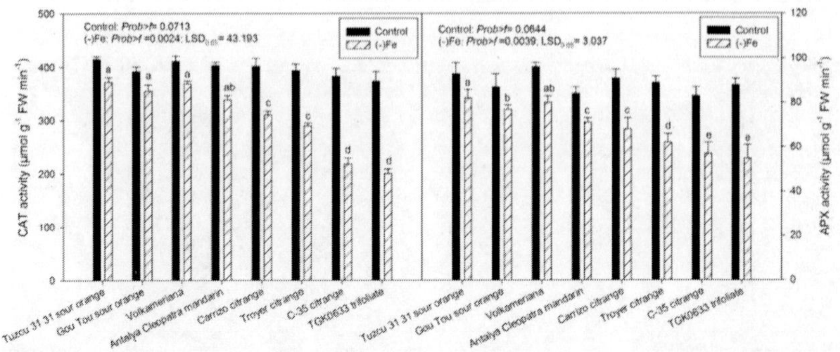

Figure 4. Catalase and ascorbate peroxidase activities of Navelina leaves of different rootstocks under Fe sufficient and deprived conditions. The bars show the standard deviation. Data presented from Cimen et al. [**55**].

The intensity of iron chlorosis can be quantified by total Fe, active Fe, leaf chlorophyll meter, photosynthetic parameters, enzymes, plant growth parameters, and visual ratings of leaf chlorosis. One of the distinctive characteristics of iron deficiency in field crops is the lack of correlation between leaf iron content and chlorosis. This has been termed the "chlorosis paradox." Therefore, leaf chlorophyll contents are generally used to monitor iron chlorosis [61]. The use of visual ratings and readings of a portable chlorophyll meter are the most efficient approaches to define iron chlorosis in citrus [51]. Yesiloglu et al. [40] established a screening study for the physiological evaluation of global tolerance to lime-induced Fe chlorosis in rootstock collection of Çukurova University in the frame of the project CIBEWU, No: 015453 and evaluated by using modified "weighted-ranking" method based on the parameters of total Fe, active Fe, chlorophyll fluorescence (*F*v'/*F*m'), leaf chlorophyll concentration by SPAD readings, visual ratings of leaf chlorosis, and shoot and root weight.

Local trifoliate, Rubidoux trifoliate, and Benecke trifoliate exhibited severe chlorosis and were more chlorotic than other genotypes. Cleopatra mandarins among mandarin and its hybrids were more tolerant than Sunki and Calamondins. Macrophylla was the best in lemon and lemon hybrids group. Campbell [50] reported that Macrophylla was well adapted to soils of high pH. Volkameriana and Rangpur were almost the same. Gou Tou was really so tolerant to iron deficiency. All sour oranges were similar to each other. Alanya Dilimli sweet orange is a variety selected in south of Turkey. It is known to be very resistant to high pH conditions. The results confirm that Alanya Dilimli is very tolerant to high pH. According to the results, Carrizo citrange, Flhorag1, Macrophylla, Antalya Cleopatra mandarin, Tuzcu 31-31 sour orange, Gou Tou sour orange, and Alanya Dilimli sweet orange were more tolerant than others in citrus rootstock collection of Çukurova University (Table 2 – unpublished data).

Although trifoliate orange has many advantages in terms of tolerance to abiotic and biotic stresses, it is susceptible to calcareous soil conditions. However, there are some superior genotypes to improve the tolerance to high pH of present rootstocks by hybridization and crossing [44].

Several rootstock breeding programs have been carried out by different countries leading in citriculture. Forner et al. [62] reported two new citrus rootstocks named F-A 5 and F-A 13, released in Spain. These rootstocks are hybrids of Cleopatra mandarin × Rubidoux trifoliate crosses with a high level of tolerance to lime-induced iron chlorosis. Besides, Bowman and Rouse [63] mentioned a new citrus rootstock named as US-812 which is a hybrid obtained by a cross between Sunki mandarin (*Citrus reticulata*) and

TABLE 2. Classification of the rootstocks in collection of Çukurova University in terms of tolerance of iron deficiency.

Genotypes	Tolerance to iron chlorosis
Rubidoux trifoliate 08A 30.15	2
Citremon 34 12 N	3
Citrumelo 4475 SRA	3
Arec Rubidoux trifoliate	3
Arec Swingle citrumelo	2
Benecke trifoliate	2
Swingle citrumelo 4475 CRC	2
Carrizo citrange	4
Flhoragl	5
Pomeroy trifoliate SRA	2
Sacaton citrumelo I	2
Sacaton citrumelo	2
Citrumelo 1452 CRC	2
Local trifoliate	1
Troyer citrange	3
Antalya Cleopatra mandarin	4
Citrus sunki	4
Calamondin CRC	2
Calamondin 108 USDA	1
Tuzcu Cleopatra mandarin	2
Volkameriana CRC 01	3
Volkameriana CRC 02	3
Macrophylla	4
Rangpur lime	3
Volkameriana	3
Citrus ichangensis CRC	3
Citrus sulcata	2
Severinia buxifolia SRA	2
Alanya Dilimli sweet orange	5
Gou Tou sour orange SRA 506	5
Smooth Seville sour orange	3
Taiwanica	2
Tuzcu 31-31 sour orange	5
Tuzcu 891 sour orange	4

*1: very sensitive, 2: sensitive, 3: acceptable, 4: tolerant, 5: very tolerant

Benecke trifoliate orange in USDA Indio Research Station, California. They have reported that using Valencia trees budded on to US-812 resulted in some tolerance to high alkalinity under pH conditions 8.1– 8.3. Moreover, Federici et al. [64] indicated that three citrus rootstocks released in August 2009 by the University of California named as "Bitters," "Carpenter," and "Furr" trifoliate hybrids, tested as C22, C54, and C57, respectively, by crossing Sunki mandarin × Swingle trifoliate orange. Bitters were found to be very tolerant to calcareous soil, whereas Carpenter and Furr were found to be moderately tolerant.

2.3. Drought

Mediterranean region has a subtropical climate and is an important region for citriculture. A significant amount of high-quality citrus fruits is produced in Mediterranean countries such as Spain, Turkey, Italy, Greece, Egypt, Morocco, and Tunisia. Fruit yield and quality are affected by genetic traits together with environmental factors. Fruit crops are frequently exposed to environmental stresses spontaneously or by conventional agronomic conditions. Some of these conditions such as high temperature may last only for a short period of time, whereas lack of water in soil may last for longer periods. Global warming is a type of greenhouse effect which is defined as the increase of Earth's average surface temperature due to the effect of greenhouse gases, such as carbon dioxide emissions from burning fossil fuels or from deforestation, which trap heat that would otherwise escape from Earth.

Yaacoubi et al. [65] indicated that Mediterranean fruit tree production is facing major changes that have environmental and socioeconomic consequences. Climatic changes related to temperature warming have been reported worldwide.

Drought stress, as one of the most ominous abiotic factors limiting the productivity of horticultural crops, is increasingly growing in dimension of severity in many regions of the world [66]. In general, the mechanism of drought resistance in plants can be explained as drought escape, drought avoidance, and drought tolerance. These traits consist of osmotic adjustments, cell membrane stability, epicuticular wax, partitioning and stem reserve mobilization, manipulation and stability of flowering processes, and seedling drought traits.

Drought tolerance is a complex trait that is important at different growth stages and involves multiple adaptations. Fundamental to this is the ability to maximize the extraction of water from the soil while minimizing loss from the leaves. Morphological adaptations include the development of deep roots and alterations in leaf morphology and cuticle structure, while physiological adaptations involve changes in stomatal density to maximize water uptake and retention [67, 68, 69].

Drought stress effects on the plant may range from slight suppression of growth and yield to temporary wilting, in which leaves flag but recover after transpiration demands decrease, to permanent wilting in which the plant suffers injury and death [70]. A plant responds to a lack of water by halting growth and reducing photosynthesis and other plant processes in order to reduce water use. As water loss progresses, leaves of some species may appear to change color, usually to blue-green. Foliage begins to wilt and, if the plant is not irrigated, leaves will fall off and the plant will eventually die. Drought lowers the water potential of a plant's root and upon extended exposure, abscisic acid is accumulated, and as a result stomatal closure occurs. This reduces a plant's leaf relative water content. The time required for drought stress to occur depends on the water-holding capacity of the soil, environmental conditions, stage of plant growth, and plant species [71]. Plants growing in sandy soils with low water-holding capacity are more susceptible to drought stress than plants growing in clay soils. A limited root system will accelerate the rate at which drought stress develops. A plant with a large mass of leaves in relation to the root system is prone to drought stress because the leaves may lose water faster than the roots can supply it. The root system has a great importance when the plant faces drought. For instance, Rough lemon rootstocks are very drought tolerant because of their extensive, deep root systems. Newly installed plants and poorly established plants may be especially susceptible to drought stress because of the limited root system or the large mass of stems and leaves in comparison to roots.

Citrus, a perennial crop with a long orchard life, is likewise a globally important fruit crop responsible for world trade and often exposed to the vagaries of soil and atmospheric drought stress [72]. Drought stress is known to restrict the vegetative growth and yield of citrus, in addition to adversely affecting fruit quality and incurring huge economic loss to the citrus growers [73]. Therefore, screening and selection of germplasm are of great importance in terms of drought tolerance.

Pedrosoa et al. [74] reported that citrus rootstocks have differential capacities for supplying shoot tissues with water and carbon, improving the resistance to biotic and abiotic stresses and affecting plant water status and photosynthesis. Water relations have been well studied in citrus trees, showing that rootstocks alter the physiological performance under water deficit through variations in plant hydraulic conductance, leaf water potential, and stomatal conductance [75, 76, 77]. In addition, several studies have found that citrus rootstocks showed different performances when they are exposed to drought [13, 73].

Treeby et al. [78] investigated irrigation management and rootstock effects on navel orange and reported that irrigation management is far more critical for external fruit quality for trees on sweet orange and, to a lesser extent, trees on the citranges compared to trees on trifoliate orange and Cleopatra mandarin.

Some studies also indicate that using tetraploid rootstocks increases the drought tolerance in comparison to their diploid clones in citrus. Allario et al. [79] reported that polyploidy is common in many plant species and often leads to better adaptation to adverse environmental conditions. The authors examined the drought tolerance in diploid (2x) and autotetraploid (4x) clones of Rangpur lime (*Citrus limonia*) rootstocks grafted with 2x Valencia Delta sweet orange (*Citrus sinensis*) scions, named V/2xRL and V/4xRL, respectively. The results of the authors showed that using tetraploid clones of Rangpur lime had increased the drought tolerance in grafted sweet orange.

3. GERMPLASM AND GENETIC VARIABILITY

The origin of citrus is believed to be southeast Asia, including south China, northeastern India, and Burma. Commercial citrus species and related genera belong to the order Geraniales, family Rutaceae, and subfamily Aurantoidea. Fruit crops is a very heterogeneous group of plants including trees, shrubs, climbing vines, and perennial herbs. They inhabit different climates, ranging from tropical to subarctic zones and altitudes from sea level to higher mountains. Fruit crops also differ considerably with respect to their origin, taxonomy, and breeding systems. The biodiversity in plants has been progressed by natural and artificial hybridization and mutation which are the basic resources of biological evolution. The deterioration of ecosystem directly or indirectly by human beings has always been causing the destruction of biodiversity and many advanced genotypes [80].

There is extensive genetic diversity in citrus. Aydin and Yesiloglu [81] reported that the genus *Citrus* L. belongs to the subtribe Citrineae, the tribe Citreae within the subfamily Aurantioideae of the Rutaceae family [82]. The Aurantioideae is one of seven subfamilies of Rutaceae which consists of two tribes and 33 genera. Each of tribes Clauseneae and Citreae is composed of three subtribes. Clauseneae includes Micromelinae, Clauseninae, and Merrillinae, and Citreae has Triphasiinae, Citrinae, and Balsamocitrinae. The Citrinae is distinct from all the other subtribes in the subfamily by having pulp vesicles in the fruit. This subtribe contains three groups: primitive citrus fruit, near citrus fruit, and true citrus fruit trees. True citrus fruits have six genera: Clymenia, Eremocitrus, Microcitrus, Poncirus, Fortunella, and Citrus [83].

All rootstocks and varieties used are included in the genus Citrus, except for kumquats (*Fortunella* spp.) and trifoliate orange (*Poncirus trifoliata* L. Raf.), the latter is used exclusively as a rootstock. Trifoliate orange [*Poncirus trifoliata* (L.) Raf.] is an important citrus relative for breeding new rootstocks. In addition to its tolerance to citrus tristeza virus, citrus nematodes, and cold weather and edaphic conditions, the dominant nature of the "trifoliate" leaves of trifoliate orange is a useful morphological marker in the visual identification of hybrids from crosses using trifoliate orange as male parent. Some hybrids of commercial interest, including citranges (sweet orange × trifoliate orange) and citrumelos (grapefruit × trifoliate orange), are used as rootstocks.

In general, the diversity of genetic structures in the subfamily of Aurantioideae, which occurred within the steps of biological evolution, is the genetic resource that has been lost before the determination of their characteristics during the rapidly vanishing process. Preservation of this kind of material is a challenge particularly for "conservation breeding" branch of plant breeding [80]. The determination of citrus genus and close relative genus and species collections with *in situ* and *ex situ* structure in the selected countries and regions, describing and establishing international legal tender for them, are very important.

The richness of germplasm has benefited genetics and breeding research in the countries that have a long history of citriculture. China is the most important place of origin for citrus. The long history and diversified climates enable China to harbor the most citrus varieties. Southern China is one of the centers of diversity for *Citrus* and related genera such as *Fortunella.* A National Citrus Germplasm Repository was established in China in the early 1980s, and a record in 1996 reported 1041 accessions [84, 85]. In India, there is an *in situ* germplasm including 627 accessions as it was reported by Singh [86], and eight *ex situ* citrus conservation collections have been established. Three *in situ* collections were established in Malaysia. In addition, there are three collections in Indonesia and Thailand, two collections in Philippines. The orchards of The Federal Fruit Crops Research Station in Tsukuba, Okitsu, and Kuchinotsu have the widest diversity collection of citrus and the relative types. It is declared that there are totally more than 1300 accessions in these three stations [84]. The original natural dispersion areas of Microcitrus and Eremocitrus as well as many relative types of citrus are located in Australia and the main collection in Australia is located at the Biological and Chemical Research Institute under the NSW Agriculture & Fisheries Department in Rydalmere in New South Wales [84]. Tuzcu [80] reported that the collection of citrus in California began in the 1890s, and the conservation and utilization system of these were structured in 1910, shortly after the establishment of Citrus Experiment Station (CES) in Riverside [87]. H.J. Webber, the first manager of the station, ensured the establishment of Citrus Variety Collection in 1917 in the field which is currently in the Riverside Campus Area of California University. United States Department of Agriculture (USDA) National Clonal Germplasm Repository for Citrus and Dates (NCGRCD) was established as a top organization for the coordination of citrus genetic resources studies in 1974. It is working in close cooperation with other two establishments. There are about 350 virus-free accessions under its conservation [84]. The most important collection is located at National Research Center for Cassava and Fruit Crops, CNPMF in Cruz das Almas/Bahaia, which includes 1858 accessions in Brazil. Valencia Agricultural Research Institute [Instituto Valenciano de Investigaciones Agrarias (IVIA)] is responsible for all the actions regarding citrus genetic resources. According to the last records, there are 478 elite accessions in total including 13 genera (16 accessions) of the Aurantioidea subfamily in addition to Citrus genus in IVIA [84, 88]. In France, 1300 accessions at Agricultural Research Station SRA tied to National Agricultural Research Institute (Institut National de la Recherches Agronomiques – The National Agronomic Research Institute (INRA) at San Nicola in Corsica Island exist in "SRA Citrus Collection" [84]. In terms of citrus genetic resources, mainly, there is one established citrus germplasm in Turkey. This establishment is named as Tuzcu Citrus Collection (TCC) and consists of 964 accessions in Çukurova University Faculty of Agriculture [80].

4. SOME DEVELOPED CITRUS ROOTSTOCKS AND THEIR TOLERANCE TO ABIOTIC STRESS

The first use of rootstocks in citriculture was in 1842 against Phytophthora in Azores Islands through the use of resistant rootstocks. Since then, commercial citrus orchards are established by combining scion and rootstocks in order to achieve the highest quality for the scion. The use of rootstocks in citrus decreases the long juvenility period and allows the cultivation of citrus under several abiotic and biotic stress conditions. The physiology of the whole tree is affected by rootstock, including traits of economic relevance such as fruit yield, fruit size, juice quality, tree vigor, and resistance against biotic and abiotic stresses [89]. Fruit maturation, fruit holding on tree, and postharvest preservation are also affected by rootstock [90].

There is no ideal rootstock in order to manage all abiotic and biotic stress conditions. For instance, sour orange (*Citrus aurantium* L.) which has a high adaptation capability to different soil conditions has been the most commonly used rootstock in commercial citrus trees. Also, sour orange *(Citrus aurantium* L.) is tolerant to root rot, citrus blight disease, calcareous soils, water deficit, and cold, inducing high yield and high fruit quality. However, sour orange is susceptible to citrus tristeza virus, and the usage of this rootstock is decreasing in some countries due to the existence of CTV. On the contrary, Rangpur lime (C. *limonia*), Volkamer lemon (*C. volkameriana* V. Ten. and Pasq.), and rough lemon (C. *jambhiri* Lush.) are drought resistant and increase the fruit yield of the grafted scion. Besides, they reduce the fruit quality as compared to fruits obtained from sour orange, trifoliate orange, Carrizo citrange, and Troyer citrange. However, they are sensitive to cold.

The search for new rootstock in citrus production as well as many fruit species is necessary to sustain production under the inevitable abiotic stresses in many different ecological conditions of the citrus-growing areas of the world. Also, new diseases, the spread of the known diseases, and different environmental conditions affected by the climatic change force the demand for developing new citrus rootstocks. Below, we have tried to cover and explain some of the newly released citrus rootstocks obtained by breeding programs carried out in several countries by several researchers.

Swingle citrumelo is a hybrid of Duncan grapefruit and trifoliate orange produced in 1907 by Swingle and released by the United States Department of Agriculture in 1974. Since then, it has been used successfully as a rootstock in many countries. Most varieties produce very satisfactory yields of good to excellent quality fruit on Swingle citrumelo. Fruits produced on Swingle citrumelo are similar to sour orange, Carrizo, and Troyer citranges in terms of fruit quality. The trees have good cold-hardiness only slightly inferior to that of trees on trifoliate oranges. Scions on Swingle citrumelo are very tolerant to CTV, blight, and root rot as well as being resistant to citrus nematodes. They also have good exocortis and xyloporosis tolerance. The trees grow well on most soils and are reportedly especially tolerant of waterlogged conditions. In contrast, Swingle citrumelo is an unacceptable choice in heavy clay, calcareous soils, and high pH soils. Swingle is sensitive to high pH soils and is unsuitable for highly calcareous soils. Therefore, it is not so popular in many Mediterranean countries. Tolerant to environmental and soil conditions, Swingle citrumelo is sensitive to high chloride levels in soil and irrigation water but is more salt tolerant than other trifoliate hybrids such as Carrizo and Troyer citranges. Swingle has moderate drought tolerance [91].

Citranges are known as hybrids of sweet orange and trifoliate orange. The main purpose of the citrange development in Florida was to combine good traits of sweet orange with the cold-hardiness of the trifoliate orange in order to create cold hardy scions. Although unsuccessful, a most significant source of new rootstocks was produced instead. There are several named selections, the more important of which are Carrizo and Troyer citranges.

These are hybrids of Washington navel orange and *Poncirus trifoliata*. The original crosses were made in the early 1900s by the United States Department of Agriculture with the intention of producing cold-tolerant scion varieties. They were later identified as being suitable for use as rootstocks. Fruit quality of the scion on Carrizo and Troyer is excellent. Trees on both grow moderately vigorous on a range of soil types but have poor salt tolerance and are sensitive to calcareous soils and exocortis virus. They have intermediate frost tolerance but are less cold-hardy than those on Cleopatra mandarin and trifoliate orange. In Turkey, Troyer citrange and trifoliate orange show superior fruit quality for Satsuma mandarins, and they are the mainly used rootstock in Aegean region of Turkey. However, using Carrizo citrange as a rootstock to especially mandarins and oranges is more common due to high soil pH levels in the Mediterranean region of Turkey. In general, the performance of Carrizo citrange is slightly better than Troyer in cancerous soils of Çukurova. However, sour orange is the main rootstock in Turkey. Ninety percent of citrus varieties in Turkey are grafted on the sour orange rootstock.

Benton citrange is a hybrid of Ruby Blood orange and trifoliate orange. It was bred in the late 1940s by the Department of Agriculture, New South Wales, Australia. The seed was first released to the industry in 1984. There are some commercial plantings using this rootstock which were established in 1990, but poor seed production in the seed source trees has been an impediment to its widespread usage. Because of its erratic performance under orange and mandarin scions, it is only recommended for Eureka lemons in Australia and only replant situations in Queensland. In Florida, Benton citrange is recommended for small-scale commercial trials with oranges and grapefruit. Trees on Benton rootstock are reported to be moderately cold tolerant and higher yielding.

C-35 citrange was bred by the University of California and released in 1987 and is a hybrid obtained by crossing Ruby Blood orange × Webber-Fawcett trifoliate. C-35 is tolerant to *Phytophthora* and CTV and resistant to citrus nematodes. Frost tolerance is good as or slightly better than Carrizo. Trees grow 25% smaller than Carrizo, making C-35 a candidate for closer spacing plantings. Trees grown in sandy, loam, and clay soils are satisfactory, but they are more sensitive to calcareous soils than Carrizo. C-32 has the same parentage as C-35 citrange and is a hybrid between Ruby orange and Webber-Fawcett trifoliate. Its very low seed production makes this citrange's seedling propagation difficult in order to use as rootstock [91].

Several rootstock breeding programs have been carried out by different leading countries in citrus industry in order to handle increasing problematic issues by abiotic and biotic stress factors.

Forner et al. [62] reported two new rootstocks released in Spain. Forner-Alcaide 5 (F-A 5) and Forner-Alcaide 13 (F-A 13) are two interspecific hybrids obtained through traditional hybridization by a senior author in a program for breeding citrus rootstocks at the IVIA in Moncada (Valencia), Spain. The researchers aimed to obtain new rootstocks tolerant to CTV, salinity, and lime-induced chlorosis and resistant to Phytophthora. They reported the resistance of F-A 5 and F-A 13 to CTV. In addition, F-A 5 was found to be more tolerant to lime-induced chlorosis than Carrrizo citrange, whereas F-A 13 is less tolerant [92]. Besides, both rootstocks have good tolerance to salinity and an excellent tolerance to flooding, as reported. Gonzalez-Mas et al. [53] conducted a rootstock field study in order to investigate rootstock effects on leaf photosynthesis in "Navelina" trees grown in calcareous soil. Authors have used seven new citrus rootstocks with Carrizo citrange obtained by J. Forner at the IVIA: F-A 5, F-A 13, F-A 418, F-A 517, 030116 (Cleopatra mandarin × *P. trifoliata*), 020324 (Troyer citrange × Cleopatra mandarin), and 230164 (*C. volkameriana* Ten. and Pasq. × *P. trifoliata*). Trees grafted on F-A 5 performed best under these calcareous soil conditions, whereas those on Carrizo citrange were poorly adapted, regarding the parameters investigated.

Bowman and Rouse [63] reported the release of US-812 citrus rootstock in May 2001 by the Agricultural Research Service of the USDA and is the result of a cross between Sunki mandarin and Benecke trifoliate. The rootstock was found to be highly productive of good quality fruit with a moderate vigor (standard medium tree size) as it was reported. The US-812 shows tolerance or resistance to CTV and citrus blight. It was reported that US-812 has good soil adaptability and disease resistance. Valencia orange grafted on US-812 performed well under high pH conditions in calcareous soils. Bowman [93] also introduced US-802 and US-897 and reported high productive per tree size, good soil adaptability and disease resistance, tolerance of Diaprepes and Phytophthora complex, large contrast in vigor, and tree size as the forthcoming features of the rootstocks. Besides, US-942 rootstock is very highly productive with good fruit quality and soil adaptability.

According to the report of Federici et al. [64], three new citrus rootstocks were released in 2009 by the University of California, Riverside. These three rootstocks were named as "Bitters," "Carpenter," and "Furr" trifoliate hybrids, tested as C22, C54, and C57, respectively, and obtained by sexual hybridization of Sunki mandarin × Swingle trifoliate orange. The main character of these three hybrids is they all show good tolerance to citrus tristeza virus.

As reported by Federici et al. [64], "Bitters" showed good tolerance to freezing. It is tolerant to CTV, moderately tolerant to *Phytophthora parasitica*, not very tolerant of citrus nematode, and very tolerant of calcareous soil. "Carpenter" showed moderate tolerance to freezing. It is tolerant to CTV, moderately tolerant to *P. parasitica*, very tolerant of citrus nematode, and moderately tolerant of calcareous soil. "Furr" also showed good tolerance to freezing, and it is tolerant to CTV, very tolerant to *P. parasitica*, very tolerant of citrus nematode, and moderately tolerant of calcareous soil [64].

5. BREEDING TECHNIQUES FOR TOLERANCE TO ABIOTIC STRESS: TRADITIONAL AND BIOTECHNOLOGICAL APPROACHES

Nearly all commercial citrus in the world are grown as grafted trees, with the scion cultivar budded on a selected rootstock cultivar. A good scion and rootstock combination supports the development of trees that bear large quantities of high-quality fruit. However, many available rootstocks are inadequate to meet the emerging needs and challenges. A large proportion of the problems faced by the citrus industry could be overcome by the use of improved rootstocks [93, 94]. However, developing an improved rootstock is a long-term approach because of the several difficulties, mainly the complexity of citrus biology. Typically, it takes at least 15 years from the begging of a cross-hybridization program until a new selected rootstock is released to the industry for a commercial use.

5.1. Traditional Breeding

Although somatic hybridizations via *in vitro* culture methods and genetic transformation via the regeneration process of plant tissues have opened new enthusiastic prospects for citrus genetic improvement, classical breeding techniques still remain important for citrus breeding. Genetic variations have a great importance in terms of plant breeding. These variations can be obtained either spontaneously or artificially by mutations and sexual hybridizations in order to achieve specific breeding objectives. Planned or unplanned sexual hybridizations have been responsible for the evolution of the new genotypes for using either rootstock or scion. Many intergenic hybrids were produced by controlled pollination. For example, citranges (*C. sinensis* × *P. trifoliate*), citrumelos (*C. paradisi* × *P. trifoliata*), citremons (*C. lemon* × *P. trifoliate*), citradia (*C. aurantium* × *P. trifoliate*), citrumquat (*Fortunella* spp. × *P. trifoliate*), and Eremoradia (*Eremocitrus glauca* × *C. aurantium*).

Citrus flowers usually bloom in the spring in one great flush, except acid limes and lemons which are noted for flowering throughout the year in cold subtropical climates. In tropics, flowering may occur more than once throughout the year. In addition, drought and excessive fertilizing may induce flowering. Citrus flowers are mostly hermaphrodite and release pollen when the stigma is receptive. However, there are some exceptions such as staminate and pistillate flowers occurring in lemons Satsumas, Shamouti, and sour orange. Besides, late harvest of the fruits instead of optimum harvest time and fertilization deficiencies can increase the ratio of staminate and pistillate flowers on trees. Besides, W. Navels are known to have the ability to set parthenocarpic fruits due to their pollenless flowers. The flowers never close; the petals merely shed a few days later. The stigma becomes receptive just before the bud breaks open, but the stamens usually do not release pollen until several hours later, after the flower is fully open. This should be considered in terms of collecting flowers for pollens to be used as male parents. Most pollination in citrus is done by insects except for varieties showing parthenocarpy that no pollination is required for fruit development.

Many citrus cultivars are known to be self-incompatible and, in some cases, cross-incompatible. With such cultivars, an appropriate pollen supply and pollinating agents is needed. Pollination requirements vary among the species and cultivars. For example, open pollinated flowers of grapefruit result in significant increases in both fruit number and seed numbers. When lemons are protected from insect visitations, a set of fruits decreases. Pummelos are known to be self-incompatible as well as Clementines, Lee, Page, Nova, and Robinson. In contrast, no pollination problems have been observed in citron, kumquat, Meyer lemon, and trifoliate orange, but there have been problems of seed set in "Morton" end "Troyer" citrange [95].

We consider the apomixes as one of the major problems in citrus rootstock breeding. Nucellar embryony is the most unusual feature that exists in the reproductive biology of citrus. This mechanism limits crossing and selfing in many varieties. Most of the genotypes that can play important roles as female parents in traditional crossing studies are highly apomictic (for example, see Table 3, unpublished data recorded at the Çukurova University, Faculty of Agriculture, Department of Horticulture, Citrus Germplasm Orchards). Hence, citrus breeding is limited by nucellar embryony of most diploid genotypes [96, 97]. The nucellar tissue which surrounds the megagametophyte can produce additional embryos (polyembryony) which are genetically identical to the parent plant. In contrast, zygotic seedlings are sexually produced and inherit genetic material from both parents. Zygotic and nucellar embryos can occur in the same seed. Not all citrus species exhibit the characteristic of polyembryony, a some produce only zygotic embryos (pummelo, citron, Clementine,

Temple, and Persian lime, for example). Others produce only nucellar embryos. Many citron and lime varieties produce a significant percentage of zygotic seedlings but oranges, grapefruit, and many mandarins usually have a low percentage.

Table 3. Polyembryony ratio of some potential genotypes in CU Citrus Germplasm Orchards to be used in breeding studies.

Genotype	Polyembryony (%)
Tuzcu 01-21 sour orange	60
Carrizo citrange	92
Troyer citrange	71
Swingle citrumelo	50
Local trifoliate (*P.trifoliata*)	80
Cleopatra mandarin	95
Volkameriana	60
Rough lemon	56
Florida Rough lemon	93
Sunki mandarin	50
Yuzu	60
Taiwanica	50
King mandarin	81
C. myrtifolia	85
C. obovidea	76

In addition, sexual hybridization faces some constrains in citrus, due to high heterozygosity, long juvenility, and polyembryony of most citrus cultivars. Moreover, it is difficult to identify sexual hybrid embryos in their early stage. In this case, using trifoliate oranges, which are valuable rootstocks due to their characteristics such as cold-hardiness and resistance to root rot, CTV, and nematodes, gains another importance for citrus rootstock breeding against polyembryony. Since the trifoliate character is dominant, progenies exhibiting the trifoliate phenotype of the pollen parent can be considered as putative hybrids. So using trifoliate trait as a morphological marker is useful for early separation and characterization in citrus rootstock breeding studies. In contrast, progenies obtained by crossing combinations using polyembryonic genotypes as females have to be identified via molecular markers systems (RAPD and SSR) in order to speed up the separation of zygotic hybrids from nucellar seedlings.

Controlled cross-pollination in citrus is mainly performed for combing desirable traits from different genotypes or species and inducing heterosis. Based on this method, many hybrid rootstocks between citrus and *Poncirus* have been developed (see Section 4).

Controlled pollination is relatively easy in citrus. Seed parent and pollen parent flowers should be protected against contamination. Emasculation is generally easy and less effortless at the flowers that are nearly ready to open. Emasculation is accomplished by gently separating the petals, pulling off the anthers while avoiding contact with stigma. Pollination should be carried out immediately after emasculation. A special storage of the pollen is seldom necessary while crossing within the genus *Citrus*. The genera *Poncirus* and *Fortunella* can be crossed with *Citrus*. Trifoliate orange naturally bloom earlier than citrus, so the pollens must be stored until the flowering time of Citrus. Pollens should be collected from unopened flowers from the branches of trees. After a waiting period of 24 h in the room temperature, a high quantity of pollen grains can be collected from anthers. Calcium chloride can be used as a drying agent just before storage of the pollen grains in a cold condition. On the contrary, *Fortunella* bloom much later than citrus in many areas. Figure 5 presents a traditional breeding procedure in citrus at Çukurova University, Citrus Germplasm Orchards.

Figure 5. Traditional Cross-Hybridization In Citrus. (A) A Large Unopened Bud, (B) Emasculation, (C) Pollination Of The Emasculated Flower, (D) Cotton Pad Wrapping Around The Twig, (E) Bagged Twig, (F) General View Of The Seed Parent After Crossing.

Fruit breeding, especially using classical breeding methods, is a difficult work taking a lot of time. In terms of citrus, chance seedlings were the main source for the cultivars, and spontaneous mutations on branches were used to select new cultivars. Current breeding projects in the present day is crossing superior selections and inducing mutations for seedlessness as well as crossing at different ploidy levels for seedless triploids.

5.2. Current Biotechnologies Applied in Rootstock Breeding

Genetic improvement of citrus through conventional breeding is limited by their genetic and reproductive characteristics. Citrus species have a complex reproductive biology, with many cases of cross- and self-incompatibility, apomixis, and high heterozygosity, and most of them have very long juvenile periods. Most species are highly heterozygous and produce progeny segregating widely many characters when crosses are made. In addition, juvenile periods are often extensive and most significantly, the presence of adventitious embryos in the nucellus of developing ovules of most citrus types greatly inhibits hybrid production [97, 98].

Plant somatic hybridization via protoplast fusion has become an important tool in plant improvement, allowing researchers to combine somatic cells (whole or partial) from different cultivars, species, or genera resulting in novel genetic combinations including symmetric allotetraploid somatic hybrids, asymmetric somatic hybrids, or somatic cybrids [99].

Briefly, the development of hybrid plants through the fusion of somatic protoplast derived from different sources of two different plant cultivars, species, and genera is called somatic hybridization. The technique of somatic hybridization involves the following steps: (1) isolation of protoplasts, (2) fusion of the protoplasts obtained from desired genotype, (3) culturing the hybrid cells, and (4) regeneration of hybrid plant. Mechanical or enzymatic methods can be used for the separation of protoplasts from plant tissue. However, the mechanical method is a laborious process that has some disadvantages such as low yield of protoplast and low protoplast viability. A plant cell consists of cell wall which has to be degraded if the protoplasts of the cell have to be manipulated as required. For this purpose, the plant cell is treated with enzymes, such as pectinase, macerozyme, cellulase, etc., that hydrolyze the plant cell wall. Since protoplasts are present in every plant cell, it can be theoretically isolated from all parts of the plant. But most successful isolations are made possible from the leaf of the plants.

Once purified protoplasts have been obtained from two different plant or tissue sources, various treatments can be given to induce them to fuse together. Generally, chemical agents or electrical manipulation is necessary to induce membrane instability that leads to protoplast fusion.

Polyethylene glycol (PEG) is used most frequently in conjunction with alkaline pH and high calcium concentrations. There are a number of steps in the fusion of plant protoplasts using PEG as a chemical facilitator. Another type of cell fusion that has emerged in recent years involves the manipulation of cell membranes by electrical currents. This process involves passing low-voltage electric pulses in a solution of protoplasts to be fused so that they line up for fusion. The protoplasts can be fused by subjecting it to brief exposure to high-voltage electric current which leads to alteration of membrane so that the adjacent protoplasts fuse. Electrofusion of plant protoplasts is often preferred over PEG fusion because it does not employ reagents that are toxic to the cells being fused. As with all other procedures, the conditions for electrofusion must be optimized for specific cell types to achieve maximum effectiveness. Typically, a yield of 20% or greater fusion products can be obtained by electrofusion of protoplasts compared to less than 1% fusion products with PEG [100].

Somatic cell fusion could overcome sexual incompatibility and long juvenility and may play a potential role in citrus genetic improvement, including producing directly or indirectly superior varieties, improving citrus scion and rootstock, or creating allopolyploids for triploid breeding [101, 102]. As Grosser and Gmitter [103] reported, this technique can facilitate conventional breeding, gene transfer, and cultivar development by bypassing some problems associated with the conventional sexual hybridization including sexual incompatibility, nucellar embryogenesis, and male or female sterility. Conversely, somatic hybridization is very promising for citrus rootstock breeding for combining genotypes having different tolerance to abiotic stress [104]. Citrus rootstock differs in terms of tolerance/resistance to abiotic stress conditions. Incompatibility between some genotypes that have high level of tolerance to abiotic stress conditions limits the usage of traditional hybridization [105]. Oigawara et al. [106] reported the first intergeneric citrus hybrids obtained by combining embryonic callus of sweet orange and *Poncirus trifoliate* leaves via protoplast fusion. Grosser et al. [107] indicated that the regeneration of more than 300 plants obtained by protoplast fusion of Hamlin sweet orange and Flying Dragon trifoliate. The regenerated plants were determined as tetraploids. Kobayashi and Ohgawara [108] recovered tetraploid somatic hybrids by fusing the protoplast obtained from the embryonic callus of Trovita orange and leaf mesophyll protoplasts of Troyer citrange. Grosser et al. [109] reported tetraploid somatic hybrids obtained by fusing the protoplast via PEG method. They used several manipulations such as Cleopatra mandarin (*Citrus reshni*) + trifoliate orange (*Poncirus trifoliata* (L) Raf.), Acidless orange (*Citrus sinensis* (L) Osb.) + trifoliate orange (*Poncirus trifoliata* (L) Raf.), sour orange (*Citrus aurantium* L) + Flying Dragon trifoliate (*Poncirus trifoliata*), sour orange (*Citrus aurantium* L) + Rangpur lime (*Citrus limonia* Osb.), and Milam lemon + Sun Chu Sha mandarin (*Citrus reticulata* Blanco). Tetraploid plants were identified and propagated for further rootstock experiments. Ollitrault et al. [104] had reported the first intergeneric somatic hybrid obtained from protoplast fusion between *Citrus reticulata* + *Fortunella japonica* in France. The authors have regenerated approximately 100 plantlets by several manipulations (*C. reticulata* + *C. sinensis*, *C. reticulata* + *C. paradisi*, *C. reticulata* + *C. limon*, *C. reticulata* + *C. aurantifolia*, *C. reticulata* + *Poncirus trifoliata*, and *Citrus aurantium* + *Eremocitrus glauca*) and reported the possible use of these population as parental germplasm for both scion and rootstock breeding programs in citrus. Grosser et al. [99] indicated that the somatic hybrids obtained by *Citrus* + *Severinia* and *Citrus* + *Fortunella crassifolia* had lower performance as rootstocks, whereas promising performance was recorded from the scion grafted on somatic hybrids obtained from the manipulations of Acidless orange + *Atalantia ceylanica* and Nova mandarin + *Citropsis gilletiana*. Also, the researchers reported the dwarfing effects of somatic hybrids obtained by fusing the protoplast of sour orange + Flaying Dragon and Cleopatra mandarin + Flying Dragon. Ollitrault et al. [16] had selected 11 allotetraploid somatic hybrids by using flow cytometry and molecular markers and propagated them for rootstock trials in order to investigate their tolerance to abiotic and biotic stress. Mourão Filho et al. [110] reported the root rot tolerance of the somatic hybrids ("Cleopatra" mandarin + "Volkamer" lemon, "Cleopatra" mandarin + sour orange, "Caipira" sweet orange + "Volkamer" lemon, and "Caipira" sweet orange + "Rangpur" lime). Somatic hybrid combinations involving sour orange or *Fortunella obovata* as one of the progenitors were intolerant to CTV. They suggested future filed evaluations with somatic hybrids, especially those with tolerance to CTV.

In addition to these findings, tetraploid rootstocks usually have a built-in tree-size control mechanism due to some unknown physiological reaction with the diploid scion. Mourão Filho et al. [110] indicated that plants budded on tetraploid rootstocks are generally smaller, which could lead to reduced harvest costs and greater production efficiency. In Florida, more than 70 somatic hybrids that can potentially be used as rootstocks have already entered into commercial field trials. Preliminary results from these trials have shown that somatic hybrid rootstocks can produce adequate yields of high-quality sweet oranges (*Citrus sinensis* L. Osbeck) on small trees [111]. Ollitrault et al. [112] reported an intergeneric somatic hybrid between Willow leaf mandarin and Pomeroy trifoliate named as "Flhorag1."

Dambier et al. [44] reported the agronomic evaluation of the Flhorag1 in Morocco. Valencia orange trees on Flhorag1 displayed the lowest growth followed by Carrizo citrange and Volkamer lemon in an agreement regarding tetraploid rootstocks controlling the tree size [109]. Flhorag1 also proved highly tolerant to iron deficiency (unpublished data of Çukurova University obtained within the framework of the INCO "CIBEWU" project).

Genetic transformation is also an attractive alternative technique for citrus genetic improvement. Almeida et al. reported that genetic transformation in Citrus has been obtained mainly from juvenile material such as embryogenic cells, epicotyl segments from *in vitro* germinated seedlings, and internodal segments from plants cultivated in the greenhouse due to a higher morphogenic ability compared to that of mature tissues [113, 114, 115, 116]. Peña et al. [18] concluded that the transformation efficiencies are generally low, and protocols are dependent on species, or even cultivar dependent. One of the limitations within this technology is low plant regeneration frequencies especially for many of the economically important citrus species [117, 118].

Another big area of biotechnology is DNA marker technology, derived from research in molecular genetics and genomics, which offers great promise for plant breeding. Owing to genetic linkage, DNA markers can be used to detect the presence of allelic variation in the genes underlying these traits. By using DNA markers to assist in plant breeding, efficiency and precision could be greatly increased. The use of DNA markers in plant breeding is called marker-assisted selection (MAS) and is a component of the new discipline of "molecular breeding" [119].

Genomic research in recent years led to the development of screening tools via marker-assisted selection, which enables much more efficient selection of superior recombinants improved for multiple traits from conventional breeding efforts. MAS can increase the efficiency of citrus breeding and may speed the release of new cultivars. In this section, the possibilities of using MAS method for early selection in citrus rootstock breeding programs will be discussed.

MAS can be very useful to efficiently select for traits that are difficult or expensive to measure, exhibit low heritability, and are expressed late in development. However, it is usually essential to confirm at certain points in the breeding process that the selected individuals or their progeny do in fact express the desired phenotype or trait. Marker types can be classified as morphological, biochemical, cytological, and DNA based (molecular). The successful application of MAS relies on the tight association between the marker and the major gene or quantitative trait locus (QTL) analysis responsible for the trait [120].

Carillo et al. [121] reported that many studies have focused on mapping QTLs for salt tolerance-related traits in rice because of its requirement for irrigation for maximum yield, its sensitivity to salinity, and its relatively small genome. Gmitter et al. [122] reported that a localized genetic linkage map of the region surrounding the citrus tristeza virus resistance gene was developed from *P. trifoliate.* The authors indicated that the identification of markers tightly linked to CTV will enable citrus breeders to identify plants likely to be CTV resistant by indirect, marker-assisted selection, rather than by labor-intensive direct challenge with the pathogen. For early selection in rootstock breeding program in citrus, Xu et al. [123] suggested that the feasible application of MAS in citrus rootstock breeding for citrus nematode resistance needs at least two genetic markers, each corresponding to related locus, in order to pyramiding the multi-genes associated or cofunctioned in controlling the citrus nematode resistance.

6. CONCLUSIONS

Citrus species are the most produced fruit crops and one of the most imported fruit groups subjected in both domestic and export markets. Citrus fruits can be produced in the tropical, semi-tropical, and subtropical climates with such a great market value. Factors limiting citrus growth in these climates significantly vary. Epidemic levels of disease and pest regarding the ecological conditions (temperature, relative humidity, and solar radiation) also remarkably differ. On the contrary, the existence of common abiotic stresses such as drought, salinity, and iron chlorosis in citrus-producing countries differs depending on the countries and the production areas of the countries. Rootstock is used for a successful citrus production against the abiotic stresses discussed.

In addition, searching for dwarfing rootstock, which enables high-density planting, has gained importance in terms of increasing the fruit yield per area, fruit quality, use of mechanical pruning, harvesting and decreasing pest, and disease management in citriculture. With this knowledge, we can say that there is no ideal rootstock which can combine all these traits. Hence, countries with different ecological conditions have to manage individual rootstock programs. In contrast, abiotic and biotic factors as well as the producer demands change depending on market conditions and new trends. Thus, breeding studies have to be persistent and managed by the current production situation and future demands.

In several breeding programs, existing rootstocks or genotypes that can be used as rootstocks in citrus production have been screened for abiotic stress conditions, and their current tolerance/resistance levels had been reported. These rootstocks currently have been used in citriculture depending on the ecological conditions of the producer countries. In addition, traditional breeding studies were carried out with these genotypes in order to combine their different tolerance/resistance characteristics in one genotype. Several successful rootstocks have been obtained so far by breeding programs. But traditional breeding takes at least 15 years from the begging of a cross-hybridization program until a new selected rootstock is released to the industry for a commercial use and limited by the complex reproductive biology of citrus. Therefore, rootstock breeders have begun to benefit using biotechnological methods in citrus breeding such as somatic hybridization, genetic transformation, and marker-assisted selection in recent years.

Genetic capacities of the parents or genotypes have great importance in breeding studies whichever method (traditional or biotechnological) is used, since the aim of breeding is to expose or combine the existing genetic capacity of parents. Therefore, screening both *ex situ* and *in situ* germplasm in the world regarding problems of citrus producer is necessary. Using promising genotypes obtained in screening studies carried out in germplasms in addition to current citrus rootstock will result in an increase in breeding programs. As mentioned before, stress conditions are not stable and differ regarding the ecological conditions. Therefore, the sustainability of germplasm is necessary as long as human beings and citriculture exist.

Practical experience shows that abiotic stresses occur at high or low intensity in about all citricultural growing areas around the world. In this study, we have tried to explain the current situation of using citrus rootstocks for tolerance to abiotic stresses and breeding studies resulted and ongoing against abiotic stress in citrus production.

REFERENCES

1. FAO. Food and Agriculture Organization. The database of annual production. FAOSTAT. Statistical database. 2015. Available from: http://faostat.fao.org [accessed 15 August 2015].
2. Iglesias DJ, Cercós M, Colmenero-Flores JM, Naranjo MA, Ríos G, Carrera E, Ruiz-Rivero O, Lliso I, Morillon R, Tadeo FR, Talon M. Physiology of citrus fruiting. Braz. J. Plant Physiol. 2007; 19(4): 333–362.
3. Ladaniya MS. Citrus Fruit: Biology, Technology and Evaluation. 1st ed. San Diego CA: Elsevier Academic Press; 2008. 558 p.
4. Tuzcu Ö, Yeşiloğlu T, Yildirim B. Citrus Rootstocks in Mediterranean and Some Suggestions for Their Future. Book of Abstracts of the 7th International Congress of Citrus Nurserymen. Book of Abstracts of the 10th International Citrus Congress; 17 – 21 September 2005; Cairo, Egypt, pp. 82–84.
5. Davies FS, Albrigo LG. Citrus Crop Protection Science in Horticulture, No: 2, CAB International, Wallington, 1994. p. 254.

6. Lawrence FP, Bridges D. Rootstocks for Citrus in Florida. Florida Cooperative Extension Service, Institute of Food and agricultural Sciences, University of Florida, Gainesville, 1974. p. 13.
7. Rouphael Y, Cardarelli M, Rea E, Colla G. Improving melon and cucumber photosynthetic activity, mineral composition, and growth performance under salinity stress by grafting on to Cucurbita hybrid rootstocks. Photosynthetica. 2012; 50: 180–188.
8. Levy Y, Syvertsen JP. Irrigation water quality and salinity effects in citrus trees. Hort. Rev. Amer. Soc. Hort. Sci. 2004; 30: 37–82.
9. Garcia-Sanchez F, Syvertsen JP. Salinity tolerance of Cleopatra mandarin and Carrizo citrange rootstock seedlings is affected by CO2 enrichment during growth. J. Am. Soc. Hortic. Sci. 2006; 131: 24–31.
10. Marschner H. Mineral Nutrition of Higher Plants. 2nd ed. London: Academic Press; 1995. 889p.
11. Pestana M, De Varennes A, Abadía J, Faria EA. Differential tolerance to iron deficiency of rootstocks grown in nutrient solution. Sci Hortic. 2005; 104: 25–36.
12. Cimen B, Yesiloglu T, Incesu M, Yilmaz B. Growth and photosynthetic response of young 'Navelina' trees budded on to eight citrus rootstocks in response to iron deficiency. New Zeal. J. Crop Hort. Sci. 2014; 42: 170–182.
13. Garcia-Sanchez F, Syvertsen JP, Gimeno V, Botia P, Perez-Perez JG. Responses to flooding and drought stress by two citrus rootstocks seedlings with different water-use efficiency. Physiol. Plantarum. 2007; 130: 532–542.
14. Castle WS, Gmitter FG. Rootstock and scion selection. In: Timmer LW, Duncan LW, editors. Citrus Health Management. Minnesota: The American Phytopathological Society; 1999. pp. 21–34.
15. Castle WS. Citrus Rootstocks In: Rom RC, Carlson RF, editors. Rootstocks for Fruit Crops. New York: John Wiley and Sons; 1987. pp. 361–399.
16. Ollitrault P, Froelicher Y, Dambier D, Seker M. Rootstock breeding by somatic hybridisation for the Mediterranean citrus industry. Acta Hortic. 2000; 535: 157–162.
17. Jacobsen T, Adams RM. Salt and silt in ancient Mesopotamian agriculture. Science. 1958; 128: 1251–1258.
18. Peña L, Cervera M, Fagoaga C, Romero J, Juarez J, Pina JA, Navarro L. Citrus. In: Pua EC, Davey MR, editors. Transgenic crops V. Biotechnology in Agriculture and Forestry, vol. 60. Heidelberg: Springer; 2007 pp. 35–50.
19. Boyer JS. Plant productivity and environment. Science. 1982; 218: 443–448.
20. Munns R. Comparative physiology of salt and water stress. Plant Cell Environ. 2002; 25: 239–250.
21. Sudhir P, Murthy SDS. Effects of salt stress on basic processes of photosynthesis. Photosynthetica. 2004; 42: 481–486.
22. Levitt J. Responses of Plants to Environmental Stresses: Chilling, Freezing, and High Temperature Stresses. Vol. 1. London, New York, Toronto: Academic Press; 1980. p. 698.
23. Munns R, Termaat A. Whole-plant responses to salinity. Aust. J. Plant Physiol. 1986; 13: 143–160.
24. Grattan ST, Grieve CM. Salinity mineral relations in horticultural crops. Sci. Hortic. 1999; 78(1–4): 127–157.
25. Carvajal M, Del Amor FM, Fernandez-Ballester G, Martinez V, Cerda A. 1998. Time course of solute accumulation and water relations in muskmelon plants exposed to salt during different growth stages. Plant Sci. 1998; 38: 103–112.
26. Del Amor FM, Martinez V, Cerda A. Salinity duration and concentration affect fruit yield and quality, and growth and mineral composition of melon plants grown in perlite. Hort. Science. 1999; 34: 1234–1237.
27. Grattan ST, Grieve CM. Salinity mineral relations in horticultural crops. Sci. Hortic. 1999; 78(1–4): 127–157.

28. Caro M, Cruz V, Cuartero J, Estan MT, Bolarin MC. Salinity tolerance of normal-fruited and cherry tomato cultivars. Plant Soil. 1991; 136: 249–255.
29. Oztekin GB, Tuzel Y. Comparative salinity responses among tomato genotypes and rootstocks. Pak. J. Bot. 2011; 43(6): 2665–2672.
30. Storey R, Walker RR. Citrus and salinity. Sci. Hortic. 1999; 78: 39–81.
31. Hasegawa PM, Bressan RA, Zhu JK, Bohnert HJ. Plant cellular and molecular responses to high salinity. Ann. Rev. Plant Physiol. Plant Mol. Biol. 2000; 51: 463–499.
32. Li LY, Stanghellini C. Analysis of the effect of EC and potential transpiration on vegetative growth of tomato. Sci. Hortic. 2001; 89: 9–21.
33. Heuvelink E, Bakker M, Sthangellini C. Salinity effects on fruit yield in vegetable crops: a simulation study. Acta Hort. 2003; 609: 133–140.
34. Flexas J, Diaz-Espejo A, Galmés J, Kaldenhoff R, Medrano H, Ribas-Carbo M. Rapid variations of mesophyll conductance in response to changes in CO2 concentration around leaves. Plant, Cell & Environment. 2007; 30: 1284–1298.
35. Hussain S, Luroa F, Costantinoa G, Ollitrault P, Morillon R. Physiological analysis of salt stress behaviour of citrus species and genera: low chloride accumulation as an indicator of salt tolerance. S. Afr. J. Bot. 2012; 81: 103–112.
36. Almansa MS, Hernandez JA, Jimenez A, Botella MA, Sevilla F. Effect of salt stress on the superoxide dismutase activity in leaves of *Citrus limonum* in different rootstock-scion combinations. Biol. Plant. 2002; 45(4): 545–549.
37. Garcia-Sanchez F, Carvajal M, Cerda A, Martinez V. Response of 'Star Ruby' grapefruit on two rootstocks to NaCl salinity. J. Hortic. Sci. Biotechnol. 2003; 78(6): 859–865.
38. Romero-Aranda R, Moya JL, Tadeo FR, Legaz F, Primo-Millo E, Talon M. Physiological and anatomical disturbances induced by chloride salts in sensitive and tolerant citrus: beneficial and detrimental effects of cations. Plant Cell Environ. 1998; 21: 1243–1253.
39. Nieves, M, Cerda A, Botella M. Salt tolerance of two lemon scions measured by leaf chloride and sodium accumulation. J. Plant Nutr. 1991; 14: 623–636.
40. Yesiloglu T, Tuzcu O, Aka-Kacar Y, Yilmaz B, Uysal-Kamiloglu M, Incesu M. Citrus rootstock breeding for efficient water and nutrient use (CIBEWU, No: 015453). Third Report for CIBEWU Project. Partner 6, Cukurova University. Unpublished project report. 2009.
41. Khoshbakht D, Ramin AA, Baninasab B. Effects of sodium chloride stress on gas exchange, chlorophyll content and nutrient concentrations of nine citrus rootstocks. Photosyntheti̇ca. 2015; 53(2): 241–249.
42. Cimen B, Yesiloglu T, Yilmaz B, Incesu M. Effects of different salinity levels on photosynthetic performances of some citrus rootstocks. Res. J. Agri. Sci. 2013; 6(2): 13–18. Available on http://www.nobel.gen.tr/Makaleler/TABAD-Issue%202-ad7928a81cee4f69873a7141db68a383.pdf.
43. Römheld V. The chlorosis paradox: Fe inactivation as a secondary event in chlorotic leaves of grapevine. J. Plant Nutr. 2000; 23(11–12): 1629-1643.
44. Dambier D, Benyahia H, Pensabene-Bellavia G, Aka Kaçar Y, Froelicher Y, Belfalah Z, Lhou B, Handaji N, Printz B, Morillon R, Yesiloglu T, Navarro L, Ollitrault P. Somatic hybridization for citrus rootstock breeding: an effective tool to solve some important issues of the Mediterranean citrus industry. Plant Cell Rep. 2011; 30(5): 883–900.
45. Mengel K. Iron availability in plant tissues, iron chlorosis on calcareous soils. Plant Soil. 1994; 165: 275–283.
46. Byrne DH, Rouse RE, Sudahono S. Tolerance to citrus rootstocks to lime-induced iron chlorosis. Subtropical Plant Sci. 1995; 47: 7–11.
47. Alvarez-Fernández A, Abadía J. Abadía A. Iron deficiency, fruit yield and fruit quality. In: Barton LL, Abadía J, editors. Iron Nutrition in Plants and Rizospheric Microorganisms. Dordrecht: Springer; 2006. pp. 437–448.

48. Castle WS, Nunnallee J. Screening citrus rootstocks and related selections in soil and solution culture for tolerance to low-iron stress. Hort. Sci. 2009; 44(3): 638–645.
49. Pestana M, Correia PJ, David M, Abadía A, Abadía J, de Varennes A. Response of five citrus rootstocks to iron deficiency. J. Plant Nutr. Soil Sci. 2011; 174: 837–846.
50. Hamze M, Ryan J, Zaabout N. 1986. Screening of citrus rootstocks for lime-induced chlorosis tolerance. J. Plant Nutr. 1986; 9: 459–469.
51. Campbell CW. Rootstocks for the Tahiti lime. Proc. Fla. State Hort. Soc. 1991; 104: 28–30.
52. Sudahono, Byrne DH, Rouse RE. Greenhouse screening of citrus rootstocks for tolerance to bicarbonate-induced iron chlorosis. HortScience. 1994; 29(2): 113–116.
53. Forner-Giner MA, Alcaide A, Primo-Millo E, Forner JB. Performance of Navelina orange on 14 rootstocks in Northern 'Valencia' (Spain). Sci. Horti. 2003; 98: 223–232.
54. González-Mas MC, Llosa MJ, Quijano A, Forner-Giner A. Rootstock effects on leaf photosynthesis in 'Navelina' trees in calcareous soil. 2009; HortScience. 44(2): 280–283.
55. Ferrarezi RS, Bataglia OC, Furlani PR, Schammass E. Iron sources for citrus rootstock development grown on pine bark/vermiculite mixed substrate. Sci. Agric. 2007; 503 64: 520–531.
56. Cimen B, Yesiloglu T, Incesu M, Yilmaz, B. Physiological investigation of tolerance to iron chlorosis of Navelina orange budded on different citrus rootstocks. Acta Hortic. 2015; 1065: 1423–1430.
57. Lawlor DW. 1995. The effect of water deficit on photosynthesis. In: Smirnoff N, editor. Environment and Plant Metabolism, Flexibility and Acclimation London: BIOS Scientific Publisher; 1995. pp. 129–160.
58. Bollivar DW, Beale SI. The chlorophyll biosynthetic enzyme-Mg protoporphyrin IX monomethylester (oxidative) cyclase. Plant Physiol. 1996; 112: 105–114.
59. Terry N. Limiting factors in photosynthesis: iron stress mediated changes in light-harvesting and electron transport capacity and its effects on photosynthesis in vivo. Plant Physiol. 1983; 71: 855–860.
60. Davis T, Jolley V, Walser R, Brown J, Blaylock A. Net photosynthesis of Fe-efficient and Fe-inefficient soybean cultivars grown under varying iron levels. J. Plant Nutr. 1986; 9: 671–681.
61. Chouliaras V, Therios I, Molassiotis A, Patakas A, Diamantidis G. Effect of iron deficiency on gas exchange and catalase and peroxidase activity in citrus. J. Plant Nutr. 2004; 27: 2085–2099.
62. Bertamini M, Nedunchezhian N, Borghı B. Effect of iron deficiency induced changes on photosynthetic pigments, ribulose-1,5-bisphosphate carboxylase, and photosystem activities in field grown grapevine (*Vitis vinifera* L. cv. Pinot noir) leaves. Photosynthetica, 2001; 39(1): 1–160.
63. Forner JB, Forner-Giner MA, Alcaide A. Forner-Alcaide 5 and Forner-Alcaide 13: Two new citrus rootstocks released in Spain. HortScience. 2003; 38(4): 629, 630.
64. Bowman KD, Rouse, RE. US-812 citrus rootsock. HortScience. 2006; 41(3): 832–836.
65. Federici CT, Kupper RS, Roose ML. 'Bitters', 'Carpenter' and 'Furr' Trifoliate Hybrids: Three New Citrus Rootstocks [Internet]. 2009. Available from: https://plantbiology.ucr.edu/faculty/new%20citrus%20rootstocks%202009.pdf. [Accessed: 2015-07-12].
66. Yaacoubi AE, Malagi G, Oukabli A, Hafidi M, Legave JM. Global warming impact on floral phenology of fruit trees species in Mediterranean region. Sci. Hortic. 2014; 180: 243–253.
67. Shukla N, Awasthi RP, Rawat L, Kumar J. Biochemical and physiological responses of rice (*Oryza sativa* L.) as influenced by *Trichoderma harzianum* under drought stress. Plant Physiol. Biochem. 2012; 54: 78–88.
68. Pardales JR, Kono Y, Yamauchi A. Response of the different root system components of sorghum to incidence of waterlogging. Environ. Exp. Bot. 1991; 31: 107–115.

69. Cruz RT, Jordan WR, Drew MC. Structural changes and associated reduction of hydraulic conductance in roots of Sorghum hicolor L. following exposure to water deficit. Plant Physiology 1992; 99: 203–212.
70. Carmo-Silva AE, Keys AJ, Beale MH, Ward JL, Baker JM, Hawkins ND, Arrabaca MC, Parry MAJ. Drought stress increases the production of 5-hydroxynorvaline in two C-4 grasses. Phytochemistry. 2009; 70: 664–671.
71. Levitt J. Response of plants to environmental stress. Responses of Plants to Environmental Stresses. New York: Academic Press; 1972. p. 698.
72. Ogbaga CC, Stepien P, Johnson GN. Sorghum (Sorghum bicolor) varieties adopt strongly contrasting strategies in response to drought. Physiol. Plant. 2014; 152: 389–401.
73. Molinari HBC, Marur CJ, Filho JCB, Kaboyashi AK, Pileggi M, Junior RPL, Pereira LFP, Vieira LGE. Osmotic adjustment in transgenic citrus rootstock Carrizo citrange (*Citrus sinensis* Osb. × *Poncirus trifoliata* L. Raf.) overproducing proline. Plant Sci. 2004; 167: 1375–1381.
74. Rodriguez-Gamir J, Primo-Millo E, Forner JB, Forner-Giner MA. Citrus rootstock responses to water stress. Sci. Hortic. 2010; 126: 95–102.
75. Pedrosoa F, Prudentea DA, Buenoa ACR, Machadoa EC, Ribeiroa RV. Drought tolerance in citrus trees is enhanced by rootstock-dependent changes in root growth and carbohydrate availability. Environ. Exp. Bot. 2014; 101: 26–35.
76. Romero P, Navarro JM, Pérez-Pérez J, García-Sánchez F, Gómez-Gómez A, Porras I, Martinez V, Botía P. Deficit irrigation and rootstock: their effects on water relations, vegetative development, yield, fruit quality and mineral nutrition of Clemenules mandarin. Tree Physiol. 2006; 26: 1537–1548.
77. Magalhães Filho JR, Amaral LR, Machado DFSP, Medina CL, Machado ECM. Water deficit, gas exchange and root growth in 'Valencia' orange tree budded on two rootstocks. Bragantia. 2008; 67: 75–82.
78. Rodríguez-Gamir J, Ancillo G, Aparicio F, Bordas M, Primo-Millo E, Forner-Giner MA. Water-deficittolerance in citrus is mediated by the down-regulation of PIP gene expression in the roots. Plant Soil. 2011; 347: 91–104.
79. Treeby MT, Henrioda RE, Bevington HB, Milne DJ, Storey R. Irrigation management and rootstock effects on navel orange [*Citrus sinensis* (L.) Osbeck] fruit quality. Agr. Water Manag. 2007; 91(1–3): 24–32.
80. Allario T, Brumos J, Colmenero-Flores J, Iglesias DJ, Pina JA, Navarro L, Talon M, Ollitrault P, Morillon R. Tetraploid Rangpur lime rootstock increases drought tolerance via enhanced constitutive root abscisic acid production. Plant, Cell & Environment. 2013; 36: 856–868.
81. Tuzcu O, Anil S, Yesiloglu T, Kafa G, Turgutoglu E. Turkey citrus genetic resources. Mersin: Selim Offset; 2013. p. 914. ISBN 978-605-4672-30-1.
82. Uzun A, Yeşiloğlu T. Genetic diversity in citrus. In: Genetic Diversity in Plats, Caliskan, editors. Rijeka: Intech; 2012. pp. 213–230.
83. Webber HJ. History and development of the itrus industry. In: Reuther W, Webber HJ, Batchelor LD, editors. The Citrus Industry, vol. 1. Berkeley: University of California Press; 1967. pp. 1–39.
84. Swingle WT, Reece PC. The botany of citrus and its wild relatives. In: Reuther W, Webber HJ, Batchelor LD, editors. The Citrus Industry, vol. 1. Berkeley: University of California Press; 1967. pp. 389–390.
85. Kahn TL, Krueger RR, Gumpf DJ, Roose ML, Arpaia ML, Batkin TA, Bash JA, Bier OJ, Clegg MT, Cockerham ST, Coggins CW, Durling D, Elliott G, Mauk PA, Mcguire PE, Orman C, Qualset CO, Roberts PA, Soost RK, Turco J, Van-Gundy SG, Zuckerman B. Citrus genetic resources in California: analysis and recommendations for long-term conservation. GRCP Report No. 22. University of California Division of Agriculture and Natural Resources, Genetic Resources Conservation Program, Davis, CA, USA; 2001. p. 74.
86. Khan IA. 2007. Citrus - Genetics, Breeding and Biotechnology. CAB International, Oxfordshire, UK.

87. Singh B. Establishment of first gene sanctuary in India for Citrus in Garo Hills. New Delhi, India: Concept Publishing; 1981.
88. Soost RK, Cameron JW, Bitters WP. Citrus germplasm collection is widely used. Calif. Agriculture. 1997; 31(9): 38, 39.
89. Navarro L, Juarez J, Pina JA, Ballester JF, Arregui JM. The citrus variety improvement program in Spain after eleven years. Proc. Int. Org. Citrus Virologists. 1988; 1: 400–406.
90. Castle WS, Tucker DPH, Krezdorn AH, Youtsey CO. Rootstocks for Florida citrus. University of Florida, Gainesville. 1993.
91. Hodgson RW. Horticultural varieties of citrus. In: Reuther W, Webber HJ, Batchelor LD, editors. The Citrus Industry, vol. 1. Berkeley: University of California Press; 1967. p. 431–591.
92. Saunt James. Citrus Varieties of the World. Norwich, England: Sinclair International Limited; 2000. p. 68.
93. Forner JB, Alcaide A, Verdejo-Lucas S, Sorribas FJ. New hybrids as citrus rootstocks in Spain. In: Manicom B, Robinson J, du Plessis SF, Joubert P, van Zyl JV, du Preez S, editors. Proceedings of 8th International Citrus Congress. vol. 1. Sun City, South Africa. 1997. pp. 167–170.
94. Bowman KD. New hybrid citrus rootstocks developed by US Department of Agriculture. In: Proceedings of the International Society for Citriculture, IX Congress, 2000. p. 51.
95. Wutscher HK, Hill LL. Performance of 'Hamlin' orange on 16 rootstocks in east-central Florida. HortScience. 1995; 30: 41–43.
96. Soost RK. Citrus Pollination. Calif. Citrog. 1963; 48: 447–452.
97. Aleza P, Juárez J, Hernández M, Ollitrault P, Navarro L. Implementation of extensive citrus triploid breeding programs based on 4x×2x sexual hybridisations. Tree Genetics & Genomes. 2012; 8: 1293–1306.
98. Frost HB, Soost RK. Seed reproduction, development of gametes and embryos. In: Reuther W, Batchelor LD, Webber HB, editors. The citrus industry, vol. 2. Berkeley: University of California Press; 1968. pp. 290–324.
99. Peña L, Cervera M, Juarez J, Navarro A, Pina JA, DuranVila N, Navarro L. Agrobacterium-mediated transformation of sweet orange and regeneration of transgenic plants. Plant Cell Rep. 1995; 14: 616–619.
100. Grosser JW, Calović M, Louzada ES, Davey MR, Anthony P. Protoplast Fusion Technology - Somatic Hybridization and Cybridization. Plant Cell Culture, John Wiley & Sons, Ltd: 2010. pp. 175–198.
101. Compton ME, Saunders JA, Veilleux RE. Use of protoplast for plant improvement. In: Trigiano RN, Gray DJ, editors. Plant Tissue Culture Concepts and Laboratory Exercise. USA: CRC; 2000. pp. 249–261.
102. Guo WW, Deng XX, Yi HL. Somatic hybrids between navel orange and grapefruit for seedless triploid breeding. Euphytica. 2000; 16: 281–285.
103. Guo WW, Deng XX. Wide somatic hybrids of citrus with its related genera and their potential in genetic improvement. Euphytica. 2001; 118: 175–183.
104. Grosser JW, Gmitter FG. Somatic hybridization of citrus with wild relatives for germplasm enhancement and cultivar development. HortiScience. 1990; 25: 147–151.
105. Ollitrault P, Dambier D, Cabasson C, Teisson C, Luro F. Protoplast fusion in citrus. Fruit. 1994; 49: 5, 6.
106. Nito N, Akihama T. Prospect of citrus and related genera for disease resistant rootstocks. In: Proceeding IV Asia-Pasific Conference on Citrus Rehabilitation, February, 1990, Chiang Mai, Thailande, Hongkong, Chine, B. Aubert ed., UNDP-FAO. Regional Project RAS/86/022, pp.39–47.
107. Oigawara I, Kobayashi S, Ohgawara E, Uchimiya H, Ishii S. Somatic hybrid plants obtained by protoplast fusion between *Citrus sinensis* and *Poncirus trifoliata*. Theor. Appl. Gen. 1985; 71: 1–4.

108. Grosser JW, Gmitter FG, Chandler JL. Intergeneric somatic hybrid plants of *Citrus sinensis* cv. Hamlin and *Poncirus tirfoliata* cv. Flaying dragon. Plant Cell Rpt. 1988; 7: 5–8.
109. Kobayashi S, Ohgawara T. Production of somatic hybrid plants through protoplast fusion in citrus. Jour. Agr. Rev. Quart. 1988; 22: 181–188.
110. Grosser JW, Louzada ES, Gmitter FG, Chandler JL. Somatic hybridization of complementary citrus rootstocks: five new hybrids. HortScience. 1994; 29(7): 812–813.
111. Mourão Filho FAA, PIO R, Mendes BMJ, Alves de Azevedo F, Schinor EH, Entelmann FA, Alves ASR, Cantuaris-Aviles T. Evaluation of citrus somatic hybrids for tolerance to phytophthora nicotiana and citrus tristeza virus. Sci. Hortic. 2008; 115: 301–308..
112. Grosser JW, Chandler JL, Ling, Barthe GA. New Somatic Hybrid Rootstock Candidates for Tree-Size Control and High Juice Quality. Proc. Fla. State Hort. Soc. 2011; 124: 131–135.
113. Ollitrault P, Dambier D, Sudahono, Luro F. Somatic hybridization in citrus: some new hybrid and alloplasmic plants. Proc. Int. Soc. Citric. 1996; 2: 907–912.
114. Hidaka T, Omura M, Ugaki M, Tomiyama M, Kato A, Ohshima M, Motoyoshi F. Agrobacterium-mediated transformation and regeneration of Citrus spp. from suspension cells. Jpn. J. Breed. 1990; 40: 199–207.
115. Cervera M, Ortega C, Navarro A, Navarro L, Peña L. Generation of transgenic citrus plants with the tolerance-tosalinity gene HAL2 from yeast. J. Hortic. Sci. Biotechnol. 2000; 75: 26–30.
116. Yang ZN, Ingelbrecht IL, Louzada E, Skaria M, Mirkov TE. Agrobacterium-mediated transformation of the commercially important grapefruit cultivar Rio Red (*Citrus paradisi* Macf.). Plant Cell Rep. 2000; 19: 1203–1211.
117. Pena L, Cervera M, Juarez J, Navarro A, Pina JA, DuranVila N, Navarro L. Agrobacterium-mediated transformation of sweet orange and regeneration of transgenic plants. Plant Cell Rep. 1995; 14: 616–619.
118. Peña L, Pérez RM, Cervera M, Juárez JA, Navarro L. Early events in Agrobacterium-mediated genetic transformation of citrus explants. Ann Bot. 2004; 94(1): 67–74.
119. Donmez D, Simsek O, Izgu T, Aka-Kacar Y, Yalcin-Mendi Y. Genetic transformation in Citrus. Sci. World J. 2013; Article ID 491207.
120. Collard BCY, Mackill DJ. Marker-assisted selection: an approach for precision plant breeding in the twenty-first century. Philos. Trans. R. Soc., B. 2008; 363: 557–572.
121. [120]Pérez-de-Castro AM, Vilanova S, Cañizares J, Pascual L, Blanca JM, Díez MJ, Prohens J, Picó B. Application of Genomic Tools in Plant Breeding. Curr. Genomics. 2012; 13: 179–195.
122. Carillo P, Annunziata MG, Pontecorvo G, Fuggi A, Woodrow P. Salinity Stress and Salt Tolerance, Abiotic Stress in Plants - Mechanisms and Adaptations, Prof. Arun Shanker, editor, 2011. InTech,
123. Gmitter FG, Xiao SY, Huang S, Hu XL, Garnsey SM, Deng Z. A localized linkage map of the citrus tristeza virus resistance gene region. Theor. Appl. Genet. 1996; 92(6): 688–695.
124. Xiang X, Deng Z, Chen C, Gmitter FG, Bowman K. Marker assisted selection in citrus rootstock breeding based on a major gene locus 'Tyr1' controlling citrus nematode resistance. Agric. Sci. China. 2010; 9(4): 557–567.

CHAPTER 15

Verbesina alternifolia Tolerance to the Holoparasite *Cuscuta gronovii* and the Impact of Drought

Bethany Evans and Victoria Borowicz

BEES Section, 4120/School of Biological Sciences, Illinois State University, Normal, IL 61790, USA

ABSTRACT

Holoparasites are nonphotosynthetic plants that acquire all resources from hosts. The holoparasite *Cuscuta gronovii* is native to much of the US with a broad host range including *Verbesina alternifolia*, an understory perennial. Both species grow in moderate to moist soils and occur in habitats that may experience prolonged or episodic drought. We applied the Wise-Abrahamson Limiting Resource Model (LRM) developed for plant-herbivore relations to examine the effects of pattern of drought stress on tolerance of *V. alternifolia* to parasitism by *C. gronovii*. Individual plants were assigned one of six treatments that were combinations of parasite (none or addition of parasite) and drought stress (well-watered, continuously-stressed, or pulse-stressed). After pulse-stressed plants had experienced two wet-dry cycles all plants were harvested. Parasitism strongly reduced both shoot and root mass and well-watered hosts exhibited the greatest decline, indicating reduced tolerance to parasitism when water was readily available. This is consistent with the LRM if parasitism limits photosynthates available to the host. However, parasitism increased allocation to shoot and this effect did not differ between well-watered and drought-stressed plants, indicating equal tolerance. This outcome is in accord with an alternative prediction of the LRM if hosts are not carbon limited. Total pot productivity was reduced by parasitism and drought stress, and this effect was greater for pulse-stressed than for continuously-stressed hosts. We discuss the applicability of the LRM for understanding the effects of drought on tolerance to parasitism.

Keywords: *Cuscuta gronovii*; focal resource; parasitism; pulse stressed; resource limitation; *Verbesina alternifolia*

1. INTRODUCTION

Being non-photosynthetic, holoparasites obtain all needed resources from hosts via root-like haustoria [1]. Haustoria bridge the vascular tissues of parasite and host, enabling the parasite to extract water, minerals, and organic compounds produced or acquired by the host at some cost. Parasites can greatly reduce host growth but the impact is likely to be a product of resource supply for the host. Because water is a key resource affecting the ability of a plant to acquire and transport minerals [2], to fix carbon [3], and distribute photosynthates, water availability is likely to affect plant-parasite interactions.

Plants transport assimilates in the phloem by loading solutes at the source, osmotically raising turgor pressure and unloading solutes at the sink, thereby lowering the turgor pressure [4]. In general, stress reduces source strength by reducing photosynthesis, and sink strength by inhibiting growth which diminishes translocation [5]. Consequently, plants experiencing moderate to severe drought stress may show not only slowed nutrient uptake and transport but also a reduction in photosynthate translocation [6] and thus a change in nutrient availability to a consumer.

Although pattern of drought stress is likely to affect plants and the holoparasites attached to them, literature is lacking on parasitic plant-host responses to water stress. Nonetheless, drought-mediated interactions between herbivores and host plants have been examined extensively [7,8,9,10,11,12]. Because herbivores and parasitic plants possess similar feeding preferences [13] and the interactions between parasitic plants and their hosts resemble herbivore-host interactions [14], theory developed for explaining plant-insect interactions may provide insight into holoparasite-plant interactions.

Tolerance describes the ability of a plant to grow and reproduce despite being under attack [15,16,17]. A tolerant plant can sustain damage or injury or support an enemy without showing a significant reduction in growth. While it is often assumed that plants exhibit lower tolerance in stressful environments [16,18], many examples show greater tolerance to herbivory under stressful conditions [19,20,21]. With the goal of providing a general explanation with clear predictions for the relationship between resources and tolerance to herbivory, Wise and Abrahamson [22] proposed the Limiting Resource Model (LRM) of plant tolerance. In our study we use this model as a framework for examining the effects of drought stress on growth of a host supporting a holoparasite.

In this novel application of the LRM, water is the "focal" (manipulated) resource [22]. If parasitism affects the ability of the host plant to acquire enough water, the impact of the parasite should be small when water availability is high and the host plant should express greater tolerance to the parasite when not drought stressed (Figure 1A). Although shoot parasites may attach to xylem and affect water availability, holoparasites can strongly limit host growth by siphoning off photosynthates transported in the phloem [23,24]. We assume carbon is the resource most affected by the holoparasite, *i.e.*, the "alternate" resource in the LRM. If a host plant does not become carbon limited when water availability is high, the parasite's effect is independent of water and the plant is predicted to be equally tolerant to parasitism at all levels of water (Figure 1B). If carbon becomes the limiting resource when the host plant has abundant water, the parasite's effect depends upon whether it increases or decreases this carbon limitation for the host. If the parasite increases carbon limitation of the host, then the impact of the parasite is relatively larger when water is readily available than when the host is drought-stressed, leading to lower tolerance to the parasite when water availability is high (Figure 1C). Because they are drawing upon the host's photosynthates, holoparasites are unlikely to relieve host carbon limitation and so the last outcome outlined in the LRM (greater tolerance at higher focal resource level due to decreased alternate resource limitation) would not apply.

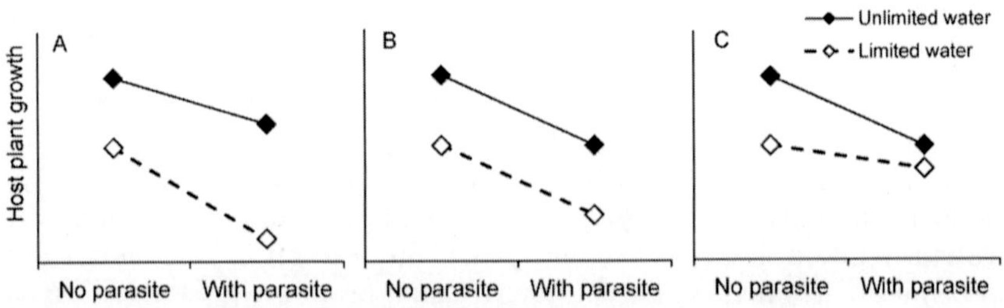

Figure 1. Tolerance slopes describing host plant growth relative to presence of the parasite. (**A**) Greater tolerance to parasite when water is not limiting. (**B**) Equal tolerance to parasite regardless of water stress. There is no interaction of water stress and parasite. (**C**) Reduced tolerance to parasite when water is not limiting.

Our experiment was intended to distinguish among three possible responses to parasitism as a function of water stress: greater tolerance, equal tolerance, or reduced tolerance. In addition to observing the effect on tolerance, we also measured effect of drought stress on total productivity (host mass + parasite mass) to determine whether reduced host growth was balanced by increased parasite growth, and whether any depression in overall productivity was altered by the temporal pattern of drought stress.

Our greenhouse experiment focused on *Verbesina alternifolia* (L.) Britton ex Kearney (wingstem) as host and *Cuscuta gronovii* Willd. Ex Schultes (dodder) as parasite. The holoparasite *C. gronovii* is native to much of the US and has a broad host range that includes *V. alternifolia*, an understory perennial. *Cuscuta gronovii* and *V. alternifolia* occur in areas that may experience prolonged or repeated episodes of drought. More severe and extended periods of drought are predicted consequences of an increase in global temperatures [25]. While continuous water stress is characteristic of many experimental studies, pulsed stress treatments may better mimic many natural situations [26,27,28]. Responses of plants to episodic water limitation and underlying mechanisms are not well studied [29]. To our knowledge, how temporal pattern of drought affects plant-holoparasite relations has not been explored. This stands in contrast to studies of plant-herbivore relations, which have focused on herbivore performance (see [27] for review). In particular, phloem-feeding insects are hypothesized to perform better on pulse-stressed plants as a result of increased concentrations of nitrogen that become available after stress is relieved [27]. We conducted a greenhouse experiment with two objectives: (1) to test the pulse-stress hypothesis with dodder as the consumer (data not presented); and (2) to examine how the combined effects of this vascular feeder and temporal pattern of drought affect growth of the host, a native herbaceous perennial. In this paper we present the results of host growth within the framework of the Limiting Resource Model of plant tolerance.

2. RESULTS AND DISCUSSION

2.1. Host Growth

Drought stress and parasitism significantly reduced the growth of the host roots and shoots, but the effect of the parasite differed among drought treatments (Table 1A) and this effect was more strongly expressed in the shoots (larger standardized canonical coefficient). Parasitism more strongly reduced root and shoot growth in control plants (high water availability) than in plants given either the pulse-stressed or continuously-stressed drought treatments (Table 1B, Figure 2A,B). Among plants that experienced drought, parasitism more strongly reduced root mass of pulse-stressed plants (Table 1B, Figure 2A). The impact of parasitism did not differ between pulse-stressed and continuously-stressed plants in terms of shoot mass (Table 1B, Figure 2B).

Table 1. (**A**) Multivariate analysis of covariance for effect of drought stress (control (=not stressed), pulsed, continuous), presence or absence of parasite (*Cuscuta gronovii*), and their interaction on total *Verbesina alternifolia* host shoot and root mass using the length of the longest leaf as a pretreatment estimate of host variation, and (**B**) contrasts testing whether the effect of parasitism is consistent across water treatments for host shoot and root dry mass.

A	Pillai's trace			Standardized canonical coefficient			
	df	*F*	*P*	Shoot	Root	Shoot 2	Root 2
Leaf length	2,161	18.78	<0.0001	1.844	0.956	-	-
Water	4,324	42.13	<0.0001	2.935	−0.463	−1.520	2.836
Parasite	2,161	352.01	<0.0001	1.494	1.281	-	-
Parasite*Water	4,324	19.84	<0.0001	1.891	0.910	−2.711	2.725
B	***V. alternifolia* shoot mass contrast**			**df**	***F***	***P***	
	Control *vs.* Pulsed			1,162	52.29	<0.0001	
	Control *vs.* Continuous			1,162	69.78	<0.0001	
	Pulsed & Continuous			1,162	1.53	0.2172	
	***V. alternifolia* root mass contrast**			**df**	***F***	***P***	
	Control *vs.* Pulsed			1,162	23.89	<0.0001	
	Control *vs.* Continuous			1,162	61.66	<0.0001	
	Pulsed & Continuous			1,162	9.39	0.0026	

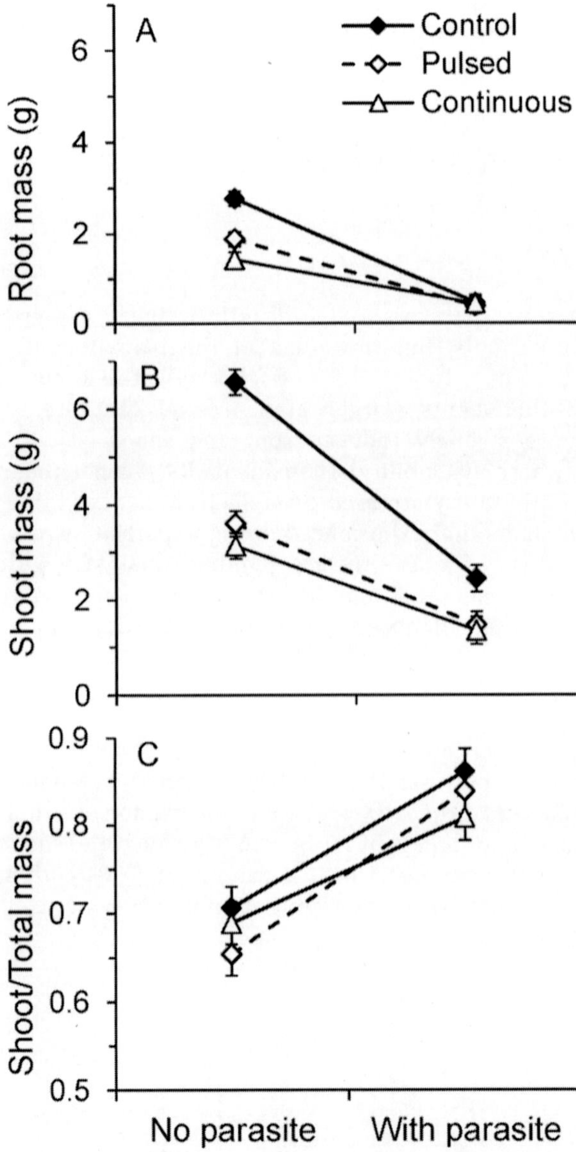

Figure 2. Effect of parasitism on (**A**) root mass, (**B**) shoot mass, and (**C**) the proportion of total mass consisting of shoot in *V. alternifolia* maintained under well-watered (=control), pulse-stressed, or continuously drought-stressed conditions. Values are least squares means (±2 se).

The LRM was originally presented as a flowchart [22] and subsequently as a dichotomous key [30] where outcomes depend upon answers to a series of questions. A pivotal question is whether higher focal resource levels cause the alternate resource to become limiting. We assumed the answer is yes, when water is plentiful photosynthesis limits plant growth. This is based on the observations that *V. alternifolia* grow well in full sun, and that light levels in our greenhouse are less than full sun. Given this assumption, the host is predicted to exhibit lower tolerance to the parasite (greater reduction in growth) when water is readily available, as indicated by a statistical interaction. When expressed as dry shoot mass or dry root mass, our results are in accord with this prediction.

Although our results suggest greater tolerance to the parasite when the plants are drought stressed, the interpretation may be less straightforward. We evaluated the raw data because they fit the assumptions of analysis of variance much better than did transformed data. However, biomass is frequently log transformed due to the allometric scaling relationship of plant growth, which is multiplicative rather than additive [31,32]; proportionally similar effects on large *vs.* small plants yield larger absolute differences for the larger plants. The comparatively large reduction in growth of well-watered plants due to parasitism is partly a product of the vigorous growth of plants in this treatment and consequent large absolute effects.

According to optimal partitioning models, plants respond to variation in the environment by partitioning biomass among various organs to capture nutrients, light, water, and carbon dioxide to maximize growth rate [33]. Well-nourished *Cuscuta*-parasitized hosts exhibit reduced root: shoot ratios because the parasite acts as a strong sink relative to the host root [24]. Therefore, demand by parasites for photosynthates would be predicted to shift host allocation to shoot in order to gather more light. Likewise, water limitation would shift allocation to roots in order to gather more water. If photosynthesis does not limit host growth then the LRM predicts no difference in the effect of parasitism at high *vs.* low resource levels, *i.e.*, these effects of parasitism and water availability should be independent. Parasitism increased ($F_{1,162} = 219.75$, $p < 0.0001$) and drought stress decreased allocation to shoot mass ($F_{2,162} = 5.43$, $p = 0.0052$) and the effect of parasitism depended upon the drought treatment ($F_{2,162} = 3.43$, $p = 0.0348$). Although parasitism and water stress were not independent, this interaction arises from the difference between the two stress treatments. Control plants, which were not drought stressed, did not differ significantly from either drought-stress treatment ($p > 0.05$ for each). However, parasitism shifted allocation to shoots more strongly in pulse-stressed plants, which occasionally experienced relief from water limitation, than in continuously-stressed plants ($F_{1,162} = 6.83$, $p = 0.0098$; Figure 2C). Thus, from the perspective of allocation, our results provide some support for two different predictions. The outcome is consistent with the prediction of no effect of drought stress on tolerance to parasitism when the standard is well-watered controls, but consistent with the prediction of reduced tolerance to parasitism with drought stress when drought stress is intermittent.

In sum, our conclusions regarding reduced tolerance *vs.* no change in tolerance to parasitism as a function of water supply remain equivocal. Growing conditions in the greenhouse did not definitively place well-watered hosts in a light environment that limited growth, and growth response of the hosts can be interpreted as meeting either of two predictions of the LRM when water is the focal resource and parasites limit an alternate resource. However, we find no support for a third potential outcome of the LRM, *i.e.*, reduced tolerance to parasitism when plants are drought stressed (Figure 1A), which is predicted if either parasitism affects the host's ability to acquire water or if parasitism alleviates carbon limitation.

2.2. Pot Productivity

Largest leaf length, a measure of pre-treatment variability among hosts, significantly affected final pot productivity, as measured by parasite + host biomass ($F_{1,162} = 36.44$, $p < 0.0001$). Nonetheless, parasitism strongly reduced total pot biomass ($F_{1,162} = 162.95$, $p < 0.0001$) indicating that the parasite not only shifted resources from the host to itself, but reduced the host's rate of growth. Drought treatment also reduced total pot biomass ($F_{2,162} = 146.98$, $p < 0.0001$) and altered the effect of parasitism ($F_{2,162} = 4.24$, $p = 0.0160$). The reduction in productivity due to parasitism was significantly greater for well-watered plants than for continuously-stressed plants ($F_{1,162} = 8.43$, $p < 0.0042$; Figure 3) but no other contrasts were significant ($p > 0.05$ for each). These data were log-transformed prior to analysis and so the effect of parasitism was in part a proportional reduction in whole pot productivity.

Parasitism can depress photosynthesis and growth [34], or cause a sink-dependent increase in photosynthesis [35] with reduced productivity when the uptake of photosynthates by the parasite does not balance loss from the host. Parasitism by *Cuscuta* can also produce a total pot yield similar to that from uninfected pots [24]. In our experiment, *C. gronovii* grew especially large on plants well supplied with water [36], and reduced total productivity of these control pots to a greater extent than pots in which the hosts were continuously stressed. Given that a holoparasite imports all photosynthates, robust parasite growth increases sink strength and the demand on the host. Increasing drought stress leads to dehydration avoidance responses, e.g., closure of stomata, decreased rates of transpiration, and depressed photosynthesis, and in turn, reduction of plant growth [37,38]. Compared to plants that were not drought stressed, near shutdown by continuously-stressed plants provided little opportunity for exploitation by the parasite.

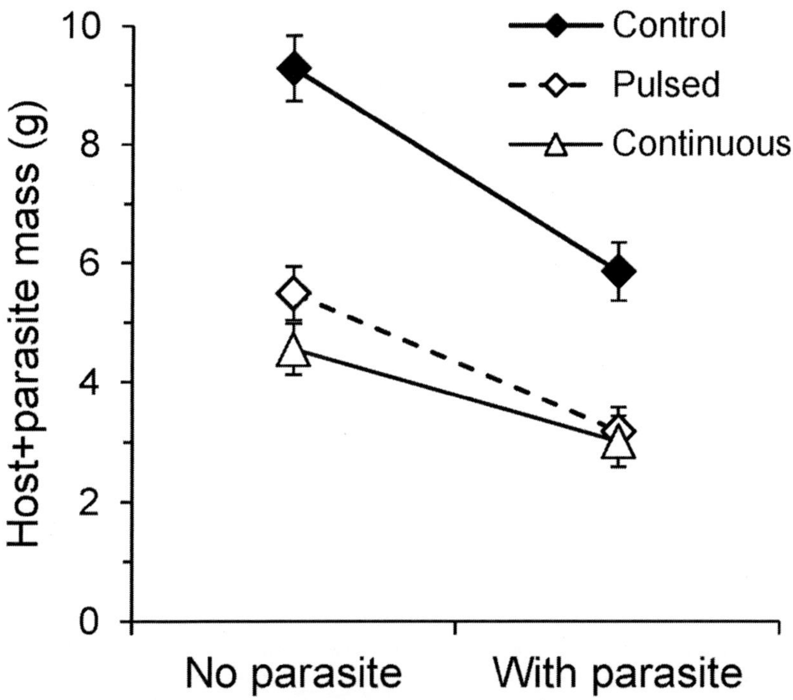

Figure 3. Total productivity from pots of individual *V. alternifolia* with or without dodder and well-watered, pulse-stressed, or continuously drought stressed. Values are back-transformed least squares mean (±2 se) dry mass of host roots + shoots + parasite.

2.3. Requirements and Utility of the LRM for Holoparasite-Host Interactions

Wise and Abrahamson [22] outline four experimental requirements for testing the LRM. Our study met the first requirement of a full factorial experimental design, where plant performance was evaluated under high and low-resource level and high and low-consumer pressure. A second requirement calls for the focal resource (water) to ultimately limit host fitness. We evaluated tolerance in this experiment by examining host plant biomass as an indicator of fitness. Based on the growth of the host in response to the treatments, we are confident that reduced water availability would indeed have limited fitness [39,40]. A third requirement entails measuring physiological parameters to confirm the resource limited by the consumer. Although we assumed that the parasite limited carbon by obtaining photosynthates via haustorial connections to the host's phloem, other resources may have been affected as well. Haustorial connections to the host's xylem could have led to nitrogen or even water limitation by the parasite [6]. However, parasitism increased investment in shoot relative to root which indicates that the parasite limited aboveground processes more than it limited belowground resource acquisition. Research on the congener *C. campestris* has demonstrated that this holoparasite reduces growth of the host both through acquisition of photosynthates and reduction in host photosynthesis [34].

Our experiment did not meet a fourth requirement of the LRM, that consumer-damage levels be consistent across environments. Although hosts were infested with a single parasite, parasites grew significantly better on well-watered hosts, to the extent that parasite growth exceeded host growth [36]. Thus, apparently lower tolerance of parasitism by well-watered plants is, at least in part, an expression of lower resistance to and consequently greater consumption by the parasite. Unlike insect herbivores, which are generally a small fraction of the size of the host, parasitic plants such as *Cuscuta* spp. can greatly exceed the size of an individual host as they grow across multiple hosts. Short of trimming the parasite, this requirement of the LRM would be difficult to meet in many plant-parasite systems. Nonetheless, we believe the LRM provides a useful framework for examining the impact of the abiotic environment on responses of plants to parasites.

3. EXPERIMENTAL

3.1. Study Organisms

The host species, *Verbesina alternifolia* (Asteraceae) grows from 1–3 m tall in full sun to light shade and moderate to moist soils. This native perennial is frequently found in thickets, woods, and bottomlands throughout the Midwestern and Eastern United States including Illinois [41]. Its common name of wingstem derives from the distinct wing-like structures running the length of the main stem. Known as common or swamp dodder, *Cuscuta gronovii* (Cuscutaceae) is the most common and abundant species of *Cuscuta* in North America. Found in moist thickets, roadsides, and fields through the Midwestern and Eastern United States as well as adjacent Canada [42,43], this annual holoparasitehas also been introduced to Europe [44]. The broad host range of *Cuscuta gronovii* includes at least 175 native and crop plants [45] and it is a particularly noxious weed in cranberry production, where heavy infestation can reduce yield up to 100% [46].

Cuscuta spp. twine around a host and form haustoria that penetrate the host's vascular tissue [47]. Following attachment, the part of the seedling that existed prior to loop formation withers away and growth is based on resources acquired from the host. Searching hyphae differentiate into either xylem or phloem elements upon contact with the host plant's vascular bundles [48] through cell recognition at the point of contact [49]. Water and minerals are acquired through connections to both the host xylem and phloem but carbon and nitrogen are obtained mainly through phloem attachments [23,35,50] by means of symplastic transfer [48]. The parasite creates a very efficient sink that deprives the host roots of carbon, leading to increased net photosynthesis of the host and a nitrogen deficit in host shoots and roots [35].

3.2. Experimental Procedure

Seeds of the host plant and the parasite were collected in October 2010 adjacent to a prairie near the Mackinaw River in Lexington, IL, USA. *V. alternifolia* seeds were cold stratified for 278 days in a refrigerator in bags containing damp Perlite™. Stratified seeds were subsequently germinated in pairs in small pots. The dodder seeds were cold stratified similarly for 188 days and then placed on the surface of a damp mixture of soil, sand, and Perlite™ to germinate. All seeds were germinated in a greenhouse and were frequently misted and watered to maintain adequate moisture.

V. alternifolia seedlings (22 days post germination) were transplanted singly into 180 1.7-L pots filled with 1 L of 1:1 potting medium (MetroMix™ 510:Perlite™). Host plants were randomly assigned one of six treatments that were combinations of parasite (no parasite or addition of a *C. gronovii* seedling 12 days after host transplant) and drought stress (well-watered, continuously drought stressed, or pulse stressed). Details are provided below.

Three pots of each of the six treatments were randomly assigned to 10 tables totaling 18 pots per table. The positions of all pots were rotated on tables and the tables were rotated around the room. Bamboo skewer sticks were placed in all 180 pots. Velcro™ strips were used when necessary to tie aggressively growing parasites to the stick in order to prevent spread to other pots. The greenhouse lights (600 watt high pressure sodium) were on a 14:10 L:D schedule and the temperature range was controlled to remain between 18–24 °C. Eleven and 22 days after transplanting, *V. alternifolia* pots were fertilized with 1 g of Scott's Rose and Bloom™ (12-4-8 N-P-K), a slow-release granular fertilizer. Total leaf count and length of the longest leaf on each host plant were recorded 13 days after transplant to be used as possible covariates in the analysis.

3.2.1. Parasite Treatment

Six days after ending parasite cold stratification and 13 days prior to initiation of drought treatments, one 3–6 cm *C. gronovii* seedling was haphazardly selected from the tray of seedlings and placed near the host in a shell vial filled with water. These seedlings were monitored twice daily to record date of attachment as well as to replace any *C. gronovii* seedlings that did not survive. Attachment was defined as one entire tight loop around the stem of the *V. alternifolia* host plant.

3.2.2. Drought Stress Treatment

Drought stress treatments began 25 days after host transplant. An initial saturation mass was determined for each pot by flooding the pot with 500 mL of water, waiting 30 min for excess water to drip, and then recording the mass. This saturation mass was used as a benchmark throughout the rest of the experiment. Each of the 120 plants was then randomly assigned one of 6 treatments that were combinations of two factors: drought stress and parasitism. Drought treatments consisted of a control (high water availability), a continuously stressed treatment, and a pulse (intermittently) stressed treatment. Control plants were maintained at >85% of initial saturation (Figure 4). For the control treatment, sentinel pots on each table were weighed daily and all 60 pots in this treatment group received water when one of the sentinel pots fell below 85% initial saturation. Each control was then watered to 100% of mass at initial saturation. Continuously-stressed treatment pots were weighed daily and maintained at 40%–45% of initial saturation. When a pot's mass fell below 40% saturation it was watered to attain 45% saturation. Pulsed (intermittently stressed) pots received no water until sentinel pots dropped below the lower limit experienced by continuously stressed plants (40% of initial saturation). Severe wilting occurred at

35% of initial saturation and was determined to be the point when recovery was necessary. When one sentinel pot reached 35% initial saturation then all 60 pots received water until they were back to 100% initial saturation. Due to the extended period without water, the pulse-stressed pots were extremely dry and the soil had constricted making watering procedures more difficult. To ensure all water was available for saturation and not lost through the bottom, trays were placed underneath these pots to allow for absorption of runoff. The intermittently-stressed pots were pulsed back to 100% saturation twice, and the experiment was terminated when they were at the point of requiring a 3rd pulse of water. They were not watered at this point in order to keep the pulsed and continuous drought treatments relatively similar to each other in terms of amount of water received throughout the experiment. The water added to each pot in the experiment was recorded daily.

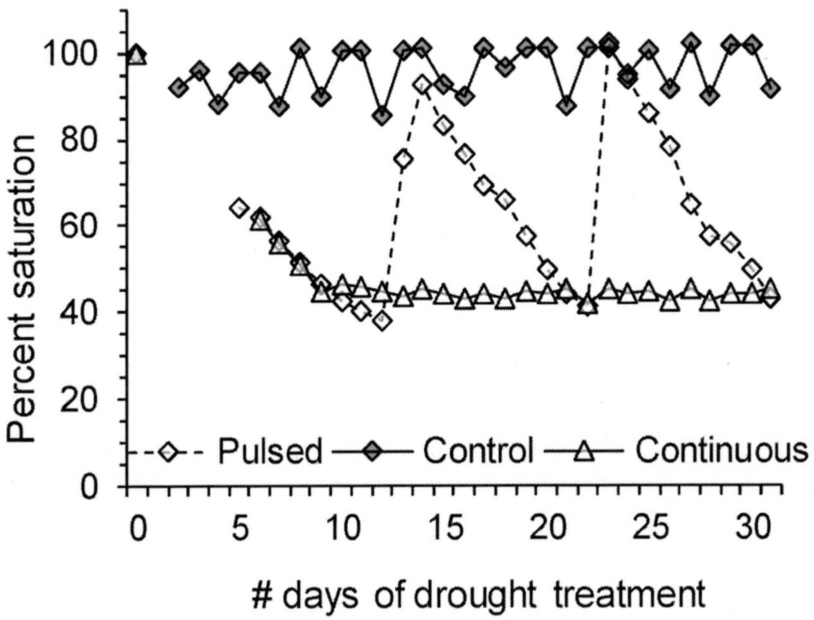

Figure 4. Percent saturation for pots containing single *V. alternifolia* throughout application of water treatments. Values for the two parasite treatments were averaged and the water treatments are expressed as a percent of initial mass of pots at field capacity.

3.2.3. Harvest

The experiment was terminated 57 days after hosts were transplanted, which was 32 days after water stress began. Parasites were carefully detached from host shoots and dried. *Verbesina alternifolia* shoots were cut at ground level, roots were washed to remove all debris from the potting medium, and each part was dried at 60 °C to constant weight.

3.3. Statistical Analysis

Because the duration of parasitism could affect final biomass of the host we examined the attachment date of *C. gronovii* prior to analysis and excluded 11 pots in which the parasite took longer than 8 days to tightly loop around the host. Among the pots that were not excluded, over 45% of the attachments occurred within one day and 90% within 3 days. We used days to attachment as a covariate in MANCOVA with water as the main effect to test whether this important event in the host-parasite interaction significantly affected final root and shoot mass of parasitized hosts. The covariate was not significant ($p = 0.1275$) indicating that timing of parasite establishment was not a significant source of variation. The final data set consisted of 26 control, 28 pulse-stressed, and 25 continuously-stressed replicates with parasites, and 30 replicates in each of the drought treatments free of parasites.

To determine the effects of water stress and parasitism on host tolerance, dry root and dry shoot mass of the hosts were analyzed with MANCOVA. Total number of leaves prior to the onset of treatments ranged from 4 to 10 and 90% of plants had 6 or 8 leaves. Preliminary analysis showed that total number of leaves did not explain significant variance of shoot and root biomass ($p > 0.05$). Length of the longest leaf from each host plant did not interact with main effects and explained significant variance in final host biomass measures. Because measures of leaf size are correlated with variation in leaf structure and function [51] we used this variable to account for pre-treatment variation among hosts. MANCOVA was followed by bivariate contrasts to test whether the effects of parasitism varied among water treatments. Generally biomass measurements are log-transformed because treatments that cause proportionately similar responses in large and small plants yield different absolute amounts [28,29]. We analyzed untransformed data because these best met the assumptions of the model. However, analysis of untransformed mass can bias towards finding significant interactions due to the problem of scale [32]. To evaluate this possibility we subsequently analyzed shoot mass divided by total mass and performed contrasts. Finally, to determine whether the parasite reduced overall productivity or merely shifted resources to itself, total host + parasite dry mass was analyzed with ANCOVA using length of the longest host leaf as a covariate to account for pre-treatment variation among hosts. All statistical analyses were performed using SAS 9.2 © 2008 for Windows Version 6.1.7601.

4. CONCLUSIONS

In a comparison of competing models of plant tolerance to herbivory, Wise and Abrahamson [30] determined that 95% of the studies produced results consistent with predictions of the LRM. When water was the focal resource, plants were more tolerant of herbivory when water was in short supply. Our results with a holoparasite as consumer and water limitation as the focal resource suggest that the LRM is broadly applicable. However, the LRM makes no predictions regarding temporal pattern of stress. While drought-stress treatments imposed in experiments of herbivory are far from standardized, stress is generally applied at a single constant level or a single pulse, with examination of different intensities the exception [28]. Plants may exhibit rapid photosynthetic and growth spurts when watered after drought, but the ability to recover may be affected by previous history of drought or compromised if drought severity impairs biochemical and physiological processes [29]. Some data suggest the severity of drought may be more important to a plant than pulse frequency [26,28] but much depends on how treatments are administered. In our study, overall limitation, rather than temporal pattern of the stress, drove the host response, but pattern of drought affected allocation in response to parasitism.

In addition to affecting individual host growth, parasites significantly reduce crop productivity [52,53] and alter community structure and function [54,55,56]. Consequently, a change in the abiotic environment is likely to alter the magnitude of a parasitic species' effects at several levels [14]. Climate change models predict more frequent and variable drought conditions [25,57] which will affect plant communities [58,59,60] and parasite-host interactions [60]. Shifts in phenologies and geographic distributions of species in the direction predicted by these models indicate that climate change is indeed occurring [61]. To better understand how climate change will impact holoparasite-host and community relations, future experiments should continue to manipulate parasitism against a background of changing resource availability.

ACKNOWLEDGMENTS

The authors are grateful to John Evans Campbell for assistance in the greenhouse. We thank Steven Juliano for helpful comments and statistical knowledge, Joseph Armstrong for his discussion and advice, and Don Schmidt for assistance in the greenhouse.

CONFLICTS OF INTEREST

The authors declare no conflict of interest.

REFERENCES

1. Watling, J.R.; Press, M.C. Impacts of infection by parasitic angiosperms on host photosynthesis. Plant Biol. 2001, 3, 244–250.
2. Bollard, E.G. Transport in the xylem. Annu. Rev. Plant Physiol. 1960, 11, 141–166.
3. Mooney, H.A. The carbon balance of plants. Annu. Rev. Ecol. Syst. 1972, 3, 315–346.
4. Thompson, M.V. Phloem: The long and the short of it. Trends Plant Sci. 2006, 11, 26–32.
5. Wardlaw, I.F. The control and pattern of movement of carbohydrates in plants. Bot. Rev. 1968, 34, 79–105.
6. Hsiao, T.C. Plant responses to water stress. Annu. Rev. Plant Physiol. 1973, 24, 519–570.
7. Gutbrodt, B.; Mody, K.; Dorn, S. Drought changes plant chemistry and causes contrasting responses in lepidopteran herbivores. Oikos 2011, 120, 1732–1740.
8. Hawkes, C.V.; Sullivan, J.J. The impact of herbivory on plants in different resource conditions: A metaanalysis. Ecology 2001, 82, 2045–2058.
9. Joern, A.; Mole, S. The plant stress hypothesis and variable responses by blue grama grass (Bouteloua gracilis) to water, mineral nitrogen, and insect herbivory. J. Chem. Ecol. 2005, 31, 2069–2090.
10. Robinson, E.A.; Ryan, G.D.; Newman, J.A. A meta-analytical review of the effects of elevated CO2 on plant-arthropod interactions highlights the importance of interacting environmental and biological variables. New Phytol. 2012, 194, 321–336.
11. Sadras, V.O.; Wilson, L.J.; Lally, D.A. Water deficit enhanced cotton resistance to spider mite herbivory. Ann. Bot.1998, 81, 273–286.
12. Sun, Y.; Ding, J.; Frye, M.J. Effects of resource availability on tolerance of herbivory in the invasive Alternanthera philoxeroides and the native Alternanthera sessilis. Weed Res. 2009, 50, 527–536.
13. Marquardt, E.S.; Pennings, S.C. Constraints on host use by a parasitic plant. Oecologia 2010, 164, 177–184.
14. Pennings, S.C.; Callaway, R.M. Parasitic plants: Parallels and contrasts with herbivores. Oecologia 2002, 131, 479–489.
15. Núñez-Farfán, J.; Fornoni, J.; Valverde, P.L. The evolution of resistance and tolerance to herbivores. Annu. Rev. Evol. Syst. 2007, 38, 541–566.
16. Rosenthal, J.P.; Kotanen, P.M. Terrestrial plant tolerance to herbivory. Trends Ecol. Evol. 1994, 9, 145–148.

17. Strauss, S.Y.; Agrawal, A.A. The ecology and evolution of plant tolerance to herbivory. Trends Ecol. Evol. 1999, 14, 179–185.
18. Belsky, A.J. Does herbivory benefit plants? A review of the evidence. Am. Nat. 1986, 127, 870–892.
19. Ferraro, D.O.; Oesterheld, M. Effect of defoliation on grass growth. A quantitative review. Oikos 2002, 98, 125–133.
20. Meyer, G.A.; Root, R.B. Effects of herbivorous insects and soil fertility on reproduction of goldenrod. Ecology 1993, 74, 1117–1128.
21. Rand, T.A. Competition, facilitation, and compensation for insect herbivory in an annual salt marsh forb. Ecology2004, 85, 2046–2052.
22. Wise, M.J.; Abrahamson, W.G. Beyond the compensatory continuum: Environmental resource levels and plant tolerance of herbivory. Oikos 2005, 109, 417–428.
23. Hibberd, J.M.; Quick, W.P.; Press, M.C.; Scholes, J.D.; Jeschke, W.D. Solute fluxes from tobacco to the parasitic angiosperm Orobanche cernua and the influence of infection on host carbon and nitrogen relations. Plant Cell Environ.1999, 22, 937–947.
24. Jeschke, W.D.; Baig, A.; Hilpert, A. Sink-stimulated photosynthesis, increased transpiration and increased demand dependent stimulation of nitrate uptake: Nitrogen and carbon relations in the parasitic association Cuscuta reflexa–Coleus blumei. J. Exp. Bot. 1997, 48, 915–925.
25. Allen, C.D.; Macalady, A.K.; Chenchouni, H.; Bachelet, D.; McDowell, N.; Vennetier, M.; Kitzberger, T.; Rigling, A.; Breshears, D.D.; Hogg, E.H.; et al. A global overview of drought and heat-induced tree mortality reveals emerging climate change risks for forests. For. Ecol. Manag. 2010, 259, 660–684.
26. Novoplansky, A.; Goldberg, D.E. Effects of water pulsing on individual performance and competitive hierarchies in plants. J. Veg. Sci. 2001, 12, 199–208.
27. Huberty, A.F.; Denno, R.F. Plant water stress and its consequences for herbivorous insects: A new synthesis. Ecology2004, 85, 1383–1398.
28. Mody, K.; Eichenberger, D.; Dorn, S. Stress magnitude matters: Different intensities of pulsed water stress produce non-monotonic resistance responses of host plants to insect herbivores. Ecol. Entomol. 2009, 34, 133–143.
29. Xu, Z.; Zhou, G.; Shimizu, H. Plant responses to drought and rewatering. Plant Signal. Behav. 2010, 5, 649–654.
30. Wise, M.J.; Abrahamson, W.G. Effects of resource availability on tolerance of herbivory: A review and assessment of three opposing models. Am. Nat. 2007, 169, 443–454.
31. Kerkhoff, A.J.; Enquist, B.J. Multiplicative by nature: Why logarithmic transformation is necessary in allometry. J. Theor. Biol. 2009, 257, 519–521.
32. Mascaro, J.; Litton, C.M.; Hughes, R.F.; Uowolo, A.; Schnitzer, S.A. Minimizing bias in biomass allometry: Model selection and log-transformation of data. Biotropica 2011, 43, 649–653.
33. McConnaughay, K.D.M.; Coleman, J.S. Biomass allocation in plants: Ontogeny or optimality? A test along three resource gradients. Ecology 1999, 80, 2581–2593.
34. Shen, H.; Hong, L.; Ye, W.; Cao, H.; Wang, Z. The influence of the Holoparasitic plant Cuscuta campestris on the growth and photosynthesis of its host Mikania micrantha. J. Exp. Bot. 2007, 58, 2929–2937.
35. Jeschke, W.D.; Rath, N.; Baumel, P.; Czygan, F.C.; Proksch, P. Modeling the flow and partitioning of carbon and nitrogen in the holoparasite Cuscuta reflexa Roxb and its host Lupinus albus L. 1. Methods for estimating net flows. J. Exp. Bot. 1994, 45, 791–800.
36. Evans, B.A. Pattern of Drought and Host Plant-Holoparasite Relations: An Examination of Resistance and Tolerance. Master's Thesis, Illinois State University, Normal, IL, USA, 2012.
37. Chaves, M.M.; Maroco, J.P.; Pereira, J.S. Understanding plant responses to drought—From genes to the whole plant. Funct. Plant Biol. 2003, 30, 239–264.
38. Ghannoum, O. C4 photosynthesis and water stress. Ann. Bot. 2009, 103, 635–644.

39. Heschel, M.S.; Riginos, C. Mechanisms of selection for drought stress tolerance and avoidance in Impatiens capensis. Am. J. Bot. 2005, 92, 37–44.
40. Sletvold, N.; Agren, J. Variation in tolerance to drought among Sandinavian populations of Arabidopsis lyrata. Evol. Ecol. 2012, 26, 559–557.
41. Gleason, H.A.; Cronquist, A. Manual of Vascular Plants of Northeastern United States and Adjacent Canada, 2nd ed.; New York Botanical Garden: New York, NY, USA, 1991.
42. Costea, M.; Tardif, F.J. The biology of Canadian weeds. 133. Cuscuta campestris Yuncker, C. gronovii Willd. Ex Schult., C. umbrosa Beyr. Ex Hook., C. epithymum (L.) L. and C. epilinum Weihe. Can. J. Plant Sci. 2006, 86, 293–316.
43. Sandler, H.A. Dodder—Cuscuta gronovii Willd; Cranberry Experiment Station, University of Massachusetts: Wareham, MA, USA, 2001.
44. Parker, C.; Riches, C.R. Parasitic Weeds of the World. Biology and Control; CAB International: Wallingford, UK, 1993.
45. Gaertner, E.E. Studies of seed germination, seed identification, and host relationships in dodders, Cuscuta spp. Cornell Exp. Stn. Mem. 1950, 294, 1–56.
46. Devlin, R.M.; Deubert, K.H. Control of swamp dodder (Cuscuta gronovii) on cranberry bogs with butralin. Proc. Annu. Meet. Northeast. Weed Sci. Soc. 1980, 11, 112–113.
47. Kelly, C.K. Resource choice in Cuscuta europaea. Proc. Natl. Acad. Sci. USA 1992, 89, 12194–12197.
48. Birschwilks, M.; Haupt, S.; Hofius, D.; Neumann, S. Transfer of phloem-mobile substances from the host plants to the holoparasite Cuscuta sp. J. Exp. Bot. 2006, 57, 911–921.
49. Vaughn, K.C. Conversion of the searching hyphae of dodder into xylic and phloic hyphae: A cytochemical and immunocytochemical investigation. Int. J. Plant Sci. 2006, 167, 1099–1114.
50. Hibberd, J.M.; Jeschke, D. Solute flux into parasitic plants. J. Exp. Bot. 2001, 52, 2043–2049.
51. Niinemets, Ü.; Portsmuth, A.; Tena, D.; Tobias, M.; Matesanz, S.; Valladares, F. Do we underestimate the importance of leaf size in plant economics? Disproportional scaling of support costs within the spectrum of leaf physiognomy. Ann. Bot. 2007, 100, 283–303.
52. Dawson, J.H.; Musselman, L.J.; Wolswinkel, P.; Dörr, L. Biology and control of Cuscuta. Rev. Weed Sci. 1994, 6, 265–317.
53. Lanini, T.; Kogan, M. Biology and management of Cuscuta in crops. Cien. Inv. Agr. 2005, 32, 165–179.
54. Grewell, B. Parasite facilitates plant species coexistence in a coastal wetland. Ecology 2008, 89, 1481–1488.
55. Press, M.C.; Phoenix, G.K. Impacts of parasitic plants on natural communities. New Phytol. 2005, 166, 737–751.
56. Yu, H.; Yu, F.-H.; Miao, S.-L.; Dong, M. Holoparasitic Cuscuta campestris suppresses invasive Mikania micrantha and contributes to native community recovery. Biol. Conserv. 2008, 141, 2653–2661.
57. Parry, M.L.; Canziani, O.F.; Palutikof, J.P.; van der Linden, P.J.; Hanson, C.E. IPCC Fourth Assesment Report (AR4). Climate Change 2007: Impacts, Adaptation and Vulnerability. Contribution of Working Group II to the Fourth Assessment Report of the Intergovernmental Panel on Climate Change; Cambridge University Press: Cambridge, UK, 2007.
58. Milnes, K.J.; Davies, W.J.; Rodwell, J.S.; Francis, B.J. The responses of Briza media and Koeleria macrantha to drought and re-watering. Funct. Ecol. 1998, 12, 665–672.
59. Rennenberg, H.; Dannenmann, M.; Gessler, A.; Kreuzwieser, J.; Simon, J.; Papen, H. Nitrogen balance in forest soils: Nutritional limitation of plants under climate change stresses. Plant Biol. 2009, 11, 4–23.
60. Phoenix, G.K.; Press, M.C. Effects of climate change on parasitic plants: The root hemiparasitic Orobanchaceae. Folia Geobot. 2005, 40, 205–216.

61. Parmesan, C. Ecological and evolutionary responses to recent climate change. Annu. Rev. Ecol. Evol. Syst. 2006, 37, 637–669.

CHAPTER 16

Drought Adaptation in Millets

Zerihun Tadele[1, 2]

[1] *Institute of Plant Sciences, University of Bern, Bern, Switzerland*

[2] *Addis Ababa University, Institute of Biotechnology, Addis Ababa, Ethiopia*

ABSTRACT

Millets are major food and feed sources in the developing world especially in the semi-arid tropical regions of Africa and Asia. The most widely cultivated millets are pearl millet [Pennisetum glaucum (L.) R. Br.], finger millet [Eleusine coracana (L.) Gaertn], foxtail millet [Setaria italica (L.) P. Beauvois], Japanese barnyard millet [Echinochloa esculneta (A. Braun) H. Scholz], Indian Barnyard millet [Echinochloa frumetacea Link], kodo millet [Paspalum scrobiculatum L.], little millet [Panicum sumatrense Roth.ex.Roem. & Schult.], proso millet [Panicum miliaceum L.], tef [Eragrostis tef (Zucc.) Trotter] and fonio or acha [Digitaria exilis (Kippist) Stapf and D. iburua Stapf]. Millets are resilient to extreme environmental conditions especially to inadequate moisture and are rich in nutrients. Millets are also considered to be a healthy food, mainly due to the lack of gluten (a substance that causes coeliac disease) in their grain. Despite these agronomic, nutritional and health-related benefits, millets produce very low yield compared to major cereals such as wheat and rice. This extremely low productivity is related to the challenging environment in which they are extensively cultivated and to the little research investment in these crops. Recently, several national and international initiatives have begun to support the improvement of diverse millet types.

Keywords: Abiotic stress, drought avoidance, drought escape, drought tolerance, millet

1. INTRODUCTION

Millets are among the major cereal crops in the developing world especially in the semi-arid tropical regions of Africa and Asia where they are used both as human food and livestock feed. Millets represent small grain crops that are mainly cultivated in marginal environments. Exceptional to this definition is pearl millet [*Pennisetum glaucum* (L.) R. Br.] that has a large seed size. Among the widely cultivated millets, those traditionally considered as millet are pearl millet, finger millet [*Eleusine coracana* (L.) Gaertn], foxtail millet [*Setaria italica* (L.) P. Beauvois], Japanese barnyard millet [*Echinochloa esculneta* (A. Braun) H. Scholz], Indian Barnyard millet [*Echinochloa frumetacea* Link], kodo millet [*Paspalum scrobiculatum* L.], little millet [*Panicum sumatrense* Roth ex Roem. & Schult.] and proso millet [*Panicum miliaceum* L.] (Table 1). Tef [*Eragrostis tef* (Zucc.) Trotter] was included in the millet group at the First Small Millets Workshop held 30 years ago in Bangalore, India [1], while both tef and fonio or acha [*Digitaria exilis* (Kippist) Stapf and *D. iburua* Stapf] were grouped under small millets by international agricultural organizations in the mid-1990s [2]. The inclusion of tef and fonio to the millet family is justifiable due to the close relationship of the two species with other millets. The genetic difference between some traditional millets is as large as that between tef or fonio and other millets. Due to this substantial variability among themselves, millets are grouped into two subfamilies, namely Panicoideae, which includes pearl millet, foxtail millets, Japanese barnyard millet and Indian millet, and to Chloridoideae, which includes finger millet and tef, and eight genera (Table 1). This indicates that finger millet which is normally grouped under millet is more closely related to tef than to other millets [3]. The divergence among traditional millets is also exhibited in the chromosome number and ploidy level which range from the diploid pearl millet ($2n = 2x = 14$) to the hexaploid fonio ($2n = 6x = 54$) [4, 5]. Millets are also divergent in the size and colour of seeds, seed weight, plant stature and shape of their panicles (Table 1). The geographical distributions of small millets were recently summarized by Goron and Raizada [6]. Except for finger millet, which is extensively cultivated in Africa and Asia, other small millets are mainly grown in Asia.

Table 1. Description and benefits of millets

Common name	Traditional millets				
	Pearl millet	**Foxtail millet**	**Japanese Barnyard millet**	**Indian Barnyard millet**	**Kodo millet**
Other names	Bulrush millet	Italian millet	Japanese millet	Billion dollar grass	Koda millet
Botanical names	*Pennisetum glaucum*	*Setaria italica*	*Echinochloa esculneta*	*Echinochloa frumetacea*	*Paspalum scrobiculatum*
Subfamily	Panicoideae	Panicoideae	Panicoideae	Panicoideae	Panicoideae
Tribe	Paniceae	Paniceae	Paniceae	Paniceae	Paniceae
Distribution			Japan, Korea, China	India, Pakistan, Nepal	
Ploidy level	Diploid	Diploid	Hexaploid	Hexaploid	Tetraploid
Chromosome number	2n = 2x = 14	2n = 2x = 18	2n = 6x = 36	2n = 6x = 36	2n = 4x = 40
Purpose	Food, feed	Food, biofuel	Food, feed	Food	Food, feed
Agronomic benefits	Drought & heat tolerance	Drought tolerance	Early maturity, anti-fungal	Early maturity	Drought tolerance
Nutritional benefits	High protein, starch & minerals		anti-diabetic	High protein content	High-quality protein
Health benefits	No gluten	No gluten	No gluten	No gluten	Low glycaemic index, anti-oxidant
Reference	[4, 20, 27]	[16, 27]	[5, 18, 27]	[6]	[27, 36]
Botanical names	*Panicum sumatrense*	*Panicum miliaceum*	*Eleusine coracana*	*Eragrostis tef*	*Digitaria exilis, D. Iburua*
Subfamily	Panicoideae	Panicoideae	Chloridoideae	Chloridoideae	Panicoideae
Tribe	Paniceae	Paniceae	Eragrostideae	Eragrostideae	Paniceae
Distribution					
Ploidy level	Tetraploid	Tetraploid	Tetraploid	Tetraploid	Diploid or hexaploid
Chromosome number	$2n = 4x = 36$	$2n = 4x = 36$	$2n = 4x = 36$	$2n = 4x = 40$	$2n = 2x = 30$ or $2n = 6x = 54$
Purpose	Food	Food, feed	Food, feed	Food, feed	Food, feed
Agronomic benefits	Abiotic stress tolerance	Drought tolerance, early maturity	Drought and salt tolerance	Water-logging tolerance, storage pest tolerance	Drought tolerance
Nutritional or	High in phytochemicals, fibre	Rich in amino acids	Rich in calcium, methionine and tryptophan	Rich in protein	Rich in amino acids
Health benefits	Anti-diabetic	Anti-cancer	Low glycaemic index, anti-oxidant	No gluten	
Reference	[38–40]	[6, 23, 27]	[6, 20, 27, 35, 37]	[33, 47]	[20, 24, 25]

2. IMPORTANCE OF MILLETS IN GLOBAL AGRICULTURE

2.1. Economic Benefits

Millets play a key role in the economy of the developing world especially in countries with extensive areas of marginal land used for crop cultivation. In 2013, the global area under millet cultivation was 34.9 million hectares, corresponding to 4.7 % of the global area for all cereals including wheat, maize and rice [7] (Table 2). On the other hand, the global production of millets in the same year was estimated to be 36.7 million tons, which contributes only 1.2% to the total cereal production. This lower production was due to the inferior average yield of millets (only 0.9 t ha^{-1}) compared to other cereals (3.8 t ha^{-1}). However, the contribution of India to global millet production is significant. In 2013, India produced over 30% of the global millet yield from only 25% of the global millet area, mainly due to improved productivity. In the same year, while the mean seed yield of millet in India was 1.2 t ha^{-1}, it was only 0.8 t ha^{-1} for other countries. This 50% production advantage in India over other countries especially African countries was due to the widespread use of improved varieties and techniques. A decade ago, the rate of adoption of improved pearl millet cultivars by farmers was 65% in India but below 10% in some African countries [8].

Tef and fonio are exclusively cultivated in Africa. While fonio is cultivated on a total of half a million hectares in West Africa mainly in Guinea, Guinea-Bissau and Côte d'Ivoire [7], tef is grown in the Horn of Africa especially in Ethiopia where it is annually cultivated on three million hectares of land and is a staple food for about 50 million people [9]. In the last two decades, the productivity of tef was raised by 100%, from just 0.7 t ha^{-1} in 1994 to 1.4 t ha^{-1} in 2013 mainly due to an increase in the use of improved cultivars.

In general, millets play a key role in food security in Asia and Africa. Together with sorghum, millets account for about half of the total cereal production in Africa [10]. Millets, are therefore considered as a poor man's crop due to their significant contributions to the diet of resource-limited farmers and consumers.

Table 2. The top 20 millet-producing countries in the world in 2013.

Country	Traditional millets[a]	Tef	Fonio (Acha)	Total
Production (ton)				
India	10,910,000			10,910,000
Ethiopia	848,956[b]	4,418,642		5,267,598
Nigeria	5,000,000		90,000	5,090,000
Niger	2,995,000		6,000	3,001,000
China	1,746,000			1,746,000
Mali	1,152,331		22,090	1,174,421
Burkina Faso	1,078,570		19,887	1,098,457
Sudan (former)	1,090,000			1,090,000
Guinea	215,000		429,000	644,000
Chad	582,000			582,000
Senegal	572,155		1,030	573,185
Russia	418,844			418,844
USA	418,145			418,145
Tanzania	322,731			322,731
Pakistan	310,000			310,000
Nepal	305,588			305,588
Uganda	228,000			228,000
Myanmar	185,000			185,000
Ghana	155,131			155,131
Cameroon	97,000			97,000
others	1,233,696		19,000	1,252,696
Total production	29,864,147	4,418,642[c]	587,007	34,869,796
Total area (ha)	33,118,792	3,016,521	554,451	36,689,764
Yield (ton ha^{-1})[d]	0.90	1.47	1.06	0.95
Reference	[7,9]	[9]	[7]	

[a] Traditional millets include finger millet, foxtail millet, Indian barnyard millet, Japanese barnyard millet, kodo millet, little millet, pearl millet and proso millet.

[b] Only for finger millet.

[c] Only for Ethiopia.

[d] Average global yield except for tef where it is the national average yield for Ethiopia.

2.2. Agronomic Benefits

Millets are resilient to the extreme climatic and soil conditions prevalent in the semi-arid regions of Asia and Africa. The similarities of millets are that they are all grown under extreme environmental conditions, especially those of inadequate moisture and poor soil fertility which are poorly suited to the major crops of the world [11] (Table 1). Proso millet is considered to have been domesticated before rice in China, based on the extreme resistance of this millet to drought [12, 13]. In addition to its resistance to drought, proso millet escapes the terminal drought that normally occurs late in the growing season since it matures in only three months; hence, proso millet is considered to be a millet with low water requirements [6].

Similar to maize and sorghum, millets possess a C4 photosynthesis system [14, 15]; hence, they prevent photorespiration and, as a consequence, efficiently utilize the scarce moisture present in the semi-arid regions. Since C4 plants are able to close their stomata for long periods, they can significantly reduce moisture loss through the leaves. In addition to its tolerance to drought, tef is tolerant to waterlogging especially in poorly drained soils where other crops such as maize and wheat could not survive. Foxtail millet is also considered to be a model plant for biofuel studies [16]. A novel peptide isolated from foxtail millet and barnyard millet has shown strong antifungal properties as has one from finger millet which is especially effective and works against four fungus species, namely *Alternaria, Trichoderma, Botrytis* and *Fusarium* [17, 18].

2.3. Nutritional Benefits

Millets are rich sources of nutrients for both humans and animals. Saleh et al. [19] have compiled detailed information on the nutritional advantages of several millets. The grains of most millets possess levels of protein comparable to those of wheat but higher than those of rice [20] (Table 1). In addition, the seeds of finger millet contain valuable amino acids especially methionine [20], which is lacking in the diets of hundreds of millions of the poor who live on starchy staples such as cassava. Other reports indicate that finger millet is rich in lysine, threonine and valine [21, 22] while proso millet has plentiful leucine, isoleucine and methionine [23]. The seeds of fonio are also nutritious, especially in amino acids such as leucine, methionine and valine [24, 25]. Since proso millet is rich in essential amino acids including leucine, isoleucine and methionine, the protein quality of the grain is higher than that of wheat [23].

The grains of extensively cultivated pearl millet contain high amounts of starch, fibres and minerals [26, 27]. In general, millets have high amounts of vitamins, calcium, iron, potassium, magnesium and zinc [28].

The straws and crop residues of millets are also the main source of livestock feed for farmers in developing countries. In Ethiopia, compared to the straw from other cereals, the straw of tef is the most palatable to livestock and fetches the highest price [29].

2.4. Health-Related Benefits

In addition to being nutritious, millets are also considered to be a healthy food. Two recent reviews examined the health-related benefits associated with millets [19, 6]. A number of leading newspapers and media have recently indicated the potential of millets particularly tef as a global lifestyle crop [30–32]. This is particularly due to the lack of gluten in the grain of tef [33] (Table 1). Gluten is a substance present in wheat and other grains that causes celiac disease or other forms of allergies. Similar to tef, several other millets, particularly foxtail millet, do not contain gluten.

Six millet species (namely kodo, finger, proso, foxtail, little and pearl millets) were shown to have an anti-proliferative property and might have a potential in the prevention of cancer initiation [34, 35]. The anti-proliferative property of these millets is associated with the presence of phenolic extracts. Among the first four millets indicated above, the maximum phenolic content was obtained in kodo millet while the minimum was in foxtail millet [36].

Finger millet is also a popular food among diabetic patients because of its low glycaemic index and slow digestion due to high fibre content [37]. The glycaemic index of little millet was also lower than that of rice, wheat and sorghum; hence, it is considered to be an anti-diabetic grain [38]. The composition of useful antioxidants and related products could be enhanced through processing the grain. A study in little millet showed that the levels of phenolics, flavonoids and tannins were substantially increased by germinating, steaming and roasting soaked grains [39].

3. DROUGHT: A MAJOR CHALLENGE TO MILLET CULTIVATION

Biotic stresses such as insect pests and diseases are a cause for substantial yield losses to diverse types of millets. However, abiotic stresses are the biggest contributor to losses every year. Although, in general, millets perform better than cereals such as wheat and rice in semi-arid environments, these challenging climatic and soil conditions are by no means an optimum environment for millet cultivation. In semi-arid and arid environments where millets are the dominant crop, drought or inadequate moisture is the major abiotic stress affecting productivity. Studies in pearl millet showed that drought impacts include growth, yield, membrane integrity, pigment, osmotic adjustment, water relations and photosynthetic activity [40].

3.1. Prevalence of Drought

Drought is defined as a temporary reduction in moisture availability in which the amount of available water is significantly below normal for a specified period. In general, drought can be explained as meteorological, hydrological or agricultural drought [41]. Agricultural drought occurs when there is not enough soil moisture to meet the needs of a particular crop at a particular time. Drought is also commonly expressed as a shortage or absence of rainfall causing a loss in rain-fed agriculture. For example, the decline in the level of rainfall during severe drought years in Ethiopia was accompanied by serious reductions in rain-fed agricultural outputs; this is because a 10% drop in rainfall (below the long-term national averages) results in an average drop of 4.2% in cereal yields [42].

As indicated above, millets are crops of dry land areas of the world. According to the United Nations, dry lands, which cover 40% of the world's land area or one-third of the global arable land, support two billion people, of which 90% live in the developing world [43]. Dry lands are classified into four, namely hyper-arid deserts, arid, semi-arid and dry subhumid. Millets are extensively cultivated in the semi-arid region, which is characterized by low and erratic rainfall and periodic drought. Climate change is expected to worsen the situation in this part of the world by reducing the grassland productivity by 49–90% by 2020 [43]. The Sahel Region in Africa, covering over three million km^2in 10 countries (namely northern Senegal, southern Mauritania, central Mali, northern Burkina Faso, the extreme south of Algeria, Niger, the extreme north of Nigeria, central Chad, central and southern Sudan and northern Eritrea) is the typical semi-arid region situated between the Sahara desert in the north and the tropical or savanna climate in the south [44].

The frequency and intensity of drought has increased in recent times. In Ethiopia, severe droughts used to occur periodically every 6–8 years [45], but recently, they have happened every 1–2 years especially in the south of the country [46].

Similar to other millets, drought is implicated among the major yield limiting factors in tef production [47]. Although tef grows in a wide variety of agro-ecological conditions ranging from semi-arid areas with low rainfall to areas with high rainfall, the rainfall pattern in most tef growing regions is not consistent enough to support the normal growth of the crop during the crop cycle. In most tef growing regions, greater rainfall variability exists over the growing period than over the year-cycle [48, 49] which results in poor agricultural outputs. The Water Requirement Satisfaction Index (WRSI), a crop-specific performance indicator taking rainfall and soil characteristics into account, indicates extreme and increasing variability in Ethiopia. A recent study also confirmed that climate will have a negative impact on the acreage and productivity of tef unless urgent interventions are implemented which favours mitigation and adaptation strategies [50].

3.2. Yield Losses Due to Drought

Various yield loss studies made for millets treated with drought conditions are summarized in Table 3. Using polyvinylchloride (PVC) tubes filled with sandy soil, Matsuura and colleagues [51] investigated the effect of moisture deficit before and after flowering on four millets, namely proso millet, little millet, foxtail millet and wild millet [*Setaria glauca* (L.) Beauv.]. Compared to the well-watered plants, a significant yield reduction was obtained in all four millets when the drought treatment was implemented at early developmental stage, that is, before flowering (or heading). However, terminal drought, which occurs from the flowering stage to the harvesting of the crop, contributed to a significant yield loss only in proso and little millets while the effect on foxtail and wild millets was negligible.

A study by Winkel et al. [52] in Niger where the annual rainfall is around 200 mm investigated the impact of water deficit at three stages of pearl millet development. The three stages were prior to flowering, at flowering and at the end of flowering. According to the findings of the work, the grain yield of pearl millet was severely reduced when moisture was limited prior to and at the flowering stage but not at the end of flowering. On the other hand, in pearl millet, terminal drought in which irrigation was terminated from the flowering until crop maturity, was severe, as it resulted in 60% yield loss [53]. The mid-season stress, which occurred from one month before flower initiation to full flowering, resulted in only 7% yield loss.

The study in two landraces of finger millet in which a drought treatment was imposed four weeks after sowing, resulted in 100% yield loss and over 30% biomass damage [54]. Similarly, yield loss reached up to 77% when the tef plant experienced drought at the flowering stage [55].

Although yield loss studies were not exhaustively made for most millets as they are considered drought tolerant, substantial damage occurs to these crops depending on the severity of drought. However, millets produce at least some grain and straw even in bad years unlike drought-intolerant cereals such as wheat and rice which completely fail to produce any yield.

Table 3. The magnitude of yield loss due to moisture scarcity in millets.

Millet type	Yield loss (%)			Critical stage	Reference
	Early drought[a]	terminal drought[b]	Long-term drought[c]		
Proso millet	30.1*	34.6*	64.0*	Before and after heading	[51]
Little millet	62.6*	80.1*	80.5*		[51]
Foxtail millet	19.2*	3.4[NS]	20.3*	Before heading	[51]
Wild millet (*Setaria gluaca*)	27.3*	15.3[NS]	30.1*		[51]
	Mid-season stress[d]	Terminal stress[e]			
Pearl millet	6.6	60.1		Flowering	[53]
Finger millet		109.8*[f]		Flowering	[54]
	Prior to flowering	Beginning flowering	End of flowering		
Pearl millet	72	61	Insignificant	From four weeks to flowering	[52]
	Early stress[g]				
Tef	69–77				[55]

[a] Early drought: water stress from 25 days after sowing till flowering.

*Indicates statistically significant difference from the well-watered samples.

[b] Terminal drought: water stress from flowering till harvesting.

[c] Long-term drought: water stress from 25 days after sowing till harvesting.

[d] Mid-season stress: water stress for 30 days from floral initiation to flowering.

[e] Terminal stress: water stress at flowering.

[f] Water stress from 28 days after sowing to harvest.

[g] Early stress: water stress from two weeks after emergence until symptoms of stress observed.

4. ADAPTATION OF MILLETS TO DROUGHT

4.1. Strategies to Drought Adaptation or Tolerance

Plants cope with drought using three main strategies, namely, drought escape, drought avoidance and drought tolerance, although a fourth strategy, known as drought recovery, has also been identified [56–60].

Drought escape: Drought escape refers to the condition in which plants reach maturity before the drought occurs. Traits associated with drought escape are rapid growth, early flowering, high leaf nitrogen level and high photosynthetic capacity [58]. The study in West Africa indicated that pearl millet matches its phenology to the mean distribution of the rainfall where precipitation is limited and erratic [61]. In this case, the development of the main panicle coincided with an increasing period of rain, thus reducing the risks associated with drought events occurring prior to or at the beginning of flowering.

Drought avoidance: Drought avoidance refers to the ability of the plant to maintain a favourable water balance under moisture stress in order to avoid water deficit in the plant tissue. Two types of drought avoidance mechanisms have been identified: (i) those that reduce water loss through transpiration (e.g. low stomata conductance and reduced leaf) and (ii) those that maintain water uptake during drought period (e.g. high root-to-shoot ratio) [56, 58, 62].

Drought tolerance: Drought tolerance refers to the ability of the plant to produce some yield by withstanding low water potential [62]. Traits associated with drought tolerance are increased osmoprotectants (or compatible solutes such as betaines and amino acids), and osmotic adjustment (i.e. reducing osmotic potential through accumulation of organic and inorganic substances) [58, 60].

Drought recovery: Drought recovery refers to a condition in which plants recover from the adverse effects of drought in order to provide some yield and/or biomass. Desiccation-tolerant or resurrection plants particularly the wild *Eragrostis nindensis* is the typical example of drought recovery since it stabilizes its cells or membranes at desiccated state [63].

These strategies which are devised by plants to cope with drought are manifested through changes in some phenotypic traits. In a recent review, Kooyers [58] showed for each strategy the path followed by plants in terms of life cycle, altered phenotypes and to the type of drought the plant fits itself. This indicates that the strategies and mechanisms of drought tolerance are interrelated.

4.2. Mechanism Of Drought Tolerance

Table 4 summarizes various mechanisms of drought tolerance in diverse millet types. These inherent properties of plants which include agronomical, morphological and physiological traits are briefly discussed below.

Table 4. Traits associated to diverse drought tolerance mechanisms in millets.

Parameter	Millet type	Response to drought	Reference
Agronomy-related traits			
Seed number and biomass	Pearl millet	Unaffected under drought	[64]
Seed yield	Pearl millet	High for drought-tolerant genotypes	[65]
Flowering time	Pearl millet	Adjust phenology to rainfall pattern	[53]
Morphology-related traits			
Shoot length	Little millet	Decreased under drought	[40]
Root length	Little millet	Increased under drought	[40]
Leaf tensile strength	Tef	Increased in drought-tolerant plants	[68]
Physiology-related traits			
Water extraction	Pearl millet	Less extraction before flowering; more extraction after flowering	[65]
Chlorophyll content	Little millet	Decreased under drought	[40]
Biochemical-related traits			
Anti-oxidants	Little millet	Accumulated under drought	[40]
ROS scavenging enzymes	Little millet, tef	Accumulated under drought	[40, 71]
Free proline	Tef, little millet	Increased concentration	[40, 71]
GB (glycine betaine)	Little millet	Accumulated under drought	[40]
Superoxide	Little millet	Accumulated under drought	[40]
AP (ascorbate peroxidase)	Tef, little millet	Increased specific activity	[40, 71]
CAT (catalase)	Little millet	Accumulated under drought	[40]
GR (glutathione reductase)	Tef	Increased concentration	[71]
MDAR (monodehydro-ascorbate reductase)	Tef	Increased concentration	[71]
Total free amino acid	Little millet	Increased concentration	[40]

Agronomy-related traits: These refer to the traits that are commonly known as yield and yield components. Among these, number of tillers, number and size of panicle, seed and biomass yield, seed weight and harvest index are the major ones. However, conclusions could not be made from the two studies using drought-tolerant pearl millet cultivars since drought did not affect the shoot biomass in the first case [64] while it boosted the seed yield in the second case [65].

Morphology-related traits: Morphological or anatomical traits which play important roles in drought tolerance include root- and shoot length and leaf area [66]. However, changes in the morphological and biochemical properties of the flag leaf play a key role in drought tolerance as flag leaves are the primary source of photosynthesis [67]. Mechanical properties of the plant also affect drought tolerance in millets. Balsamo et al. [68] studied the leaf tensile strength or also known as force to tear in three *Eragrostis* species with different levels of tolerance to drought. According to their findings, drought-tolerant *E. curvula* had higher tensile strength values than the moderately drought-tolerant *E. tef*, which in turn had higher values than the drought-susceptible *E. capensis*, indicating a positive correlation between drought tolerance and leaf tensile strength [68]. Structural investigations of leaves from the three species revealed the presence of extensive lignification of bundle sheath extensions in *E. tef* and *E. curvula* unlike in *E. capensis*. A study in maize indicated that lignification of the midrib parenchyma and epidermis was directly correlated with increased tensile strength [69].

Physiology-related traits: Among the several physiological traits that are differentially regulated during moisture deficit, osmotic adjustment is a major mechanism that increases drought avoidance to enable the plant produce some yield. Osmotic adjustment, which refers to the lowering of the osmotic potential in the cytoplasm due to the accumulation of compatible solutes such as proline, glycine betaine and organic acids, contributes to turgor maintenance of shoots and roots [40]. In little millet, drought stress increased the amount of proline and glycine betaine in both the root and leaf [40]. According to the authors, the accumulation of free amino acids in this millet during drought might be related to the disruption in protein synthesis, induced proteolysis or its partial hydrolysis [40].Water-use efficiency of the plant is also important as moisture is mostly limited in the areas where millets are extensively cultivated. The experiment using drought-sensitive and drought-tolerant pearl millet genotypes showed that under moisture deficit conditions, the total amount of water extracted by both genotypes was comparable [65]. However, compared to susceptible genotypes, tolerant genotypes extracted less water prior to flowering and more water after flowering, enabling these genotypes to support the tillers and maintain the stay-green phenotype.

Biochemical-related traits: Reactive oxygen species (ROS) are chemically reactive molecules that are useful in cell signalling at low concentrations but are damaging to cells when present at high concentrations. The main causes for the high production of ROS are environmental stresses such as drought and salinity [70]. In order to reduce the damaging effects of ROS, plants produce antioxidants, which include glutathione, ascorbate and carotenoids and ROS-scavenging enzymes which include superoxide dismutase (SOD), peroxidase (POD), catalase (CAT) and ascorbate peroxidase (AP or APX) [40]. In little millet, the activity of SOD, POD and CAT were elevated under drought conditions to enable the plant cope with unfavourable ROS accumulation [40]. Similarly, the activity of AP and monodehydro-ascorbate reductase (MDAR) increased in tef plants treated with drought compared to control plants receiving normal watering [71].

4.3. Genes involved in drought tolerance

The sequence of the genome and transcriptome of plants provides information important to the understanding of the types of genes involved in the regulation of drought tolerance, particularly in plants with increased resistant to moisture scarcity. So far, the genome of foxtail millet [72, 73] and tef [3] has been sequenced.

Transcriptome sequencing of millets after exposure to moisture-deficit condition provides information on genes differentially regulated under exposure to abiotic stresses particularly to drought. A transcriptome-wide study of finger millet plants exposed to drought obtained 2824 genes that were differentially expressed under these conditions [74].

Genes known to be involved in drought response and/or tolerance of selected millets are presented in Table 5. Wang et al. [75] indicated that the overexpression of SiLEA14, a type of LEA gene from foxtail millet, increased the tolerance of *Arabidopsis* plants to salt and osmotic stress. Parvathi et al. [76] reported the induction of several genes when finger millet was exposed to drought. The up-regulated genes include metallothionein, farnesylated protein ATFP6, Farnesyl pyrophosphate synthase and protein phosphatase 2A.

Traits associated with drought tolerance were investigated using a genome scan and association mapping methods [77, 78]. A single gene known as β-carbonic anhydrase (*Pg*CA) was consistently up-regulated in pearl millet exposed to multiple abiotic stresses including drought, salinity and heat [79]. Hence, this particular gene is useful in adapting the plant to diverse abiotic stresses. Other genes known to be involved in drought response or tolerance in millets were EcDehydrin 7 [80], mt1D [81] and Ec-apx1 [82] from finger millet, and SiARDP [83] from wild foxtail millet.

Although not yet reported for millets, the suppression of two genes, namely, SAL1 and ERA1, increased the drought tolerance of the model plant *Arabidopsisthaliana* [84, 85]. The *era1* mutants develop tolerance to drought through a mechanism involving closing of the stomata [85].

Table 5. Differentially Regulated Drought-Related Genes In Millets.

Gene name	Source of the gene	Test organism (type)	Reference
SiLEA	Foxtail millet	Overexpression in foxtail millet and Arabidopsis increased drought tolerance	[75]
SiARDP	Foxtail millet	Overexpression in foxtail millet and Arabidopsis increased drought tolerance	[83]
EcDehydrin7	Finger millet	Overexpression of EcDehydrin7	[80]
Ec-apx1	Finger millet	Expression increased under drought	[82]
Mt1D	bacteria	Finger millet expressing mt1D had better osmotic adjustment and chlorophyll retention under drought	[81]
Metallothionein,	Finger millet	Induced under drought	[76]
Farnesylated protein ATFP6	Finger millet	Induced under drought	[76]
Farnesyl pyrophosphate synthase	Finger millet	Induced under drought	[76]
Protein phosphatase 2A	Finger millet	Induced under drought	[76]
RISBZ4	Finger millet	Induced under drought	[76]
β-carbonic anhydrase (*Pg*CA)	Pearl millet	Up-regulated when exposed to drought	[79]

5. BREEDING MILLETS FOR EXTREME DROUGHT TOLERANCE

5.1. Germplasm acquisition and utilization

National and international efforts have been made to collect and maintain landraces of various millets types. The recent review by Goron and Naizanda [6] indicates the institutions involved in the preservation efforts and the amount of germplasm available at each institution. In general, India and China dominate the collections of millets. While institutions in India maintain 67% of the total of 33650 finger millet accessions, a single institution in China called the Chinese National Gene Bank preserves 61.2% of the total of 43,580 foxtail collections. Similarly, in the Ethiopian Institute of Biodiversity (EIB), over 5000 tef landraces collected from various tef-growing regions in the country are available [86]. Although these germplasm collections might not be exhaustive, they can play a key role in improving the productivity of respective crops. Further, large-scale expeditions need to be made for other millets in order to fully survey and bank the existing diversity in millets.

5.2. Breeding for drought tolerance

Breeding for drought tolerance is the major objective of many crop-breeding programmes due to the widespread prevalence of the moisture-deficit problem in global agriculture. A number of crops with drought tolerance have been developed. There are two options for the management of crops in water-limiting environments: the genetic and agronomic [87]. The genetic approach requires robust and reproducible screening methods for the identification of traits of drought tolerance in germplasm and breeding materials, and incorporation of the same into high-yielding varieties using conventional and biotechnological tools.

Crop breeding has relied for many years on conventional and ancient techniques such as selection and hybridization. Mutation breeding, the process of using chemicals or radiation to generate mutant plants with desirable traits, has also been used for several decades and has been a key in the release of over 2000 crop varieties to the farming community among which drought-tolerant cultivars are included [88]. Crop improvement techniques that apply modern genetic and omics (genomics, transcriptomics, proteomics and metabolomics) tools include the following: (i) *marker-assisted selection* (MAS) which refers to the utilization of molecular markers located near genes of interest to breed for traits that are difficult to observe, (ii) *TILLING* (targeting induced local lesions in genomes) [89] or EcoTILLING [90], the high-throughput and non-transgenic techniques which rapidly detect point mutations in mutagenized populations, and (iii) *Gene* targeting that relies on the following three tools to increase the efficiency of gene targeting: *zinc-finger nucleases* [91, 92], *TALEN* (transcription activator-like effector nuclease) [93] and *CRISPR/Cas* (clustered regularly interspaced short palindromic repeats)/(CRISPR associated), type II prokaryotic adaptive immune system [94, 95].

5.3. Improved crop management

The wise use of crop management practices which include the time of planting, frequency of tillage and the rate and time of fertilizer application is important particularly in the semi-arid regions where moisture is scarce. Flexibility to change from late maturing crops to early maturing crops when the rainfall arrives late in the season is important. In the central semi-arid regions of Ethiopia farmers start their season by planting sorghum in April. When sorghum fails due to late arrival of rain, they sow wheat in June. However, if the rain is still late or not enough for wheat plant establishment, farmers sow tef in July or early August as the last option.

Compared to sorghum and wheat, tef requires less moisture and matures early.

Suggestions have been earlier given on the type of technologies to be adopted in the semi-arid regions of Southern Africa [96] and West Africa [97]. According to Mir and colleagues, these technologies should include genomics, physiology and breeding [98].

5.4. Agricultural inputs and insurance

Access to agricultural inputs such as improved seeds, fertilizer and chemicals as well as credit and markets is important for farmers. In semi-arid areas where millets are dominantly cultivated, the amount and pattern of rainfall is erratic. Due to this, an insurance system known as Weather Index Drought Insurance has been implemented for the last decade in several African countries including Niger [99], Ghana [100], Kenya [101] and Burkina Faso [102] as well as India [103]. The successful insurance organization called 'Kilimo Salama' which was initially established by Syngenta Foundation for Sustainable Agriculture (SFSA) and implemented in several East African countries has been recently transferred to the Agriculture and Climate Risk Enterprise Ltd. (ACRE) [104, 105].

5.5. Partnership in research and development

Collaborations among national and international institutions are required in both research and development, in order first to develop improved millet cultivars and later to disseminate them to the farming community. Among the institutions with a global mandate to improve millets, ICRISAT (International Crops Research Institute for Semi-Arid Tropics) has recently added tef to the list of its mandate crops [106]. With its headquarters in Patancheru, India and regional officers in Nairobi (Kenya) and Bamako (Mali), it has been focusing on the improvement of diverse millets. The centre is among the 15 international agricultural research centers that belong to the CGIAR (Consultative Group for International Agricultural Research), the global partnership that unites organizations engaged in research for food security. Hence, the research and development of tef, a vital crop in the Horn of Africa that feeds over 50 million people in Ethiopia alone, will receive a global partnership towards its improvement and development. In general, suggestions given to the improvement of understudied or orphan crops [107, 108] could also be applied to the research and development of millets.

6. CONCLUSIONS

Millets play a significant role in the livelihood of the population of developing world especially due to their enormous contribution to the food security of these countries. However, these crops have not been sufficiently studied and hence have been named orphan crops. Both conventional and modern improvement techniques have not yet been adequately implemented. It is believed that the changing climate will have significant effects on the types of crops cultivated in the next century. Currently, widely cultivated crops that provide the daily diet for many (such as wheat) might not be extensively cultivated in the future due to environmental stresses, especially the increase in global temperature. Millets might provide alternative climate-smart crops, as their adaptations to challenging environment are better than the current major crops of the world. Enhancing the productivity of millets requires concreted efforts of breeders, agronomists, policy makers and donors at both individual and institutional capacities.

8. ACKNOWLEDGEMENTS

I would like to thank Syngenta Foundation for Sustainable Agriculture, SystemsX.ch and the University of Bern for supporting the Tef Improvement Project based at the Institute of Plant Sciences, University of Bern, Switzerland.

REFERENCES

1. Seetharam, A., K.W. Riley, and G. Harinarayana, eds. *Small millets in global agriculture: Proceedings of the First International Small Millets Workshop, 29 October to 2 November 1986, Bangalore, India*. 1989, Oxford & IBH Publishing Co. Pvt. Ltd: New Delhi 413.
2. ICRISAT-FAO, *The world sorghum and millet economies: facts, trends and outlook.* 1996: International Crops Research Institute for the Semi-Arid Tropics, India & Food and Agriculture Organization of the United Nations, Rome, Italy.
3. Cannarozzi, G., et al., *Genome and transcriptome sequencing identifies breeding targets in the orphan crop tef (Eragrostis tef).* BMC Genomics, 2014. 15: p. 581.
4. Martel, E., et al., *Genome size variation and basic chromosome number in pearl millet and fourteen related Pennisetum species.* Journal of Heredity, 1997. 88(2): p. 139–143.
5. Mann, D.G.J., et al., *Switchgrass (Panicum virgatum L.) polyubiquitin gene (PvUbi1 and PvUbi2) promoters for use in plant transformation.* BMC Biotechnology, 2011. 11: p. 74.

6. Goron, T.L. and M.N. Raizada, *Genetic diversity and genomic resources available for the small millet crops to accelerate a New Green Revolution.* Frontiers in Plant Science, 2015. 6.
7. FAOSTAT. *Crop production.* 2015 [cited 2015 September 7]; Available from: http://faostat3.fao.org/home/E.
8. Bantilan, M.C.S. and U.K. Deb, *Impacts of genetic enhancement in pearl millet,* in *Crop Improvements and Its Effect on Productivity: The Impact of International Agricultural Research,* R.E. Evenson and D. Gollin, Editors. 2003, CABI International: UK. p. 215–240.
9. CSA, *Agricultural Sample Survey for 2013/14,* in *Statistical Bulletin 532.* 2014: Addis Ababa, Ethiopia.
10. Belton, P.S. and J.R.N. Taylor, *Sorghum and millets: protein sources for Africa.* Trends in Food Science & Technology, 2004. 15(2): p. 94–98.
11. Baker, R.D., *Millet production. Guide A-414* 2003: Cooperative Extension Service, College of Agriculture and Home Economics, New Mexico University, Las Cruces, USA.
12. Lu, H.Y., et al., *Earliest domestication of common millet (Panicum miliaceum) in East Asia extended to 10,000 years ago.* Proceedings of the National Academy of Sciences of the United States of America, 2009. 106(18): p. 7367–7372.
13. Jones, M.K. and X. Liu, *Archaeology. Origins of agriculture in East Asia.* Science, 2009. 324(5928): p. 730–731.
14. Brutnell, T.P., et al., *Setaria viridis: a model for C4 photosynthesis.* The Plant cell, 2010. 22(8): p. 2537–2544.
15. Warner, D.A. and G.E. Edwards, *C-4 Photosynthesis and Leaf Anatomy in Diploid and Autotetraploid Pennisetum-Americanum (Pearl-Millet).* Plant Science, 1988. 56(1): p. 85–92.
16. Doust, A.N., et al., *Foxtail Millet: A Sequence-Driven Grass Model System.* Plant Physiology, 2009. 149(1): p. 137–141.
17. Xu, W., et al., *A novel antifungal peptide from foxtail millet seeds.* Journal of the Science of Food and Agriculture, 2011. 91(9): p. 1630–1637.
18. Nolde, S.B., et al., *Disulfide-stabilized Helical Hairpin structure and activity of a novel antifungal peptide EcAMP1 from seeds of barnyard grass (Echinochloa crus-galli).* Journal of Biological Chemistry, 2011. 286(28): p. 25145–25153.
19. Saleh, A.S.M., et al., *Millet grains: nutritional quality, processing, and potential health benefits.* Comprehensive Reviews in Food Science and Food Safety, 2013. 12: p. 281–295.
20. NAS, *Lost crops of Africa. Vol. I. Grains.* 1996, Washington, DC: National Academy of Science.
21. Ravindran, G., *Studies on Millets – Proximate Composition, Mineral-Composition, and Phytate and Oxalate Contents.* Food Chemistry, 1991. 39(1): p. 99–107.
22. Sripriya, G., U. Antony, and T.S. Chandra, *Changes in carbohydrate, free amino acids, organic acids, phytate and HCl extractability of minerals during germination and fermentation of finger millet (Eleusine coracana).* Food Chemistry, 1997. 58(4): p. 345–350.
23. Kalinova, J. and J. Moudry, *Content and quality of protein in proso millet (Panicum miliaceum L.) varieties.* Plant Foods for Human Nutrition, 2006. 61(1): p. 45–49.
24. Temple, V.J. and J.D. Bassa, *Proximate Chemical-Composition of Acha (Digitaria-Exilis) Grain.* Journal of the Science of Food and Agriculture, 1991. 56(4): p. 561–563.
25. Vodouhè, R., *Promoting fonio production in West and Central Africa through germplasm management and improvement of post-harvest technology,* 2004: Benin. p. 18.
26. Ragaee, S., E.M. Abdel-Aal, and M. Noaman, *Antioxidant activity and nutrient composition of selected cereals for food use.* Food Chemistry, 2006. 98(1): p. 32–38.
27. FAO, *Sorghum and millets in human nutrition.* FAO Food and Nutrition Series No 27. 1995., Rome: FAO (Food and Agriculture Organization).

28. Léder, I., *Sorghum and millets*. Füleky, G. Editors, Cultivated Plants Primarily as Food Sources. 2004, UNESCO, Eolss Publishers, Oxford, UK.
29. Yami, A., *Tef straw: a valuable feed resource to improve animal production and productivity.*, in *Achievements and prospects of tef improvement.*, K. Assefa, S. Chanyalew, and Z. Tadele, Editors. 2013, EIAR-Uni. Bern: Bern. p. 233–251.
30. Provost, C. and E. Jobson, *Move over quinoa, Ethiopia's teff poised to be next big super grain*, in *The Guardian*. 2014.
31. NYDailyNews. *Ethiopia's teff grain set to be the world's next superfood. New York Daily News.* 2014 [cited 2015 September 14]; Available from: http://www.nydailynews.com/life-style/health/teff-world-grain-superfood-article-1.1716593.
32. Jeffrey, J. *Will Ethiopia's teff be the next 'super grain'? BBC Business.* 2015 [cited 2015 September 14]; Available from: http://www.bbc.com/news/business-32128441.
33. Spaenij-Dekking, L., Y. Kooy-Winkelaar, and F. Koning, *The Ethiopian cereal tef in celiac disease.* The New England journal of medicine, 2005. 353(16): p. 1748–1749.
34. Chandrasekara, A. and F. Shahidi, *Antiproliferative potential and DNA scission inhibitory activity of phenolics from whole millet grains.* Journal of Functional Foods, 2011. 3(3): p. 159–170.
35. Bhatt, D., et al., *Responses to drought induced oxidative stress in five finger millet varieties differing in their geographical distribution.* Physiology and molecular biology of plants: an international journal of functional plant biology, 2011. 17(4): p. 347–353.
36. Rao, B.R., M.H. Nagasampige, and M. Ravikiran, *Evaluation of nutraceutical properties of selected small millets.* Journal of Pharmacy & Bioallied Sciences, 2011. 3(2): p. 277–279.
37. Chandrashekar, A., *Finger Millet Eleusine coracana.* Advances in food and nutrition research, 2010. 59: p. 215–262.
38. Itagi, S., R. Naik, and N. Yenag, *Versatile little millet therapeutic mix for diabetic and non-diabetics.* Asian Journal of Science and Technology, 2013. 4: p. 33–35.
39. Pradeep, S.R. and M. Guha, *Effect of processing methods on the nutraceutical and antioxidant properties of little millet (Panicum sumatrense) extracts.* Food Chemistry, 2011. 126(4): p. 1643–1647.
40. Ajithkumar, I.P. and R. Panneerselvam, *ROS Scavenging System, Osmotic Maintenance, Pigment and Growth Status of Panicum sumatrense Roth. Under Drought Stress.* Cell Biochemistry and Biophysics, 2014. 68(3): p. 587–595.
41. Ramamasy, S. and S. Baas, *Climate variability and change: adaptation to drought in Bangladesh: A resource book and training guide.* 2007, Rome: Food and Agriculture Organization of the United Nations (FAO). 66.
42. Webb, P. and J. Von Braun, *Famine and Food security in Ethiopia.* 1994: Wiley, New York.
43. UN, *Global Drylands: A UN system-wide response. Prepared by the Environment Management Group.* 2011: United Nations.
44. Wikipedia. *Sahel.* Available from: https://en.wikipedia.org/wiki/Sahel.
45. Gebrekirstos, A., et al., *Climate-growth relationships of the dominant tree species from semi-arid savanna woodland in Ethiopia.* Trees-Structure and Function, 2008. 22(5): p. 631–641.
46. Riché, B., et al., *Climate-related vulnerability and adaptive-capacity in Ethiopia's Borana and Somali communities. Final assessment report of CARE International and Save the Children UK.* 2009, International Institute for Sustainable Development (IISD). p. 82.
47. Assefa, K., et al., *Breeding tef [Eragrostis tef (Zucc.) trotter]: conventional and molecular approaches.* Plant Breeding, 2011. 130(1): p. 1–9.
48. Rosell, S. and B. Holmer, *Rainfall change and its implications for belg harvest in South Wollo, Ethiopia.* Geografiska Annaler Series a-Physical Geography, 2007. 89A(4): p. 287–299.

49. Cheung, W.H., G.B. Senay, and A. Singh, *Trends and spatial distribution of annual and seasonal rainfall in Ethiopia*. International Journal of Climatology, 2008. 28(13): p. 1723–1734.
50. ABCIC, *Effects of climate change on Eragrostis tef in Ethiopia: a call for action to avert food security crisis, in ABCIC Policy Brief No.1*. 2011.
51. Matsuura, A., et al., *Effect of pre- and post-heading water deficit on growth and grain yield of four millets*. Plant Production Science, 2012. 15(4): p. 323–331.
52. Winkel, T., J.F. Renno, and W.A. Payne, *Effect of the timing of water deficit on growth, phenology and yield of pearl millet [Pennisetum glaucum (L.) R Br] grown in Sahelian conditions*. Journal of Experimental Botany, 1997. 48(310): p. 1001–1009.
53. Bidinger, F.R., V. Mahalakshmi, and G.D.P. Rao, *Assessment of Drought Resistance in Pearl-Millet [Pennisetum-Americanum (L) Leeke].2. Estimation of Genotype Response to Stress*. Australian Journal of Agricultural Research, 1987. 38(1): p. 49–59.
54. Maqsood, M. and A.N.A. Ali, *Effects of drought on growth, development, radiation use efficiency and yield of finger millet (Eleucine coracana)*. Pakistan Journal of Botany, 2007. 39(1): p. 123–134.
55. Takele, A., *Genotypic variability in dry matter production, partitioning and grain yield of tef [Eragrostis tef (Zucc.) Trotter] under moisture deficit*. SINET: Ethiopian Journal of Science, 1997. 20: p. 177–188.
56. Fang, Y.J. and L.Z. Xiong, *General mechanisms of drought response and their application in drought resistance improvement in plants*. Cellular and Molecular Life Sciences, 2015. 72(4): p. 673–689.
57. Kholová, J., *Understanding of terminal drought tolerance mechanisms in pearl millet [Pennisetum glaucum (L.) R. Br.]* in *Faculty of Science* 2010, Charles University in Prague Prague, p. 115.
58. Kooyers, N.J., *The evolution of drought escape and avoidance in natural herbaceous populations*. Plant Science, 2015. 234: p. 155–162.
59. Monneveux, P. and J.P. Ribaut, *Secondary traits for drought tolerance improvement in cereals*, in *Drought adaptation in cereals*, J.M. Ribaut, Editor. 2006, Food Products Press: New York. p. 97–143.
60. Blum, A., *Drought resistance, water-use efficiency, and yield potential - are they compatible, dissonant, or mutually exclusive?* Australian Journal of Agricultural Research, 2005. 56(11): p. 1159–1168.
61. Sivakumar, M.V.K., *Empirical-analysis of dry spells for agricultural applications in West Africa*. Journal of Climate, 1992. 5(5): p. 532–539.
62. Agriinfo. *Plant Breeding for Drought Resistance*. 2015 [cited 2015 September 16]; Available from: http://www.agriinfo.in/default.aspx?page=topic&superid=3&topicid=2152.
63. Vander Willigen, C., et al., *Mechanical stabilization of desiccated vegetative tissues of the resurrection grass Eragrostis nindensis: does a TIP 3;1 and/or compartmentalization of subcellular components and metabolites play a role?* Journal of Experimental Botany, 2004. 55(397): p. 651–661.
64. Aparna, K., et al., *Seed number and 100-seed weight of pearl millet [Pennisetum glaucum L.) respond differently to low soil moisture in genotypes contrasting for drought tolerance*. Journal of Agronomy and Crop Science, 2014. 200(2): p. 119–131.
65. Vadez, V., et al., *Small temporal differences in water uptake among varieties of pearl millet [Pennisetum glaucum (L.) R. Br.] are critical for grain yield under terminal drought*. Plant and Soil, 2013. 371(1-2): p. 447–462.
66. Shao, H.B., et al., *Water-deficit stress-induced anatomical changes in higher plants*. C. R. Biologies, 2008. 331(3): p. 215–225.
67. Biswal, A.K. and A. Kohli, *Cereal flag leaf adaptations for grain yield under drought: knowledge status and gaps*. Molecular Breeding, 2013. 31(4): p. 749–766.
68. Balsamo, R.A., et al., *Drought tolerance of selected Eragrostis species correlates with leaf tensile properties*. Annals of Botany, 2006. 97(6): p. 985–991.

69. Balsamo, R.A. and J.A.J. Orkwiszewski, *Leaf architecture, lignification, and tensile strength during vegetative phase change in Zea mays* Acta Societatis Botanicorum Poloniae, 2008. 77(3): p. 181–188.
70. Sharma, P. and R.S. Dubey, *Drought induces oxidative stress and enhances the activities of antioxidant enzymes in growing rice seedlings.* Plant Growth Regulation, 2005. 46(3): p. 209–221.
71. Smirnoff, N. and S.V. Colombe, *Drought influences the activity of enzymes of the chloroplast hydrogen-peroxide scavenging system.* Journal of Experimental Botany, 1988. 39(205): p. 1097–1108.
72. Zhang, G., et al., *Genome sequence of foxtail millet (Setaria italica) provides insights into grass evolution and biofuel potential.* Nature Biotechnology, 2012. 30(6): p. 549–554.
73. Bennetzen, J.L., et al., *Reference genome sequence of the model plant Setaria.* Nature Biotechnology, 2012. 30(6): p. 555–561.
74. Qi, X., et al., *Genome-wide annotation of genes and noncoding RNAs of foxtail millet in response to simulated drought stress by deep sequencing.* Plant Molecular Biology, 2013. 83(4-5): p. 459–473.
75. Wang, M.Z., et al., *SiLEA14, a novel atypical LEA protein, confers abiotic stress resistance in foxtail millet.* BMC Plant Biology, 2014. 14: p. 290.
76. Parvathi, M.S., et al., *Expression analysis of stress responsive pathway genes linked to drought hardiness in an adapted crop, finger millet (Eleusine coracana).* Journal of Plant Biochemistry and Biotechnology, 2013. 22(2): p. 193–201.
77. Mariac, C., et al., *Genetic basis of pearl millet adaptation along an environmental gradient investigated by a combination of genome scan and association mapping.* Molecular Ecology, 2011. 20(1): p. 80–91.
78. Yadav, R.S., et al., *Mapping and characterisation of QTL x E interactions for traits determining grain and stover yield in pearl millet.* Theoretical and Applied Genetics, 2003. 106(3): p. 512–520.
79. Kaul, T., et al., *Biochemical and molecular characterization of stress-induced beta-carbonic anhydrase from a C(4) plant, Pennisetum glaucum.* Journal of Plant Physiology, 2011. 168(6): p. 601–610.
80. Singh, R.K., et al., *Isolation and characterization of drought responsive EcDehydrin7 gene from finger millet [Eleusine coracana (L.) Gaertn.].* Indian Journal of Genetics and Plant Breeding, 2014. 74(4): p. 456–462.
81. Hema, R., et al., *Stable expression of mtlD gene imparts multiple stress tolerance in finger millet.* Plos One, 2014. 9(6): p. e99110.
82. Bhatt, D., et al., *Cloning, expression and functional validation of drought inducible ascorbate peroxidase (Ec-apx1) from Eleusine coracana.* Molecular Biology Reports, 2013. 40(2): p. 1155–1165.
83. Li, C., et al., *An ABA-responsive DRE-binding protein gene from Setaria italica, SiARDP, the target gene of SiAREB, plays a critical role under drought stress.* Journal of Experimental Botany, 2014. 65(18): p. 5415–5427.
84. Wilson, P.B., et al., *The nucleotidase/phosphatase SAL1 is a negative regulator of drought tolerance in Arabidopsis.* The Plant Journal: For Cell and Molecular Biology, 2009. 58(2): p. 299–317.
85. Wang, Y., et al., *Molecular tailoring of farnesylation for plant drought tolerance and yield protection.* The Plant Journal: For Cell and Molecular Biology, 2005. 43(3): p. 413–424.
86. Tesema, A., *Genetic diversity of tef in Ethiopia*, in *Achievements and Prospects of Tef Improvement*, A. Assefa, S. Chanyalew, and A. Tadele, Editors. 2013, EIAR-University of Bern: Bern, Switzeralnd. p. 15–20.
87. Saxena, N.P. and C. John, eds. *Field screening for drought tolerance in crop plants with emphasis on rice.* 2002, ICRISAT, Patancheru, India. 211.
88. Ahloowalia, B.S., M. Maluszynski, and K. Nichterlein, *Global impact of mutation-derived varieties*. Euphytica, 2004. 135(2): p. 187–204.

89. McCallum, C.M., et al., *Targeting induced local lesions IN genomes (TILLING) for plant functional genomics*. Plant Physiology, 2000. 123(2): p. 439–442.
90. Comai, L., et al., *Efficient discovery of DNA polymorphisms in natural populations by Ecotilling*. Plant Journal, 2004. 37(5): p. 778–786.
91. Shukla, V.K., et al., *Precise genome modification in the crop species Zea mays using zinc-finger nucleases*. Nature, 2009. 459(7245): p. 437–441.
92. Townsend, J.A., et al., *High-frequency modification of plant genes using engineered zinc-finger nucleases*. Nature, 2009. 459(7245): p. 442–445.
93. Cermak, T., et al., *Efficient design and assembly of custom TALEN and other TAL effector-based constructs for DNA targeting (vol 39, pg e82, 2011)*. Nucleic acids research, 2011. 39(17): p. 7879–7879.
94. Miao, J., et al., *Targeted mutagenesis in rice using CRISPR-Cas system*. Cell Research, 2013. 23(10): p. 1233–1236.
95. Jiang, W., et al., *Demonstration of CRISPR/Cas9/sgRNA-mediated targeted gene modification in Arabidopsis, tobacco, sorghum and rice*. Nucleic Acids Research, 2013. 41(20): p. e188.
96. Gwata, E.T. and J. Mzezewa, *Optional crop technologies at a semi-arid ecotope in southern Africa*. Journal of Food Agriculture & Environment, 2013. 11(2): p. 291–295.
97. Haussmann, B.I.G., et al., *Breeding strategies for adaptation of pearl millet and sorghum to climate variability and change in West Africa*. Journal of Agronomy and Crop Science, 2012. 198(5): p. 327–339.
98. Mir, R.R., et al., *Integrated genomics, physiology and breeding approaches for improving drought tolerance in crops*. Theoretical and Applied Genetics, 2012. 125(4): p. 625–645.
99. Leblois, A., et al., *Weather Index drought insurance: An ex ante evaluation for millet growers in Niger*. Environmental & Resource Economics, 2014. 57(4): p. 527–551.
100. Molini, V., et al., *Safety nets and index-based insurance: historical assessment and semiparametric simulation for Northern Ghana*. Economic Development and Cultural Change, 2010. 58(4): p. 671–712.
101. Chantarat, S., et al., *Improving humanitarian response to slow-onset disasters using famine-indexed weather derivatives*. Agricultural Finance Review, 2008. 68(1): p. 169–195.
102. Berg, A., P. Quirion, and B. Sultan, *Can weather index drought insurance benefit to Least Developed Countries' farmers? A case study on Burkina Faso*. Weather, Climate and Society, 2009. 1: p. 7184.
103. Zant, W., *Hot stuff: Index insurance for Indian smallholder pepper growers*. World Development, 2008. 36(9): p. 1585–1606.
104. SFSA. *Kilimo Salama is ACRE*. 2014; Available from: http://www.syngentafoundation.org/index.cfm?pageID=562.
105. ACRE. *Acre Africa: Agriculture and Climate Risk Enterprise Ltd. (ACRE)*. Available from: http://acreafrica.com/.
106. ICRISAT. *Tef: New superfood crop in ICRISAT's portfolio*. 2015 [cited 2015 September 7]; Available from: http://www.icrisat.org/newsroom/latest-news/happenings/happenings1689.htm.
107. Tadele, Z. and K. Assefa, *Increasing Food Production in Africa by Boosting the Productivity of Understudied Crops*. Agronomy, 2012. 2(4): p. 240–283.
108. Tadele, Z., *Role of crop research and development in food security of Africa*. International Journal of Plant Biology and Research, 2014. 2(3): p. 1019.

CHAPTER 17

Histone Methylation - A Cornerstone for Plant Responses to Environmental Stresses?

Roberto Bobadilla[1] and Alexandre Berr[1]

[1] Institut de Biologie Moléculaire des Plantes, CNRS UPR2357, Université de Strasbourg, Strasbourg, France

ABSTRACT

Plants are sessile organisms and, as such, their survival relies on their ability to respond quickly all along their life cycle to any kind of environmental stimuli, including abiotic and biotic stresses. In this respect, plants have developed efficient mechanisms of protection and/or adaptation to minimize deleterious effects of stress on their growth and development. In a stress type-dependent manner, external signals are firstly sensed. This step is then followed by the activation of particular signalling pathways, resulting ultimately in the rapid and specific modulation of the plant transcriptome. Currently, transcriptional regulation is considered as a central process in the build-up of plant responses to both abiotic and biotic stresses. Among mechanisms involved in transcriptional regulation, the combined effect of different histone tail post-translational modifications (PTMs; e.g. acetylation and methylation) through the activity of particular histone-modifying enzymes can lead to changes in the local chromatin structure environment and hence the underlying DNA accessibility.By focusing on histone lysine methylation, in this chapter we highlight our current understanding of the transcriptional roles played by chromatin-remodelling mechanisms in regulating plant response/adaptation to different biotic and abiotic stresses. Based on recent advances, we further discuss the stability and transmission of such methylation marks to subsequent generations, with the underlying idea of an epigenetically based transcriptional memory of stresses in plants.

Keywords: Histone methylation and demethylation, histone methy ltransferases and demethylases, biotic and abiotic stresses

1. INTRODUCTION

Stress, as we currently think of it, is a highly subjective phenomenon defined as a state of threatened homeostasis. Depending on their nature, external stresses are usually divided into biotic (i.e. herbivorous insects and pathogens such as fungi, bacteria and viruses) or abiotic (i.e. including, among others, high or low temperature, submergence or drought and salinity). During their lifetime, all living organisms inevitably and constantly face all sorts of environmental stresses that often occur suddenly and/or simultaneously. Classically, different strategies can be applied to minimize deleterious effects of stresses, such as resistance, tolerance, avoidance or escape. Being sessile, plants cannot escape and are therefore more prone to the deleterious effect of unfavourable environmental growth conditions. Because responses are critical to ensure their survival, plants have developed specific and efficient strategies that allow them to precisely perceive different environmental stresses and respond and/or adapt to them [1, 2]. In addition to preformed defence traits, plants have evolved inducible defence strategies. Indeed, upon perception, each stress will raise a complex and more or less specific repertoire of cellular and molecular responses implemented by the plant to minimize or prevent damage. Particularly, the stimulation of a given stress-signalling pathway after pathogen detection will be integrated into the plant cell nucleus through a set of regulatory transcription factor cascades, which prioritizes defence over growth-related cellular functions, while conserving enough valuable resources for survival and reproduction [3, 4]. Supporting the idea that the capacity of a plant to rapidly reprogramme its gene expression at the transcriptional level is an essential and common component of all plant response strategies to stress and disturbance; more than 1,000 transcription factors were found to be involved in stress responses [5, 6]. Because eukaryotic genes function in the context of chromatin, modifications and remodelling of the chromatin configuration from permissive for transcription to restrictive, and vice versa, may be an integral part of mechanisms involved in this vital transcriptional reprogramming. In this chapter, we review and discuss the current knowledge about the functional impact of chromatin changes on the transcriptional regulation of genes under different stress conditions, with particular emphasis on histone methylation/demethylation.

2. CHROMATIN STRUCTURE AND HISTONE METHYLATION/DEMETHYLATION

In eukaryotes, genomic DNA in the cell nucleus is packaged in a complex and evolutionarily conserved structure named chromatin, with nucleosome as the basic unit. The nucleosome complex contains about 160–241 base pairs (bp) of DNA, a nucleosome core particle and the H1 linker histone. The nucleosome core particle is composed of an octamer of core histones, consisting of two H3–H4 dimers associated with two H2A–H2B dimers. About 146 bp of DNA is wrapped in ~1.65 negatively supercoiled circles around the histone octamer, while the linker DNA associated with H1 varies in length from 8 to 114 bp [7]. At first sight, the chromatin as it is described appears as a barrier, restricting the access of all kinds of enzymes that process the DNA. However, nucleosomes are not merely static but highly dynamic entities. Indeed, nucleosomes can be moved, stabilized/destabilized, disassembled/reassembled at particular genome locations in response to specific environmental signals or developmental cues [8]. This dynamic leads to a wide range of chromatin condensation states modulating the DNA accessibility, with euchromatin, being relaxed, and heterochromatin, being compacted. Therefore, in eukaryotic cells, an intimate connection exists between the structural organization of the genome and its functioning. For this reason, the level of chromatin condensation is directly related to all aspects of DNA metabolism, thus playing a major role in regulating transcription, DNA replication, DNA repair, recombination, transposition and chromosome segregation. In plants, changes in the chromatin structure were reported to affect various biological processes such as root growth, flowering, organogenesis, gametophyte or embryo formation [9–11].

In the nucleosome core particle, histones H2A, H2B, H3 and H4 possess two common regions, a histone-fold domain and a histone tail. The histone-fold domain is the most conserved region and the main element of histone dimerization [12]. The tail protrudes from the nucleosome core particle and is more variable and unstructured than the fold [13]. All four core histones have an N-terminal tail domain, but only histone H2A has an additional long

C-terminal tail. Histone tails are extremely basic due to their particularly high content in basic amino acid residues, such as lysine and arginine [14]. Resulting positive charges allow them to closely associate with the negatively charged nucleosomal DNA through electrostatic interactions [15]. In addition, histone tails, especially N-terminal ones, may undergo diverse types of post-translational modifications such as acetylation or methylation. The great diversity of these modifications as well as the high number of amino acid residues that can be modified within histone tails, and the correlation between these modifications and various nuclear processes, lead to the hypothesis that the specific combination of histone modifications constitute a histone 'code' [16].

Technically, these histone marks can be localized by chromatin immunoprecipitation (ChIP) using specific antibodies against the modification [17]. Briefly, protein–DNA interactions are stabilized by cross-linking with formaldehyde; chromatin is sheared into small pieces to facilitate analysis and then immunoprecipitated using an antibody raised against a specific histone modification. Following enrichment, cross links are reversed to release DNA, which is then quantified by polymerase chain reaction (PCR) to measure the relative amount of the specific histone mark on selected plant genes. ChIP can also be combined with microarray hybridization (ChIP-chip) or high-throughput sequencing (ChIP-Seq), allowing the genome-wide discovery of DNA–histone modification interactions.

Methylation is the most abundant one compared with other histone PTMs. It can occur at both lysine (K) and arginine (R) residues of core histone tails. Further extending the indexing potential of this modification, mono-, di- and trimethylation of lysine and mono- and dimethylation (symmetric or asymmetric) of arginine are common at N-terminal tails of H2A, H2B, H3 and H4. Although histone acetylation is generally associated with active gene transcription, histone methylation can be associated with either active or silent gene expression, depending upon the histone, the methylated residue or the level (mono-, di- or tri-) of methylation. In *Arabidopsis*, genome-wide analyses revealed that trimethylations of H3K4 and H3K36 (H3K4me3 and H3K36me3) are generally enriched at actively transcribed genes, whereas H3K27me3 is associated with repressed genes and H3K9me2 and H4K20me1 are enriched at constitutive heterochromatin and silenced transposons [18]. For histone arginine methylation, a definitive role has not yet been clearly established. However, because the level of symmetric H3R2me2 and H4R3me2 was negatively correlated with the level of H3K4me3, a well-known mark reflecting active transcription, high levels of H3R2me2 and H4R3me2 are thought to cause transcriptional repression [19–21]. In contrast, asymmetric H4R3me2 was associated with gene activation [22, 23].

Histone methylation is relatively stable and can be established on lysine and arginine by two distinct families of enzymes, the histone lysine methyltransferases (HKMTs), all containing the evolutionary conserved catalytic SET domain in plants [24], and the protein arginine methyltransferases (PRMTs) [25], respectively. As a counterpart, methyl groups on histone can also be removed by at least two evolutionarily conserved classes of histone demethylases, the lysine-specific demethylase1 (LSD1) type and the Jumonji C (JmjC) domain-containing demethylases [26]. Histone methyltransferases and demethylases are well conserved in angiosperms and have been identified and classified on the basis of phylogenetic analyses and domain organization in several plants, including *Arabidopsis*, maize, tomato, rice, grapevine and *Brassica rapa*, [27–32]. However, cellular and molecular functions of many of these modifiers have not yet been addressed.

Although histone acetylation can directly modulate the chromatin structure, arginine and lysine methylation of histone tails can promote or prevent the docking of key transcriptional effector molecules, named readers, needed to 'translate' the code in order to determine the functional and structural outcome of the corresponding PTMs. Just as there are a large number of PTMs on histone tails, there are also numerous protein domains that recognize and bind to particular PTMs on these tails. For example, PTM-recognition domains such as plant homeodomain (PHD) fingers, chromodomains and Tudor domains all recognize methylated lysine residues [33].

3. HISTONE METHYLATION CHANGES ASSOCIATED WITH BIOTIC STRESS CONDITIONS

Biotic stress is the result of the damage done to plants by insects or pathogens, such as bacteria or fungi. Plant pathogens are generally divided into two distinct categories: biotrophs, which colonize living plant tissue and obtain nutrients from living host cells, and necrotrophs, which depend on dead host tissue for nutrients and reproduction. To fend off pathogens with different infection strategies, plants have evolved complex defence mechanisms. Classically, the pathogen-sensing machinery induces signalling cascades that promote the accumulation of hormones such as salicylic acid (SA) or jasmonic acid (JA)/ethylene (ET) [34]. These hormones then orchestrate the overall plant defence reaction locally and systemically by inducing the transcriptional activation of defence genes through an intricate signalling network. In this part, we highlight recent examples illustrating how histone methylations condition major steps leading to immunity, ranging from initial pathogen perception to hormonal homeostasis changes for antimicrobial effector expression.

3.1. Histone Methylation/Demethylation in The Defence Against Biotrophic Pathogens

The phytohormone SA plays an important role in plant defence, from the induction of pathogen resistance (*PR*) genes against biotrophic bacteria (e.g. *Pseudomonas syringae*) to the establishment of systemic acquired resistance (SAR) [35]. Several studies suggested that the SA signalling pathway is notably controlled by histone methylation. Under normal growth conditions, *Arabidopsis* mutants for *SNI1* (*Suppressor of NPR1, Inducible*), a negative regulator of SAR required to dampen the basal expression of *PR* genes, presented an increased H3K4me2 on *PR1* [36]. Rather than being a constitutive mark of transcription, H3K4me2 was proposed to be involved in the fine-tuning of tissue-specific expression [37]. Using the functional SA-analogue S-methyl benzo [1,2,3] thiadiazole-7-carbothioate (BTH), an increased level of H3K4me2 on *PR1* was observed in wild-type plants 48 h after treatment and was not detected in mutants. Interestingly, when expressed in yeast, SNI1 also repressed transcription, suggesting a highly conserved mechanism of transcriptional repression. These results together with the structural similarity of SNI1 with armadillo repeat (ARM) proteins (i.e. a motif known to mediate protein–protein interactions) imply that SNI1 may form a scaffold for interaction with proteins that modulates the chromatin structure of *PR* genes, thus repressing their transcription. In addition, the presence of H3K4me2 detected on *PR1* before induction suggested that this mark is readily in place, providing the appropriate chromatin configuration for the efficient induction of *PR1* upon need. Using a similar approach, Alvarez-Venegas et al. [38] reported no significant changes in levels of H3K4me2 and H3K4me3 on *PR1* 24 h after the SA treatment [38]. This discrepancy may reflect differences in experimental conditions. Indeed, the action of the so-called 'SA-analog' BTH on gene transcription is significantly broader than the action of SA itself [39]. Moreover, samplings were performed 48 h versus 24 h after treatment. Together, because the H3K4 methylation increase does not occur immediately after the induction of *PR1*, this mark may not be directly related to the transcriptional induction itself, but later, for the maintenance/reinforcement of *PR1* expression.

The ARABIDOPSIS HOMOLOG OF TRITHORAX (ATX1) is a H3K4 trimethyltransferase providing basal resistance against *Pseudomonas syringae pv. tomato* (*Pst*); [40]. Despite being not induced by either *Pst* infection or SA, ATX1 positively and directly regulates the expression of the transcription factor WRKY70 through H3K4 trimethylation at the WRKY70 promoter. In addition, *atx1* mutant shows induced expression of the JA-inducible *THI1.2* gene and the reduced *PR1* expression without detectable changes in their chromatin, resulting in impaired resistance to *Pst* infection. Since the transcriptional factor WRKY70 was positioned at the convergence nod of the SA and JA signalling pathways, activating the SA-responsive *PR1* gene and repressing the JA inducible genes [41], ATX1 was proposed to indirectly regulate *PR1* and *THI1.2* through *WRKY70*. SET DOMAIN GROUP 8 (SDG8), another HKMT encoding the major *Arabidopsis* H3K36 di- and trimethyltransferase [42], was also involved in the plant-defence against *Pst*, but it was more upstream than ATX1 [43]. Indeed, SDG8 sustains the basal transcription of particular *R* genes (*RPM1* or *LAZ5*) by maintaining a basal level of H3K36me3, another histone mark tightly associated with active transcription. SDG8 is also required for the transcriptional induction of these *R* genes upon BTH treatment or *Pst* inoculation. However, this induction occurs without any detectable increase of H3K36me3. Therefore, in resting plants, SDG8 may establish a 'permissive' chromatin structure at some *R* genes by methylating H3K36, thus ensuring their basal expression and their transcriptional inducibility upon need. Similarly as *atx1* and *sdg8* mutants, loss-of-function mutants for the putative HKMT *SDG7* were also found to be more sensitive to *Pst* infection than wild-type plants [44]. The expression of other *R* genes seems to be under the control of histone methylation. Indeed, enhanced downy mildew 2 (EDM2) impacts disease resistance by controlling levels of H3K9me2 at an alternative polyadenylation site in the immune receptor gene *RPP7*, thus regulating the balance between full-length *RPP7* transcripts and prematurely polyadenylated transcripts, which do not encode the *RPP7* immune receptor [45, 46]. EDM2, as an epigenetic 'reader', contains two stretches of atypical PHD-finger motifs known to dock specifically several forms of methylated or unmethylated lysine residues on histones [47]. Besides this, EDM2 was also proposed to cooperate within a large protein complex with EMSY-like (AtEML) members, harbouring an Agenet domain related to the Tudor domain family of epigenetic 'readers' [48].

Apart from *Arabidopsis*, little is known about the regulatory role of histone methylation in the defence against pathogen attack in other plant species. In rice, the JmjC protein gene *JMJ705* encoding a histone lysine demethylase that specifically reverses H3K27me2/3 was found induced during infection with the bacterial pathogen *Xanthomonas oryzae*. JMJ705 was further involved in the dynamic removal of the basal H3K27me3 over defence-related genes, thereby increasing their basal expression and/or potentiating their higher expression upon biotic stress. Interestingly, the *JMJ705* overexpression resulted in an enhanced resistance to the bacterial pathogen, while its mutation reduces the plant resistance [49].

3.2. Histone Methylation/Demethylation in The Defence Against Necrotrophic Pathogens

While to combat biotrophic pathogens the plant activates mainly the SA signalling pathway, the activation of the JA/ET signalling pathway is prominent to mediate defences against necrotrophic pathogens and herbivorous insect attacks [50]. The involvement of histone methylation in the defence against necrotrophic pathogens is far less documented as compared with the defence against biotrophic pathogens. Besides being more susceptible to *Pst* [43], *sdg8* mutants were also reported to be more sensitive to necrotrophic fungal pathogens such as *Alternaria brassicicola* (*Alt*) and *Botrytis cinerea* [51]. This increased susceptibility was the consequence of the inefficient transcriptional induction of different genes along the JA/ET signalling pathway that was correlated with a stably weak level of H3K36me3 at these genes. Inversely, in wild-type plants, H3K36me3 together with gene expression were increased upon *Alt* infection or stimulation with exogenous MeJA. Under resting conditions, a similarly weak level of H3K36me3 was correlated with a reduced basal expression in *sdg8*. On that account, H3K36 methylation was proposed to act as a 'permissive' mark correlated with gene activity and readily in place at a subset of JA/ET signalling-related genes to raise their rapid and efficient transcriptional induction when required [52]. Interestingly, a stable and very low level of H3K27me3 was detected in defence effector genes. Because H3K27me3 is often associated with epigenetic silencing [53], this low H3K27me3 level may provide these genes with a reduced probability for undesired silencing, thus participating in the reactivity of plants to pathogen infections.

4. HISTONE METHYLATION CHANGES ASSOCIATED WITH ABIOTIC STRESS CONDITIONS

Abiotic stresses such as heat, cold, drought, salinity and nutrient deficiency are inherent to every ecosystem and essentially unavoidable. Abiotic stresses are considered the most harmful factors in terms of growth and productivity of crops worldwide [54, 55], especially when they occur in combinations [56]. Here, we summarize and discuss various studies in order to clarify the functional involvement of different histone methylation marks in setting up plant responses to adverse environmental growth conditions.

4.1. Histone Methylation/Demethylation and The Plant Stress Hormone ABA

The phytohormone abscisic acid (ABA) is a crucial signalling molecule playing versatile functions in regulating many developmental processes, including seed dormancy and germination [57, 58]. ABA also plays a pivotal role in adaptive stress processes, integrating both biotic and abiotic environmental constraints in a complex network of interacting pathways with crosstalks at different levels [59–61]. Currently, ABA is considered as a global regulator of stress responses that can dominantly control the switch in priority between the responses to biotic or abiotic stress, allowing plants to respond to the most severe threat [62].

The transition from heterotrophic to autotrophic development at the post-germinative stage (i.e. embryonic state) is highly vulnerable to osmotic stress [63]. During a period of osmotic stress, ABA promotes the expression of transcription factors such as ABI3 and ABI5, which in turn delay germination and lead to osmotolerance and survival [64]. In *Arabidopsis*, mutation in *PICKLE* (*PKL*), encoding a putative chromatin modifier, results in increased and abnormally sustained expression of *ABI3* and *ABI5* in response to exogenous ABA treatment. This sustained expression was correlated with reduced levels of H3K9me2 and H3K27me2, two methylation marks found in the chromatin of silent genes [65]. Based on these results, it was suggested that PKL might act on *ABI3* and *ABI5* to promote directly or indirectly the formation of a repressed chromatin state through a so-far-unknown mechanism.

In adult plants, the establishment of a response and tolerance to drought stress by ABA has been extensively studied and is well discussed in several outstanding reviews [58, 66]. Briefly, under drought conditions, water stress perception triggers ABA biosynthesis and increased tissue ABA accumulation, resulting in stomatal closure and reduced transpiration. Among major enzymes involved in the ABA biosynthesis pathway, NINE-CIS-EPOXYCAROTENOID DIOXYGENASE 3 (NCED3) is thought to be the rate-limiting enzyme [67]. In *Arabidopsis*, the increased transcription of *NCED3* upon dehydration was correlated with the binding of the HKMT ATX1 and the increased level of H3K4me3 at *NCED3* [68]. Therefore, the loss-of-function *atx1* mutant showed less tolerance to dehydration, notably because of the lower enrichment of RNA Polymerase II (RNAPII) and H3K4me3 at *NCED3* under stress. ATX1-modified H3K4me3 may thus have an important function in the transcriptional regulation of *NCED3*. However, it is still unclear whether this function is directly linked with the transcriptional induction or is more related to the reinforcement of the increased transcription upon stress perception.

4.2. Histone Methylation/Demethylation in Response to Water Stresses

Water stresses including drought or submergence are major environmental factors limiting plant growth and crop productivity worldwide [69, 70]. Consequently, plants have evolved a variety of biochemical and physiological mechanisms to respond/adapt to these stresses [71, 72]. In the following section, we distinctly address the involvement of histone methylation in responses to drought and submergence.

4.2.1. Drought Stress

Using *Arabidopsis*, the molecular response to water deficit was found to rely notably on the transcriptional regulation of stress-inducible genes with products thought to function in drought tolerance and response [73]. Using ChIP analyses in 15-day-old *Arabidopsis* seedlings, the level of the active mark H3K4me3 was found gradually enriched in response to dehydration stress, preferentially on the coding region of four drought-inducible genes (*RD29A*, *RD29B*, *RD20* and *RAP2.4*), and was correlated with their upregulation [74]. Consistent with this result, ATX1 was reported to be required for the efficient transcriptional induction of *RD29A* and *RD29B* during a dehydration stress response in an ABA-independent manner [68]. Also, for unknown reasons, it is worth noting that upon exogenous ABA treatment the transcriptional induction of *RD29A* and *RD29B* was stronger in *atx1* mutants than in wild-type plants (their basal transcript levels were lower in *atx1* than in wild-type, while their transcript levels upon ABA treatment were identical [68]. Next, for the ABA-dependent *RD29A* and ABA-independent *RAP2.4* genes, a time lag was observed between their transcriptional induction and the increase in H3K4me3 [74]. Based on these findings and the rapid saturation in RNAPII enrichment compared with H3K4me3 (i.e. already saturated 1 h after stress exposure for RNAPII, while H3K4me3 was still increasing up to 5 h), Kim et al. [74] concluded that the H3K4me3 enrichment may be established, gradually, in response to drought stress after full transcriptional activation of *RD29A* and *RAP2.4*. Because the timing of H3K4me3 enrichment followed subsequent to the RNAPII enrichment, H3K4me3 might be dispensable for the initiation of transcription. Finally, the gradual increase of H3K4me3 further indicates that the longer the stress lasts, the more H3K4me3 will be enriched, suggesting that the epigenetic responsiveness must depend on the intensity of a stress [75].

In a similar approach but using ChIP-Seq, van Dijk et al. [76] established the whole-genome distribution patterns of H3K4me1, H3K4me2 and H3K4me3 in 4-week-old rosette *Arabidopsis* leaves under dehydration stress conditions. They also observed a strong correlation between H3K4me3 abundance and transcripts levels from responding genes. Indeed, among the most strongly downregulated genes, an increase in H3K4me1 and a decrease in H3K4me3 were detected, suggesting the involvement of a histone demethylase in modulating the expression of a subset of stress-responsive genes. Supporting this finding, a putative PKDM7 subfamily-like H3K4 demethylase homologue and two putative demethylase enzymes containing a JmjC domain were found to be drought-inducible in two barley cultivars and in young peanut plants, respectively [77, 78]. Surprisingly, in contrast to the classical genome-wide H3K4me3 enrichment around the transcriptional start site observed in all other eukaryotes [79], H3K4me3 displayed a broader distribution on dehydration and ABA-inducible genes. Such an unusual profile may reflect a function not strictly related to transcription initiation, as will be discussed later in the stress memory section.

Through a genome-wide approach in rice seedling, Zong et al. [80] also uncovered a weak but positive correlation between H3K4me3 enrichment and the transcript level of some drought-responsive genes under drought stress. This correlation was extended to many genes involved in stress-related metabolite and hormone signalling pathways, further supporting the role played by H3K4me3 in the stress response [80]. However, because H3K4me3 is not the only histone mark for gene activation, this weak correlation may reflect that other active histone marks may also play important roles in regulating gene expression in response to stress in rice. Although these large data sets have provided much information on drought responses in rice, more detailed analyses will be required to elucidate whether the observed variations in H3K4 methylation are a cause or a consequence of the transcriptional changes triggered by water stress. Moreover, identifying key histone modification enzymes is indispensable to better understand the transcriptional regulatory network of the abiotic stress response.

4.2.2. Submergence

Submergence is a complex stress that encompasses many changes in environmental factors, including light intensity, pH and dissolved oxygen concentration. Alcoholic fermentation is important for the survival of plants especially under anaerobic environments [81]. In rice, *alcohol dehydrogenase 1* (*ADH1*) and *pyruvate decarboxylase 1* (*PDC1*) genes are involved in this anaerobic metabolism and their expression is reversibly induced (i.e. activated upon submergence and repressed upon re-aeration)

[82]. Using these two genes as a model, Tsuji et al. [83] observed that the level of H3K4 methylation, specifically at both the 5′- and 3′-coding regions of *ADH1* and *PDC1*, was changed from a dimethylated state to a trimethylated state upon their transcriptional upregulation in response to submergence. This change was reverted back to its initial level following re-aeration, indicating that in this particular case, H3K4me3 does not serve as a memory mark of a prior transcriptional activity. Similarly to drought, these results highlight the dynamic and reversible change of histone H3K4 methylation at stress-related genes in response to the occurrence and disappearance of a stress.

4.3. Histone Methylation/Demethylation in Response to Salt Stress

Salinity is also a serious factor affecting plants in several ways (i.e. water stress, ion toxicity, nutritional disorders, oxidative stress, alteration of metabolic processes, membrane disorganization, genotoxicity, reduction of cell division and expansion), thus limiting plant growth, development and survival [84]. In *Arabidopsis*, a NaCl or ABA treatment has been shown to induce the transcription of a range of abiotic stress-responsive genes (*ABI1*, *ABI2*, *KAT1*, *KAT2*, *DREB2A*, *RD29A* and *RD29B*). Using ChIP, this induction was found significantly correlated with an increase in the active mark H3K4me3 and a decrease in the repressive mark H3K9me2 [85]. Also, suggesting a link between different histone PTMs, mutations in the histone deacetylase HDA6 partially suppressed the H3K4me3 increase observed in response to stress, while the H3K9me2 decrease was not affected. Since histone acetylation and H3K4 trimethylation are often associated with gene activation [86], the repressive function of HDA6 may suggest that acetylation on histone H3 is required for K4 methylation to occur [87]. Unfortunately, because ChIP experiments were not normalised against the total H3 density, it is difficult to attribute the decrease in H3K9me2 to either the active removal of methyl groups by a histone demethylases or the active removal of nucleosome in response to a stress. Supporting the second possibility, the nucleosome density of two other drought stress-inducible genes (*RD20* and *RAP2.4*) was found to gradually decrease in response to drought stress [74].

In soya bean and in response to a high NaCl concentration, these histone methylation marks were also found altered at some salinity-induced transcription factors (i.e. MYB, b-ZIP and AP2/DREB family members) that were primarily identified by microarray analysis [88]. For some genes, their transcriptional induction was correlated with an increased level of histone acetylation and H3K4me3, accompanied or not with a reduced level of DNA methylation and H3K9me2 in various parts of the promoter or coding regions. For other genes, DNA methylation had no influence on histone methylation. This work perfectly reflects the heterogeneity of the effect of salinity on histone methylation and DNA methylation, and supports the role(s) of histone methylation changes in the expression of some transcription factors important for salinity tolerance.

As mentioned above, H3K4me3 was found to be involved in the transcriptional induction of stress-responsive gene upon salt stress exposure. In plants, the JmjC-domain-containing histone demethylases JMJ14, JMJ15 and JMJ18 have been reported to display an H3K4me2/3 demethylase activity as well as to regulate diverse aspects of chromatin function and development [89–95]. Recently, the overexpression of *JMJ15* was reported to preferentially downregulate many stress-related genes preferentially marked by H3K4me2/3 and to enhance salt stress tolerance [96]. In contrast, the loss-of-function mutant was more sensitive to salt. Despite the fact that an increased JMJ15 level may regulate stress-responsive gene transcription programmes in *Arabidopsis*, the role of H3K4me3 resetting in these processes is still very elusive.

Besides histone lysine methylation, arginine methylation was also involved in establishing the transcriptional response to salt stress. The protein arginine methyltransferase 5 (PRMT5), also named Shk1 kinase-binding protein1 (SKB1), is a type II methyltransferase that catalyses symmetric H4R3 dimethylation, a repressive mark known to promote flowering through the repression of the floral repressor *FLOWERING LOCUS C* (*FLC*) in *Arabidopsis* [20, 97]. Interestingly, gain-of-function *SKB1* mutants showed an enhanced salt stress tolerance and sensitivity to ABA [98]. As a consequence of PRMT5 disassociating from chromatin, the H4R3sme2 level at stress-responsive genes was reduced during salt stress, resulting in their induced expression. Suggesting an additional function of PRMT5 on non-histone proteins, the methylation level of the U6 small nuclear ribonucleoprotein Sm-like4 (LSM4, a core protein of the spliceosome) was increased in response to salt stress and ABA. Since splicing defects were observed in the *prmt5/skb1* and *lsm4* mutants, with both of them being hypersensitive to salt stress, authors proposed that PRMT5 might mediate plant development and salt response by altering the methylation status of H4R3me2 and LSM4, linking transcriptional regulation to pre-mRNA splicing [98].

4.4. Histone Methylation/Demethylation in Response to Temperature

In plants, temperature stresses are classically classified into different types according to temperature exposure, which may be warm, high, chilling or freezing temperature. Due to global warming and because temperature stress greatly affects plant growth and development, immunity and circadian rhythm, and poses a serious threat to the global food supply, the genetic mechanisms of plant responses to heat have been well studied. Plants exposed to temperature stresses modulate the transcription of a large number of genes involved in distinct biochemical and physiological response pathways and networks of phytohormones or secondary metabolites, ultimately leading to increased tolerance to hazardous temperature stresses [99–102]. The role played by histone methylation during the plant response to a heat or a cold stress is discussed separately hereafter.

4.4.1. Heat Stress

Heat stress during seed development decreases the seed size in many cereals, resulting in severe yield losses [103, 104]. In rice, a molecular mechanism involving the putative rice polycomb repressive complex 2 (PRC2) gene *Fertilization-Independent Endosperm1* (*OsFIE1*) was suggested as being a potential key component involved in regulating the thermal sensitivity of seed enlargement during endosperm development [105]. When developing seeds were exposed to a heat stress, both DNA methylation and H3K9 methylation were reduced on *OsFIE1* resulting in its derepression. Under heat stress, syncytial stage-specific MADS-box genes involved in seed size regulation were precociously repressed, due to the increased deposition of H3K27me3 silencing marks by the PRC2 complex [105]. In the unicellular green alga *Chlamydomonas reinhardtii*, histone modification was also affected by heat stress [106]. Indeed, after heat stress, the level of H3K4me1 was found decreased and the level of histone acetylation increased at promoter regions of active genes compared with inactive regions. As a hypothesis, authors proposed that upon heat stress, the heat shock transcription factor HSF1 might promote chromatin remodelling and RNAPII recruitment for transcription initiation/elongation [106]. Finally, while addressing molecular mechanisms of the response of cotton anthers to high temperature, two jmjC domain-containing genes, putatively involved in histone demethylation, were found significantly repressed during anther development under heat stress [107].

Because euchromatin is gene rich and usually transcriptionally active, investigation about the role of histone methylation in temperature stress acclimation was largely centred on euchromatin-associated coding regions. Focusing on the transcriptionally silent heterochromatin, mainly constituted of repetitive DNA sequences, some works demonstrated the transcriptional activation of normally silent transposable element embedded within heterochromatic regions under stress conditions [108].

Intriguingly, such activation under heat stress can occur without alteration of DNA methylation and with only minor changes in both H3K9me2 and H3K4me3 [109, 110]. In summary, these works suggest that temperature stress-mediated transcription of tandem-repeat elements might play a vital role in the adaptation of plants to temperature stimuli, offering an efficient mechanism by which heat or cold could promote the expression of some stress-responsive genes. Upon activation and when inserted into or very close to a gene, such transposable elements could interfere with the expression of this gene, giving rise to deleterious mutations, genetic instability or positive contribution to gene regulation and adaptation [111].

4.4.2. Cold Stress

The increased tolerance of plants to cold is referred as 'cold acclimation'. Cold acclimation differs from vernalization, as the last one requires a long-term exposure to cold temperatures, while cold acclimation can be achieved in a couple of days under non-freezing low temperatures [112]. Locally, histone methylation changes in cold-responsive genes were addressed in *Arabidopsis*. Upon cold stress, the repressive mark H3K27me3 decreased, in both a histone occupancy-dependent (i.e. arising from the lowering nucleosome density) and -independent (i.e. as the result of the activity of a not yet identified histone demethylase) manner, on the cold-responsive genes, cold-regulated 15A (*COR15A*) and galactinol synthase 3 (*ATGOLS3*; [113]). Interestingly, the decrease in H3K27me3 upon stimulation occurred more gradually than their rapid transcriptional induction, so that their activation may not be inhibited by H3K27me3 itself but rather lead to the removal of H3K27me3. Also, while the transcription of *COR15A* and *ATGOLS3* was completely repressed to the initial level upon returning plants to normal growth conditions, the H3K27me3 decrease was maintained. Given that this decrease does not affect the transcriptional induction of *COR15A* and *ATGOLS3* upon re-exposure to cold temperatures, such chromatin change can so far only be view as a 'reminiscence' of a recent transcriptional activity and not as a stress memory implicated in a gene priming process.

In maize during cold stress, changes in histone modifications, including the heterochromatic marks H3K9me2 and DNA methylation, were assessed through a genome-wide approach [114]. The more detailed analysis of the two knob-associated tandem-repetitive sequences, the 180-bp repeat and the 350-bp repeat termed TR-1, demonstrated that their selectively and transiently cold-activated transcription was correlated with a decreased H3K9me2 and DNA methylation, together with an increased H3K9 acetylation. Such cold-induced transcriptional activation of tandem repeats is selective and transient, and the silencing state is recovered as the treatment continues.

5. HISTONE METHYLATION AS A MEMORY MARK OF STRESS

In animals, the formation of memory immune cells after primary antigen recognition confers long-lasting resistance, resulting in an accelerated and a more effective immune response in case of second exposure. Despite the absence of such memory immune cells, plants often acquire a systemic immunity to further infections after a primary localized infection [115]. This requires the accumulation of the plant hormone SA in systemic tissues and is called systemic acquired resistance (reviewed in [116]. The SAR is also associated with gene priming in systemic tissues, in which defence genes will be expressed more rapidly and robustly in case of a second attack [117]. At the transcriptional level, gene expression is primarily influenced by the chromatin structure, which in turn is controlled partly by processes, often referred to as 'epigenetic' processes, which can be transmitted through mitosis and/or inherited through meiosis [118]. Therefore, chromatin remodelling through histone methylation offers a potential mechanism for short-/long-term stress memory within the lifespan of an individual, referred to as somatic memory, and/or across generations, referred to as transgenerational memory.

5.1. Somatic Stress Memory

In *Arabidopsis*, a priming event, either treatment with BTH or infection with *Pseudomonas syringae pv. maculicola*, systematically resulted in an increase in the level of H3K4me2/me3 at defence gene promoters (namely the WRKY transcription factors *WRKY6*, *WRKY29* and *WRKY52*) that are normally found on active genes, while the genes remain inactive [119]. More interestingly, this increase also occurs in leaves distal to localized foliar infection. Hence, even if the histone-modifying enzyme involved in this process remains unidentified, results from Jaskiewicz et al. [119] clearly suggest that histone methylation might create a 'memory' of the primary infection that is associated with an amplified reaction to a second stress stimulus. Further, Luna et al. [120] observed that promoters of SA-inducible *PR1*, *WRKY6* and *WRKY53* in the progeny of *Pst*-inoculated *Arabidopsis* plants were enriched with acetylated histone, while the promoter of the JA-inducible gene *PLANT DEFENSIN1.2* (*PDF1.2*) showed an increased level of H3K27me3. For decades, the signalling protein *NON EXPRESSOR OF PR1* (NPR1) has been implicated in mediating SAR induction [115] and also the crosstalk between SA- and JA/ET-dependent defence pathways, enabling plants to mount an appropriate defence reaction, depending on the nature of the attacker and the stage of infection [121, 122]. More recently, NPR1 has been proposed to play a critical role in the expression of the transgenerational SAR as progeny from *npr1* failed to develop transgenerational defence phenotypes and failed to present enrichment for H3K27me3 at the *PDF1.2* promoter [120]. Together, these findings suggested that one or more systemic signals are stored as an immune memory on defence-related gene promoters in the form of histone modifications, thus providing the plant with a life-long protection, which can be transmitted to subsequent generations.

Besides being involved in defence priming related to biotic stress, histone methylation was also proposed as a priming strategy against drought. To further explore the functional impact of histone methylation on biotic stress responses in *Arabidopsis*, Kim et al. [123] followed chromatin dynamics of several drought genes (*RD20, RD29A* and *AtGOLS2*) and a rehydration-inducible gene (*ProDH*) during drought and rehydration. As previously discussed, a strong correlation was observed between H3K4me3 enrichment (i.e. especially in gene bodies) and transcription for drought genes upon drought [74, 123]. Such a correlation was also detected for the rehydration gene upon rehydration [123]. Suggesting a memory role for H3K4me3, RNAPII rapidly disappeared after rehydration at drought genes, while H3K4me3 was gradually decreased. Concomitantly, by training plants with up to four successive drought treatments, Ding et al. [124] uncovered the existence of two distinct subsets of genes within the dehydration stress–response gene fraction. The 'non-trainable' genes (e.g. the ABA-independent *RD29A* and *COR15A*) have repetitively similar transcription rates during each stress treatment, while 'trainable' ones (e.g. the ABA-dependent *RD29B* and *RAB18*) increased the magnitude of their subsequent transcriptional response, relative to their initial stress response. Using ChIP, Ding et al. [124] observed that the H3K4me3 enrichment at 'trainable' genes, especially in gene bodies, was atypically retained from the preceding transcription after rehydration. Even more interestingly, the RNAPII phosphorylated at C-terminal domain (CTD) repeat serine 5 (Ser5P; associated with transcription initiation) was found stalled on these genes as a memory mark from a previously transcribed state. In contrast to 'trainable' genes, the stress-induced H3K4me3 and Ser5P enrichment at 'non-trainable' genes was decreased to its basal level during recovery [124]. Moreover, this transcriptional memory can persist in the absence of inducing signals at least for 5 days, but is lost after 7 days. Supporting a specific role for H3K4me3 in stress memory, other active chromatin marks such as acetylation of histones H3 and H4 were found rapidly increased at drought genes upon stress and decreased at comparable levels as before induction quickly after recovery [123, 124]. Consistent with an activating role of ATX1 at dehydration stress–response genes [68, 125, 126], dehydration-induced transcript levels were diminished in *atx1* plants [127]. However, ATX1 does not seem to have a critical impact on drought stress memory in *Arabidopsis*. Indeed, while being less increased than in wild-type plants, trainable genes still produced increased transcripts in trained, relative to untrained, *atx1* plants and retained high H3K4me3 levels during the watered recovery states. ATX1, ATX2, SDG25 and SET DOMAIN GROUP2/ATX-RELATED3 (SDG2/ATXR3) belong to the same class III of H3K4me3 methyltransferases and are thought to act, partially redundantly, as H3K4 methyltransferase [128–134]. More recently, SDG2 has been found to be essential for the full transcriptional activation of various hormone-responsive genes upon hormone treatment (i.e. including the ABA-dependant *RD29A*) via its H3K4 trimethyltransferase activity [135]. It is, therefore, likely that the other class III HKMT might also contribute to the drought stress memory. In summary, these results suggest that in addition to be a good marker of gene activation when found around promoter and 5′ regions of genes, H3K4me3 might also play a role in establishing a transcriptional short-term somatic memory of drought stress when found in gene bodies.

Using a large-scale approach, the distribution of H3K4me2, H3K4me3, H3K9me2 and H3K27me3 was analysed in *Arabidopsis* seedlings, which have been treated with mild salt stress in the seedling stage, resulting in an increased tolerance upon an additional salt stress application [136]. At low resolution in primed seedlings, H3K4me2 and H3K4me3 most commonly consisted of higher peaks of pre-existing enriched histone modification domains, named islands, whereas H3K9me2 produced the least differences. By contrast, majority of differences in H3K27me3 resulted from a higher number of islands with lower genome coverage. At high resolution, changes in H3K27me3 were already detectable a few hours after salt addition, suggesting that demethylation of H3K27me3 operates at a speed that is comparable to that of transcriptional regulation. Interestingly, this effect fades over time; however, it is still clearly visible after a 10-day-growth period in control conditions. In response to a second stress treatment, genes with high responsiveness, such as *HKT1* (i.e. encoding a root-specific Na transporter) and *PIP2E* (i.e. encoding a plasma membrane aquaporin), experienced a decrease of H3K27me3, whereas genes with lower responsiveness, such as *GH3.1* and *GH3.3* (i.e. encoding auxin and JA-amino acid-conjugating enzymes, respectively), experienced an increase of H3K27me3. Conversely, another group reported that in plants that have experienced several exposures to dehydration stress no significant change in the level of H3K27me3 could be detected on trainable and non-trainable genes, or during transcriptionally active/inactive gene states [137]. However, the high H3K27me3 level present at inactive dehydration stress memory genes did not interfere with the transition to an active transcription and with the accumulation of H3K4me3 [138]. Together, the function of H3K27me3 in genes that dynamically change transcription seems to depend on the type of environmental stimuli.

In contrast to H3K27me3, the higher level of H3K4me3 retained at the trainable gene *RD29B*, when its transcription is low, further supports the idea that H3K4me3 works as a 'memory' histone mark of a previously active state [138]. Generally, H3K4me3 and H3K27me3 play antagonistic roles in gene transcription and are therefore mutually repulsive at developmental genes [37, 139]. Seemingly, the presence of both marks, referred as 'bivalent domains', was first described in mammalian stem cells and was proposed to represent a pluripotent chromatin state that poises genes for activation upon appropriate developmental cues [140, 141]. Further work is required to determine whether H3K4me3 and H3K27me3 co-exist at certain genes whose expression is rapidly altered in response to environmental stimuli.

Because plant stress research has traditionally focused on single stresses, we separately described priming in response to abiotic or biotic environmental cues. However, in nature, plants are constantly exposed to mild environmental stresses during their lifetime. While testing how different environmental histories can affect the response of the plant to a subsequent biotic stress, Singh et al. [142] reported that *Arabidopsis* plants exposed to a recurrent abiotic stress (i.e. heat, cold or salt) were more resistant to *Pst* than plants grown in a more stable environment. This enhanced resistance was due to the priming of commonly used marker genes of pattern-triggered immunity (PTI; *WRKY53, FLG22-INDUCED RECEPTOR KINASE1* (*FRK1*) and *NDR1/HIN1-LIKE10* (*NHL10*)). Indeed, enrichment for epigenetic marks associated with transcriptional activation, such as H3K4me2 and H3K4me3, at PTI-responsive genes was observed after the exposure to recurrent stress, resulting in an enrichment of RNA polymerase II and a primed transcription in response to a subsequent bacterial infection. Collectively, these works on somatic stress memory promote the idea that, in plant, the environmental history can shape/modulate the response to stress, providing a mechanistic link between histone methylation and gene priming.

5.2. Transgenerational Stress Memory

The transgenerational stress memory refers to the transmittance of certain environmental responses from one generation to the next, thus providing the offspring of environmentally challenged plants with an adaptive advantage for better fitness (i.e. improve plant stress tolerance and impart developmental flexibility; [143]). Compared with DNA methylation and RNA interference (RNAi), very few studies suggest the involvement of histone methyltransferases and histone methylation changes in this process [144]. In *Arabidopsis*, changes in DNA methylation, histone modifications and gene expression were followed in the progeny of plants exposed to salt stress over one generation [145]. Although the DNA from the progeny of plants exposed to salt stress was globally hypomethylated, the majority of genes and promoters causing methylation changes were hypermethylated and lowly expressed. In addition, DNA hypermethylation was correlated with an increased level of the repressive mark H3K9me2. Among these hypermethylated genes, a large number was encoding different histone methyltransferases, such as the *Arabidopsis* *SU(VAR)3-9* homologues *SUVH2, SUVH5, SUVH8*, involved in H3K9 methylation, or the PRC2 subunit *CURLY LEAF* (*CLF*), involved in H3K27 methylation. Following this work, the progeny of heat-stressed plants was used to explore epigenetic variations under both normal and stressed conditions, in comparison to the progeny of control plants [146]. Similarly to salt stress, the progeny of plants exposed to heat stress had a global decrease of genomic DNA methylation and a reduced expression of several *SUVH* genes, which correlated with their enrichment in H3K9me2. Together, the hypermethylation of SUVH genes in the progeny of stressed plants may represent a protective mechanism against hypermethylation of the entire genome. Interestingly, in both works [145, 146], the transposon expression was elevated in the progeny of stressed plants. Because main targets of the SUVH pathway are transposable elements [24], the authors proposed that a decrease in the expression of *SUVH* genes might contribute to transposon activation, which at opportune times can create intragenomic potential upon transposition to facilitate adaptation in response to environmental changes [147]. In summary, these works suggested a role for histone methylation in the inheritance of stress memory; however, whether histone methylation changes are heritable through multiple generations and whether they sustain the acquisition of adaptive traits is still a matter of debate [148]. However, the evidence to date favours the view that stress-induced transgenerational changes in chromatin might increase the survival chances of the plant species, rather than each individual, by broadening the phenotypic plasticity and the genetic variation within the population [149, 150].

6. DISCUSSION AND PERSPECTIVES

Recent advances, especially in *Arabidopsis*, have uncovered that chromatin remodelling through histone methylation changes are not only restricted to developmental needs but also an integral part of the very complex cascade of events that lead to abiotic/biotic stress tolerance, resistance and short-/long-term memory. Currently, a preliminary view is emerging, indicating that histone methylation changes, providing specific chromatin configurations, can be classified into several interrelated categories when involved in stress responses (Fig. 1): (*i*) histone methylation changes that

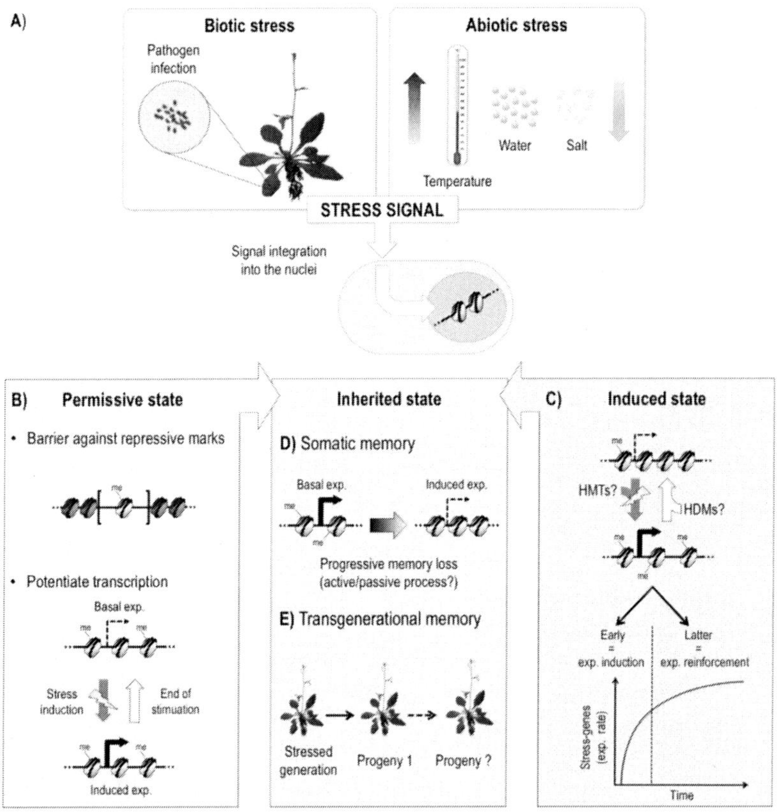

Figure 1. Hypothetical model depicting the role of histone methylation/demethylation in regulating plant stress responses. (A) Plants exposed either to biotic or abiotic stresses integrate the signal stress into the cell nuclei, where it affects the chromatin structure through histone methylation changes. According to our knowledge, these histone methylation changes can be classified into several interrelated categories. (B) The permissive state represents a more loosened chromatin state that will either offer a protection against repressive marks (represented with nucleosomes in red) and/or potentiates a rapid transcriptional induction upon stress induction. (C) The induced state represents histone methylation changes that are transiently induced by a stress signal. If methylation changes occur early during the stress-response process, they might participate to the transcriptional induction of stress-responding genes, while if they occur later, they might reinforce the transcription of stress-responding genes. Both permissive and the induced states can be maintained allowing a faster and/or stronger transcriptional induction of stress-responding genes upon a subsequent challenge. The memorized chromatin state can be maintained (D) for a certain time during the lifespan of an individual and referred as the somatic memory, or (E) transmitted to one or more subsequent generations and referred as the transgenerational memory.

are basally present on stress-related genes to establish a 'permissive' chromatin state that may either limit the spreading of repressive chromatin marks and/or potentiate a rapid transcriptional induction upon need; (*ii*) histone methylation changes that are transiently induced from an inactive or a permissive chromatin state by stress, to either facilitate the transcriptional initiation and/or reinforce transcription of stress-responding genes, and finally, histone methylation changes that are established in response to a stress; (*iii*) maintained for a certain time during the lifespan of an individual (i.e. somatic memory) or (*iv*) transmitted to one or more subsequent generations (i.e. transgenerational memory).

Nonetheless, this emerging view is facing many gaps, inaccuracies and divergences, mainly related to numerous difficulties inherent to the study of such a dynamic and acute process. In this respect, plants in nature are usually challenged simultaneously by different kinds of stresses. Responses to these stress combinations are largely controlled by different signalling pathways that can interact in a non-additive manner, producing effects that could not have been predicted from the study of either stress individually [151, 152]. The occurrence of simultaneous biotic and abiotic stresses introduces an added degree of complexity that requires stresses to be imposed simultaneously and to treat each set of environmental conditions as an entirely new stress. For this reason and to clarify the mechanism behind the regulation of stress responses by histone methylation changes, there is a strong necessity to intensify our investigations. For instance, the correlation between histone methylation/demethylation and stress responses remains elusive and clarifications will require in-depth dynamic approaches based on comparative analyses of both epigenomes and transcriptomes during stress responses. In parallel, current knowledge about the corresponding histone-modifying enzymes is still largely missing. This lack of knowledge is pending on the identification of different stress-responsive histone modifiers and will require large-scale screens and genetic analyses for the sensitivity of different histone methyltransferases/demethylases mutants to various stresses, combined or not. Among other factors governing stress-induced chromatin changes, almost nothing is known about the specific reader/effector that will recognize particular histone methylation sites in order to determine their functional and structural outcome. An effort in this direction will most likely benefit the comprehensive understanding of the fundamental mechanisms connecting histone methylation changes with the modulation of transcription of stress-responsive genes, subsequently enabling plant to withstand stress. Higher-resolution chromatin studies are undoubtedly required to reveal the targeted stress-responsive genes and the specific sites of histone methylation/demethylation. Nevertheless, investigation of the direct effects of histone methylation/demethylation in plants is difficult. One reason is that plant genomes harbour high-copy number of histone genes (e.g. the *Arabidopsis* genome comprises 47 genes that encode 33 different core histone proteins; www.chromdb.org) and the incorporation/modification of such variants can result in the formation of chromatins with particular properties and functions [153–155]. Although ChIP assays have proven valuable in helping to identify histone methylation changes, many antibodies used to detect these changes have been so far unable to distinguish between different variants. New technologies (e.g. generation of mutants with point mutations targeting amino acid in the N-terminal tail of histone using the clustered regularly interspaced short palindromic repeats (CRISPR)/CRISPER-associated (Cas) system) [156] will need to be explored to unravel histone methylation changes of specific histone variants and their functions. Another challenge is that plants consist of many functionally specialized tissues and cell types, each with its own unique epigenome, transcriptome and proteome. Until now, histone methylation changes induced by stresses were exclusively addressed in entire plant or organs, meaning that the obtained profiles most likely reflect the consensus of multiple tissue- or cell-specific profiles that may differ. New methods allowing the mapping of chromatin features in specific tissue/cell types such as the one described by Wang and Deal [157] will be decisive for determining the cell-/tissue-specific chromatin alterations involved in a particular stress response. Finally, as plants have finite resources that must be balanced between growth and defence against stresses, often resulting in a growth or yield penalty, histone methylation changes in response to stress should be integrated in a more global developmental view, taking into account the involvement of several histone methyltransferases/demethylases in various processes such as root growth, flowering time, floral organogenesis, gametophyte or embryo formation [33]. Finally, understanding such regulatory network is an essential step to provide both novel paradigms and potential tools for further exploitation towards sustainable agriculture.

7. ACKNOWLEDGEMENTS

This work was supported by the Centre National de la Recherche Scientifique, the Agence Nationale de la Recherche (ANR-12-BSV2-0013-02), the European Commission (FP7-PEOPLE-2013-ITN, grant number 607880), the National Council of Science and Technology of Mexico (CONACYT) and the Program 'Estancias Posdoctorales y Sabaticas al Extranjero' (Fellowship number 237557. BOBADILLA-LANDEY Roberto; CVU:162391).

REFERENCES

1. Osakabe Y, Yamaguchi-Shinozaki K, Shinozaki K and Tran LS. (2013) Sensing the environment: key roles of membrane-localized kinases in plant perception and response to abiotic stress. J Exp Bot. 64(2):445–458.
2. Pieterse CMJ, Leon-Reyes A, Van der Ent S and Van Wees SCM. (2009) Networking by small-molecule hormones in plant immunity. Nat Chem Biol. 5(5):308–316.
3. Nakashima K, Ito Y and Yamaguchi-Shinozaki K. (2009) Transcriptional regulatory networks in response to abiotic stresses in *Arabidopsis* and grasses. Plant Physiol. 149(1):88–95.
4. van Verk MC, Gatz C and Linthorst HJM. (2009) Transcriptional regulation of plant defense responses. Adv Bot Res. 51:397–438.
5. Riechmann JL and Ratcliffe OJ. (2000) A genomic perspective on plant transcription factors. Curr Opin Plant Biol. 3(5):423–434.
6. Moore JW, Loake GJ and Spoel SH. (2011) Transcription dynamics in plant immunity. Plant Cell. 23(8):2809–2820.
7. Luger K, Mäder AW, Richmond RK, Sargent DF and Richmond TJ. (1997) Crystal structure of the nucleosome core particle at 2.8 Å resolution. Nature. 389(6648):251–260.
8. Andrews AJ and Luger K. (2011) Nucleosome structure(s) and stability: variations on a theme. Annu Rev Biophys. 40:99–117.
9. Nelissen H, Boccardi TM, Himanen K and Van Lijsebettens M. (2007) Impact of core histone modifications on transcriptional regulation and plant growth. Crit Rev Plant Sci. 26(5-6):243–263.
10. Shen WH and Xu L. (2009) Chromatin remodeling in stem cell maintenance in *Arabidopsis thaliana*. Mol Plant. 2(4):600–609.
11. Berr A and Shen WH. (2010) Molecular mechanisms in epigenetic regulation of plant growth and development. In: Pua E. C. and Davey M. R. (Eds.) Plant Dev Biol – Biotechnol Perspect. 2:325–344.
12. Arents G and Moudrianakis EN. (1995) The histone fold: a ubiquitous architectural motif utilized in DNA compaction and protein dimerization. Proc Natl Acad Sci USA. 92(24):11170–11174.
13. Davey CA, Sargent DF, Luger K, Maeder AW and Richmond TJ. (2002) Solvent mediated interactions in the structure of the nucleosome core particle at 1.9 Å resolution. J Mol Biol. 319(5):1097–1113.
14. Marks P, Rifkind RA, Richon VM, Breslow R, Miller T and Kelly WK. (2001) Histone deacetylases and cancer: causes and therapies. Nat Rev Cancer. 1(3):194–202.
15. Mutskov V, Gerber D, Angelov D, Ausio J, Workman J and Dimitrov S. (1998) Persistent interactions of core histone tails with nucleosomal DNA following acetylation and transcription factor binding. Mol Cell Biol. 18(11):6293–6304.
16. Strahl BD and Allis CD. (2000) The language of covalent histone modifications. Nature. 403(6765):41–45.
17. Jaskiewicz M, Peterhansel C and Conrath U. (2011) Detection of histone modifications in plant leaves. J Vis Exp. (55).
18. Zhang X. 2012. Chromatin modifications in plants. In: Wendel JF, editor. Plant genome diversity. Vienna: Springer. p. 237–56.
19. Guccione E, Bassi C, Casadio F, Martinato F, Cesaroni M, Schuchlautz H, Lüscher Band Amati B. (2007) Methylation of histone H3R2 by PRMT6 and H3K4 by an MLL complex are mutually exclusive. Nature. 449(7164):933–937.
20. Wang X, Zhang Y, Ma Q, Zhang Z, Xue Y, Bao S and Chong K. (2007) SKB1-mediated symmetric dimethylation of histone H4R3 controls flowering time in *Arabidopsis*. EMBO J. 26(7):1934–1941.
21. Zhao Q, Rank G, Tan YT, Li H, Moritz RL, Simpson RJ, Cerruti L, Curtis DJ, Patel DJ, Allis CD, Cunningham JM and Jane SM. (2009) PRMT5-mediated methylation of

histone H4R3 recruits DNMT3A, coupling histone and DNA methylation in gene silencing. Nat Struct Mol Biol. 16(3):304–311.

22. Wang H, Huang ZQ, Xia L, Feng Q, Erdjument-Bromage H, Strahl BD, Briggs SD, Allis CD, Wong J, Tempst P and Zhang Y. (2001) Methylation of histone H4 at arginine 3 facilitating transcriptional activation by nuclear hormone receptor. Science. 293(5531):853–857.
23. Strahl BD, Briggs SD, Brame CJ, Caldwell JA, Koh SS, Ma H, Cook RG, Shabanowitz J, Hunt DF, Stallcup MR and Allis CD. (2001) Methylation of histone H4 at arginine 3 occurs in vivo and is mediated by the nuclear receptor coactivator PRMT1. Curr Biol. 11(12):996–1000.
24. Thorstensen T, Grini PE and Aalen RB. (2011) SET domain proteins in plant development. Biochim Biophys Acta. 1809(8):407–420.
25. Di Lorenzo A and Bedford MT. (2011) Histone arginine methylation. FEBS Lett. 585(13):2024–2031.
26. Pedersen MT and Helin K. (2010) Histone demethylases in development and disease. Trends in Cell Biol. 20(11):662–671.
27. Springer NM, Napoli CA, Selinger DA, Pandey R, Cone KC, Chandler VL, Kaeppler HF and Kaeppler SM. (2003) Comparative analysis of SET domain proteins in maize and *Arabidopsis* reveals multiple duplications preceding the divergence of monocots and dicots. Plant Physiol. 132(2):907–925.
28. Ng DW, Wang T, Chandrasekharan MB, Aramayo R, Kertbundit S and Hall TC. (2007) Plant SET domain-containing proteins: structure, function and regulation. Biochim Biophys Acta. 1769(5-6):316–329.
29. Pontvianne F, Blevins T and Pikaard CS. (2010) *Arabidopsis* histone lysine methyltransferases. In: Delseny M. And Kader J. C. (eds) Advances in botanical research. London: Academic Press Ltd-Elsevier Science Ltd. 1–22.
30. Aquea F, Vega A, Timmermann T, Poupin MJ and Arce-Johnson P. (2011) Genome-wide analysis of the SET DOMAIN GROUP family in grapevine. Plant Cell Rep. 30(6):1087–1097.
31. Huang Y, Liu C, Shen WH and Ruan Y. (2011) Phylogenetic analysis and classification of the *Brassica rapa* SET-domain protein family. BMC Plant Biol. 11:175.
32. Aiese-Cigliano R, Sanseverino W, Cremona G, Ercolano MR, Conicella C and Consiglio FM. (2013) Genome-wide analysis of histone modifiers in tomato: gaining an insight into their developmental roles. BMC Genomics. 14:57.
33. Berr A, Shafiq S and Shen WH. (2011) Histone modifications in transcriptional activation during plant development. Biochim Biophys Acta. 1809(10):567–576.
34. Hammond-Kosack KE and Jones JDG. (2000) Response to plant pathogens. In: Buchannan B. Gruissem W. Jones R. (eds). Biochemistry and molecular biology of plants. Rockville, MD: American Society of Plant Physiologists. 1102–1157.
35. Jones JDG and Dangl JL. (2006) The plant immune system. Nature. 444(7117):323–329.
36. Mosher RA, Durrant WE, Wang D, Song J and Dong X. (2006) A comprehensive structure-function analysis of *Arabidopsis* SNI1 defines essential regions and transcriptional repressor activity. Plant Cell. 18(7):1750–1765.
37. Roudier F, Ahmed I, Bérard C, Sarazin A, Mary-Huard T, Cortijo S, Bouyer D, Caillieux E, Duvernois-Berthet E, Al-Shikhley L, Giraut L, Després B, Drevensek S, Barneche F, Dèrozier S, Brunaud V, Aubourg S, Schnittger A, Bowler C, Martin-Magniette M L, Robin S, Caboche M and Colot V. (2011). Integrative epigenomic mapping defines four main chromatin states in *Arabidopsis*. EMBO J. 30:1928–1938.
38. Alvarez-Venegas R, Al Abdallat A, Guo M, Alfano JR and Avramova Z (2007). Epigenetic control of a transcription factor at the cross section of two antagonistic pathways. Epigenetics. 2(2):106–113.
39. Gruner K, Griebel T, Návarová H, Attaran E and Zeier J. (2013) Reprogramming of plants during systemic acquired resistance. Front Plant Sci. 4:252.
40. Alvarez-Venegas R, Sadder M, Hlavacka A, Baluska F, Xia YN, Lu GQ, et al. (2006) The *Arabidopsis* homolog of trithorax, ATX1, binds phosphatidylinositol 5-

phosphate, and the two regulate a common set of target genes. Proc Natl Acad Sci USA. 103(15):6049–6054.

41. Li J, Brader G, Kariola T and Palva ET. (2006) WRKY70 modulates the selection of signaling pathways in plant defense. Plant J. 46(3):477–491.
42. Xu L, Zhao Z, Dong A, Soubigou-Taconnat L, Renou JP, Steinmetz A and Shen WH. (2008) Di- and tri- but not monomethylation on histone H3 lysine 36 marks active transcription of genes involved in flowering time regulation and other processes in *Arabidopsis thaliana*. Mol Cell Biol. 28(4):1348–1360.
43. Palma K, Thorgrimsen S, Malinovsky FG, Fiil BK, Nielsen HB, Brodersen P, et al. (2010) Autoimmunity in Arabidopsis *acd11* Is Mediated by Epigenetic Regulation of an Immune Receptor. PLoS Pathog 6(10): e1001137.
44. De-La-Peña C, Rangel-Cano A and Alvarez-Venegas R. (2012) Regulation of disease-responsive genes mediated by epigenetic factors: interaction of *Arabidopsis-Pseudomonas*. Mol Plant Pathol. 13(4):388–398.
45. Tsuchiya T and Eulgem T. (2011) EMSY-like genes are required for full RPP7-mediated race-specific immunity and basal defense in *Arabidopsis*. Mol Plant Microbe Interact. 24(12):1573–1581.
46. Tsuchiya T and Eulgem T. (2013) An alternative polyadenylation mechanism coopted to the *Arabidopsis* RPP7 gene through intronic retrotransposon domestication. Proc Natl Acad Sci U S A. 110(37):E3535-E3543.
47. Musselman CA and Kutateladze TG. (2011) Handpicking epigenetic marks with PHD fingers. Nucleic Acids Res. 39(21):9061–9071.
48. Tsuchiya T and Eulgem T. (2013) Mutations in EDM2 selectively affect silencing states of transposons and induce plant developmental plasticity. Sci Rep. 3:1701.
49. Li T, Chen X, Zhong X, Zhao Y, Liu X, Zhou S, Cheng S and Zhou DX. (2013) Jumonji C domain protein JMJ705-mediated removal of histone H3 lysine 27 trimethylation is involved in defense-related gene activation in rice. Plant Cell. 25(11):4725–4736.
50. Pieterse CM, Poelman EH, Van Wees SC and Dicke M. (2013) Induced plant responses to microbes and insects. Front Plant Sci. 4:475.
51. Berr A, McCallum EJ, Alioua A, Heintz D, Heitz T and Shen WH. (2010). *Arabidopsis* histone methyltransferase SET DOMAIN GROUP8 mediates induction of the jasmonate/ethylene pathway genes in plant defense response to necrotrophic fungi. Plant Physiol. 154(3):1403–1414.
52. Berr A, Ménard R, Heitz T and Shen WH. (2012) Chromatin modification and remodelling: a regulatory landscape for the control of *Arabidopsis* defence responses upon pathogen attack. Cell Microbiol. 14(6):829–839.
53. Grossniklaus U and Paro R. (2014) Transcriptional silencing by polycomb-group proteins. Cold Spring Harb Perspect Biol. 6(11):a019331.
54. Wang W, Vinocur B and Altman A. (2007) Plant responses to drought, salinity and extreme temperatures towards genetic engineering for stress tolerance. Planta. 218(1):1–14.
55. Gao JP, Chao DY and Lin HX. (2007) Understanding abiotic stress tolerance mechanisms: recent studies on stress response in rice. J Integr Plant Biol 49(6):742–750.
56. Mittler R. (2006) Abiotic stress, the field environment and stress combination." Trends Plant Sci. 11(1):15–19.
57. Santner A, Calderon-Villalobos LI and Estelle M. (2009) Plant hormones are versatile chemical regulators of plant growth. Nat Chem Biol. 5(5):301–307
58. Cutler SR, Rodriguez PL, Finkelstein RR and Abrams SR. (2010) Abscisic acid: emergence of a core signaling network. Annu Rev Plant Biol. 61:651–679.
59. Fujita M, Fujita Y, Noutoshi Y, Takahashi F, Narusaka Y, Yamaguchi-Shinozaki K and Shinozaki K. (2006) Crosstalk between abiotic and biotic stress responses: a current view from the points of convergence in the stress signaling networks. Curr Opin Plant Biol. 9(4):436–442.

60. Asselbergh B, De Vieesschauwer D and Hofte M. (2008) Global switches and fine-tuning – ABA modulates plant pathogen defense. Mol Plant Microbe Interact. 21(6):709–719.
61. Yasuda M, Ishikawa A, Jikumaru Y, Seki M, Umezawa T, Asami T, Maruyama-Nakashita A, Kudo T, Shinozaki K, Yoshida S and Nakashita H. (2008) Antagonistic interaction between systemic acquired resistance and the abscisic acid-mediated abiotic stress response in *Arabidopsis*. Plant Cell. 20(6):1678–1692.
62. Atkinson NJ and Urwin PE. (2012) The interaction of plant biotic and abiotic stresses: from genes to the field. J Exp Bot. 63(10):3523–3543.
63. Rajjou L, Duval M, Gallardo K, Catusse J, Bally J, Job C and Job D. (2012) Seed germination and vigor. Annu Rev Plant Biol. 63:507–533.
64. Lopez-Molina L, Mongrand S and Chua NH. (2001) A postgermination developmental arrest checkpoint is mediated by abscisic acid and requires the ABI5 transcription factor in *Arabidopsis*. Proc Natl Acad Sci USA. 98(8):4782–4787.
65. Perruc E, Kinoshita N, Lopez-Molina L. (2007). The role of chromatin-remodeling factor PKL in balancing osmotic stress responses during *Arabidopsis* seed germination. Plant J. 52(5):927–936.
66. Raghavendra AS, Gonugunta VK, Christmann A and Grill E. (2010) ABA perception and signalling. Trends Plant Sci. 15(7):395–401.
67. Koornneef M, Leon-Kloosterziel K, Schwartz SH and Zeevaart JAD. (1998) The genetic and molecular dissection of abscisic acid biosynthesis and signal transduction in *Arabidopsis*. Plant Physiol Biochem. 36:83–89.
68. Ding Y, Avramova Z and Fromm M. (2011) The Arabidopsis trithorax-like factor ATX1 functions in dehydration stress responses via ABA-dependent and ABA-independent pathways. Plant J. 66(5):735–744.
69. Ramachandra RA, Chaitanya KV and Vivekanandan M. (2004) Drought induced responses of photosynthesis and antioxidant metabolism in higher plants. J Plant Physiol. 161(11):1189–1202.
70. Araus JL, Slafer GA, Royo C and Serret MD. (2008) Breeding for yield potential and stress adaptation in cereals. Crit Rev Plant Sci. 27:377–412.
71. Hirayama T and Shinozaki K. (2010) Research on plant abiotic stress responses in the post-genome era: past, present and future. Plant J. 61(6):1041–1052.
72. Fukao T and Xiong L. (2013) Genetic mechanisms conferring adaptation to submergence and drought in rice: simple or complex? Curr Opin Plant Biol. 16(2):196–204.
73. Huang D, Wu W, Abrams SR and Cutler AJ. (2008) The relationship of drought-related gene expression in *Arabidopsis thaliana* to hormonal and environmental factors. J Exp Bot. 59(11):2991–3007.
74. Kim JM, To TK, Ishida J, Morosawa T, Kawashima M, Matsui A, Toyoda T, Kimura H, Shinozaki K and Seki M. (2008). Alterations of lysine modifications on the histone H3 N-tail under drought stress conditions in *Arabidopsis thaliana*. Plant Cell Physiol. 49:1580–1588.
75. Kim JM, Sasaki T, Ueda M, Sako K and Seki M. (2015) Chromatin changes in response to drought, salinity, heat, and cold stresses in plants. Front Plant Sci. 6:114.
76. van Dijk K, Ding Y, Malkaram S, Riethoven JJ, Liu R, Yang J, Laczko P, Chen H, Xia Y, Ladunga I, Avramova Z and Fromm M. (2010) Dynamic changes in genome-wide histone H3 lysine 4 methylation patterns in response to dehydration stress in *Arabidopsis thaliana*. BMC Plant Biol. 10:238.
77. Papaefthimiou D and Tsaftaris A. (2012) Significant induction by drought of HvPKDM7-1, a gene encoding a jumonji-like histone demethylase homologue in barley (H. Vulgare). Acta Physiol Plant. 34(3):1187–1198.
78. Govind G, Vokkaliga Thammegowda H, Jayaker Kalaiarasi P, Iyer D, Muthappa S, Nese S and Makarla U. (2009) Identification and functional validation of a unique set of drought induced genes preferentially expressed in response to gradual water stress in peanut. Mol Genet Genomics. 281(6):591–605.

79. Rando OJ and Chang HY. (2009) Genome-wide views of chromatin structure. Annu Rev Biochem. 78:245–271.
80. Zong W, Zhong X, You J and Xiong L. (2013). Genome-wide profiling of histone H3K4-tri-methylation and gene expression in rice under drought stress. Plant Mol Biol 81(1-2):175–188.
81. Hsu FC and Shih MC. (2013) Plant defense after flooding. Plant Signal Behav. 8(11):e26922.
82. Tsuji H, Nakazono M, Saisho D, Tsutsumi N and Hirai A. (2000). Transcript levels of the nuclear-encoded respiratory genes in rice decrease by oxygen deprivation: evidence for involvement of calcium in expression of the alternative oxidase 1a gene. FEBS Lett. 471(2-3):201–204.
83. Tsuji H, Saika H, Tsutsumi N, Hirai A and Nakazono M. (2006) Dynamic and reversible changes in histone H3-Lys4 methylation and H3 acetylation occurring at submergence-inducible genes in rice. Plant Cell Physiol. 47(7):995–1003.
84. Carillo P, Annunziata MG, Pontecorvo G, Fuggi A and Woodrow P. (2011) Salinity stress and salt tolerance. In: Arun S, editor. Abiotic Stress in Plants - Mechanisms and Adaptations. Croatia: InTech. pp. 22–38.
85. Chen LT, Luo M, Wang YY and Wu K. (2010). Involvement of *Arabidopsis* histone deacetylase HDA6 in ABA and salt stress response. J Exp Bot. 61(2):3345–3353.
86. He G, Elling A and Deng X. (2011) The epigenome and plant development. Rev Plant Biol. 62(1):411.
87. Nightingale KP, Gendreizig S, White DA, Bradbury C, Hollfelder F and Turner BM. (2007) Cross-talk between histone modifications in response to histone deacetylase inhibitors: MLL4 links histone H3 acetylation and histone H3K4 methylation. J Biol Chem. 282(7):4408–4416.
88. Song Y1, Ji D, Li S, Wang P, Li Q and Xiang F. (2012) The dynamic changes of DNA methylation and histone modifications of salt responsive transcription factor genes in soybean. PLoS One. 7(7):e41274.
89. Deleris A, Greenberg MV, Ausin I, Law RW, Moissiard G, Schubert D and Jacobsen SE. (2010) Involvement of a Jumonji-C domain-containing histone demethylase in DRM2-mediated maintenance of DNA methylation. EMBO Rep. 11(12):950–955.
90. Lu F, Cui X, Zhang S, Liu C and Cao X. (2010) JMJ14 is an H3K4 demethylase regulating flowering time in *Arabidopsis*. Cell Res. 20(3):387–390.
91. Searle IR, Pontes O, Melnyk CW, Smith LM and Baulcombe DC. (2010) JMJ14, a JmjC domain protein, is required for RNA silencing and cell-to-cell movement of an RNA silencing signal in *Arabidopsis*. Genes Dev. 24(10):986–991.
92. Le Masson I, Jauvion V, Bouteiller N, Rivard M, Elmayan T and Vaucheret H. (2012) Mutations in the *Arabidopsis* H3K4me2/3 demethylase JMJ14 suppress posttranscriptional gene silencing by decreasing transgene transcription. Plant Cell. 24(9):3603–3612.
93. Yang H, Mo H, Fan D, Cao Y, Cui S and Ma L. (2012) Overexpression of a histone H3K4 demethylase, JMJ15, accelerates flowering time in *Arabidopsis*. Plant Cell Rep. 31(7):1297–1308.
94. Yang H, Han Z, Cao Y, Fan D, Li H, Mo H, Feng Y, Liu L, Wang Z, Yue Y, Cui S, Chen S, Chai J and Ma L. (2012) A companion cell-dominant and developmentally regulated H3K4 demethylase controls flowering time in *Arabidopsis* via the repression of FLC expression. PLoS Genet. 8(4):e1002664.
95. Cui X, Jin P, Cui X, Gu L, Lu Z, Xue Y, Wei L, Qi J, Song X, Luo M, An G and Cao X. (2013) Control of transposon activity by a histone H3K4 demethylase in rice. Proc Natl Acad Sci USA. 110(5):1953–1958.
96. Shen Y, Conde E, Silva N, Audonnet L, Servet C, Wei W and Zhou DX. (2014) Over-expression of histone H3K4 demethylase gene JMJ15 enhances salt tolerance in *Arabidopsis*. Front Plant Sci. 5:290.
97. Schmitz RJ, Sung S and Amasino RM. (2008) Histone arginine methylation is required for vernalization-induced epigenetic silencing of FLC in winter-annual *Arabidopsis thaliana*. Proc Natl Acad Sci USA. 105(2):411–416.

98. Zhang Z, Zhang S, Zhang Y, Wang X, Li D, Li Q, Yue M, Li Q, Zhang YE, Xu Y, Xue Y, Chong K and Bao S. (2011) *Arabidopsis* floral initiator SKB1 confers high salt tolerance by regulating transcription and pre-mRNA splicing through altering histone H4R3 and small nuclear ribonucleoprotein LSM4 methylation. Plant Cell. 23(1):396–411.
99. Thomashow MF. (1999) PLANT COLD ACCLIMATION: freezing tolerance genes and regulatory mechanisms. Annu Rev Plant Physiol Plant Mol Biol. 50:571–599.
100. Chinnusamy V, Zhu J and Zhu JK. (2007) Cold stress regulation of gene expression in plants. Trends Plant Sci. 12(10):444–451.
101. Bokszczanin KL and Fragkostefanakis S. (2013). Perspectives on deciphering mechanisms underlying plant heat stress response and thermotolerance. Front Plant Sci. 4:315.
102. Qu AL, Ding YF, Jiang Q and Zhu C. (2013). Molecular mechanisms of the plant heat stress response. Biochem Biophys Res Commun. 432(2):203–207.
103. Peng S, Huang J, Sheehy JE, Laza RC, Visperas RM, Zhong X, Centeno GS, Khush GS and Cassman KG. (2004) Rice yields decline with higher night temperature from global warming. Proc Natl Acad Sci USA 101(27):9971–9975.
104. Lobell DB, Schlenker W and Costa-Roberts J. (2011) Climate trends and global crop production since 1980. Science. 333:616–620.
105. Folsom JJ, Begcy K, Hao X, Wang D and Walia H. (2014) Rice fertilization-independent endosperm1 regulates seed size under heat stress by controlling early endosperm development. Plant Physiol. 165(1):238–248.
106. Strenkert D, Schmollinger S and Schroda M. (2013) Heat shock factor 1 counteracts epigenetic silencing of nuclear transgenes in *Chlamydomonas reinhardtii*. Nucleic Acids Res. 41(10):5273–5289.
107. Min L, Li Y, Hu Q, Zhu L, Gao W, Wu Y, Ding Y, Liu S, Yang X and Zhang X. (2014). Sugar and auxin signalling pathways respond to high-temperature stress during anther development as revealed by transcript profiling analysis in cotton. Plant Physiol. 164(3):1293–1308.
108. Grandbastien MA. (1998) Activation of plant retrotransposons under stress conditions. Trends Science. 3(5):181–187. Plants
109. Pecinka A, Dinh HQ, Baubec T, Rosa M, Lettner N and Mittelsten Scheid O. (2010) Epigenetic regulation of repetitive elements is attenuated by prolonged heat stress in *Arabidopsis*. Plant Cell. 22(9):3118–3129.
110. Tittel-Elmer M, Bucher E, Broger L, Mathieu O, Paszkowski J and Vaillant I. (2010) Stress-induced activation of heterochromatic transcription. PLoS Genet. 6(10):e1001175.
111. Lippman Z, Gendrel AV, Black M, Vaughn MW, Dedhia N, McCombie WR, Lavine K, Mittal V, May B, Kasschau KD, Carrington JC, Doerge RW, Colot V and Martienssen R. (2004) Role of transposable elements in heterochromatin and epigenetic control. Nature. 430(6998):471–476.
112. Sung S and Amasino RM. (2005) Remembering winter: toward a molecular understanding of vernalization. Annu Rev Plant Biol. 56:491–508.
113. Kwon CS, Lee D, Choi G and Chung WI. (2009) Histone occupancy-dependent and -independent removal of H3K27 trimethylation at cold-responsive genes in *Arabidopsis*. Plant J. 60(1):112–121.
114. Hu Y, Zhang L, He S, Huang M, Tan J, Zhao L, Yan S, Li H, Zhou K, Liang Y and Li L. (2012) Cold stress selectively unsilences tandem repeats in heterochromatin associated with accumulation of H3K9ac. Plant Cell Environ. 35(12):2130–2142.
115. Durrant WE and Dong X. (2004) Systemic acquired resistance. Annu Rev Phytopathol 42:185–209.
116. Mishina TE and Zeier J. (2007) Pathogen-associated molecular pattern recognition rather than development of tissue necrosis contributes to bacterial induction of systemic acquired resistance in *Arabidopsis*. Plant J. 50(3):500–513.
117. Conrath U. (2011) Molecular aspects of defense priming. Trends Plant Sci. 16(10):524–531.

118. Heard E and Martienssen RA. (2014) Transgenerational epigenetic inheritance: myths and mechanisms. Cell. 157(1):95–109.
119. Jaskiewicz M, Conrath U and Peterhansel C. (2011) Chromatin modification acts as a memory for systemic acquired resistance in the plant stress response. Embo Reports. 12(1):50–55.
120. Luna E, Bruce TJ, Roberts MR, Flors V and Ton J. (2012) Next-generation systemic acquired resistance. Plant Physiol. 158(2):844–853.
121. Spoel SH, Koornneef A, Claessens SM, Korzelius JP, Van Pelt JA, Mueller MJ, Buchala AJ, Métraux JP, Brown R, Kazan K, Van Loon LC, Dong X and Pieterse CM. (2003) NPR1 modulates cross-talk between salicylate- and jasmonate-dependent defense pathways through a novel function in the cytosol. Plant Cell. 15(3):760–770.
122. Koornneef A and Pieterse CM. (2008) Cross talk in defense signaling. Plant Physiol. 146(3):839–844.
123. Kim JM, To TK, Ishida J, Matsui A, Kimura H and Seki M. (2012) Transition of chromatin status during the process of recovery from drought stress in *Arabidopsis thaliana*. Plant Cell Physiol. 53:847–856.
124. Ding Y, Fromm M and Avramova Z. (2012) Multiple exposures to drought 'train' transcriptional responses in *Arabidopsis*. Nat Commun. 3:740.
125. Ding Y, Lapko H, Ndamukong I, Xia Y, Al-Abdallat A, Lalithambika S, Sadder M, Saleh A, Fromm M, Riethoven JJ, Lu G and Avramova Z. (2009) The *Arabidopsis* chromatin modifier ATX1, the myotubularin-like AtMTM and the response to drought. Plant Signal Behav. 4(11):1049–1058.
126. Ding Y, Avramova Z and Fromm M. (2011) Two distinct roles of ARABIDOPSIS HOMOLOG OF TRITHORAX1 (ATX1) at promoters and within transcribed regions of ATX1-regulated genes. Plant Cell. 23(1):350–363.
127. Ding Y, Ndamukong I, Xu Z, Lapko H, Fromm M and Avramova Z. (2012) ATX1-generated H3K4me3 is required for efficient elongation of transcription, not initiation, at ATX1-regulated genes. PLoS Genet. 8(12):e1003111.
128. Saleh A, Alvarez-Venegas R, Yilmaz M, Le O, Hou G, Sadder M, Al-Abdallat A, Xia Y, Lu G, Ladunga I and Avramova Z. (2008) The highly similar *Arabidopsis* homologues of trithorax ATX1 and ATX2 encode proteins with divergent biochemical functions. Plant Cell. 20(3):568–579.
129. Avramova Z. (2009) Evolution and pleiotropy of TRITHORAX function in *Arabidopsis*. Int J Dev Biol. 53(2-3):371–381.
130. Berr A, Xu L, Gao J, Cognat V, Steinmetz A, Dong A and Shen WH. (2009) SET DOMAIN GROUP25 encodes a histone methyltransferase and is involved in FLOWERING LOCUS C activation and repression of flowering. Plant Physiol. 151(3):1476–1485.
131. Tamada Y, Yun JY, Woo SC and Amasino RM. (2009) ARABIDOPSIS TRITHORAX-RELATED7 is required for methylation of lysine 4 of histone H3 and for transcriptional activation of FLOWERING LOCUS C. Plant Cell. 21(10):3257–3269.
132. Berr A, McCallum EJ, Ménard R, Meyer D, Fuchs J, Dong A and Shen WH. (2010) *Arabidopsis* SET DOMAIN GROUP2 is required for H3K4 trimethylation and is crucial for both sporophyte and gametophyte development. Plant Cell. 22(10):3232–3248.
133. Guo L, Yu Y, Law JA and Zhang X. (2010) SET DOMAIN GROUP2 is the major histone H3 lysine [corrected] 4 trimethyltransferase in *Arabidopsis*. Proc Natl Acad Sci USA. 107(43):18557–18562.
134. Yun JY, Tamada Y, Kang YE and Amasino RM. (2012) *Arabidopsis* trithorax-related3/SET domain GROUP2 is required for the winter-annual habit of *Arabidopsis thaliana*. Plant Cell Physiol. 53(5):834–846.
135. Kim S, Lee J, Yang JY, Jung C and Chua NH. (2013) *Arabidopsis* histone methyltransferase SET DOMAIN GROUP2 is required for regulation of various hormone responsive genes. J Plant Biol. 56(1):39–48.

136. Sani E, Herzyk P, Perrella G, Colot V and Amtmann A. (2013). Hyperosmotic priming of *Arabidopsis* seedlings establishes a long-term somatic memory accompanied by specific changes of the epigenome. Genome Biol. 14(6):R59.
137. Liu N, Ding Y, Fromm M and Avramova Z. (2014) Different gene-specific mechanisms determine the 'revised-response' memory transcription patterns of a subset of *A. thaliana* dehydration stress responding genes. Nucleic Acids Res. 42(9):5556–5566.
138. Liu N, Fromm M and Avramova Z. (2014) H3K27me3 and H3K4me3 chromatin environment at super-induced dehydration stress memory genes of *Arabidopsis thaliana*. Mol Plant. 7(3):502–513.
139. Bouyer D, Roudier F, Heese M, Andersen ED, Gey D, Nowack MK, Goodrich J, Renou JP, Grini PE, Colot V and Schnittger A. (2011) Polycomb repressive complex 2 controls the embryo-to-seedling phase transition. PLoS Genet. 7(3):e1002014.
140. Azuara V, Perry P, Sauer S, Spivakov M, Jorgensen HF, John RM, Gouti M, Casanova M, Warnes G, Merkenschlager M and Fisher AG. (2006). Chromatin signatures of pluripotent cell lines. Nat Cell Biol. 8(5):532–538.
141. Bernstein BE, Mikkelsen TS, Xie X, Kamal M, Huebert DJ, Cuff J, Fry B, Meissner A, Wernig M, Plath K, Jaenisch R, Wagschal A, Feil R, Schreiber SL and Lander ES. (2006) A bivalent chromatin structure marks key developmental genes in embryonic stem cells. Cell. 125(2):315–326.
142. Singh P, Yekondi S, Chen PW, Tsai CH, Yu CW, Wu K, Zimmerli L. (2014) Environmental history modulates *Arabidopsis* pattern-triggered immunity in a HISTONE ACETYLTRANSFERASE1-Dependent manner. Plant Cell. 26(6):2676–2688.
143. Bruce TJA, Matthes MC, Napier JA and Pickett JA. (2007) Stressful 'memories' of plants: evidence and possible mechanisms. Plant Sci. 173(6):603–608.
144. Liu J, Feng L, Li J and He Z. (2015) Genetic and epigenetic control of plant heat responses. Front Plant Sci. 6:267.
145. Bilichak A, Ilnystkyy Y, Hollunder J and Kovalchuk I. (2012) The progeny of *Arabidopsis thaliana* plants exposed to salt exhibit changes in DNA methylation, histone modifications and gene expression. PLoS One. 7(1):e30515.
146. Migicovsky Z, Yao Y and Kovalchuk I. (2014) Transgenerational phenotypic and epigenetic changes in response to heat stress in *Arabidopsis thaliana*. Plant Signal Behav. 9(2):e27971.
147. Oliver KR, McComb JA and Greene WK. (2013) Transposable elements: powerful contributors to angiosperm evolution and diversity. Genome Biol Evol. 5(10):1886–1901.
148. Iwasaki M and Paszkowski J. (2014) Epigenetic memory in plants. EMBO J. 33(18):1987–1998.
149. Becker C and Weigel D. (2012) Epigenetic variation: origin and transgenerational inheritance. Curr Opin Plant Biol. 15(5):562–567.
150. Weigel D and Colot V. (2012) Epialleles in plant evolution. Genome Biol. 13(10):249.
151. Rizhsky L, Liang HJ, Shuman J, Shulaev V, Davletova S and Mittler R. (2004) When defense pathways collide. The response of *Arabidopsis* to a combination of drought and heat stress. Plant Physiol. 134(4):1683–1696.
152. Asselbergh B, Achuo AE, Hofte M and Van Gijsegem F. (2008) Abscisic acid deficiency leads to rapid activation of tomato defence responses upon infection with *Erwinia chrysanthemi*. Mol Plant Pathol. 9(1):11–24.
153. She W and Baroux C. (2014) Chromatin dynamics during plant sexual reproduction. Front Plant Sci. 5:354.
154. Talbert PB and Henikoff S. (2014) Environmental responses mediated by histone variants. Trends Cell Biol. 24(11):642–650.
155. Probst AV and Mittelsten Scheid O. (2015) Stress-induced structural changes in plant chromatin. Curr Opin Plant Biol. 27:8–16.
156. Sander JD and Joung JK. (2014) CRISPR-Cas systems for editing, regulating and targeting genomes. Nat Biotechnol. 32(4):347–355.

157. Wang D and Deal RB. (2015) Epigenome profiling of specific plant cell types using a streamlined INTACT protocol and ChIP-seq. Methods Mol Biol. 1284:3–25.

Index